THE PLANETARY SYSTEM

THIRD EDITION

DAVID MORRISON

TOBIAS OWEN

Addison Wesley

San Francisco Boston New York
Capetown Hong Kong London Madrid Mexico City
Montreal Munich Paris Singapore Sydney Tokyo Toronto

Acquisitions Editor:	Adam Black
Project Editor:	Nancy G. Benton
Managing Editor:	Joan Marsh
Marketing Manager:	Christy Lawrence
Media Producer:	Claire Masson
Manufacturing Coordinator:	Vivian McDougal
Production Services:	Joan Keyes, Dovetail Publishing Services
Cover Illustrator:	Lynette Cook
Cover Design:	Blakeley Kim

Library of Congress Cataloging-in-Publication Data

Morrison, David, 1940–
     The planetary system / David Morrison, Tobias Owen.—3rd ed.
       p. cm.
     Includes bibliographical references and index.
     ISBN 0-8053-8734-X
     1. Planets. I. Owen, Tobias C. II. Title.

QB601 .M76 2003
523.2—dc21   2002034522

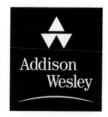

1 2 3 4 5 6 7 8 9 10—VHP—03 02
www.aw.com

CONTENTS

PREFACE

THE FINAL THIRD of the twentieth century will be remembered as the golden age of planetary exploration, the period in which we launched the first spacecraft missions to the planets. Within three extraordinary decades, the wanderers in the night sky that have fascinated humans for millennia became individual worlds, with moons, rings, and fantastic landscapes that have given us a new perspective on our own place in the universe. We can now compare our blue and white Earth with other planets in ways that would have been impossible without the space program. What has emerged from this exciting period is a new scientific discipline, usually termed *planetary science*. This book, *The Planetary System*, is an introduction to this discipline.

In 1988 we published the first edition of *The Planetary System* as a text for a one-semester introductory college course in planetary science. Such courses are taught in many universities, offered by departments of astronomy, physics, and geology. As the first comprehensive introductory text that covered the entire field, giving roughly equal weight to both solid surfaces and atmospheres, the book was well received. We therefore published a second edition in 1996, incorporating the many new discoveries that had been made during the intervening years. In that edition, we were barely able to report the first discovery of an extrasolar planet as we went to press. Seven years later, the number of known extrasolar planets stands at more than 100, with many more on the way. The Galileo orbiter has competed its exploration of Jupiter. We have obtained remarkable new close-up images of both an asteroid and a comet, intriguing evidence for subsurface liquid water has been found on Mars, and a 1000-kilometer–diameter object has been discovered in the Kuiper belt, more than a billion kilometers beyond Pluto. The new multidisciplinary study of astrobiology is also providing us with different perspectives on planets—both within our solar system and beyond—as potential abodes of life. It is time for a new edition that describes these and other new discoveries and explains their impact on our vision of the system in which we live.

The general subject matter and target audience are unchanged in the third edition. As before, we intend this book to be accessible to the nonscience undergraduate with no prior training in either astronomy or geology. At the same time, we realize that in many universities the first course in planetary science is not offered until the upper-tier undergraduate level, and we have tried to provide material that is suitable for these courses as well. For the more advanced audience, most chapters include optional quantitative supplements that use simple algebra, together with problems that can be solved with a handheld calculator.

While other introductory texts in solar-system science have appeared recently, most are written by astronomers and are derivative of longer general astronomy texts. This book remains unique as a stand-alone volume focusing on planetary science as understood and practiced by planetary scientists, with a clear focus on the processes by which the planetary system formed and evolved.

As an interdisciplinary subject, planetary science defies easy description. Traditionally, the Earth and planets have been studied within the fields of astronomy and the geosciences (geology, geophysics, meteorology, and so on). Only recently have components of these standard disciplines come together to generate the new perspectives of comparative planetology—the study of the Earth as a planet and of the other planets as worlds. Such a synthesis would not have been possible without the spectacular success of the planetary spacecraft launched by the United States and the former Soviet Union, spurred in part by competition between the capitalist and socialist systems. With the collapse of the USSR, the urgency of competition has evaporated, and the pace of spacecraft exploration has slowed accordingly. However, the European Space Agency (ESA) and Japan have joined the United States as players, compensating in part for these cutbacks. Also, we have seen a resurgence of discoveries about the solar system using more traditional astrophysical methods enhanced by new detectors and carried out with both ground-based and orbiting telescopes. Today both space-based and ground-based planetary research contribute to a healthy and growing field.

The primary subject of this book is the solar system. There is some difference of opinion among teachers as to whether it is appropriate to include the Earth and Sun in an introductory course on the planets. We have always felt that the study of the Earth is essential for a proper perspective on the other planets, but we omitted the Sun from the first edition. At the urging of a number of colleagues, we added a chapter on the Sun in the second edition, and we include it here. However, we note that the treatment of this nearest star, as well as of our own planet Earth, is less detailed than that found in most astronomy and geology texts, respectively. As indicated by the title, the focus of this book is on the entire solar system of planets, satellites, comets, and asteroids, of which the Sun and Earth are only two members. For the third edition, we introduce additional perspectives based on the discovery of other planetary systems and the growing interest in life on other worlds. Both of these perspectives stimulate new ways of thinking about the objects in our planetary system.

The order of presentation of topics is designed to emphasize the processes that formed and continue to influence the members of the planetary system. The first three chapters provide an introduction to the solar system itself, a brief discussion of the Sun, and an explanation of some of the basic physical and chemical concepts used later. Students who have already had a course in astronomy or the Earth sciences may wish to go directly to Chapter 4, which begins the study of the simplest, most primitive solid objects in the

system: comets, asteroids, and meteorites. We like to introduce the meteorites at the beginning of our discussion, since these samples of extraterrestrial material provide an essential "ground truth" for remote studies. Chapter 7 moves us one step up the scale of complexity to examine the small cratered worlds, Mercury and the Moon, as an introduction to planetary geology. Chapters 9–12 deal with the Earth-Venus-Mars triad of terrestrial planets, and their active geological evolution and complex atmospheres. In Chapter 12, we pay special attention to the question of life on planets, introducing concepts from the new discipline of astrobiology. The outer solar system is the subject of the next four chapters, which cover the giant planets, their large and small satellites, and their ring systems.

In Chapter 17 we review the structure of the entire system in terms of its origin and evolution. At that point, we are ready for something new and different: How does our system compare with other planetary systems in the Milky Way Galaxy? How well do our models for the origin and evolution of the solar system in which we live explain the characteristics of these new systems? These are challenging questions indeed, and 7 years ago we could not even begin to answer them. But now we can, and the answers take up the first part of Chapter 18. The final part of this chapter takes us though the next logical steps after the discovery of other planets by examining the future quest for Earth-like planets and the probabilities for finding life on some of the brave new worlds that circle other stars.

Throughout the book we emphasize comparative studies and the common processes acting on the members of the system, but we do so in the context of descriptions of individual objects. Generalizations are important, but we feel they need to be derived from understanding of the specific planets, satellites, asteroids, and comets that we have explored. We have enjoyed including the most important new discoveries and interpretations in the third edition of this book. We have not reported novel results for their news value alone, however, but rather have made every effort to blend the new discoveries with what was already known in order to present a more complete and balanced picture of the planetary system and its evolution. We have made changes in every section of every chapter to provide comprehensive, current perspectives.

For this new edition, we have compressed the discussion of small satellites and rings to make room for the new chapter on extrasolar planets and for expanded discussions of astrobiology. Thus the book has not increased in size. We are delighted to be able to use color throughout the book, as we do not live in a black and white universe and the use of color has a real pedagogical advantage. We are also pleased to be using a larger format page to display the wealth of images. We have introduced a new feature in this edition called *Perspective*—essays that deal with the nature of science (and pseudoscience) and relationships between science and society. While these Perspective essays are optional (that is, no basic concepts or vocabulary are introduced in them), we expect most students will find them as interesting as we do.

Any scientific discipline has specialized jargon. Planetary science, which draws from several disciplines, may be worse than most. Some of this jargon is useful and even necessary, but much of it is superfluous in an introductory course. We feel that technical vocabulary should be used as a tool, not learned as an end in itself—this is not a text for a foreign language course! We continue to practice "jargon control," introducing only those specialized terms (about 200 in all) that we feel are really necessary for understanding the concepts in this book. These terms are printed in boldface where first explained in the text and are defined again in the Glossary.

Another problem for authors concerns the practice of making reference to the work of individuals. We decided that a book at this level should not include specific references to the literature, although lists of additional reading are presented at the end of each chapter. Yet we did not want to neglect the work of individuals entirely, since it would give a false impression of the nature of science to imply that it is an impersonal endeavor. Of necessity, however, we have been able to mention by name only a few of the scientists who have made important contributions to this field, and we apologize to the many others for what may seem to be a rather arbitrary selection.

This book would not have been possible without the encouragement and assistance of our colleagues who have shared their new discoveries and interpretations, explained to us what sort of book they need for their undergraduate classes, and provided constructive criticisms of the previous editions. We are especially grateful for the advice and assistance we have received from Michael Allison, Jeffery Brown, Joe Cain, Chris Chyba, Clark Chapman, Gene Clough, Dale Cruikshank, Jeff Cuzzi, David DesMarais, Frank Drake, Jack Farmer, Andrew Fraknoi, Don Goldsmith, Ron Greeley, Woody Harrington, Alan Harris, Tom Hockey, David Hogenboom, Jim Houck, David Hughes, Andy Ingersoll, Bruce Jakosky, Bill Kaula, Larry Lebofsky, Jean Pierre Lebreton, Jack Lissauer, Paul Mahaffy, Frank Maloney, Geoff Marcy, Dennis Matson, Chris McKay, Bill McKinnon, Denise Meeks, Jeff Moore, Deane Peterson, Lawrence Pinsky, Tricia Reeves, Carl Sagan, Bill Schopf, Gene Shoemaker, Seth Shostak, Dale Smith, Ted Snow, Jill Tarter, Karen Teramura, Don Yeomans, and Kevin Zahnle. In addition, Janet Morrison provided welcome editorial advice. Author Owen thanks Leonid Slutsky for his generous hospitality in Moscow during completion of the final chapters. Diane Tokumura and Louise Good provided critical help with the preparation of the manuscript. We are also grateful for the consistent support we have received from our publisher, and especially from our editors: Adam Black, who had the faith to encourage a full-color book for this third edition; Nancy Benton, who has overseen the preparation of the book; Carol Reitz for her careful copyediting; and especially Joan Keyes, who has masterfully organized the entire effort, including editing, design, artwork, and layout. We thank them all, and we hope that you, the reader, will find the product of these efforts to be both useful and enjoyable.

David Morrison   Saratoga, California
Tobias Owen   Honolulu, Hawaii
    October 2002

ABOUT THE AUTHORS

DAVID MORRISON is Senior Scientist at the NASA Astrobiology Institute and former Director of Space at NASA Ames Research Center. He participated in the Mariner mission to Mercury, the Voyager mission to the outer planets, and the Galileo mission to Jupiter. With colleagues, he has used astronomical observations to discover that Uranus has no internal energy source, to make the first measurements of the volcanic heat flow from Io, and to formulate the classification of the asteroids into the C-S-M taxonomic groups. He received the Klumpke-Roberts award of the Astronomical Society of the Pacific for education and public service in astronomy, as well as NASA Outstanding Leadership Medals for his contributions to the Galileo mission and to protection of the Earth from comet and asteroid impacts. In addition to this book, Morrison is the author of ten other texts and popular books on space science. He considers himself lucky to have participated in the birth of two new scientific fields: planetary science and astrobiology.

TOBIAS OWEN is professor of astronomy at the Institute for Astronomy of the University of Hawaii. He was a member of the NASA Viking mission to Mars, Voyager to the outer planets, and Galileo to Jupiter. He is currently participating in the Japanese Nozomi Mars mission, the ESA rendezvous with Comet Wirtanen, and is an interdisciplinary scientist and member of several experiment teams on the Cassini-Huygens mission to the Saturn system. NASA awarded him an Exceptional Scientific Achievement Medal for his work on the martian atmosphere. In Hawaii, he pursues a program of spectroscopic studies of the members of our planetary system, using the great telescopes on Mauna Kea. This work has included the discovery and evaluation of deuterium on Mars and in comets Hale Bopp and Hyakutake, hydrogen cyanide on Neptune, and ices of nitrogen and carbon monoxide on Pluto. He has written over 250 scientific articles and co-authored *The Search for Life in the Universe* with Donald Goldsmith, currently in its third edition.

THE PLANETARY SYSTEM

THIRD EDITION

1 FINDING OUR PLACE IN SPACE

The Great Galaxy in Andromeda, a nearby spiral galaxy that looks very much as our own Milky Way would appear if seen from the outside.

Standing under the stars on a clear, dark night, you are looking out from the surface of our rocky planet into the vast depths of the universe—the grandest assemblage of matter and space that we know. "This majestical roof, fretted with golden fire" (Shakespeare, *Hamlet*) has been made familiar to us by the efforts of our ancestors, who gave names to the stars and the patterns they form. Throughout history, humans have struggled to try to understand the place our planet occupies in this starry realm.

The planetary system that is the principal subject of this book consists of nine planets, nearly 100 satellites discovered so far, and the innumerable asteroids, comets, and other objects that orbit the Sun. We are also beginning to glimpse the properties of other planetary systems as we discover planets orbiting other stars, as we will discuss later. We first want to know the shape and size of our own planetary system and its internal motions. What moves around what, and why? These puzzles, which have fascinated humans since they first recognized the existence of the planets, are the subject of Chapter 1.

1.1 WATCHING THE SKY: SUN, MOON, AND STARS

Most of us live in cities these days. We can see little in the sky, especially at night when city lights obscure all but the brightest stars. We have thus lost contact with an important window on the universe, one we can look through with our own eyes, unaided by any optical or electronic devices. Even so, some celestial events play an important part in our lives, setting the cycle of the seasons and giving us various measures of time. Occasionally, our attention is focused by the news media on some rare event: a solar eclipse or the appearance of a comet. Most of the time, we have only random views of the Sun and Moon and pay little enough attention to these.

For many of us, even the time-keeping aspect of the sky is remote, since we have calendars, watches, radios, television sets, newspapers, and magazines to remind us of daily, monthly, and yearly timescales and events. But to ancient civilizations without such assistance, the sky was a very important part of the natural environment. It was a source of pleasure as well as fear, and its changing aspects gave these peoples the ability to predict seasonal changes that were vital to their livelihood. In trying to understand the reasons for what they were seeing, the ancients initiated the subject of astronomy. We are still struggling with some of the problems they could not solve, such as the origins of the Earth, Moon, Sun, and stars, and the question of whether or not there are other inhabited worlds.

Many of the phenomena these early observers found mysterious, however, we can understand. Indeed, we take these solutions so much for granted that it is useful to precede a review of the modern perspective by returning to earlier problems in order to see how they were solved. In the same spirit, we encourage you to go out and make some simple, basic observations yourselves—watching seasonal changes in the altitude of the Sun at noon, following the Moon's apparent motion through the sky, learning the constellations, and identifying the planets. All this can be done with no optical aid whatsoever, and these simple activities help you establish a personal connection with the planetary system we inhabit.

THE CONSTELLATIONS

To become acquainted with the night, you need to find a location that is far from any city and has a really dark sky. This is the kind of environment that was always at hand to the ancients. Such a night sky is filled with stars of different brightness and colors. With its usual quest for order, the human mind found certain patterns among these apparently random points of light. These patterns are called **constellations** and were given names associated with myths, legends, and historical events in various cultures. Astronomers still use some of these names today. Probably most of you at one time have identified the Big Dipper, part of the Roman "Great Bear" or the Old English "Hay Wain," and Orion, the brilliant constellation of cold winter nights (Fig. 1.1). (All these descriptions are given for observers who live at latitudes from 25° to 50° north of the equator. The sky looks different to people in the Southern Hemisphere.)

If you watch for a while, you will see some constellations setting in the west while others emerge above the eastern horizon. In addition, the constellations change with the season. In summer, Orion is replaced by Lyra and Cygnus nearly overhead and by Scorpius and Sagittarius toward the south. While this grand seasonal cycle is taking place, the positions of the stars within each constellation remain fixed. The Big Dipper looked the same to Charlemagne in the year 800 as it does to you. One of the early mysteries the ancients tried to solve was the reason for this daily and annual repetition in the sky. The obvious solution was to suggest that the Earth was fixed in the center of a set of giant spheres that slowly revolved about it, giving rise to day and night and the seasonal change in the visible constellations. Today we know that it is the Earth that moves, but this is a relatively new discovery in the history of human thought.

APPARENT MOTION OF THE SUN AND STARS

Like the Sun, the stars rise in the east and set in the west. This diurnal, or daily, motion can be noticed in an hour or less, and by the end of the

Figure 1.1 The constellation Orion, showing the brightest stars and the location of a giant gas cloud, the Orion Nebula, where new stars are forming at the present time.

night it has brought many new constellations into view to replace the stars present at dusk. A slower seasonal cycle is also apparent, however.

Try going outside each evening at the same time for several weeks, and you will see that the eastern constellations are farther up in the sky and the western ones farther down than when you first began looking. This seasonal shift ultimately brings the same constellations back to the same place at the same time each year (Fig. 1.2). The movement is the manifestation in the night sky of the seasonal motion of the Sun through the background of stars.

The Sun's daily motion through the sky is also subject to seasonal changes. In winter the days are shorter than in summer. The Sun is lower in the sky at noon and casts long shadows. Thus, a careful observer can detect the seasonal motion of the Sun by plotting its height above the horizon at noon. This motion comes to a stop and reverses at the summer **solstice** ("Sun-stationary"), when the Sun is highest in the sky, and again six months later at the winter solstice, when the Sun is at its lowest point (Fig. 1.3). Halfway between we have the spring (or vernal) and autumn

equinoxes, when days and nights are equal in length. These apparent motions were well known to most ancient civilizations, even though they did not understand their ultimate cause.

In repeating this regular cycle, the Sun appears to move through a certain set of constellations. The apparent path of the Sun through the constellations in the course of its seasonal round is an

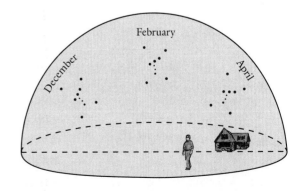

Figure 1.2 Appearance of Orion in the sky at 9 P.M. on the first of December, February, and April. The seasonal change in the appearance of the night sky is caused by the revolution of the Earth about the Sun.

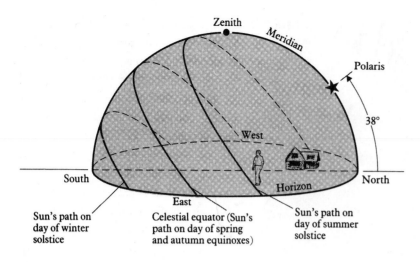

Sun's path on day of winter solstice

Celestial equator (Sun's path on day of spring and autumn equinoxes)

Sun's path on day of summer solstice

Figure 1.3 A person living at north temperate latitudes of the United States will see the Sun rise directly in the east at the spring and fall equinoxes and set directly in the west. At the summer solstice, the Sun rises north of east and sets north of west, reaching its highest noontime elevation in the sky. At the winter solstice, when days are shortest, the Sun rises south of east and sets south of west, reaching its lowest noontime altitude. The meridian is an imaginary line running north-south and passing directly overhead (through the zenith).

imaginary line in the sky known as the **ecliptic.** It may seem strange that the constellations through which the Sun moves can be determined, since it is impossible to see the stars during the day. However, it is simply a matter of watching which constellations are near the Sun in the twilight skies just after it sets and before it rises. This is something you can do yourself, if you have access to clear horizons. When you know the configuration of constellations over the entire sky, you can tell where the Sun is by figuring out the position halfway between the stars seen in the morning and evening twilight. Alternatively, you can note the constellation that is opposite to the Sun, rising when the Sun sets. Either way, you can trace out the annual path of the Sun.

It is easier to see which constellations form the backdrop for the motions of the Moon. It turns out that the Moon moves through the same constellations as the Sun. The Moon closely follows the ecliptic, deviating from it by only a few degrees, sometimes appearing above it, sometimes below. The set of constellations through which the Sun and Moon appear to move is called the **zodiac,** from the same Greek word that gives us *zoo* and *zoology*. That etymology tells you right away that most of these constellations are named for various real and imaginary animals (Fig. 1.4).

The Sun requires one year or 365.26 days to make its slow, eastward journey through the constellations of the zodiac and return to its starting point. It is this journey, in fact, that defines our year. The cycle of the Moon through the zodiac is much faster, giving us the period of time we call a month.

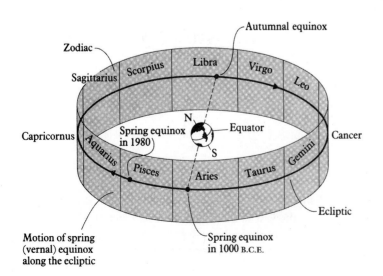

Motion of spring (vernal) equinox along the ecliptic

Spring equinox in 1000 B.C.E.

Figure 1.4 The planets, the Sun, and the Moon all appear to move through a narrow band of 12 constellations called the zodiac.

PHASES OF THE MOON

Although the Moon is a nearly perfect sphere, its apparent shape changes from night to night. These lunar phases exist because the Moon shines by reflected sunlight, and its position relative to the Sun is constantly changing. During the course of its monthly journey through the zodiac, the Moon goes through a complete set of phases, from new to full and back to new (Fig. 1.5). The new moon occurs when the Moon is approximately between us and the Sun. In this configuration, the Moon rises and sets with the Sun and is above the horizon only in the daytime sky. The side of the Moon that faces us receives no sunlight and we cannot see it. The night after new moon, we see a thin crescent in the western sky at twilight. The next night, the Moon has moved farther east through the stars and the crescent has grown. When the crescent is far enough from the Sun to be seen against a dark sky, we can sometimes see the rest of the Moon faintly illuminated by sunlight reflected from the Earth, a phenomenon called "the old Moon in the new Moon's arms."

The first-quarter moon is half-full and rises at noon. At nightfall it is halfway between the eastern and western horizons, and at midnight it sets. The full moon rises at sunset and sets at sunrise. The Moon has now moved to a position opposite the Sun, so it is fully illuminated. But we are still looking at the same hemisphere that faced us when the Moon was new and we could not see it; only the illumination has changed.

Since the full moon is opposite the Sun, in winter full moons occur with the Moon high in the sky, the position the Sun occupies in summer. Similarly, summer full moons are low, in the winter position of the Sun. This opposition of Sun and Moon creates a striking winter effect in places where it snows, since the snow that resists melting because it lies in shadow during the day is brightly illuminated by the full moon at night.

Moving still farther east through the constellations, the waning (shrinking) Moon again reaches half-full phase, rising now at midnight with its curved edge facing the eastern horizon instead of the western one. This is called the third-quarter moon. And finally we have a thin crescent in the morning sky, rising just before the Sun.

Many people in the modern world have not thought much about the phases of the Moon, and they may think that the phases result from the

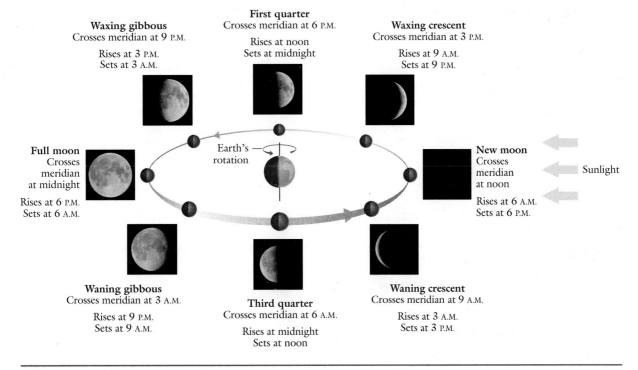

Figure 1.5 Phases of the Moon.

shadow of the Earth falling on the Moon. You might consider how you would explain that this is not the cause of the phases. If you are having trouble visualizing the lunar phases, hold a ball to represent the Moon and illuminate it with a lightbulb. As you pivot and move the ball around you, you will see it display the same phases we have described for the Moon.

The complete cycle of lunar phases happens every 29.5 days, giving us the basic period of our month. Nature was not kind enough to make the year exactly divisible by the month, but 12 is clearly the closest multiple. Thus, we have acquired our present patchwork of 12 stretched and diminished months in one year. And we have 12 constellations in the zodiac, each one corresponding approximately to one of these months (Fig. 1.4). By knowing the location of the Sun with respect to these constellations, the ancients had a good monthly calendar available. Within the months, the phase of the Moon provided a cycle of days.

It should be evident from this discussion that one can account for the apparent motions of both the Moon and the Sun by thinking of them as revolving around the Earth, just like the great sphere holding the stars. This was the popular ancient belief, later codified by Ptolemy of Alexandria and enforced by the medieval Catholic Church. But we know today that the Sun is at the center of our system and the Earth travels around it in one year, turning daily on its axis, while the Moon makes its monthly circuit of our planet.

Eclipses

The circuit of the Moon around the Earth sometimes produces rare events called **eclipses.** There are two different types of eclipse. The most dramatic, but also most rarely seen, is an eclipse of the Sun. If, at new moon, our satellite is exactly between the Earth and the Sun, its shadow falls on the Earth, blotting the Sun from the view of those in its path (Fig. 1.6). It is a coincidence of nature

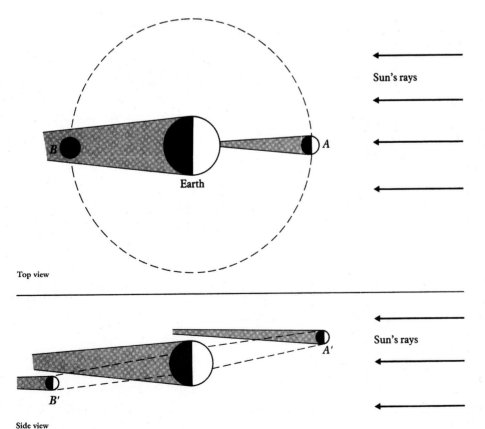

Top view

Side view

Figure 1.6   Top view: Looking down at the Moon's orbit from space. An eclipse of the Sun occurs at new moon when the Moon is directly between the Earth and the Sun (A). An eclipse of the Moon occurs at full moon when the Moon passes into the Earth's shadow (B). (This would be experienced as an eclipse of the Sun by an astronaut standing on the Moon.) Side view: Because the Moon's orbit is slightly inclined relative to the orbit of the Earth about the Sun, eclipses do not occur at each new and full moon. In the configuration shown here, both shadows miss their targets.

Figure 1.7 Multiple exposure photograph of an eclipse of the Sun. Totality (central image) lasts only a few minutes.

that the Moon's distance from the Earth is such that its apparent size is approximately equal to that of the Sun. If the Moon completely covers the Sun, the eclipse is said to be total. A total eclipse of the Sun is one of the most spectacular sights in nature.

The complement to an eclipse of the Sun is an eclipse of the Moon, which occurs when the Moon passes through the Earth's shadow. A lunar eclipse can occur only when the full moon is exactly lined up with the Earth and the Sun (Fig. 1.6). The Earth's shadow is not as dark as the Moon's, however, because our planet's atmosphere scatters sunlight into the shadow, illuminating the eclipsed Moon with a dull orange glow. Unlike an eclipse of the Sun, which is visible only along the narrow path of the Moon's shadow, a lunar eclipse can be seen from the entire night side of the Earth.

Eclipses also occur on other planets. On the giant planet Jupiter, the dark shadow of its satellite Io would produce an eclipse of the Sun for Jovians floating above that planet's colorful clouds. Similarly, the jovian satellites go into eclipse themselves when they pass into their planet's shadow. They can even eclipse each other, with the shadow of one satellite falling on another.

1.2 WATCHING THE SKY: THE PLANETS

Although most of the points of light we see in the night sky are fixed stars that form the unchanging patterns of the constellations, there are five bright ones that move. "Wild Sheep," the Babylonians called them, while the Greeks used the term *wanderer*, from which we get our word *planet*. These wild sheep do not roam over the entire heavenly meadow. Like the Sun and the Moon, they confine their wandering to the 12 constellations of the zodiac.

Our night skies also contain airplanes, whose flashing lights reveal their motion in a few seconds, and artificial satellites in orbit about the Earth, whose motion is also quickly discernible. So it is important to stress that although the planets most definitely move, it usually takes several nights to detect their motion with respect to the fixed stars.

MERCURY AND VENUS

Each ancient civilization had its own names for the planets, often associated with its pantheon of gods. The names we use today are primarily

Roman in origin. The fastest moving planet was appropriately named Mercury (Greek Hermes), after the swift messenger of the gods. This planet is visible only in morning and evening twilight. For this reason, it is not easy to see, and some cultures thought it was actually two objects seen at different times of year: one visible only in the morning and one in the evening.

This same duplicity was sometimes assumed for Venus, which also stays close to the Sun. Venus is the brightest object in the sky after the Sun and Moon, so bright that it is sometimes visible during the day if you know exactly where to look. Sailing serenely in the silent sky, Venus shines like a brilliant jewel, marking the transition between day and night. The ancient Greeks and Romans must have been very impressed, for they named this planet after their goddess of love and beauty (Greek Aphrodite). Venus appears farther from the Sun than Mercury, but never so far that it can be seen in the middle of the night.

Mars, Jupiter, and Saturn

The other three planets easily visible to the unaided eye do not exhibit such a close association with the Sun, and they appear to move through the constellations of the zodiac at a more leisurely pace. They also exhibit **retrograde motion**—a reversal of the normal direction of movement. Like the Sun and the Moon, the planets usually move eastward through the constellations of the zodiac. Each day they rise and set with the other stars, but from one night to the next, you can notice that they have shifted their position eastward. However, sometimes they reverse themselves and move toward the west, and this retrograde motion presents a challenge to any theory that purports to explain planetary motions.

Because of its red color, Mars (Greek Ares) was associated with the god of war. We might first see Mars in the late night sky, moving slowly eastward through the constellations of the zodiac. As the changing seasons cause the stars to rise earlier and earlier, Mars rises earlier with them, becoming visible shortly after sunset and riding high in the sky in the middle of the night. We now say it

is at **opposition** because it is on the opposite side of the Earth from the Sun.

A few months before Mars reaches opposition, its apparent motion gradually slows and then reverses, now going west with respect to the stars. The planet speeds up along a retrograde path, then slows down and returns to its normal direction. Its path appears as a loop in the sky, and in traversing this loop, near opposition, Mars becomes much brighter than usual. Although this pattern must have been apparent to ancient observers, its causes remained obscure. Early perceptions of Mars must have included a certain amount of uneasiness at its apparent unpredictability, an appropriate characteristic for the god of war.

The brightest of the planets after Venus, Jupiter (Greek Zeus) was named for the king of the gods. Jupiter takes 12 years to complete one circuit of the zodiac, and it exhibits only a small retrograde loop each year. The retrograde motion of Jupiter is illustrated in Figure 1.8.

Saturn (Greek Chronos) is even more dignified. This planet requires a full 30 years to travel through the zodiac. Saturn was Jupiter's father and the ancient god of time. Since 30 years was (and still is) a large fraction of a human lifetime, the name is appropriate.

Ancient Interpretations of Planetary Motion

The various apparent motions of the stars and planets provided the first observations that people could use to understand the Earth's place in the universe. The Sun, Moon, stars, and planets all apparently circle the Earth in some kind of spherical system. The circle was regarded by the ancient Greeks as the most perfect geometric figure, and casual observations of the sky indeed suggested that circular paths were entirely appropriate for the bright objects it contained. The Sun and Moon were clearly circular themselves, giving added support to this approach.

With our space-age perspective, we may regard these ideas as hopelessly naive, but a moment's reflection will indicate that all of your daily

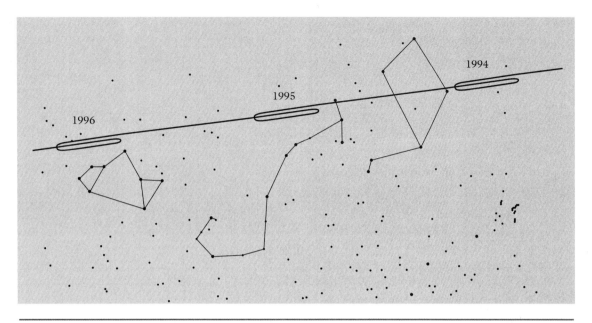

Figure 1.8   Retrograde loops made by Jupiter as it wanders among the stars. The apparent motion of the planet slows down, reverses, slows again, and continues in the original direction.

needs can be met by this simple model of the universe—a flat Earth around which everything else revolves. It is only if you take a long trip or stand at the seashore watching boats disappear over the horizon (or see a picture of Earth taken from space) that you need to question the flat Earth. And even if you understand (as did the ancient Greeks) that the world is spherical, you can still attribute celestial motions to the revolution of the sky about a stationary Earth.

As observations improved, however, the inadequacy of the simple **geocentric,** or Earth-centered, model of circular motions became apparent. Near the beginning of the Roman Empire 20 centuries ago, astronomers started to develop more elaborate models to explain the apparent motions of the planets. They kept the geocentric idea and the perfect circles, but they postulated that instead of moving on a circle with the Earth at the center, each planet moved on a small circle (called an epicycle) that was attached to a larger circle. They also allowed the Earth to be offset from the centers of the planetary circles. By the second century A.D., when Claudius Ptolemy wrote the definitive treatise on classical Greco-Roman astronomy, dozens of cycles and epicycles were employed in the calculation of planetary positions. Cumbersome as this Ptolemaic system may seem to us, it worked very well. It was not until the fifteenth and sixteenth centuries that this approach was abandoned, as a part of the European artistic and intellectual rebirth known as the Renaissance.

1.3   THE SLOW GROWTH OF REASON

Among the basic questions that have been asked in almost all human civilizations are those that concern the size and shape of the Earth, the motions of the stars and planets, and the distances to the Sun, Moon, and planets. Fundamental information about the structure of our solar system is now taught in elementary school. Nevertheless, these "facts" were only slowly established, and even today few of us could actually cite the evidence that proves that our modern perspective is correct.

The slow growth of understanding that led to our present view is a fascinating story, but we can mention only a few highlights here. It is useful to

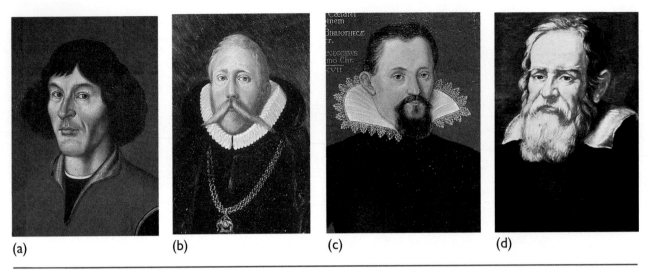

(a) (b) (c) (d)

Figure 1.9 **(a)** Copernicus, **(b)** Tycho, **(c)** Kepler, and **(d)** Galileo—a gallery of the four Renaissance scientists who changed our perception from an Earth-centered, mysterious universe to a Sun-centered solar system surrounded by distant stars.

review these because they offer some lessons about how we should approach new, unsolved problems today, and they show how the perspective we now take for granted was achieved. Here we shall consider the work of just four people: Copernicus, Tycho, Kepler, and Galileo (Fig. 1.9), moving on to Newton in the next section.

COPERNICUS AND THE HELIOCENTRIC SYSTEM

Nicholaus Copernicus (1473–1543) was a Polish astronomer and cleric. He was a contemporary of some of the great Italian painters of the Renaissance, such as Botticelli and Michelangelo, and he was 19 years old when Columbus made his famous voyage to the New World. Copernicus spent some time as a student in Italy before returning to Poland in 1503, where he was given a position in the Church.

Copernicus was taught the geocentric system that had been developed by Claudius Ptolemy 1300 years earlier (Fig. 1.10). In the Ptolemaic system, the motionless Earth stood in the center of the universe and everything else circled around it. Copernicus did not like this system because it seemed too complex with its various cycles and epicycles. Aesthetically, it made more sense to

him for the radiant, life-giving Sun to be in the center, with the planets revolving around it (Fig. 1.11). This bold hypothesis allowed him to explain the apparent motions of the planets more easily than the geocentric system could. It also allowed the stars to be motionless; this pleased Copernicus because he thought it was more reasonable for the small Earth to move about the Sun than for the great vault of heaven to turn about the Earth. This new perspective is called **heliocentric,** or Sun-centered.

It is hard for us to recapture the challenge to human thought that the Copernican system required. We are so accustomed to the modern heliocentric view that we find the older geocentric system absurd. The geocentric perspective, however, had been supported by most of the philosophers of ancient Greece and Rome as part of a broader framework that argued for a central Earth and unchangeable perfection in the heavens. These ideas had become entrenched in the medieval Christian Church, which was then the dominant intellectual and political force in Europe. The new heliocentric theory was seen by many as a direct affront to religion. Moreover, it contained the seeds of a radical alteration in our perception of ourselves. Previously, the Earth was large and the rest of the universe seemed

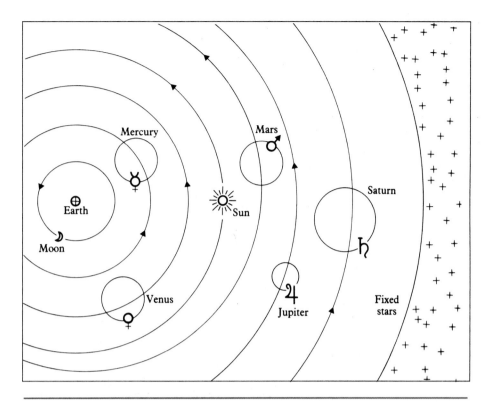

Figure 1.10 The geocentric, or Ptolemaic, system developed in antiquity to explain the apparent motions of the Sun and planets. Each object except the Earth must move in small epicycles in addition to its circular motion about the Earth.

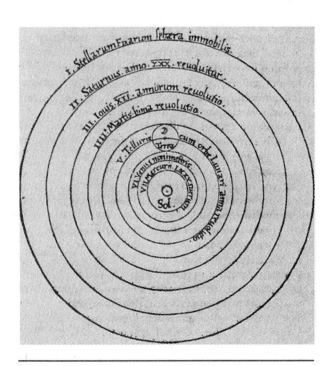

Figure 1.11 The heliocentric system: a copy of the original sketch of his plan for the solar system published by Copernicus. Note that only the Moon orbits the Earth, while all the planets (including Earth) move around the Sun.

secondary. Humans lived at the center of the cosmos. Copernicus made the Earth small and the universe big, thereby demoting humanity from its central position.

Copernicus had no proof that his heliocentric system was correct. To him it was both more clear and more pleasing than the geocentric system. Others felt differently. Our senses tell us that the Earth is stationary and the sky revolves around it—not an easy point of view to abandon. Nor, in that period of history, were people accustomed to the idea of using experiments or observations of nature to distinguish between different philosophical systems. But consider just one simplifying stroke that Copernicus achieved.

Previously it had been thought that the motions of the planets as seen in the sky were real. Thus, when a planet looked as if it were moving backward (retrograde motion), it really was. The retrograde motion was thought to be caused by the planet's motion on small circles called epicycles that were centered on their circular orbits. When a planet needed to go backward, it traversed

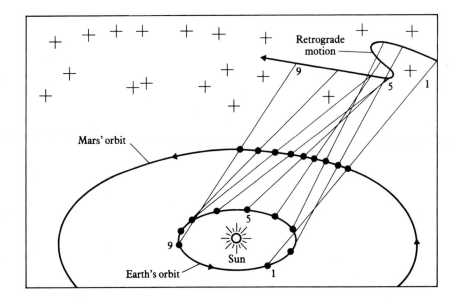

Figure 1.12 Retrograde motions as explained by the Copernican system. The outer planets appear to move backward as the faster Earth overtakes them.

the small circle in a direction opposite to that in which the big circle carried it. This complex arrangement may be compared with what happens when the Sun is placed at the center. Now the retrograde motion of the planet is simply the effect seen by an observer on Earth as we overtake the planet in our journey around the Sun (Fig. 1.12).

Note that both the geocentric and the heliocentric models for the solar system explain retrograde motion. Copernicus simply does it in a less complex, more elegant way. This is a frequent situation in science, where more than one theory can sometimes explain the same set of observations. However, when the theories are as different as these, they will make different predictions about other phenomena that have not yet been observed. These predictions can then be tested by new observations, which will provide a way of choosing between the two theories.

Tycho and Kepler

In the absence of observations to distinguish the heliocentric from the geocentric system, the ideas of Copernicus remained highly controversial. One of the astronomers who found them inter-

esting but not entirely satisfactory was Tycho Brahe (1546–1601). Born in Denmark three years after the death of Copernicus, Tycho's early education was in jurisprudence. An eclipse of the Sun he had seen as a boy of 14 made a profound impression on him, however, and he soon abandoned law to become the most accomplished observer of the heavens in the era before the invention of the telescope. Using instruments of high precision but with no optical parts, Tycho surveyed the heavens, measuring the positions of the stars and planets with an accuracy far greater than that of anyone before him.

In those days, one of the chief tasks of astronomers was the accurate prediction of planetary motions, and the old Ptolemaic tables were no longer adequate for this task. Tycho was determined to improve this situation. He was also a colorful character, rather vain, who wore a silver nose as the result of a dueling injury. Among his own discoveries, his observational proof that a "new star" that appeared in 1572 was farther away than the Moon was an important step away from ancient ideas. It showed that the heavens were not immutable, that the stars themselves could change.

It was Tycho's successor, Johannes Kepler (1571–1630), who extracted the hidden science from these remarkably precise observations of planetary positions. Kepler was taught the Copernican system at the University of Tübingen, where he was studying to become a Lutheran minister. But the faculty there, recognizing his intellectual gifts, persuaded him to take a post as a mathematician at the University in Graz, Austria, in 1594. (To place the period, note that in England Shakespeare was writing his plays and sonnets at this time.) In addition to exhibiting a real gift for mathematics, Kepler was strongly influenced by a mystical sense of the intrinsic harmony and order in the universe. It was this sense that provided his primary motivation for studying the motions of the planets: He felt that there must be an underlying set of simple principles that governed the structure of the solar system.

Kepler was fortunate in that his early work came to the attention of Tycho Brahe. Tycho recognized the younger man's mathematical abilities and invited Kepler to join him in Prague in 1600. Tycho died a year later, and Kepler took over his post.

Kepler was fascinated by the simplicity and order in the Copernican system. Because of his own mystical beliefs, he was certain that there must be a way of expressing its apparent harmony in simple mathematical relationships. This proved very difficult to do, but nine years after his arrival in Prague, Kepler published a book containing the first two principles of planetary motion that we now refer to as Kepler's laws.

KEPLER'S LAWS OF PLANETARY MOTION

Kepler had been trying to analyze Tycho's observations of Mars, fitting them to the system of orbits described by Copernicus. It simply wouldn't work. The problem lay in the circular orbits that the Copernican system still used. Mars obviously didn't move about the Sun in a circle, not even a circle in which the Sun was not at the center. So Kepler tried a variety of other geometric shapes until he found the elongated oval called an **ellipse.** The ellipse worked, not just for Mars but for all the planets.

At first Kepler was unhappy with the ellipse because it doesn't have the perfect symmetry of the circle. The Sun is at one focus, but there is nothing at the center of the ellipse and nothing at the other focus (Fig. 1.13). The ellipse is still elegantly simple, however, and it worked. This then is Kepler's first law:

1. *Each planet moves in an elliptical orbit about the Sun, with the Sun at one focus of the ellipse.*

Another problem that had vexed astronomers was the fact that at some times the planets moved faster in their orbits than at others. This seemed random and unexpected, until Kepler found another principle that made the behavior intellectually acceptable. With the planets moving in elliptical orbits, it turns out that the fastest motion occurs when they are nearest to the Sun. Consequently, an imaginary line drawn between the Sun and the planet will sweep out the same

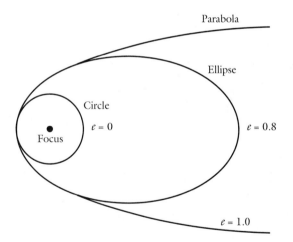

Figure 1.13   The circle, ellipse, and parabola are all possible shapes for orbits according to Kepler's laws. These curves are sometimes called conic sections, and you have probably encountered them in geometry class. The common point in this drawing is called the focus; it is the same as the center of a circle but is offset for ellipses. These curves have different shapes or eccentricities, increasing from $e = 0$ (circle) to $e = 1.0$ (parabola). The Earth's orbit has an eccentricity of only 0.02 and that of Venus (which has the most nearly circular orbit) is only 0.007. In contrast, comets often have eccentricities as high as 0.99, indicating an ellipse so elongated that it is practically a parabola.

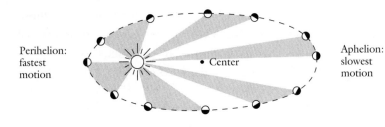

Perihelion:
fastest
motion

• Center

Aphelion:
slowest
motion

Figure 1.14   Kepler's second law. A planet moves fastest when it is closest to the Sun (perihelion), yet the area swept out by a line connecting the planet to the Sun is always the same for a given period of time. Thus, the areas of the light and dark triangles are all equal.

area within the ellipse during the same interval of time, regardless of where the planet is in its orbit (Fig. 1.14). This discovery restored a certain order to nature and became another simple mathematical description of what was observed. It is Kepler's second law:

> 2. *An imaginary line connecting the Sun with a planet sweeps out equal areas in equal times as the planet moves about the Sun.*

The best was yet to come. Kepler was still searching for some principle that would relate the motions of the planets to one another. He now had an accurate description of how they moved, but he was certain that there must be a harmony in those motions that would explain why the planets moved the way they did. Ten years after publishing his first two laws, he found that harmony. What he discovered was a simple mathematical relationship between the period of a planet's revolution about the Sun and its distance from the Sun. This is Kepler's third law (Fig. 1.15):

> 3. *The cube of the distance from the Sun divided by the square of the time required to traverse the orbit is a constant, the same for every planet.*

Written as an algebraic formula, the third law is

$$D^3 = AP^2$$

where D is the distance of the planet from the Sun and P is the orbital period. The letter A stands for a number whose value depends on the units used; if we measure the period in years and the distance in astronomical units, then $A = 1$. The **astronomical unit (AU)** is the name astronomers give to the average distance of the Earth from the

Sun, approximately 150 million km. Kepler didn't know the length of the astronomical unit very precisely, but the third law gave him a way to express the sizes of all the other planetary orbits in terms of the Earth's orbit.

This third law represented a remarkable achievement. Kepler was overjoyed to have found this underlying harmony in the solar system. His work had greatly strengthened the Copernican system, providing the predictive power it had lacked in its original form when the planets were thought to move in circles. Kepler's laws, however, are still descriptive. His work could be viewed as simply providing a much better way to calculate the positions of planets. The proof that would clearly distinguish between the geocentric and heliocentric systems was still lacking.

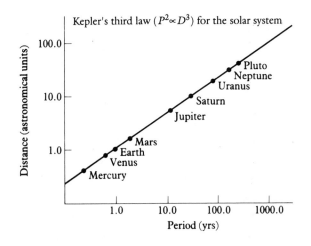

Figure 1.15   Graph illustrating Kepler's third law. The square of a planet's period (P^2) is proportional to the cube of its distance from the Sun (D^3). Periods are plotted in years and distances in units of the Earth's distance from the Sun; both scales are logarithmic.

GALILEO AND THE FOUNDATION OF MODERN SCIENCE

Kepler was a contemporary of the Italian Galileo Galilei (1564–1642), who was at this time laying the foundations of modern physical science. Galileo's experiments and observations, just as much as Copernicus's and Kepler's theories, were responsible for overturning the geocentric model of the solar system.

Galileo, the son of a musician, began his studies in medicine but was soon attracted to mathematics. In 1609, he was teaching mathematics in Padua when he heard about the invention of the telescope. Unlike many of his contemporary academics, he recognized that this combination of lenses enables one to see better, not just differently (as with a distorting mirror). He quickly made some telescopes of his own and began using them to study the night sky. What he found provided the first experimental demonstration that the geocentric universe and its philosophical underpinnings were not simply complex and inelegant; they were wrong!

One of Galileo's earliest discoveries was the presence of some small "stars" very close to Jupiter. Observing night after night, he found that these stars changed their positions with respect to the planet in an orderly, repetitive way. Evidently, they were satellites moving in four separate orbits around Jupiter. This discovery was a major blow to the geocentric theory, which required that everything must orbit the Earth. Galileo saw that what he had found resembled a miniature Copernican solar system itself, in which Jupiter played the role of the Sun and the four satellites that he had found resembled the planets.

The evidence for a heliocentric system was further strengthened when Galileo discovered that the planet Venus exhibited phases like those of our Moon. Only in a heliocentric system could Venus be fully illuminated at some part of its orbit, as seen from the Earth. If this planet orbits the Sun, it also should become larger in apparent size as it nears the Earth, at the same time shrinking to a thinner and thinner crescent. Conversely, as Venus moves around to the other side of the Sun, we can see more of it because of the favorable viewing angle, but it appears much smaller than when it is a crescent because it is now much farther away from us (Fig. 1.16). This behavior, which is exactly what Galileo observed, is impossible in the geocentric system, which would not permit an observer on Earth to see Venus fully illuminated.

The geocentric system was clearly untenable, at least to those who had looked through Galileo's telescope or who believed what he wrote, but the Catholic Church did not accept these proofs because they appeared to contradict scripture. The Church felt that the Copernican ideas could be discussed, but only as "theory," not fact. Galileo was unwilling to abide by these restrictions, and in 1633 he was arrested and forced to deny the heliocentric theory, 14 years after Kepler (who did not live in a Catholic country) had published his third law. These were perilous times for dissenters: Kepler had to interrupt his work at one point to save his mother from being executed as a witch, and only nine years before Galileo began experimenting with telescopes, Giordano Bruno was burned at the stake for his heretical views of the universe. Galileo was treated much less severely, being sentenced to house arrest for the last eight years of his life. Galileo saw no contradiction between astronomy and religion, and gradually his view prevailed, although the Roman Catholic Church did not formally exonerate him until 1992.

1.4 NEWTON AND THE LAW OF GRAVITATION

Isaac Newton (1642–1727) was born only a few months after the death of Galileo. He therefore grew up at a time when the basic discoveries we have just described should have been part of his college curriculum. As an undergraduate at Cambridge from 1661 to 1665, however, Newton found that the university was still teaching the old geocentric view of the universe. Nevertheless, he managed to learn about the new ideas, to which he added his own insights. These days undergraduates interested in science and mathematics study

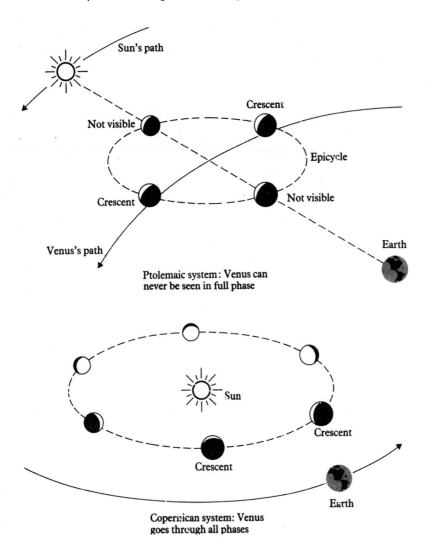

Figure 1.16 The Ptolemaic (Earth-centered) view of the solar system (bottom) could explain crescent phases for Venus but could not predict that a full phase would occur. (Venus would have to appear in the midnight sky, which it does not do.) The Copernican (Sun-centered) view (top) correctly predicts the phases and also the observed fact that Venus at full phase appears much smaller than it does as a crescent—it is simply much farther away.

calculus. Newton began inventing calculus as an undergraduate, while also formulating ideas about optics and mechanics.

Throughout a long career (he died at age 84), Newton made many contributions to physics and astronomy (Fig. 1.17). He was one of the first scientists to use quantitative, or mathematical, arguments to test scientific ideas. Because he was so successful, he fundamentally altered the human perception of the universe and its workings. The impact of his accomplishments was nicely summed up in the following couplet by a contemporary poet, Alexander Pope:

Nature and Nature's laws lay hid in night:
God said, let Newton be! and all was light.

THE LAWS OF MOTION

Newton's most important contributions to astronomy were his three laws of motion and the law of gravitation. The first law of motion may be stated as follows:

1. *Every body continues in its state of rest or of uniform motion in a straight line unless it is compelled to change that state by the action of some outside force.*

The first part of this law seems pretty obvious—things don't move unless they are forced to do so—but the second may not be so evident. There are no familiar examples of an object moving uniformly in a straight line. Indeed, the Greeks and

Figure 1.17 Isaac Newton, the scientist who discovered the physical laws that govern planetary motions, thereby bringing a sense of order to an apparently random universe.

Romans had thought that the circle was the natural form of motion of celestial objects. But Newton realized that in the heavens, as on Earth, an object moving without any force on it would go straight. In everyday life, we find that things that are moving eventually slow down and stop. Newton realized that this was because of friction. As friction is reduced, motion continues for a longer time, and one can imagine a situation with no friction at all in which a moving object coasts on forever.

How can we express the relationship between the forces that act on bodies and the motions of these bodies? Newton's second law provided the answer:

> 2. *The change of motion (acceleration) is proportional to the force that acts on the body and inversely proportional to the mass of the body.*

Newton's laws, like Kepler's description of orbits, are so powerful because they can be expressed mathematically and used for calculations. Stated in words alone they are rather obtuse, but expressed mathematically they are tremendously powerful. Written algebraically, with A as acceleration, F as force, and M as mass, the second law becomes $F = MA$. This law provides a way to calculate the motion of a body from the force exerted on it. Or, conversely, by tracking the motion, we can calculate the force. These calculations, carried out today with large computers, are still the way we navigate our spacecraft on their interplanetary journeys.

The larger the force, the greater the acceleration. Also, a given force will have more effect on a body of small mass than on a large one. All of this is well known from everyday experience, where we deal frequently with objects (autos, for example) that change speeds. What may be less obvious is that an object moving at constant speed in a circular or elliptical path is also accelerating. Changing direction is a form of acceleration, which means that a force is acting on the body to produce the curved path.

Newton's third law takes us back to the issue of balancing forces:

> 3. *To every action there is an equal and opposite reaction.*

This law says that when an apple falls toward the Earth, the Earth must also move toward the apple. But since the Earth has about 10^{26} times as much mass as the apple, yet the force acting on the two of them is the same, the Earth accelerates 10^{26} times less. That is a small number indeed! To see this law in action, you might find a partner and try pushing each other while standing on roller skates or skateboards. This law can also make getting into or out of a canoe rather wet work, while getting into and out of a big boat is easy.

THE LAW OF GRAVITATION

The framework provided by Newton's three laws of motion seems very close to providing an understanding of why the planets move around the Sun in the way that Kepler described. Obviously, a force is acting on them, or they would be moving in straight lines. But what is that force that pulls them toward the Sun? And what is the balancing force that allows them to continue this motion instead of slowing to a stop or falling into the Sun?

Newton's brilliant solution to these dilemmas was the law of gravitation. If an apple falls on Earth, it is being attracted to the Earth by a force. Perhaps that same force is attracting the Moon. The Moon doesn't fall all the way to the surface of the Earth because it is moving sideways, approximately at right angles to the force

---

PERSPECTIVE

SCIENTIFIC SKEPTICISM

What is science? On one level it is a vast collection of facts gathered from observation and experiment, facts that can be used to describe the world around us and to develop the technology that dominates our lives. Scientific knowledge and invention have made incredible progress. We know far more about the world and the way it works than did our predecessors. As we noted, the mathematics called calculus, invented by Newton, one of the smartest scientists who ever lived, is now taught to millions of undergraduates each year. Einstein's theory of relativity, which challenged the best minds of the early twentieth century, is now required knowledge for seniors majoring in physics. How has this explosive progress come about?

To answer this question, we should think about science not as a collection of facts but as a way of thinking. Science is a process that allows us to build upon the accomplishments of the past. It provides guidelines for testing ideas, so that we can keep the best ideas and discard those that fail the test. While scientists, like everyone, make many mistakes, the scientific process allows those mistakes to be exposed and superseded by more accurate concepts.

One way to look at this process is to say that scientists are skeptics. They do not accept ideas uncritically. They question each other. They debate new ideas or results and devise experiments to test them. They want to know why things work the way they do.

Often we start with observations or the results of experiments. Newton measured how fast an apple fell from a tree, and he was aware of the distance to the Moon and of its motion. If we are clever, we may come up with an idea that relates the information. In Newton's case, the idea was that both the apple and the Moon are pulled toward the Earth by the force of gravitation. He went further and put this idea in quantitative form, suggesting that the gravitational attraction declined with the square of the distance. We might call such an idea, carefully phrased in words or mathematical symbols, a *hypothesis*.

A hypothesis is an invitation to make further observations or experiments—to collect more data. Newton calculated the acceleration of the apple and the Moon and found that they matched his hypothesis. Others applied the same hypothesis to the moons of Jupiter. Again it worked. More and more applications were found, and they all confirmed the hypothesis. Eventually this hypothesis became the theory of gravitation. A *theory* is a scientific idea

➥

---

of gravity. The attractive force changes the motion of the Moon from a straight line to a closed curve, an orbit around the Earth. In effect, the Moon is falling around the Earth.

Newton's law of gravitation states:

The gravitational force between two objects is proportional to the product of their masses and inversely proportional to the square of the distance between them.

If the force is F, the masses are M_1 and M_2, and the distance is R, the same law can be written as

$$F = \frac{GM_1 M_2}{R^2}$$

where G stands for a number called the gravitational constant.

This is an extremely powerful law (Fig. 1.18). With it, we can determine the mass of a planet or the mass of the entire Galaxy. It also governs the flights of both spacecraft and polevaulters. And it explains why the planets move the way they do. The planets are "falling" around the Sun in the same way the Moon moves around the Earth. Left

➥ that has passed many tests and seems capable of tying together large amounts of otherwise disparate data. In science, to call an idea a theory or law is the highest praise. This is quite different from the common usage of the word, where *theory* seems to mean any sort of casual idea or hypothesis.

A scientific theory or law is not absolute. It can be overthrown or superseded as new data become available. Newton's theory of gravitation was superseded in the twentieth century by the theory of relativity, which provides a different way of looking at gravitation and makes slightly different predictions about the way planets move. In most applications, Newton's theory works just fine. But when we want maximum precision—for example when Jet Propulsion Laboratory engineers are navigating a spacecraft to another planet—we use more complicated formulas of relativity. Each theory has its place, but neither represents absolute truth.

A skeptical, questioning attitude is essential for a research scientist, who is on the lookout for some flaw in current theory, some new discovery or unexplained phenomenon that might lead to a more profound understanding of the universe. But this skepticism also requires an understanding of how to test a hypothesis. We need to know something about the rules of evidence. We have to learn about statistics and probability. These are the same skills that a citizen needs to make judgments about important issues of the day. We all need to have some understanding of the evidence concerning global warming, for example—is the Earth really getting hotter, and what will that mean to our lives? We need this sort of skepticism every time we hear about a new treatment for a disease—has it been shown to be effective, and what are the possible side effects? When a politician makes a statement about the relationship between tax rates and the future growth of the economy, we need to ask what is the evidence for the expected outcome. When a police department makes an arrest based on the testimony of someone under hypnosis, we should ask to see the evidence that hypnosis actually leads to valid testimony.

Carl Sagan, one of the best known planetary scientists of the late twentieth century, was passionate about the value of scientific skepticism as a kind of "baloney detector kit" for citizens. In *The Demon Haunted World*, he wrote, "The tenants of skepticism do not require an advanced degree to master, as most successful used car buyers demonstrate. The whole idea of democratic application of skepticism is that everyone should have the essential tools to effectively and constructively evaluate claims to knowledge."

to their own devices, they would move through space in straight lines. The acceleration toward the Sun that they experience from gravity is just balanced by the acceleration produced by their orbital motion. Hence, they are condemned to move forever in the orbits in which we find them.

One important consequence of the law of gravitation is the realization that the gravitational force between two objects depends on the product of their masses. It logically follows that the orbit of one object around another must also depend on their masses. Although the effect of the mass of the planet on its orbit is small, it is important, for it allows us to calculate the masses of the planets from observations of their motion.

SCIENCE SINCE NEWTON

"If I saw a little farther than others," Newton wrote in a letter, "it was because I was standing on the shoulders of giants." This modest appraisal of his extraordinary achievements is nevertheless an accurate description of how science proceeds, each step forward building on the

Figure 1.18 A couple holding each other at the intimate distance of 1 cm are exerting a gravitational attraction on each other that is about equal to the attraction of the Moon on either one of them. Fortunately, the gravitational force exerted by the Earth on both of them is about 100,000 times greater, and there are other forces operating between people that are stronger than gravity.

work of those who came before. This was particularly the case after Newton, as his carefully reasoned, mathematical approach set the tone for most subsequent scientific inquiry.

Galileo and Newton ushered in a new world of experimental and quantitative thinking. During the following decades, the pace of science quickened, and one discovery followed another. Ever more elegant and powerful mathematical tools were developed, just as scientific instruments become more complex and precise. Even though space travel was still more than two centuries in the future at Newton's death in 1727, the way toward the modern world was clearly indicated.

1.5 ESCAPING FROM EARTH

The laws of motion and of gravitation provide a framework for understanding the orbits of planets and other members of the solar system. They also permit us to understand the operation of rockets and the principles that govern Earth satellites and interplanetary probes.

ROCKETS

Whether we are launching a space telescope into low Earth orbit or sending a probe to the outer planets, we need to overcome the force of gravity exerted by the Earth. To do that, we must lift a spacecraft above the atmosphere (so friction will not slow it down) and supply it with an amount of energy sufficient to keep it from falling back. The best device to impart this energy to a spacecraft is a rocket.

A rocket accelerates a spacecraft by expelling gases backward at high speed. This is an excellent example of Newton's third law: The exhaust goes in one direction and the reaction drives the spacecraft in the opposite direction (Fig. 1.19). A jet engine works on the same principle, but it draws oxygen from the atmosphere to burn with its petroleum fuel. A rocket carries its own oxidizer as well as its fuel, and therefore it can operate in the near-vacuum of space.

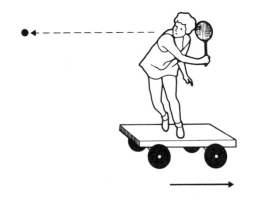

Figure 1.19 Newton's third law applies both to tennis and to rockets. When the girl puts the ball in play with a backhand shot to the left, she and the cart will move to the right.

Many kinds of rockets have been developed for spaceflight. For large, brute-force applications there are solid-fuel rockets, which look like giant firecrackers and are nearly as explosive. The external boosters of the Space Shuttle are solid-fuel rockets. More easily controlled are liquid-fuel rockets, such as those of the three shuttle main engines. Many liquid-fuel rockets are designed to be restarted in space, a requirement if we are to make in-flight adjustments on interplanetary trajectories. The most efficient liquid-fuel rockets burn hydrogen and oxygen, but these two gases must be kept extremely cold to be used in liquid form, so they are not suitable for engines that must be restarted. Less efficient fuels that remain liquid at moderate temperatures are needed for flights to deep space. A nitrogen compound called hydrazine is one of these.

GOING INTO ORBIT

Suppose that a rocket has lifted a spacecraft above most of the Earth's atmosphere and is ready to inject it into orbit. Enough horizontal speed must be imparted to keep the new artificial satellite from falling back to the Earth. To understand how the trajectory depends on the speed, imagine you are on a mountain that sticks up above the atmosphere, and ask yourself how fast you must throw a ball in the proper direction to keep it in orbit (Fig. 1.20).

The motion of the ball results from the pull of gravity in combination with the ball's forward speed. If the ball is dropped instead of thrown, it simply falls toward the center of the Earth. A moderate forward speed—say, a few thousand kilometers per hour—results in a curving trajectory that takes the rocket partway around the world, on a path similar to that of an intercontinental missile. When the speed reaches a critical value called the **orbital velocity,** the surface of the Earth curves away as fast as the ball falls, resulting in a low-altitude circular orbit. For Earth, the orbital speed is about 8 km/s, or about 28,000 km/hr.

Calculations using Newton's laws show that the square of the orbital velocity is proportional to the mass of the planet. Thus, a smaller planet with its weaker gravity requires a lower speed to achieve orbit, just as common sense dictates. In addition, the higher the satellite orbit, the slower its speed. At the distance of the Moon, the orbital speed is only about 1.3 km/s.

It is easy to calculate the orbital period of an Earth satellite. The period is the distance traveled in one orbit divided by the orbital speed. Since the circumference of the Earth is about 40,000 km, a satellite orbiting just above the atmosphere (at an altitude of about 200 km) will take P = circumference/speed = 40,000/28,000 = 1.4 hours or about 90 minutes to circle the planet. We can sometimes see these satellites moving against the pattern of fixed stars in our night sky. The International Space Station is by far the brightest such satellite. As it happens, the period for a low orbit is nearly the same for any planet. From a tiny asteroid up to giant Jupiter, the minimum orbital period is between 1 and 2 hours.

We know the Moon takes about 29 days to orbit the Earth at a distance of 384,000 km, so there must be a distance somewhere between 200 km and 384,000 km where the period of an orbit is just equal to 24 hours, the length of our day. Of course there is, and it is 35,680 km.

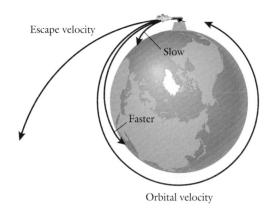

Figure 1.20   Going into orbit. Shooting a ball horizontally from a mountaintop with greater and greater velocity produces different trajectories that will ultimately put the ball into orbit.

A spacecraft put into an orbit at this altitude above the equator will appear stationary to an observer on Earth. In fact, only observers who are on the right side of the Earth could see such a satellite; it will never become visible to inhabitants of the other hemisphere. This kind of orbit is called geostationary. It is very useful for communications or meteorological satellites on Earth. We shall find when we study Pluto that this planet's natural satellite occupies just such a stationary orbit.

Escape Velocity

Return to Figure 1.20 and consider what happens when we throw a ball into orbit with greater speed. As the speed increases, the orbit becomes more and more elongated. The point on the ellipse that is closest to the Earth is called the *perigee*, and the greatest distance (on the opposite side of the Earth) is the *apogee*. (If the orbit were around the Sun, these points would be called **perihelion** and **aphelion,** respectively.) As the initial speed increases, the apogee moves farther and farther out, until we reach the point where the ball escapes completely.

The **escape velocity** described here is exactly the same as the speed required for a ball projected straight up from the surface to escape entirely from the planet. At this speed, the energy is so great that gravity, pulling backward, can slow but never quite stop the outward motion. Mathematically, the escape velocity is equal to the orbital velocity multiplied by the square root of 2, or about 1.4. For the Earth, the escape velocity is 40,000 km/hr, or just over 11 km/s.

Escape from a planet becomes easier if the mass of the planet is small. For example, the escape velocity from the asteroid Eros is only about 40 km/hr. This means that you could literally throw a ball into space from Eros if you have a strong arm, or you could ride into space on a motorcycle if you could find a smooth highway to serve as your launch pad. When the spacecraft called NEAR-Shoemaker landed on Eros in 2000 (the inverse of escape), the bump was no worse than jumping off a stepladder on Earth.

1.6 The System Revealed

It should be clear from the work of Kepler and Newton why the planets appear to move the way they do. The fact that they confine their motions to a narrow band along the ecliptic is a natural consequence of having their orbits nearly in the same plane. Imagine yourself as a letter in a word on the middle of a single page taken from this book. Stretching away from you on all sides are the letters that make up the other words you are now reading, so you would be surrounded by words, but only in the plane of the paper. Looking up or down, you would see no writing.

It is the same for us in the solar system. If we stood on the surface of Mars and studied the night sky, we would see the same constellations we see from Earth, and we would find that the planets were still moving through the zodiac. When we look out toward the planets moving in front of these constellations, we are seeing an edge-on view of the solar system. From Mars, the Earth would be a gorgeous blue-white planet in the twilight sky, exhibiting the same kind of apparent motions with respect to the Sun that Venus does from our terrestrial perspective.

The two planets that stray farthest above and below the plane defined by the Earth's orbit are Mercury and Pluto. Their orbits have **inclinations** of 7° and 17°, respectively (Fig. 1.21). All the other planets have orbital inclinations within 4° of Earth's orbit. The orbits of Mercury and Pluto also have the greatest **eccentricity**; that is, the ellipses along which these planets move are more elongated than are the paths of the other planets.

We can also gain an approximate idea of the relative distances and sizes of the planets from simple visual observations, once we have adopted the Copernican point of view. Jupiter appears to move more slowly through the zodiac than does Mars, so it must be farther away from us. The fact that it is nevertheless the brightest object in the sky after Venus suggests that it must be unusually large. Saturn, still slower, is even farther away and must also be big to be as bright as it is. Mars is much closer than these two giants and must therefore be relatively small.

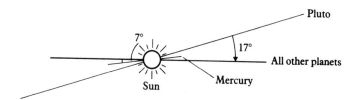

Figure 1.21 All the orbits of the planets except those of Mercury (closest to the Sun) and Pluto (most distant) lie essentially in the same plane.

DIRECTIONS OF MOTIONS

Which way is up? The convention on the Earth is to say that north is up. North in the sky is defined as the direction in space toward which the Earth's North Pole points. If we imagine ourselves located in space somewhere above the solar system in this sense, we would find when we looked down at the planets that they were all moving about the Sun in a counterclockwise direction. Looking down at the Earth from above its North Pole, we would see that our planet is also rotating on its axis in a counterclockwise direction, in the same direction that the Earth moves around the Sun.

Looking down on Venus from our lofty perch north of the system, we would find that this planet is rotating in the opposite direction, clockwise instead of counterclockwise, even though its revolution about the Sun is in the same direction as all the other planets. Uranus and Pluto would appear tipped over on their sides, with their north poles (defined by rotation) actually a little below their orbital planes. They too are technically rotating backward compared with Earth and the other planets.

The various satellite systems show a different kind of symmetry. Most of the larger satellites stay close to the equatorial planes of their parent planets. The same is true for systems of rings. In the case of Saturn, for example, the ring system and the satellite orbits all share the planet's inclination of 27° to the plane of Saturn's orbit.

Most satellites revolve around their planets in the same counterclockwise direction (as seen from the north) that the planets move around the Sun, but several of Jupiter and Saturn's small moons, and one of Neptune's, revolve in the opposite sense. (Clockwise rotation or revolution is called **retrograde**, the same word used in Section 1.2 for the apparent backward motion of a planet seen in the sky.)

The asteroids are a diverse lot of small bodies with diameters less than 1000 km that orbit the Sun primarily between Mars and Jupiter. There are tens of thousands of them, but their combined mass is less than one-tenth the mass of the Moon. Their orbits show a variety of eccentricities and inclinations, but on average they are inclined about 8° and are more eccentric than most planetary orbits.

The orbits of the comets are even more diverse. In their pristine state, most of these bodies orbit the Sun at huge distances, some as much as 1000 times the distance of Pluto. Unlike the planets, their orbits show a random distribution of inclinations. When one of these comets appears in our night skies, it is therefore rarely in a constellation of the zodiac. Its orbit may be perpendicular to the plane of the planets, or it might move around the Sun in a retrograde direction. For us to see it, the comet must have sailed in from its distant domain, perhaps perturbed by the gravity of a passing star.

A SCALE MODEL

To gain an appreciation of the size of the solar system and the distances between the various planets, we can make use of a scale model. Let's reduce every dimension in the solar system by a factor of 200 million. On this scale, the Earth is the size of an orange, and our Moon would be a grape, orbiting the orange at a distance of 2 m. The Sun would be a little more than 1 km away, and it would have a diameter of 7 m, the height of a two-story house. It is the great distance of the Sun from the Earth that makes it appear to be about the same size as the Moon when we see both of them in our skies.

Although the Earth and the other inner planets in this model are only the size of pieces of fruit, we do have some bigger planets. At this scale, Jupiter would be a large pumpkin, still small compared with the Sun but 11 times larger than Earth. Considering the Saturn system, we find that if we measure it from one edge of the rings to the other, the planet and its rings would just fit between the Earth and the Moon. Saturn's own large satellites would lie far beyond the Moon. Scales for the outer solar system are considerably grander than those for the planets close to the Sun.

Moving farther out, we pass Uranus and Neptune and finally encounter Pluto, another grape (but smaller than our Moon) at an average distance of 30 km from the Sun. This is still not the edge of the solar system, however. The edge lies at a point where the gravitational field of the Sun is challenged by the fields of passing stars. The great spherical cloud of comets extends to a distance of some 35,000 km from our scale-model Sun. This means that our scale model solar system is about four times the size of the real Earth.

This exercise illustrates how large the Sun is, how much space exists between the planets, even in the inner solar system, and to what an enormous distance the Sun's gravitational influence extends. The latter emphasizes once again how huge

the distances are between the stars. The nearest star (Alpha Centauri) is five times the distance of the comet cloud; to include it, our model would have to extend to a dimension of 200,000 km.

The real solar system is described in Table 1.1. It is useful to try to keep a few dimensions in mind. Jupiter, the closest of the giant outer planets, is about five times as far from the Sun as we are (that is, 5 AU). The next giant planet, Saturn, is about as far from Jupiter as Jupiter is from us. And the next step out, to Uranus, again nearly doubles the size of the system.

1.7 QUANTITATIVE SUPPLEMENT:
KEPLER'S LAWS IN MATHEMATICAL FORM

Kepler's laws provide a fundamental description of planetary motion that can be generalized to any system of objects of small mass in orbit around a larger object (such as the satellite systems of the giant planets). Although Newton and his successors developed many ways of improving the accuracy of orbital calculations by taking into account the gravitational influence of each planet upon the others (called perturbations), the basic utility of Kepler's three laws remains.

Kepler's first law states that orbits are ellipses with the Sun at one focus. The formula for an ellipse is

$$r = \frac{D(1 - e^2)}{1 + e \cos \theta}$$

where r is the distance from the Sun, D is the semimajor axis, e is the eccentricity, and θ is the angle that specifies the position of the planet along the ellipse, as seen from the Sun. Perihelion, the smallest value of r, corresponds to $\cos \theta = 1$ and is given by

$$\frac{r_{min}}{D} = \frac{1 - e^2}{1 + e^2} = 1 - e$$

The maximum value of r (aphelion) is similarly

$$\frac{r_{max}}{D} = 1 + e$$

Table 1.1 Dimensions of the planetary system

Planet	Distance from Sun (million km)	Distance from Sun (AU)
Mercury	58	0.39
Venus	108	0.72
Earth	150	1.00
Mars	228	1.52
Asteroid belt	330–500	2.20–3.30
Jupiter	778	5.20
Saturn	1429	9.55
Uranus	2875	19.22
Neptune	4504	30.12
Pluto	5900	39.44

Kepler's third law can be expressed as

$$D^3 = AP^2$$

when D is in AU and the period, P, is in years. The constant A equals 1 AU^3/yr^2. Newton later showed that the masses of both the Sun and the planet should be taken into account, in which case the third law is written

$$D^3 = A(M_1 + M_2)P^2$$

where M_1 is the mass of the Sun and M_2 is the mass of the planet, both expressed in units of the combined mass of the Sun and Earth. We make this choice of units so that the term in parentheses stays equal to 1 for the case of the Earth orbiting the Sun.

One application of the third law allows us to calculate the mass of a distant planet like Jupiter. We can do this as long as the planet has a satellite (natural or artificial) with a known period and distance from the planet. Ganymede, Jupiter's largest satellite, is observed to orbit the planet with a period of 7.16 days (1.96×10^{-2} yr) with a semimajor axis of 7.15×10^{-3} AU. We thus have

$$M_1 + M_2 = \frac{D^3}{P^2} = 9.5 \times 10^{-4}$$

Since the mass of Ganymede (M_2) is small compared with that of Jupiter (M_1), we conclude from this calculation that the mass of Jupiter is about 1/1000 that of the Sun.

Now consider the orbit of Comet Halley, which has a period of 76 yr and a small mass, negligible in comparison with that of the Sun. From the period, we calculate the semimajor axis of the comet's orbit as follows:

$$D^3 = (76)^2 = 5776$$
$$D = 18 \text{ AU}$$

To find the perihelion distance of the comet from the Sun, we need to know its eccentricity. For Comet Halley, the eccentricity is 0.97. Substituting for D and e in the formula for the perihelion distance, we have

$$r_{min} = 18(1 - 0.97) = 0.54 \text{ AU}$$

This places the comet's perihelion inside the orbit of Venus.

SUMMARY

Before we study the detailed nature of the individual planets as other worlds, it is useful to look up into the sky and try to reconstruct the evolution of ideas about the solar system from the period before spaceflight, or even before the invention of the telescope.

The most basic apparent motions of the lights in the sky (Sun and Moon, stars and planets) are both daily and seasonal. The stars remain in fixed patterns as they rise and set, but the more complicated motions of the Sun, Moon, and planets challenged the ingenuity of ancient peoples to find an adequate interpretation. This was not just an intellectual challenge but a practical one as well. A working knowledge of the seasons, in particular, was required for the successful development of agricultural society. By 2000 years ago, the Greco-Roman world had developed a sophisticated geocentric view of the heavens. People understood time-keeping and seasons, the phases of the Moon, and the causes of eclipses. Their view prevailed in the Western world until about 400 years ago.

The modern worldview of the planetary system was developed between the Renaissance and the eighteenth century, primarily by five extraordinary European scientists. Copernicus (1473–1543) devised a heliocentric theory of the solar system, which he advocated on grounds of simplicity and aesthetic appeal. The heliocentric theory was verified by the first telescopic observations of the planets, carried out in 1610 by Galileo (1564–1642), the founder of modern experimental science. Meanwhile, Kepler (1571–1630), using a remarkable body of pretelescopic measurements of planetary positions made by Tycho (1546–1601), placed the heliocentric theory on a sound mathematical basis by developing his three laws of planetary motion, which still form the foundation for the descriptions of the orbits of planets, satellites, comets, and other solar system bodies.

Kepler's laws are purely descriptive; they tell how the planets move but not why. The unifying concepts were developed by Newton (1642–1727), who established the physical laws that govern the motions of all bodies, developed the theory of gravitation (a force that acted equally on both falling apples and celestial objects), and invented the mathematics required to calculate trajectories and orbits.

Modern observations leave no doubt that the heliocentric model for the solar system is correct. The true distances, motions, and sizes of the planets have been determined, and these properties easily account for the observations that were so puzzling to our ancestors. Furthermore, we can use this information to understand the operations of rockets and spacecraft that have made possible the exploration of the planetary system.

KEY TERMS

aphelion	heliocentric
astronomical unit (AU)	inclination
constellation	opposition
eccentricity	orbital velocity
eclipse	perihelion
ecliptic	retrograde motion
ellipse	retrograde rotation
equinox	solstice
escape velocity	zodiac
geocentric	

REVIEW QUESTIONS

1. Take a good look at the night sky, and also visit a planetarium if there is one near you. Make sure you understand the apparent motions of the Sun and stars, both daily and seasonal.

2. What is retrograde motion of planets? Explain how this phenomenon was interpreted in both the geocentric and heliocentric systems.

3. What were the main contributions of each of the five scientists discussed in this chapter: Copernicus, Tycho, Kepler, Galileo, and Newton? To what extent was each aware of the accomplishments of the others, and how did they use previous discoveries as a basis for their own contributions?

4. Explain the sequence of phases for both the Moon and Venus as seen from the Earth. What would be predicted for the phases of Venus according to the geocentric theory? Are they any different from the phases of the Moon, since in the geocentric theory both Venus and the Moon orbit the Earth?

5. Describe Kepler's three laws of planetary motion. Write the formula for the period of a planet given its distance (in AU). Write the formula for the distance given the period.

6. How is an interplanetary spacecraft launched toward its target? Describe each step: achieving Earth orbit, escaping from the Earth, and going into orbit around the target planet.

7. What evidence could you offer to a friend to support the idea that the Earth revolves around the Sun?

QUANTITATIVE EXERCISES

1. What is the distance from the Sun (in AU) of an asteroid that has a period of revolution of 8 years?

2. What is the revolution period of a hypothetical planet that orbits the Sun at half the distance of Mercury? Of one that has twice the distance from the Sun of Pluto?

3. What would be the period of revolution for the Earth if the Sun had the same mass but twice its present diameter? If the Sun had the same diameter but twice its present mass?

4. What is the period of revolution about the Earth for a satellite in circular orbit at an altitude above the surface of 10,000 km?

5. A spacecraft on a trajectory from Earth to Saturn follows an ellipse with perihelion at the Earth's orbit (1 AU) and aphelion at Saturn's orbit (9 AU). If the semimajor axis of this transfer ellipse is halfway from Earth to Saturn (5 AU), what is the time required for the trip from the Earth to Saturn? Using similar reasoning, find the trip time to Mars.

ADDITIONAL READING

Beatty, J. K., C. C. Peterson, and A. Chaiken, editors. 1999. *The New Solar System*, 4th ed. Cambridge, MA: Sky Publishing Corp. An excellent selection of chapters by leading planetary scientists, which can be used as a companion volume to this text.

Boorstin, D. J. 1983. *The Discoverers*. New York: Random House. Historical and philosophical analysis of the human search for knowledge in astronomy, time-keeping, and geography, especially during the European Renaissance and Age of Discovery.

Krupp, E. C. 1983. *Echoes of the Ancient Skies.* New York: Harper & Row. Excellent account of archaeoastronomy, which is the study of astronomy in ancient cultures, especially among the native peoples of the New World.

Sagan, C. 1995. *The Demon-Haunted World: Science as a Candle in the Dark*. New York: Random House. Deeply felt but entertaining defense of science and scientific skepticism in the face of superstition and pseudoscience.

Sobel, D. 1999. *Galileo's Daughter: A Historical Memoir of Science, Faith and Love*. New York: Walker. Popular discussion of Galileo and his work, seen in part through the words of his cloistered daughter, Sister Marie Celeste.

2

THE SUN: AN ORDINARY STAR

The Japanese sun goddess Amaterasu comes forth from her retreat in a rocky cave in the mountains, symbolizing the return of spring.

The Earth is a planet—this is the fundamental message of Chapter 1. The demonstration by Copernicus and Galileo almost 500 years ago that our own world is one of several planets circling the Sun represented a watershed in intellectual history. This discovery involved more than just a model for planetary motion. It joined together the terrestrial and celestial realms and dethroned humanity from its central place in the universe. Without this perspective, the astronomers could not have developed such concepts as the laws of motion or the universal nature of gravitation. We take the idea of the Earth as a planet for granted today, but we should try to imagine how profound and disturbing this idea was in the time of Copernicus.

The Sun is a star—this is the second great scientific discovery that shaped our image of ourselves and our place in the universe. Galileo, Kepler, and Newton extended the reach of physical law to encompass the planetary system, but they could hardly imagine the vast and varied cosmos that we now recognize, or the insignificant place our own solar system occupies in the great order of things. To them, the Sun was the center of the universe and the stars were tiny lights of unknown (and unknowable) distance and nature. A few philosophers such as Giordano Bruno speculated on the existence of other suns with other planetary systems, but these ideas could not be evaluated scientifically. Indeed, making such bold and unsupported suggestions led to Bruno's execution in 1600.

The Sun is the centerpiece of the solar system. In this chapter, we look at this nearest star before we begin a detailed examination of the planets and other small bodies that accompany the Sun in space.

2.1 THE SUN AS A STAR

How did scientists first learn that the Sun is a star? Actually, they worked the problem the other way around. They discovered that the stars were suns.

The uniqueness and importance of the Sun for our planetary system stem from two properties: its great mass (more than a hundred times greater than that of all the planets combined) and its prodigious output of energy (a total power of about 4×10^{26} watts). It is this unvarying output of energy that warms the planets and makes life on Earth possible. In contrast, the stars appear to be very faint. Only if they are also very far away can their faintness be understood. The key to recognizing that the stars are suns was to measure their distances and thus determine their true power output—that is, their **luminosity.**

DISTANCES OF THE STARS

Early in the nineteenth century, several leading European astronomers undertook the task of measuring the distances to the stars. The only way to measure such distances was by applying the technique of triangulation (as used in surveying), but with extraordinary precision. Their approach was to measure the very slight apparent shift in the positions of nearby stars as seen from opposite ends of the Earth's orbit (Fig. 2.1). This apparent shift in position from two different viewpoints is called parallax. The stellar parallax the astronomers wanted to measure amounted to a shift of only about 1/10,000°, presenting quite an observational challenge. (The apparent diameter of the

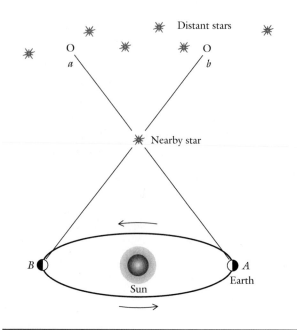

Figure 2.1 As the Earth moves around the Sun from position A to position B, an observer with a powerful telescope will see the nearby star appear to move from position *a* to position *b* among the background stars. This proves that the Earth moves around the Sun. The parallax is grossly exaggerated in this drawing.

Moon in our sky is 1/2°, and the smallest angle that can be measured without a telescope is about 1/100°.) However, in 1838 the distances were measured to three relatively nearby stars: Alpha Centauri, 61 Cygni, and Vega. Their distances were extremely large (each greater than 250,000 AU), which immediately indicated that they have luminosities similar to that of the Sun.

Nineteenth-century astronomers also succeeded in measuring the masses of some stars. They found pairs of stars, called double stars or binary stars, in orbit about each other. By measuring the orbital periods and separation of the stars, astronomers could apply Newton's laws to derive the masses. These masses are also similar to that of the Sun. The double stars provided the first evidence that Newton's laws applied outside the solar system, supporting the universality of physical law.

If the stars are suns and the Sun is a star, then the universe is truly a vast and wonderful place. Thousands of stars can be seen with the unaided eye. Tens of millions of stars can be observed with even a modest telescope. Although the stars are not exactly alike in such properties as mass, luminosity, and temperature, each of these stars has roughly the same nature as that of our own Sun. Each may therefore be the center of its own planetary system.

If the universe is filled with stars, then stars must represent one of the most abundant forms of matter. Planets, by comparison, are only small and insignificant fragments of rock, metal, and cool gas.

Although small dark objects of some sort may exist in interstellar space, the planets in our solar system formed together with the Sun, and we believe that planets generally are the result of the same process that forms stars. Understanding the origin of the planets and the Sun is a major theme of planetary research, as we will see throughout this book.

Basic Properties of the Sun

We have already noted that the mass of the Sun is more than a hundred times greater than that of all the rest of the solar system combined. This mass can be derived by applying Newton's laws to the motion of the Earth and other planets. In metric units, the mass is 2×10^{30} kg, or 2×10^{27} tons. This is a very big number even when expressed in units of the mass of the Earth: 333,000 Earth masses.

The diameter of the Sun is also very much greater than that of even the largest planet. Seen in the sky, the apparent or angular diameter of the Sun is about $1/2°$. Its distance is equal to the astronomical unit (by definition): 150 million km. To appear so large at such a great distance, the true diameter of the Sun must be greater than 1 million km, 10^9 times the diameter of the Earth. The measured value is 1.4×10^6 km. The Sun is thus large enough to hold approximately 1 million Earths within its volume.

The Sun's luminosity is the power (the energy per second) that is constantly radiated from its surface in the form of sunlight. This radiant energy fills the solar system, providing light to illuminate the surfaces of the planets and heat to maintain their temperatures. Even at the distance of the Earth from the Sun, this flood of solar energy amounts to 1370 watts per square meter of area. One of the most important questions to be asked concerning the Sun, a question that will be discussed in detail later in this chapter, is: What is the source of this luminosity?

The Sun's energy originates in its deep interior, but the sunlight we see is streaming from the "surface" of the Sun, where the temperature is about 5800 on the **Kelvin temperature scale** (abbreviated as 5800 K). The Kelvin scale uses the same size degrees as the more familiar Celsius scale, but these degrees are measured relative to the absolute zero of temperature, $-273°C$. Temperatures in the Kelvin scale are always positive numbers—degrees above absolute zero.

At this temperature of 5800 K, the material of the Sun is not solid or liquid but gaseous. When we speak of the solar surface, we refer to the apparent surface, the place where the gas in the solar atmosphere becomes opaque. This apparent surface is called the **photosphere,** meaning the sphere from which the light (photons) originates. Inside the photosphere the Sun must be even hotter because energy always flows "downhill" from hotter to cooler regions—from the interior to the surface. Thus, we conclude that the entire Sun is composed of hot, incandescent gas.

As the old rhyme goes, "The Sun is a mass / of incandescent gas." The challenge for the astronomer is to learn the composition of this gas, how the Sun's energy is generated, and how long the Sun can continue to shine at its present rate. The past and future of the planetary system, and especially of life on Earth, are intimately bound to the ability of the Sun to provide a continuing, stable source of energy to heat and illuminate the planets. Table 2.1 summarizes some of the most important facts about the Sun for ready reference.

Table 2.1   Facts about the Sun

Mass	2.0×10^{30} kg
	333,000 Earth masses
Diameter	1.4×10^9 m
	10^2 Earth diameters
Luminosity	4×10^{26} watts
Surface temperature	5800 K
Core temperature	15×10^6 K
Average density	1.4 g/cm^3
Distance from Earth	1.5×10^{11} m
	1.0 AU
Rotation period at equator	25 days

2.2 BUILDING BLOCKS: ATOMS AND ISOTOPES

The matter that makes up the Sun consists of pure substances called **elements.** An element cannot be decomposed, by chemical means, into any simpler substance; it is the basic building block of matter. The creation of the elements is a natural process in stars, resulting from nuclear reactions in the deep interior. The details of element formation are the subject for another book. But to understand the composition of the Sun and planets, we need to know something about these fundamental constituents of matter.

THE STRUCTURE OF MATTER

The 92 naturally occurring elements are composed of atoms, each consisting of a nucleus surrounded by orbiting **electrons.** The nucleus of an atom contains almost all of the mass. It is made up of particles called **protons,** which have a positive charge, and **neutrons,** which have no charge at all. The protons and neutrons have about the same mass. The electron carries negative charge and has 1/1836 the mass of a proton.

Each atom normally has as many negatively charged electrons as it has positively charged protons, so the atom itself exhibits no net charge. If an electron is removed, however, the atom is said to be an **ion;** it is **ionized** and now exhibits a positive charge. The simplest atom is hydrogen, with a single proton as its nucleus and one electron in orbit about it. If the electron is removed, we have the hydrogen ion, which is just a proton.

The electrons of an atom are capable of both emitting and absorbing light. This ability of each atom to interact with light provides a powerful tool for determining the composition of celestial materials without requiring that actual samples be brought into the laboratory.

ISOTOPES

Imagine adding a neutron to the nucleus of a hydrogen atom (Fig. 2.2). The mass of the atom is changed by a factor of two, but since the charge of the nucleus is the same, the same single electron is

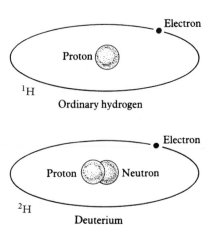

Figure 2.2 An ordinary hydrogen atom consists of one proton as the nucleus and one electron in orbit around it. Adding a neutron to the nucleus changes the mass of the atom but not its electric charge. The new isotope with one proton and one neutron is called deuterium.

all this new atom needs to remain electrically neutral. This new atom is an **isotope** of hydrogen—that is, a form of an element that differs from others only in the number of neutrons in the nucleus.

The isotope of hydrogen with one neutron in the nucleus is called deuterium, written ^{2}H or D. Since deuterium still has just one electron, its chemical properties are virtually identical to those of ordinary hydrogen (^{1}H), despite its greater mass. Thus, deuterium can combine with oxygen to form water, but we now give it the symbol D_2O instead of the familiar H_2O. The surprise comes when an ice cube of D_2O is placed in ordinary H_2O. It sinks! The greater mass of the nucleus of deuterium atoms leads to an increased density for D_2O. More mass is packed into the same volume compared with ordinary H_2O. It is for this reason that D_2O is often called "heavy water."

This is not just an idle thought experiment because deuterium is found abundantly in nature. Other elements also exhibit more than one isotope. Ordinary carbon, for example, which we find in pencil lead, diamonds, coal, and ourselves, also has two stable isotopes. One—with six neutrons and six protons—is 90 times more abundant than the other—with seven neutrons and six protons. Oxygen has three stable isotopes.

COMPOUNDS

Most of the matter we encounter is not in the form of pure elements. Water, carbon dioxide, alcohol, and quartz are all examples of **compounds,** substances that are composed of more than one element. Just as the smallest unit of an element that still preserves the element's chemical identity is called an atom, the smallest unit of a compound is called a **molecule.** A molecule is formed from the atoms of the various elements that make up the compound. Collisions between atoms can form molecules, and ultraviolet light can break molecules apart. Like atoms, molecules can be ionized by losing one or more electrons. The tails of some comets, for example, contain ionized molecules of water, written H_2O^+. The gases around the comet's head contain H and OH, fragments of H_2O molecules that have been broken apart by ultraviolet light.

Both compounds and elements can change their state, becoming gases, liquids, or solids depending on the local temperature and pressure. Of the three possibilities, the liquid state is the most rare because liquids are stable over restricted ranges of temperature. Water is liquid over a range of 100°C. Ammonia is liquid only from 233°C to 278°C, or less than half the range of water.

2.3 COMPOSITION OF THE SUN

The measurement of the composition of objects from a great distance represents one of the triumphs of astronomy. In 1835, the French philosopher August Comte speculated that we would eventually be able to determine the distances and motions of the stars but that it would never be possible to determine their chemical composition. Yet within a few decades astronomers were doing exactly that. The key to this accomplishment was the ability to interpret the spectra of the light emitted by the Sun and stars.

ELECTROMAGNETIC RADIATION

The spectrum of sunlight is the collection of its many constituent colors. The colors of the rainbow represent just the visible forms of what is called **electromagnetic radiation.** The light illuminating this page as you read it is an example of electromagnetic radiation. This radiation may be defined generally as the propagation of energy through space by varying electric and magnetic fields. It can be produced and absorbed by interactions of these fields in atoms and molecules.

Electromagnetic radiation can be understood in terms of a wavelike motion of electric and magnetic fields, with each kind of radiation having its own wavelength. But electromagnetic radiation also behaves as if it were made up of particles called **photons.** These particles have no mass, but they do carry energy. The energy of an individual photon is inversely proportional to its wavelength, so long waves correspond to low energies and short waves to high energies.

Astronomers often use the term *radiation*, which is roughly synonymous with *light*. Astronomy consists, in large part, of measuring and interpreting the radiation from distant objects that we detect with our telescopes. Don't confuse our usage with the common use of the word *radiation* to refer to various harmful emissions from radioactive materials.

THE SPECTRUM AND SPECTROSCOPY

The **electromagnetic spectrum** is a way of describing the energy range of electromagnetic radiation (Fig. 2.3). It extends from radiation with very short wavelengths (gamma rays at 10^{-10} cm, x rays at 10^{-8} cm) through the so-called visible region of the spectrum to which our eyes are sensitive (0.35×10^{-4} cm to 0.76×10^{-4} cm), and on through the infrared to the radio region, where wavelengths are measured in meters and kilometers. The spectrum can also be thought of as a progression from very high energy (gamma rays) to very low energy (long-wave radio waves).

In the part of the electromagnetic spectrum corresponding to visible light, we usually measure wavelengths in nanometers (abbreviated nm). One nanometer is 1 billionth (10^{-9}) of a meter, or 10 millionths of a centimeter. In these units, the wavelengths of visible light are from 350 nm (violet) to 760 nm (red). The longer wavelengths

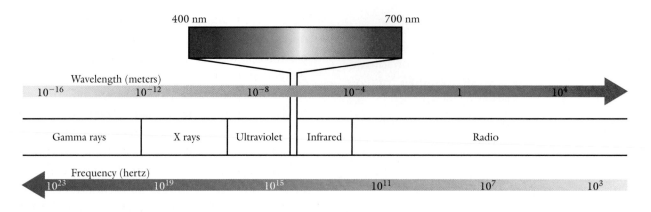

Figure 2.3 The electromagnetic spectrum extends from gamma rays with wavelengths shorter than 1 billionth of a centimeter to radio waves with wavelengths measured in millions of centimeters. All of this radiation travels through space at the speed of light. Only certain wavelengths are able to pass through a planetary atmosphere (see Fig. 3.4).

of infrared radiation are usually measured in micrometers (μm), where 1 μm is equal to 1000 nm.

The primary tool for the analysis of electromagnetic radiation is spectroscopy or spectrometry. Everyone is familiar with rainbows, the beautiful array of colors created when sunlight passes through a mist of water droplets. The yellow-white light of the Sun is split into its component colors, which are spread out for us to

admire. This is a very simple spectrum, in which radiation is sorted by energy, from violet to red.

We can achieve the same effect in our laboratories by using a prism or a diffraction grating instead of raindrops. Adding a slit to limit the overlapping of the radiation, some lenses to make images, and a detector to record them, we have constructed a **spectrometer** (Fig. 2.4). Now we can spread out sunlight and examine it in great detail.

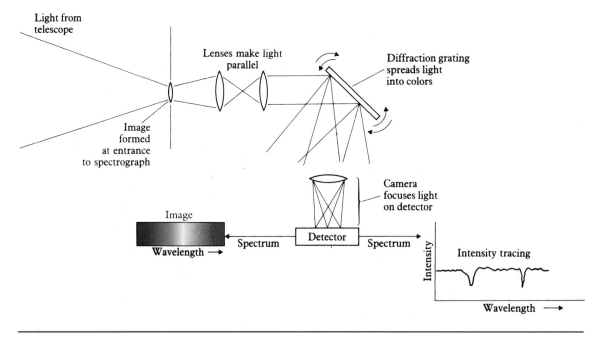

Figure 2.4 A spectrometer spreads out the light from a source according to its wavelengths, which we perceive as colors in visible light.

FORMATION OF SPECTRAL LINES

When we look at the solar spectrum in detail, we find that amidst the beautiful display of colors, which faithfully reproduces the range from violet to red exhibited by rainbows, there are many dark lines (missing colors) called **absorption lines** (Fig. 2.5). These absorption lines represent wavelengths where energy is being removed from sunlight by atoms or molecules in the Sun's atmosphere. We can think of sunlight as radiant energy produced deep in the Sun's interior. This radiation gradually works its way to the Sun's photosphere, from which it escapes into space. But during this last stage it encounters the cooler outer envelope of the Sun, and it is the effects of absorption by atoms and molecules in this solar atmosphere that we see as dark lines in the spectrum.

Why are there discrete spectral lines instead of a general decrease in the intensity of the light, as we experience on a cloudy day on Earth? In other words, why do atoms and molecules of gas absorb only specific energies or wavelengths, unlike cloud droplets, which absorb all colors? The reason is found in the quantum theory of matter. This powerful theory, which is at the foundation of most modern physics, explains mathematically the discrete nature of the absorption and emission of energy by atoms and molecules.

A violin string produces an approximate analogy to the behavior of an atom of gas. If you pluck the string, you get only one note because the string vibrates at only one frequency. An atom is a bit more complex, but it also operates at only a few specific frequencies or wavelengths, which can be calculated from quantum theory. You can, however, change the violin note (frequency) by changing the length or tension of the string. This would correspond to changing the internal configuration of an atom or molecule—for instance, by shifting an electron into a different orbit or removing it to form an ion. In such a changed state, the atom or molecule interacts with a different set of wavelengths or frequencies of electromagnetic radiation to produce a different set of absorption lines.

Atoms and molecules in the solid or liquid state also interact with electromagnetic radiation to produce a characteristic spectral signature. The atomic configuration is generally different for a solid or liquid, however, so the spectrum is different from that of the gaseous form. Because the close proximity of atoms to one another in a solid or liquid restricts the vibrations of individual atoms, the solid or liquid is less finely tuned; its spectral features are therefore often broad and fuzzy. To pursue our musical analogy, these forms respond like a bass drum rather than like a violin. Therefore, spectroscopy is a less precise tool for analyzing solids and liquids than it is for gases, and consequently spectral studies have told us less about the nature of solid planetary surfaces than about their gaseous atmospheres.

Since the spectrometer that recorded the spectrum in Figure 2.5 was located on the Earth's surface, we might expect to find absorption lines

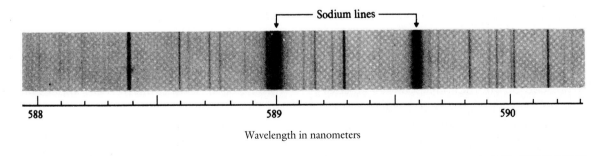

Figure 2.5 Bright or dark lines in the spectrum reveal the presence of gases in the source (or between the source and the observer) that absorb or emit specific wavelengths of light. This portion of the visible region of the Sun's spectrum shows dark lines caused by atoms in the outer atmosphere of the Sun. The two prominent absorptions at 589.0 nm and 589.6 nm are caused by atoms of sodium.

produced by gases in our planet's atmosphere also. After all, the sunlight has to pass through the Earth's atmosphere to reach the spectrometer, and along the way the photons are going to encounter a lot of atoms and molecules. It turns out that molecular nitrogen does not absorb visible light, nor does argon. But oxygen does, although not in the region shown in Figure 2.5.

Spectral Analysis

It was not until the middle of the nineteenth century that British and German physicists began to use the spectrum as an analytical tool. They found that the solar spectrum contained many absorption lines. Experimenting in the laboratory, they learned that different kinds of gases could absorb specific wavelengths to mimic the solar spectrum. The precise wavelengths that are absorbed depend only on the composition of the gas, not on its density or (over a limited range) its temperature. This is the great strength of spectroscopy as a tool: If we can measure the spectrum of light emitted by the Sun or any other source anywhere in the universe and can match the observed pattern with that of gases measured in the laboratory, we can identify the presence of those gases in the distant source.

The simplest application of spectroscopy occurs when the gas is in atomic form. Fortunately for the nineteenth-century astronomers, the Sun and stars are so hot that most of their material is in the form of individual atoms. In such circumstances, the interpretation of observed spectra is relatively straightforward, once an adequate library of laboratory spectra is available for comparison.

Many of the elements identified spectrally on the Sun are familiar. The prominent lines include the metals iron, magnesium, and sodium. Silicon, an important constituent of terrestrial rock, is clearly present. So are the four elements most essential to life: hydrogen, oxygen, carbon, and nitrogen. But some lines in the solar spectrum do not correspond to common elements on the Earth. Most notable is a prominent set of lines identified in 1899 that seemed to be produced by a simple atom that was not known on Earth. The unknown element was named helium for the

Greek word for Sun, *helios*. Shortly thereafter helium was discovered as a trace constituent of the atmosphere of the Earth, where it makes up less than one part in a billion.

Solar Abundances

While it is relatively easy to use spectroscopy to identify elements present in the Sun, it is much more difficult to estimate the relative proportions of the different elements. Many factors in addition to the abundance of the gas influence the strength of the spectral lines.

At first scientists assumed that the solar abundances were similar to those on the Earth, perhaps with the exception of a few odd gases like helium. Not until the late 1920s did a Harvard graduate student, Cecilia Payne, succeed in deriving the correct abundances for the major elements in the Sun and stars. She showed that in spite of their differences in size or temperature, the stars are all made primarily of the light gas hydrogen. Subsequently it was found that helium is the second most abundant element, and that fully 99% of the Sun and most other stars is composed of these two gases. Other more familiar elements, such as oxygen, nitrogen, carbon, silicon, and iron, are just trace constituents of the Sun. Table 2.2 lists the abundances of the major elements in the Sun. Later in this book, when we discuss the origin of the solar system, we will explain why the composition of the Sun is so different from that of the Earth and other planets.

2.4 The Sun's Energy

The composition of the Sun derived from spectroscopy is that of the outermost layers, which is where the sunlight we see originates. Indeed, almost everything we can determine about the Sun from the analysis of sunlight really applies only to this thin upper layer, consisting of gases in the photosphere and above it. However, the application of a few basic ideas helps us to understand the invisible solar material below the surface.

The Sun is radiating energy from its photosphere, energy that must originate in the interior. In order for this energy to continue to flow out,

Table 2.2 Cosmic abundances of the major elements

Element	Symbol	Atomic Number	Number of Atoms per Million Hydrogen Atoms
Hydrogen	H	1	1,000,000
Helium	He	2	97,000
Carbon	C	6	360
Nitrogen	N	7	110
Oxygen	O	8	850
Neon	Ne	10	120
Sodium	Na	11	2
Magnesium	Mg	12	40
Aluminum	Al	13	3
Silicon	Si	14	40
Sulfur	S	16	20
Argon	Ar	18	4
Calcium	Ca	20	2
Iron	Fe	26	32
Nickel	Ni	28	2

the interior of the Sun must be hotter than the surface. If the interior were not hotter, the energy would flow from the surface back into the interior, and we would all be in bad trouble. In addition, the upper layers of the Sun must be supported by gas pressure from underneath. If they were not, the Sun would collapse from its own great weight—another bad outcome!

There is evidence that the Sun has been shining for billions of years and that it has not changed much in either size or luminosity over that period of time. Its present configuration must therefore represent some sort of equilibrium, with a stable source of energy at the center and with the weight of each overlying layer supported by the pressure of the gas below. This assumption of equilibrium permits astronomers to calculate the range of pressure and temperature throughout the solar interior. They have found that the central temperature of the Sun is 15 million K (1.5×10^7 K) and that the pressure is 300 billion times greater than the surface pressure of the Earth's atmosphere (3×10^{11} bars).

Astronomers also verify that most of the solar energy originates in the core, within the innermost 10% of the volume of the Sun. But how is this energy generated?

ENERGY SOURCE: EARLY IDEAS

The observed luminosity of the Sun is 4×10^{26} watts. This is the power that the Sun is emitting into space today. If the Sun is in equilibrium, this must also be the rate of energy generation in the core.

The source of energy most familiar to us on Earth is oxidation or burning. Early in the nineteenth century, some scientists suggested that the Sun could obtain its energy by oxidation or other chemical reactions, but calculations quickly showed that this was not possible. Even if the immense mass of the Sun consisted of a burnable material like coal, oxidation could not produce energy at its present rate for more than a few thousand years. With a chemical source excluded, scientists began to look for a physical source of the Sun's power.

The simplest physical mechanism to produce large quantities of power is a slow contraction of the Sun. To see how contraction can generate power, we turn to one of the fundamental laws of nature known to nineteenth-century scientists: the law of conservation of energy. This law says that energy cannot be created or destroyed, but it can be transformed from one form to another. The steam engine, for example, relies on the transformation of thermal energy (from the boiler) to the mechanical energy of a piston or rotating wheel. The reverse is also possible, to transform mechanical motion into heat. You do this, for example, when you rub your hands together to warm them.

If an object—a planet, for example—should fall into the Sun, it would be accelerated by the Sun's gravity and would strike with great force, transforming its energy of motion into heat. In a similar way, the German physicist Hermann von Helmholtz and the English physicist William Thomson (Lord Kelvin) proposed that the outer layers of the Sun might "fall" inward and thereby produce heat. Calculations showed that a tremendous amount of power could be produced by even a very small contraction. Helmholtz and Kelvin found that the total power output of 4×10^{26} watts could be provided by an annual contraction of only 40 m. Over the time span of human history (about 10,000 years), such a contraction would amount to about 400 km, or less than 0.03% of the Sun's diameter. Such a small contraction would not be measurable. Helmholtz and Kelvin proposed contraction as the source of the Sun's power, and they calculated that this energy source could have kept the Sun shining for about 100 million years. At the time, this seemed ample, since the Earth was thought to be only a few million years old.

By the beginning of the twentieth century, the Kelvin-Helmholtz theory began to run into trouble. New evidence showed that the Earth was hundreds of millions of years old—perhaps more than a billion years. Contraction could not provide enough energy to keep the Sun shining for so long. So physicists and astronomers began to search for an alternative energy source.

THE POSSIBILITY OF NUCLEAR ENERGY

Albert Einstein was a young patent clerk in Bern, Switzerland, when he developed the special theory of relativity in 1905 (Fig. 2.6). Relativity is a fascinating concept that we cannot discuss in any detail in this text. Here we are concerned with one of the unexpected by-products of Einstein's work: the discovery of the equivalence of mass and energy. Nineteenth-century physics had established that while energy was conserved, it could be transformed from one form to another. Einstein discovered that the conservation of energy was not strictly correct: Mass and energy are different forms of the same thing, and what is conserved is a new quantity called mass-energy. In principle, at least, energy can be transformed into mass, and mass into energy.

Einstein's famous equation, $E = mc^2$, expresses the equivalence between energy (E) and mass (m). The other term in the equation is c, the speed of light (about 300,000 km/s). Because c^2 is a very large number, this equation tells us that even a small amount of mass can create a great deal of energy. Here, then, is a possible solution to the dilemma of the Sun's long-term energy source.

If 4 million tons of matter could be transformed each second into energy, then 4×10^{26} watts of power would be generated. This sounds like a lot of material, but compared with the total mass of the Sun it is not. The Sun could lose this much mass every second for tens of billions of years without altering its total mass by more than 1%. But can mass actually be transformed into energy? Is there a practical reality associated with this theoretical prediction?

THERMONUCLEAR FUSION

Several decades of work by physicists were required to identify a reaction in the Sun that might transform mass into energy. From careful laboratory measurements, they found that the nucleus of the helium atom, which can be thought of as four

Figure 2.6 Young Albert Einstein, the most eminent scientist of the twentieth century.

hydrogen atoms combined, has only 99.28% of the mass of the sum of four hydrogen atoms. Thus, if there were a way to combine or fuse four hydrogen nuclei to make one helium nucleus, excess mass would be present. This mass is converted to energy in the fusion process and is liberated primarily in the form of high-energy gamma radiation. As we have seen, the Sun and stars are composed mostly of hydrogen. It also turns out that the high temperatures and pressures in the solar interior are conducive to the fusion of hydrogen into helium. Because these reactions involve atomic nuclei and take place at high temperature, they are called **thermonuclear fusion** reactions.

Physicists have discovered two ways that hydrogen fusion can take place in the core of the Sun and other stars. As we know, they have also found ways to achieve fusion on a much smaller scale, producing the thermonuclear or hydrogen fusion bomb. But that is another story.

The most important set of reactions in the Sun is called the **proton-proton chain** (Fig. 2.7). At temperatures higher than 10 million K, the hydrogen nuclei (or protons) are moving at speeds of more than 1000 km/s. Even at these speeds, an average proton will rebound from collisions with other protons for about 14 billion years, at a rate of 100 million collisions per second. But very occasionally, two protons will stick together or fuse to form an atom of deuterium, with the release of energy in the form of gamma rays. At this point, the main barrier to the creation of helium is passed. In just 6 seconds (on average) another proton is absorbed to form light helium, a nucleus consisting of two protons and one neutron, again with the release of gamma rays. About 1 million years later, the nucleus of light helium collides with another light helium nucleus to form one nucleus of regular helium (two protons and two neutrons) and two protons, which can start this process again.

The mass lost in the proton-proton sequence of reactions is 0.71%. Thus, if 1 kg of hydrogen turns into helium, then 0.0071 kg of the mass is converted into energy. To maintain the luminosity of the Sun, 600 million tons of hydrogen must be converted to helium each second, with the net conversion of 4 million tons of mass to energy.

This energy works its way out through the layers of the Sun until it bursts forth as visible light to illuminate the rest of the planetary system—a process that we will discuss in more detail in the next section.

SOLAR NEUTRINOS

The explanation of the power source of the Sun was one of the triumphs of theoretical physics in the middle of the twentieth century. All the numbers seemed to work out fine. But of course there was no way to measure directly what was happening deep within the Sun, or to actually measure the conversion of hydrogen to helium.

In the 1980s, some experimental physicists thought they had a way to monitor the reactions in the core of the Sun. One by-product of the fusion reactions was calculated to be **neutrinos**,

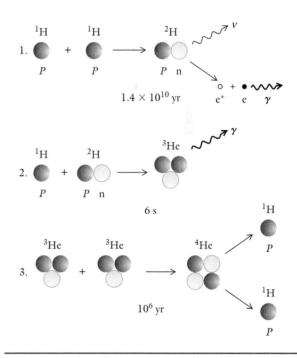

Figure 2.7 The Sun generates its energy from the fusion of four hydrogen atoms into one helium atom. The sequence of nuclear reactions illustrated here is called the proton-proton chain, because the first (and most difficult) step is the fusion of two protons (H) to produce one atom of deuterium (D). This step also releases a neutrino (ν) and a positron (e^+), which is a positively charged electron. The positron collides with an electron (e) to produce a gamma ray (γ).

nearly massless subatomic particles that move at close to the speed of light. Since neutrinos hardly interact with other atoms, the Sun is nearly transparent to them, and they escape directly from the core. This flood of neutrinos reaches the Earth within 8 minutes of the time they are formed in the core, thus (in principle) providing physicists with a way to measure the reactions taking place there. The problem is that neutrinos are extremely difficult to detect; they pass unhindered through the Earth and our detection instruments just as they do through the Sun.

In spite of the great difficulty involved, several huge neutrino detectors were built deep underground in abandoned mines. It took several years of operation to make a reliable measurement—and the results were shocking. There were only one-third as many detections as were predicted by the models for the solar fusion process. Taken at face value, this meant that the fusion reactions, as monitored by the neutrinos, were converting only 200 million tons of hydrogen, not the 600 million tons required to account for the Sun's luminosity. Here was a fundamental challenge to our understanding of the energy source of the Sun and stars.

Nearly two decades were required to solve this vexing problem. Theorists noted that the detection instruments on Earth were not sensitive to all kinds of neutrinos. They suggested that part of the neutrinos generated in the core of Sun were transformed into different (and undetectable) kinds on the 8-minute trip to the Earth. Unfortunately, there is no way to bottle up neutrinos in the lab to see whether such changes actually take place. Also, to undergo such transformations, the neutrinos must have a mass, and most physicists thought they were massless. To get out of this dilemma, an instrument was finally constructed that could measure more than one kind of neutrino. In 2000, the "missing" neutrinos were finally measured. When the transformation en route to the Earth was taken into account, the solar luminosity turned out to be exactly what was needed to produce the visible power of the Sun. And the very tiny mass of the neutrino was measured in this natural lab, something that had proved impossible previously. The resolution of the neutrino problem required two decades of patient work and the expenditure of

millions of dollars for equipment, but in the end the results provided our first direct confirmation of the power that lights the Sun and stars.

2.5 LIFE HISTORY OF THE SUN

Once scientists recognized the source of the Sun's energy, calculating an approximate limit to the Sun's lifetime was a straightforward matter. Recall that the Sun must transform 4 million tons of matter into energy each second to satisfy Einstein's equation $E = mc^2$. This quantity of excess mass is produced by the fusion into helium of 600 million tons (or 6×10^{11} kg) of hydrogen per second. Let us suppose that the entire mass of solar hydrogen could be converted to energy in this way. The mass of the Sun is 2×10^{30} kg, of which about 75%, or 1.5×10^{30} kg, is hydrogen. The time necessary to exhaust this hydrogen at the current consumption rate is 1.5×10^{30} kg divided by 6×10^{11} kg/s, or about 3×10^{18} s. Since there are 3×10^7 s in a year, this lifetime is 10^{11} years, or 100 billion years. This is a very long time, but is the result realistic? Can the Sun really convert all of its hydrogen into helium?

ENERGY GENERATION AND TRANSPORT

To understand what is happening inside the Sun, we must look more closely at the generation of energy in its core and the flow of energy from the core to the surface. First note that the rate of fusion of hydrogen into helium is sensitive to the temperature. A relatively small increase in temperature causes the protons to move faster and results in a higher fusion rate. More fusion means more power and an even hotter gas. If this situation were to continue, the core would explode. However, the Sun has a way of compensating. When the production of energy starts to increase, higher pressures in the core cause the Sun to expand and the core to cool. Conversely, if the energy generation in the core should lag, the Sun would contract, converting gravitational energy to heat and rekindling the nuclear fires in the core. In this way, equilibrium is maintained.

In a steady state, all of the energy produced in the core must find its way outward to the surface, where it is radiated into space. There are three ways for energy to be transported from one place to another. On Earth, the most familiar form of energy transport is **conduction**—the process that causes a pan placed on the stove to become hot. Conduction involves the transport of energy in a solid by molecular motion. Apply heat at one point, and the molecules increase their speed. They bump into their neighbors, and soon all are moving faster; that is, the temperature rises. Conduction plays a major role in the transport of heat inside planets, as we will see later, but inside the Sun it is overwhelmed by more efficient transport mechanisms.

One of these mechanisms is **radiation.** As we have already noted, electromagnetic radiation involves the transmission of energy through a vacuum by electric and magnetic waves. Radiation is the process by which sunlight illuminates and warms the surface of the Earth. It is also the way you are warmed by a fire in a fireplace. Radiation works best in a vacuum, but it can also play an important role in the transfer of energy within the Sun, especially in the core itself. The gamma rays released by thermonuclear fusion work their way outward primarily through the process of radiation.

The other important mechanism for the transport of energy in the Sun is **convection.** Like conduction, convection requires the presence of matter. In convection, however, it is not individual molecules that vibrate and collide with each other but large masses of gas or liquid that move from warmer to cooler regions, carrying energy with them. Convection is a macroscopic, not a microscopic, process. Where such mass motion is possible, convection becomes the most efficient of these three energy transport processes.

Within the Sun, convection is possible only in the outer layers. Within about the outer third of the Sun, the gas is in a constant state of agitation, with plumes of warm gas rising and blobs of cooler gas descending to take the place of the rising columns. Deep in the interior, radiation dominates. In either case, the time required to move energy from its source in the core up to the surface layers is very long—about a million years. The sunlight you see today originated in fusion reactions that took place near the time of the emergence of *Homo sapiens* on our planet. This

resistance to the outward flow of energy is critical to the stability of the Sun.

SOLAR EVOLUTION

Astronomers use all of the concepts we have described—the equilibrium of the Sun, hydrogen fusion in its core, and the transport of energy from the core to the surface—to calculate detailed models of the solar interior. They thus develop a quantitative picture of the structure of the Sun as it is today. These same computational tools also permit them to calculate what will happen as the hydrogen in the core is gradually converted to helium. In this way, it is possible to predict the future of the Sun.

The current structure of the Sun is illustrated in Figure 2.8, which shows the regions in which energy is carried outward by radiation and convection. After nearly 5 billion years of fusion reactions, the hydrogen is already partly used up in the core. It is the gradual depletion of hydrogen fuel that causes the Sun to change as it ages.

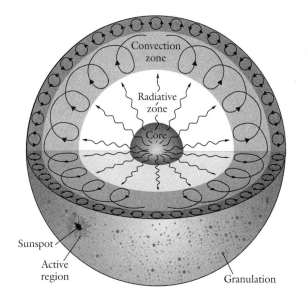

Figure 2.8 The internal structure of the Sun has been determined from computer models aided by observations of waves and vibrations in the solar atmosphere. As shown in this illustration, there are three main regions: a hot, dense core where nuclear reactions take place; a layer extending more than halfway to the surface in which the energy is transported primarily by radiation; and the outer third of the Sun, where energy is transported mainly by convection.

The response of the Sun to the exhaustion of hydrogen at its center is to readjust its structure so as to increase central temperatures. At the same time, its total energy production also increases. Thus we predict that the Sun will gradually become more luminous with time. After several billion years, the inner core will consist almost entirely of helium, and the solar luminosity will be several tens of percent greater than it is today. Then, the calculations predict that the Sun will expand rapidly to an enormous size, becoming what astronomers call a red giant. In its bloated state, the Sun will extend as far as the orbit of the Earth. Our planet will be consumed in the solar fires, and the planetary system as we know it will cease to exist.

The same calculations can be extended backward in time to determine what the Sun used to be like. We conclude that the Sun began to convert hydrogen to helium in its core between 4 billion and 5 billion years ago. Thus, the Sun is between 4 billion and 5 billion years old—a result that is in excellent agreement with the measured ages of the planets, indicating that the Sun and its retinue of planets formed together.

Early in its history, when hydrogen still made up about 75% of its core, the Sun must have been less luminous than it is today. Calculations show that the Sun has increased in luminosity by at least 35% in the past 4 billion years. Therefore, the Earth and other planets must have been cooler in the past than they are at present. We will return to this constraint on planetary history in Chapter 17, when we discuss the origin and early evolution of the planetary system.

2.6 SOLAR ACTIVITY

We depend on the steady flow of power from the Sun to warm the Earth and other planets. If the Sun's luminosity varied substantially, life would be difficult or impossible in the solar system. Fortunately for us, the Sun maintains a stable equilibrium. This equilibrium is only approximate, however, and observations show that our star is capable of changes whose influence is felt throughout the planetary system.

SURFACE ACTIVITY ON THE SUN

Beneath the visible photosphere of the Sun, convection currents transport energy upward. These convection currents are characterized by cells of rising warmer gas typically 700–1000 km across, traveling upward at speeds of several kilometers per second. These rising cells are separated by sinking columns of cooler gas. The solar surface seethes and boils like a giant witch's cauldron (Fig. 2.9).

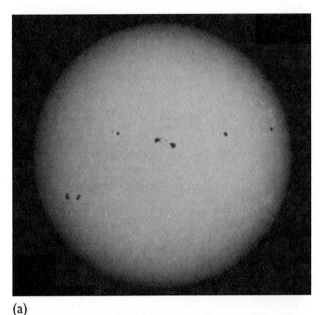

(a)

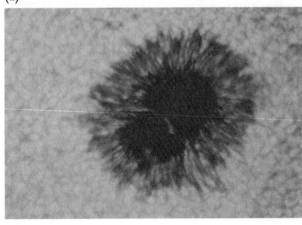

(b)

Figure 2.9 (a) The visible surface of the Sun is a seething mass of gas, with motions driven by convection currents of heat that rise from the interior. The surface includes dark regions called sunspots. **(b)** Close-up of a sunspot, surrounded by the honeycomblike structure of the photosphere. This dark spot on the Sun is about twice the size of the Earth.

Figure 2.10 The tenuous, hot outer atmosphere of the Sun is called the corona. It is most easily seen during an eclipse of the Sun, when the Moon blocks the bright surface from our view.

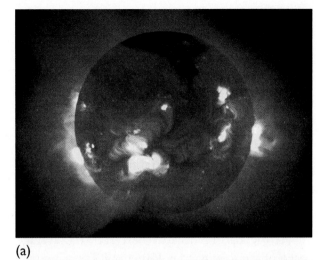

(a)

(b)

Figure 2.11 The hottest regions of the solar corona are strong sources of x-ray emission. Many centers of activity with temperatures higher than 2 million K are shown in these images. **(a)** Yohkoh space observatory image of the solar corona. **(b)** TRACE space observatory image of hot gas trapped in the Sun's magnetic field. Because these are x-ray images, the colors are false.

Like breaking waves, the hot convecting gas crashes into the outer layers of the Sun. In rising from the interior, this hot gas carries both thermal and mechanical energy. The rising gas also becomes entangled with the magnetic field of the Sun. One result of these complex interactions is the transport of a great deal of energy to the outer atmosphere of the Sun, called the **corona.** The solar corona consists of very thin gas, heated to more than a million degrees, extending for millions of kilometers into space. It is most easily seen in visible light during a total eclipse of the Sun, when the Moon blocks the brighter disk of the Sun from our view (Fig. 2.10).

Because it is so hot, the gas of the corona emits most of its radiation in the form of x rays. These x rays cannot be seen from the surface of the Earth, but x-ray images of the Sun are commonly produced from space. Figure 2.11 illustrates the appearance of the Sun in x rays and thus outlines the hottest (brightest) regions of the corona.

THE SOLAR WIND

The solar corona does not end suddenly but continues out to great distances from the Sun. Because they are so hot, the coronal gases expand rapidly and flow away from the Sun with only minimal resistance from solar gravity. The result is a **solar wind** of hot gas streaming out to fill the solar system.

The gas of the solar wind has essentially the same composition as the atmosphere of the Sun itself. Like the gas of the corona, the atoms in the solar wind are ionized. The great bulk of the solar wind is composed of individual protons (hydrogen nuclei) and electrons. An electrically charged gas of this sort, in which the protons and electrons are separate instead of combined in the form of electrically neutral atoms, is called a **plasma.** This plasma has embedded within it a part of the magnetic field of the Sun, which is also carried out to the realm of the planets.

At the Earth's distance of 1 AU, the average speed of the solar wind is 450 km/s. The wind particles travel from the Sun to the Earth in about five days. The wind speed declines slowly with increasing distance, but even at the outer planets it is still moving several hundred kilometers per second.

The solar wind was discovered before the space age from its effects on the tails of comets. This solar plasma interacts with the gas steaming outward from the center of a comet and carries this gas away from the Sun, in the direction of the motion of the solar wind (Fig. 2.12). By measuring the way individual streamers of gas move outward along the tail of a comet, astronomers can determine the speed of the solar wind and estimate its density. Since about 1960 these indirect astronomical observations have been supplemented by direct measurements of the solar wind made by interplanetary spacecraft.

(a)

(b)

Figure 2.12   (a) A photograph of Comet Halley, when it visited the inner solar system in 1986, shows a straight plasma tail with considerable structure. The tail is glowing with light from ionized carbon monoxide gas molecules that are being blown away by the solar wind. **(b)** A painting of Donati's Comet as it appeared to the naked eye on October 4, 1858, over La Cité in Paris. Note the two straight plasma tails and the curved dust tail. The bright star near the comet's head is Arcturus in the constellation Boötes. (Note that this is an artist's reconstruction, not a drawing from nature.)

SUNSPOTS

The Sun has a complex magnetic field that varies with both time and position on the surface. In some parts of the photosphere, the magnetic field can become so compressed that it inhibits the flow of hot gas from the interior. The result is a region of reduced temperature called a **sunspot.** If you look at the Sun with a small telescope equipped with a suitable filter, you can see the sunspots as small dark blotches against the bright photosphere. Although the sunspots look black, this is mostly a contrast effect. The typical temperature in a sunspot is about a thousand degrees lower than that of the surrounding photosphere, but that is still hot enough that the gas of the spot would radiate brightly if it were seen by itself (Fig. 2.9).

Sunspots vary in size from the smallest we can see with telescopes up to several times larger than the Earth (tens of thousands of kilometers across). They tend to form in groups, often with one large spot accompanied by as many as 100 smaller spots (Fig. 2.13). As the Sun rotates with a period of about 25 days, the spots are carried across the surface. A large sunspot group may have a lifetime of several solar rotations, although individual spots form and coalesce within the group on a timescale of days.

Sunspots tell us about the changing magnetic field of the Sun. The magnetic field in turn influences a number of energetic outbursts on the Sun that propagate into the corona, into the solar wind, and ultimately to the upper atmospheres of the Earth and planets.

THE SOLAR CYCLE

In 1843 Heinrich Schwabe, an amateur astronomer, noted from a long series of his observations that sunspots come and go in a regular cycle of 11 years. At sunspot maximum there may be dozens of sunspot groups visible on the solar disk, whereas at sunspot minimum the number drops nearly to zero (Fig. 2.14). The location of the spots also varies periodically. At the start of a cycle, the sunspots appear at a latitude of about 35° north and south of the equator, but as their numbers increase, the spots are found at lower and lower latitudes. The survivors of one cycle can still be seen near the equator as spots of the next cycle appear at higher latitudes.

When astronomers early in the twentieth century developed techniques for measuring directly the magnetic field of the Sun, they discovered that the 11-year sunspot cycle is closely related to variations in the solar magnetic field. They found that the sunspots in alternate cycles had reversed magnetic polarity, so that the total period for a repeat of spots with the same magnetic orientation was 22 years and not 11 years. The entire global magnetic field of the Sun also reverses itself in the

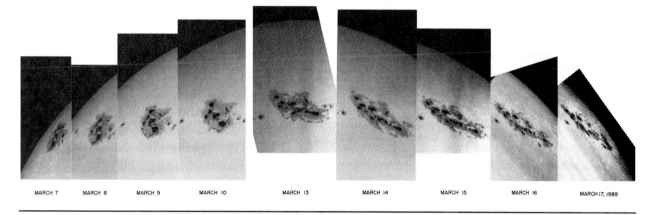

MARCH 7   MARCH 8   MARCH 9   MARCH 10   MARCH 13   MARCH 14   MARCH 15   MARCH 16   MARCH 17, 1989

Figure 2.13 Sunspots form in groups near regions of enhanced magnetism on the solar surface. This series of photos shows a large group of sunspots as it evolves over a period of nine days.

3 GETTING TO KNOW OUR NEIGHBORS

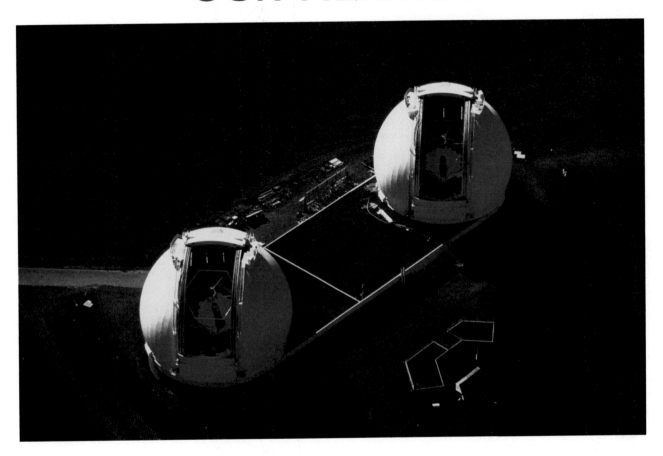

Mauna Kea Observatory is located on the summit of an extinct volcano in Hawaii. The twin 10-m Keck telescopes are shown in this aerial view.

Planets are places, made of rock and metal, ice and various gases. Some are similar to the Earth; others are very different. We begin Chapter 3 with a discussion of the materials that form the planets. We want to know how the composition varies from one object to another, and how closely individual planets represent the mixture from which the solar system was originally made. Here and throughout the rest of the book we will use a comparative approach, concentrating on the fundamental processes that determine the similarities and differences between one world and another. We also will be asking questions concerning the origin of the objects we see, and the ways they have evolved over the history of the planetary system.

The Earth has a composition very different from that of the Sun, consisting mainly of silicon, oxygen, magnesium, and iron instead of hydrogen and helium. We are living on a kind of cosmic cinder that is greatly depleted in the abundant light elements. Or, to put it more attractively, the Earth is a chocolate chip in the great galactic cookie. One of our tasks in this book is to see how this peculiar concentration of heavy elements came about and how likely it is that other chocolate chips may be out there among the stars. Another task is to ask about the role of life in planetary history, to determine whether there are likely to be many other worlds that are host to living things.

3.1 BASIC PROPERTIES: MASS, SIZE, AND DENSITY

What are the members of our planetary system really like? Before we begin the detailed answer to this question (starting in Chapter 4), let's take a quick look from a very basic point of view. We shall describe the sizes and masses of these bodies, and from these two characteristics, we can then derive their densities. Even before we study the planets with spectrometers and spacecraft, knowledge of the densities gives us an important clue to their composition.

DENSITY AS A GUIDE TO COMPOSITION

Density is a measure of the amount of mass contained in a given volume. In this book, we express density as grams per cubic centimeter or, equivalently, as tons per cubic meter. In these units, water has a density of 1.0 g/cm^3. Ice, since it floats in water, must have a lower density. In fact, it is only a little lower: 0.92 g/cm^3. That's why only the tip of an iceberg shows above the surface of the ocean in which it floats. On the other hand, a piece of pinewood has a density of 0.5 g/cm^3, and a piece of the porous volcanic rock called pumice may have a density of 0.7 g/cm^3. Both the wood and this unusual rock float better than ice. So would the planet Saturn, if we could build a big enough bathtub, since its density is only 0.7 g/cm^3. Metals, in contrast, have high densities; lead has a density of 11 g/cm^3, indicating 11 times as much mass in the same volume as a gram of water (Fig. 3.1).

Intuition tells us that we could expect a planet like the Earth to be at least as dense as the rocks we find on its surface because in effect the crust floats on the denser material beneath. (Recall that in the Sun also, density increases with depth.) Crustal rocks typically have densities between 2.5 and 3.5 g/cm^3, while the Earth's overall density is

nearly twice as great. The central core of our planet is composed of iron and nickel, obviously more dense than rock, and these metals are compressed to a greater than normal density by the weight of the overlying material. The result is an average density for the Earth as a whole of 5.5 g/cm^3. For more massive planets, we expect higher densities because the central compression is greater, and conversely for the less massive ones, if they are made of the same rocky and metallic materials as the Earth. Any deviations from this expectation must indicate differences in composition.

DENSITIES OF PLANETS

The density of a planet can be calculated once we know its mass (which can be determined from its gravitational influence on its satellites or other objects in the solar system) and its size. The technique requires Kepler's laws, Newton's laws, some simple trigonometry, and a telescope. We need no fancy astrophysical equipment, and the picture of the solar system we can now unfold was already available in the eighteenth century (except, of course, for the new planets that were discovered later). Let's see what this picture looks like.

A summary of the masses, sizes, and densities of the planets is given in Table 3.1. It is evident that the planets can be divided into three categories according to these characteristics: small high density, large low density, and small low

Table 3.1 The planets: size and density

Planet	Diameter*	Mass*	Density (g/cm^3)
Mercury	0.38	0.06	5.4
Venus	0.95	0.85	5.3
Earth	1.00	1.00	5.5
Mars	0.53	0.11	4.0
Jupiter	10.8	318	1.3
Saturn	8.9	95	0.7
Uranus	4.1	15	1.3
Neptune	3.8	17	1.6
Pluto	0.2	0.002	2.1

*Both relative to Earth = 1

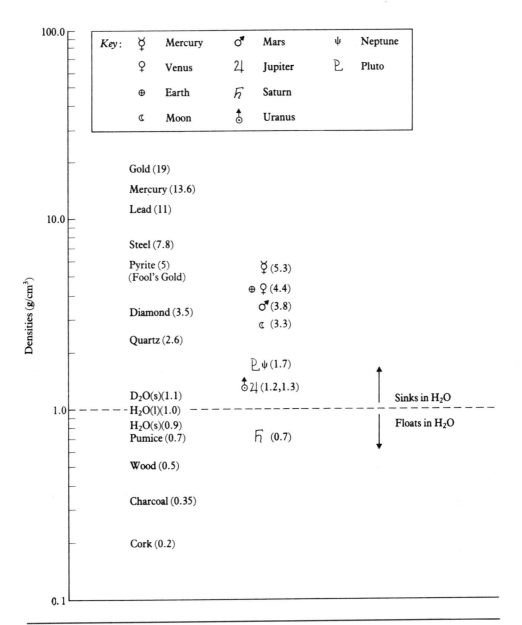

Figure 3.1   The densities of various substances compared with planetary densities. The densities of the inner planets are corrected for the effect of compression. Note that just as pumice and Saturn could float in water, steel and Mercury could float in mercury (pumice and Saturn could also).

density. These categories also turn out to be correlated with distance from the Sun: inner planets, outer planets, and Pluto. Our Moon may also be thought of as an inner planet in terms of its composition, but it is distinctly different from Earth, as we'll see in Chapter 7.

If we consider only the densities, all four inner planets are similar. Both the mass and density of Earth and Venus are nearly the same, making them twin planets (but not identical twins, as we will see later). But a closer look reveals that even

with this simple approach it is possible to find some distinguishing characteristics. We expect bodies less massive than Earth to have smaller average densities because their centers will be less compressed. This is certainly true for Mars, but not for Mercury, whose density of 5.4 g/cm^3 is nearly equal to that of our planet. As we shall see in Chapter 8, the solution to this apparent paradox is to assume that Mercury has a different composition, with a disproportionately large core of iron and nickel.

We confront the opposite problem when we consider the planets in the outer solar system. Here we are dealing with planets far more massive than the Earth, and we would expect them to have greater densities (due to compression) if they had the same composition. Instead, we find that these objects are less dense than any of the inner planets. The conclusion is inescapable: They cannot be made of rock and metal. The only way to construct massive planets with such low densities is to make them predominantly of the two lightest elements—hydrogen and helium. Thus, the outer planets are fundamentally different from the inner ones, in both size and composition. Their composition is more closely related to the Sun and stars than to the rocky bodies in our immediate neighborhood.

Pluto is an exception. This planet is about two-thirds the size of our Moon, but its lower density (2.1 vs. 3.3 g/cm^3) suggests that Pluto is probably made partly of some material with a lower density than rock or metal. Water ice is a good bet. We can reach the same conclusion if we look at the sizes and densities of many of the moons of the outer planets. Pluto resembles one of these large icy satellites more closely than it resembles any of the inner or outer planets. It is also similar to various icy objects on the outskirts of the solar system, as we will describe in Chapter 16. Some scientists think it would be less confusing if we did not classify Pluto as a planet at all.

Satellites, Asteroids, and Comets

Among the satellites of the outer planets, we find an interesting parallel to the solar system itself in the densities of the four moons of Jupiter discovered by Galileo. In order of increasing distance from the planet, the names and densities (in g/cm^3) of these objects are Io (3.3), Europa (3.0), Ganymede (1.9), and Callisto (1.9). Like the planets, these satellites exhibit decreasing densities with increasing distance from the center of their system. Evidently proximity to Jupiter has had a comparable effect to proximity to the Sun. You might also ask yourself what such low

densities for Ganymede and Callisto imply about the composition of these objects. They are far too small to be made of hydrogen and helium.

Our information about the densities of asteroids and comets is much less secure. Nevertheless, it is clear from the evidence we do have that the asteroids are composed predominantly of rock, while the comet nuclei contain substantial quantities of lower-density ice.

3.2 Chemistry of the Planets

Based on their observed densities and some hints about their surface and atmospheric composition obtained from spectroscopy, we can begin to see the outlines of planetary composition. The most fundamental distinction is between the oxygen-dominated chemistry of the inner solar system and the hydrogen-dominated chemistry of the outer planets.

Hydrogen and Oxygen

Hydrogen and oxygen are two of the most abundant and chemically reactive elements in the universe. Each is capable of forming compounds with many other elements. For example, carbon can combine with hydrogen to form methane (CH_4) or with oxygen to form carbon dioxide (CO_2). Hydrogen and oxygen can also combine with each other to form H_2O. If they outnumber oxygen atoms, the hydrogen atoms can combine with other elements to form abundant hydrogen compounds; these conditions are said to be chemically **reducing.** If oxygen predominates, we find many compounds that contain oxygen, and conditions are said to be **oxidizing.**

Hydrogen is the major constituent of the gas and dust that gave birth to the Sun and planets. In the parts of the solar system that retain abundant hydrogen, like the atmospheres of the giant planets, reducing conditions prevail, and we find such reduced compounds as methane (CH_4) and ammonia (NH_3) as well as more complex hydrocarbons (compounds containing atoms of hydrogen and carbon). In general, regions of low temperature have been able to retain more hydrogen compounds than warmer environments.

Closer to the Sun the temperatures are too high for much hydrogen gas to remain, and in the absence of hydrogen, oxygen takes control. All of the inner or terrestrial planets are at least partially oxidized, and their atmospheres are also oxidizing. The most extreme example of oxidized chemistry is our own atmosphere, which contains oxygen gas (O_2)—an exceedingly rare material in the universe because it so easily combines with almost any other element. It is only because of the presence of life on Earth that oxygen gas persists in our atmosphere.

FOUR TYPES OF MATTER

One group of ancient Greek philosophers thought matter could be divided into just four categories: Fire, Air, Water, and Earth. As we saw in Section 2.2, matter is actually composed of atoms of pure substances called elements, of which there are many more than four. But this simple Greek view is a useful tool for organizing the objects in the solar system.

For our purposes, it is helpful to change the categories somewhat. We shall consider the following four forms: gas, ice, rock, and metal. Using these four components, we can classify the various members of the solar system according to the relative proportions of these forms of matter that they contain.

GAS We all know what gas is, and we know that planetary atmospheres are composed of gas. The atmospheres of the giant planets Jupiter and Saturn are composed predominantly of hydrogen and helium gas, but at the great pressures of these planets' interiors, the hydrogen and helium are in liquid form. Jupiter and Saturn preserve the material that has been least modified chemically since the formation of the solar system.

No other planets resemble Jupiter and Saturn in having a composition similar to that of the Sun. The next closest are Uranus and Neptune. These two giants still contain a great deal of hydrogen

and helium, but they have a higher proportion of heavy elements. We see this both in the composition of their atmospheres and from their densities, which are substantially higher than the densities of Jupiter and Saturn, even though Uranus and Neptune are less massive and hence less compressed.

The solar system began as a condensing cloud of gas and dust, commonly called the **solar nebula.** The abundances of the elements we find in the atmosphere of the Sun are approximately the same as those that existed in the solar nebula (see Table 2.2). This is because when the solar nebula collapsed to form the solar system, most of the material went to form the Sun. Deep in the solar interior, nuclear reactions are changing this primordial composition, but in the outer atmosphere that we observe, the Sun can serve as a standard for the composition of the matter from which the entire solar system formed.

ICE Ices are molecules that are liquid or gaseous at moderate temperatures but form solid crystals at low temperatures. The most familiar examples on Earth are water (H_2O) and carbon dioxide (CO_2). Other ices found in the planetary system include carbon monoxide (CO), ammonia (NH_3), and methane (CH_4). These molecules are also frequently called **volatiles,** meaning that they melt or sublime at moderate temperatures. The comets and most planetary satellites are composed in large part of volatiles, predominantly water ice.

Ices formed from the original solar nebula in regions of low temperature. If it was cold enough to form ice, it was also cold enough for rock and perhaps some metal to be present, since we are considering a solar mixture from which we have simply stripped away the two lightest gases, hydrogen and helium. Since there are more molecules of water than of any other compound, the total mass of water or ice is about equal to the mass of the (much denser) rocky material. We think that the dense cores of the outer planets consist of a mixture of rock and ice, although at their high pressures and temperatures, they do not resemble the kinds of rock and ice we normally encounter.

The next step away from the primordial composition is to eliminate the hydrogen and helium gas entirely. This would leave us with ice, rock, and metal. If such bodies are large enough and not too warm, they have the potential to retain atmospheres of heavier gases. In our solar system, Pluto and the large icy satellites of Jupiter, Saturn, and Neptune seem to satisfy this description, although only Titan has a significant atmosphere.

ROCK   If we began with objects composed of mixtures of ice and rock and subsequently heated them, we could imagine the ice evaporating to leave predominantly rocky objects. The Moon is an example of a body composed almost entirely of rock. About three-quarters of the Earth is made of rock. The most common rocks are **silicates,** which are oxides of silicon, aluminum, and magnesium. We will discuss the rocks and their constituent minerals in Section 3.3.

METAL   At still higher temperatures, the rock itself undergoes chemical and structural transformations. In some circumstances, the iron, nickel, and magnesium that are common constituents of rock can separate into metallic form. Small metallic grains were a common constituent of the original solar nebula, but today most of the metal in the solar system is found in the cores of planets. Mercury is an example of a planet that is composed of about three-quarters metal, and some asteroids are nearly pure iron and nickel.

DENSITY AND COMPOSITION

We have described a progression from gas through ice and rock to metal, which corresponds to an increase in the amount of heating or other processing that the original material from the solar nebula has undergone. This progression is also correlated with the composition of the gases we find around those planets capable of holding atmospheres. In the warm inner solar system, we find oxygen compounds in planetary atmospheres, whereas hydrogen dominates the atmospheres of the outer planets.

If you recall the discussion of density in Section 3.1, you will expect a general progression from low to high density as we move to more evolved objects. This is a natural consequence of losing the abundant light elements, hydrogen and helium, and the most abundant compound, water ice. Thus, a good determination of an object's density can tell us a great deal about both its composition and history before we begin the much more difficult work required to make a detailed analysis of chemical composition.

These various trends are illustrated in Table 3.2, which provides a useful one-page introduction to solar system chemistry and planetary evolution.

3.3 ORIGIN AND CLASSIFICATION OF ROCKS

Some of the most important solids we will encounter in our exploration of the solar system are the rocks. Rocks are compounds that include the elements silicon and oxygen. Living on the Earth, we tend to take rock for granted. But as we have seen, hydrogen and helium are far more abundant than silicon in the universe at large, so rock is actually rather rare.

ROCKS AND MINERALS

Even if we restrict ourselves to solid compounds, we can see that water ice should be far more common than rock because oxygen is 11 times more abundant than silicon in the universe. What this means is that even if oxygen combined with every silicon atom in the universe to make quartz (SiO_2), there would still be plenty of oxygen left to make an even larger quantity of water ice. When we explore the outer solar system where ice is stable, we will find that ice is indeed the dominant solid. Meanwhile, let us consider the rocks.

The rock we see in the familiar landscapes of our planet is composed of assemblages of compounds or elements called **minerals.** The principal difference between a rock and a mineral is homogeneity. A mineral is composed of a single substance, whereas a rock may be made up of several different minerals. Because of their monetary value, gold and silver are probably the most famous examples of single-element minerals. Elemental sulfur, copper, and carbon in the form of graphite

Table 3.2   Planets and satellites: overview of composition

Object	Distance from Sun (AU)	Density (g/cm^3)	Bulk Composition
Mercury	0.4	5.4	iron, nickel, silicates
Venus	0.7	5.4	silicates, iron, nickel
Earth	1.0	5.5	silicates, iron, nickel
Moon	1.0	3.3	silicates
Mars	1.4	3.9	silicates, iron, sulfur
Jupiter	5.2	1.3	hydrogen, helium
Callisto	5.2	1.8	water ice, silicates
Ganymede	5.2	1.9	water ice, silicates
Europa	5.2	3.0	silicates, water ice
Io	5.2	3.4	silicates
Saturn	9.6	0.7	hydrogen, helium
Titan	9.6	1.8	water ice, silicates
Uranus	19.2	1.3	ices, hydrogen, helium
Neptune	30.1	1.6	ices, hydrogen, helium
Triton	30.1	2.1	silicates, water, other ices
Pluto	39.4	2.1	silicates, water, other ices

and diamonds are also economically important. However, only a few elements occur in nature in a pure state, so most minerals are compounds. Minerals composed of single compounds include quartz (SiO_2), hematite (Fe_2O_3), iron pyrite or "fool's gold" (FeS_2), and calcite ($CaCO_3$). These four minerals are examples of silicates, oxides, sulfides, and carbonates, respectively, the four most common types of rock-forming minerals.

IGNEOUS, SEDIMENTARY, AND METAMORPHIC ROCK

Rocks can be classified in terms of the minerals they contain, as we have just seen. However, it is also convenient to divide them into three large categories on the basis of their origins: igneous, sedimentary, and metamorphic.

Igneous rocks are those that have formed directly by cooling from a molten state. A rock picked up from the slope of a volcano in Hawaii is an obvious example (Fig. 3.2). This is a representative of the family of igneous rock called **basalt,** which we shall encounter repeatedly during our studies of the inner planets. Igneous rocks make up roughly two-thirds of the Earth's crust.

Sedimentary rocks are composed of fragments of other rocks (or of the shells of living creatures) that are cemented together. Limestone (a carbonate from seashells) and sandstone (made of silicates) are examples. On Earth, the fragments are produced by various weathering (erosion) processes that break up the parent rocks. The most effective weathering processes on our planet involve liquid water. Fragmentation can also be produced when one rock bangs into another, creating a different type of sedimentary rock on the lunar surface, for example.

Metamorphic rocks are produced from either igneous or sedimentary rocks that have been buried far beneath the Earth's surface, modified by the high pressures and temperatures they encountered there, and then returned to the surface. This process of burial and return is part of

(a)

(b)

Figure 3.2 **(a)** This ropy pahoehoe lava on the slopes of Hawaii's Kilauea volcano was still partially molten when the photo was taken. The glowing cracks are about 5 cm wide. **(b)** Astronomer Dale Cruikshank holds a 2-year-old rock (Hawaii lava) in his left hand and a 4.5-billion-year-old rock (a primitive meteorite) in his right hand.

the great cyclical movement of the Earth's continental plates, which we shall discuss in Chapter 9. At the present time, we don't know whether any other planet in our solar system exhibits this phenomenon, so Earth may be the only place where metamorphic rock exists. Marble is probably the best known metamorphic rock.

PRIMITIVE ROCK AND THE ORIGIN OF THE EARTH

In this review of rocks and minerals, we have actually been talking about reprocessed material because all the rock on Earth was melted after the planet formed. Are there more ancient rocks in the solar system? There certainly are. We can define a fourth category called **primitive** rock, which has never melted and has been affected only moderately by chemical and physical processes since the solar system began. We expect such rock to have the same abundances of the nonvolatile elements (elements that are solid rather than gaseous at low temperatures) that exist in the Sun or the giant planets. Examples of this kind of rock are found among the meteorites, as we shall see in Chapter 4.

We expect that primitive rock is lodged in the ices of the comets. Some asteroids and many of the smaller planetary satellites must also be composed of this material (see Fig. 3.2b).

We don't find primitive rocks on Earth because our planet was heated at the beginning of its existence. Current theories for the formation of the inner planets start with solid material from the solar nebula. This material is primitive in the sense described above, and the meteorites are remnants of it. But as this dust and rocky debris come together to form a planet, the object is heated. This heating results in part from the energy released by the impacting material that is bombarding the forming planet and causing it to grow, and partly from radioactivity deep inside the body.

A large planet can retain heat better than a small one can. The generation of heat usually depends on the volume of the planet, which is proportional to the cube of the radius (remember, volume = $\frac{4}{3}\pi R^3$). But the planet can lose heat only through its surface, and the surface area is proportional to the square of the radius (area = $4\pi R^2$). Thus, since heat is generated faster than it is lost, a large planet will tend to grow warmer than a small one. If the internal temperatures rise enough, the rocks melt and the central part of the planet becomes a liquid. At this stage, the denser materials are free to migrate to the center and the lighter materials rise to the top.

The process of gravitational separation according to density is known as **differentiation,** and it leads to the development of differences in internal composition. All the inner planets have undergone differentiation, as we would expect. An interesting puzzle, however, is posed by the presence of asteroids that are also differentiated, in spite of their small sizes. What processes have heated these tiny bodies? We shall return to this problem in Chapters 4 and 5.

We can see why our own planet does not have any primitive rocks. In fact, we even have trouble finding old rocks on Earth because an active geology and erosion have effectively eliminated the first billion years of our planet's geological record. To probe this ancient history, we must go to other places—the Moon, Mars, the asteroids—where less alteration of primitive conditions has taken place.

3.4 PLANETARY ATMOSPHERES

An important characteristic that distinguishes one solar system object from another is the nature of its atmosphere. Earth is 11 times smaller than Jupiter and more than 4 times as dense, with a totally different composition. Yet both planets have atmospheres, although the compositions of these atmospheres are very different. On the other hand, Jupiter's satellite Ganymede has no atmosphere, whereas Saturn's slightly smaller moon Titan has an atmosphere denser than Earth's. What accounts for these differences?

GETTING AND HOLDING AN ATMOSPHERE

There are two basic ways in which a planet can obtain an atmosphere: It can form with one (a primordial or captured atmosphere), or it can produce one from the material of which it is made (secondary or outgassed atmosphere). Often when there are two explanations, a third can be created by combining the first two. In fact, capture plus outgassing appears to be the most likely process for the formation of the atmospheres of the giant planets. They are a blend of captured solar nebula gases and gases produced from the planetary cores.

In order for a planet to hold on to an atmosphere over the 4.5-billion-year life span of the solar system, the molecules in that atmosphere must not move fast enough to escape from the planet's gravitational field. In other words, their speed must be less than what we call the escape velocity. Otherwise, they will soar off into space, eventually leaving a denuded planet behind.

The escape of gas molecules occurs from a layer that is so tenuous that a molecule moving in an upward direction will not encounter any other molecules. It is free to leave the planet if it is moving fast enough. Scientists have called this layer of the atmosphere the **exosphere** because the gases here can exit from the Earth if they have escape velocity.

Physics tells us that the velocities of molecules in a gas are determined by the temperature of the gas and its composition. If the temperature is high enough and/or the gas is sufficiently light, then individual atoms and molecules in the exosphere may reach escape velocity and be lost. For a planet to keep an atmosphere, it should have a large mass and thus a high escape velocity. It should also be cold, so that velocities of the gas molecules will be low (Fig. 3.3). Gases with large molecular or atomic weights (for example, N_2, with weight 28) are more easily retained than lighter ones (e.g., H_2, with weight 2).

These two conditions are well met in the outer solar system, where we find the giant planets. These bodies have such high escape velocities (typically tens of kilometers per second) that they can retain thick atmospheres of even the light gases hydrogen and helium.

Saturn's satellite Titan is not massive enough to retain hydrogen, but it is sufficiently massive and cold to keep an atmosphere of heavier gases, such as the nitrogen and methane that we find there. Similar conditions apply to Ganymede, so the reason this massive satellite of Jupiter does not have an atmosphere must be found elsewhere.

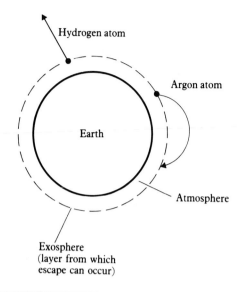

Figure 3.3 The light hydrogen atoms (atomic mass = 1) move much faster than the heavier argon atoms (atomic mass = 40) at the same temperature. Thus, hydrogen can escape from the Earth, but argon cannot.

(We will discuss it in Chapter 15.) Our Moon and the planet Mercury are both too warm and insufficiently massive to maintain atmospheres.

WHY ATMOSPHERES ARE DIFFERENT

There are thus several reasons for the diversity of atmospheres that now exist around various objects in the solar system (when they exist at all). We will come back to this discussion as we consider each planet individually, but let's briefly consider the implications of what we have learned.

Large planets form with sufficient mass to capture hydrogen and helium (and everything else) from the solar nebula, which is why they exhibit hydrogen-rich atmospheres today. Their composition resembles that of the Sun. Small planets were not able to capture and hold solar nebula gases, so they must produce their own atmospheres. The composition of these atmospheres depends on the materials that make up the planet and on the planet's geological and chemical evolution. Mars and Venus have atmospheres that are dominated by carbon dioxide. Earth should exhibit the same thing. The fact that it

doesn't is a result of the presence of liquid water and life on our planet, as we will discuss later. We can also ask why methane is much more abundant than carbon dioxide in Titan's atmosphere—another question we will return to in later chapters.

3.5 STUDYING MATTER FROM A DISTANCE

We have reviewed several basic characteristics of the planets and some properties of the matter that composes them. We now want to tie these two themes together by asking what we can learn about these distant worlds by studying the light, heat, and radio emissions that they send us across the vast emptiness of space. It is important to understand how we make the measurements of composition and other properties that are crucial to understanding how the planetary system has developed.

The process of investigating distant objects by the analysis of their radiation is often referred to as **remote sensing,** to distinguish it from *in situ* (or "on site") studies. Astronomy is almost entirely a remote sensing science, although astronomers rarely use the term. Today, we study the members of the planetary system by a combination of astronomical studies (remote sensing) from Earth, remote sensing carried out by flyby or orbiting spacecraft, and direct studies from entry probes and landers.

Often the most dramatic discoveries come from images radioed back from cameras on spacecraft either orbiting a planet or landed on its surface. We can all relate to a spectacular picture of an alien landscape. Good images require that we carry the camera to the target on a spacecraft, however; telescopes on Earth are not powerful enough to reveal landscapes on distant planets. But telescopes—on the ground or in orbit—provide wonderful tools for learning about the composition of planets.

SPECTRAL ANALYSIS

One of the most powerful techniques of planetary remote sensing is spectroscopy. In Section 2.3, we discussed the formation of spectral lines and their use in determining the composition of a gaseous object like the Sun. Similar techniques are used to study the spectra of the planets. However, in planetary remote spectroscopy, some of the most important clues are derived from the invisible parts of the electromagnetic spectrum.

Leaving the visible spectrum and considering shorter wavelengths, we find that we cannot detect sunlight at the Earth's surface in the ultraviolet region. This is not because the Sun does not radiate ultraviolet light, but because this light is absorbed by a layer of **ozone** in the Earth's atmosphere. Ozone has the chemical formula O_3. It is formed from oxygen molecules in the Earth's upper atmosphere and shields the Earth's surface from high-energy ultraviolet photons. This is beneficial for life on Earth because this high-energy electromagnetic radiation is capable of destroying many of the molecules that make up living organisms.

Moving toward longer wavelengths, past the red end of the visible region of the spectrum, we encounter the infrared, first detected by English astronomer William Herschel (1738–1822), who is best known for his discovery of the planet Uranus. Here again we encounter absorption from the Earth's atmosphere, primarily from water vapor and carbon dioxide. Despite their relatively low abundances in the Earth's atmosphere, these two gases play a dominant role in absorbing infrared radiation. Unlike the short-wavelength end of the spectrum, however, some infrared light can penetrate our planet's atmosphere and reach the Earth's surface. This penetration occurs in regions of the spectrum that lie between the strong absorption bands of water vapor and carbon dioxide (Fig. 3.4). Such regions of transparency are called **atmospheric windows,** although they are windows

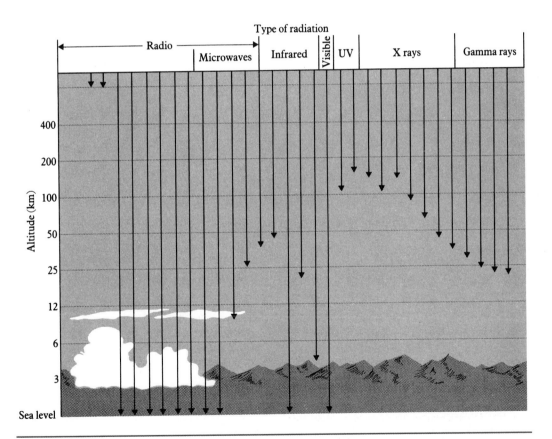

Figure 3.4 The Earth's atmosphere absorbs radiation at most wavelengths at various altitudes above our planet's surface. Only visible light, and some infrared and radio waves, can penetrate the atmosphere and reach the surface. These regions of the spectrum in which radiation can pass through the atmosphere are called atmospheric windows. The same situation (with variations depending on atmospheric mass and composition) exists on all other planets with atmospheres.

in the spectral domain, not in physical space. The atmosphere acts as a kind of color filter, letting some wavelengths (colors) through and absorbing others.

The pattern of spectral windows and absorptions continues out to longer and longer wavelengths. The atmospheric absorption grows progressively stronger until it ceases at a wavelength of about 1 mm. The atmosphere is transparent to radiation of longer wavelengths, providing another clear window on the universe. The long-wave limit on this window is at about 30 cm, the wavelength at which the atmosphere becomes opaque again. This window includes the microwaves, radar, television, and FM broadcasts, which are often lumped together as "radio." AM broadcasts occur at still longer wavelengths and are reflected back to the Earth's surface by the region of our planet's upper atmosphere called the ionosphere.

SPECTRA OF OTHER PLANETS

Orbiting the Sun, the planets shine by reflected light. Just as sunlight passing through the Earth's atmosphere is absorbed at certain wavelengths by our planet's atmospheric gases, we can expect the same thing to occur in the atmospheres of the other planets. By examining the reflected sunlight, we can then hope to discover new absorption lines that will tell us what gases are present on that planet (Fig. 3.5).

Not all the sunlight that strikes the planets is reflected, however. Astronomers call the fraction of sunlight that is reflected the **albedo** of the object. Our Moon, for example, reflects only 11% of the sunshine that illuminates it, despite the fact that it seems so bright in our night skies. This low albedo is characteristic of an exposed, rocky surface. In contrast, cloud-covered Venus reflects 75% of the incident sunlight, while icy Enceladus (a satellite of Saturn) has an albedo of nearly

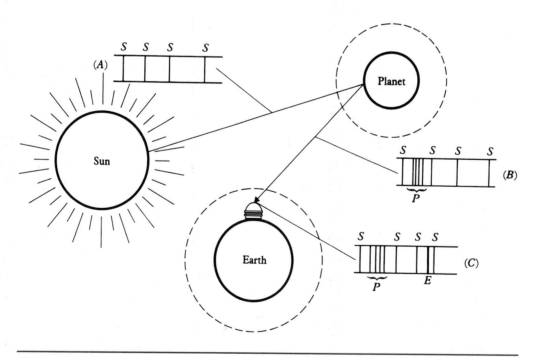

Figure 3.5 Sunlight, whose spectrum (A) contains absorptions (S) from gases in the Sun's atmosphere, penetrates the atmosphere of another planet, where some of it is absorbed by planetary gases, producing new lines (P) in the spectrum of the planet (B). Reflected by clouds or the planet's surface, the light continues its journey to Earth, where some of it may be absorbed by gases in our planet's atmosphere, producing new lines (E) before it reaches a telescope and spectrometer to form the observed spectrum (C).

100%. What happens to the radiation that is not reflected? It is absorbed by the planet, heating its surface or atmosphere. Thus, each planet, satellite, asteroid, or other object assumes a temperature that depends on its distance from the Sun, its albedo, and what kind of an atmosphere it has, if any. If there is a large internal source of heat, that is also important, as we'll see in the case of some of the giant planets.

When the surface of a planet absorbs sunlight and becomes warm, it begins to radiate energy at infrared and radio wavelengths. Imagine you are studying the spectrum of such a planet. After you look at the visible region, where you would be analyzing reflected sunlight, you might move on to the infrared. Here you detect thermal radiation from the planet itself. Since there are no solar lines in the spectrum to confuse you, the absorptions you find in the planet's infrared spectrum are caused by gases in its atmosphere or in the atmosphere of the Earth.

Spectroscopy provides a powerful tool for studying planets from a distance. By analyzing the light they reflect as well as the radiation they emit, we can learn about the composition of the planets' atmospheres and the nature of their surfaces. In fact, we can do even more than this. The strengths of absorption lines can tell us the amount of the absorbing gas. By observing many lines of the same gas, we can often derive an average atmospheric pressure and temperature. If we compare the intensities of thermal emission at several different wavelengths, it is possible to determine physical and chemical properties of the emitting surface. In short, there is a huge bag of tricks that can be played with planetary radiation, once we have enough of it to work with. To achieve that, we need telescopes.

TELESCOPES AND OBSERVATORIES

Astronomers would like to collect as much electromagnetic radiation from the planet as possible, to view the planet from all angles, and to study small regions as well as the entire globe. There are two ways to do this: (1) use a large collecting area that focuses all the radiation falling on it into an image that can then be examined (but this does not broaden the angle of view), or (2) go closer to the object. The first approach employs a telescope; for the second a spacecraft is essential. In practice, the optical instruments on spacecraft also use small telescopes, thereby further improving their performance.

The reasons scientists need to make this effort to study the planets are quite simple. First, even the brightest planets don't send us very much light. Second, when a spectrometer spreads out the planetary radiation to form a spectrum, it is essentially sorting photons by wavelength or energy. Instead of just detecting the planet by adding up all the photons in the visible spectrum, which is what our eyes do, the detector in the spectrometer is sampling only a few discrete energies. Unless a lot of radiation is collected before it is spread out in this way, there won't be enough photons to stimulate the detector.

An optical telescope performs two critical functions for the astronomer: It collects light, and it forms a magnified image of the object being studied. Light is collected by the mirror or lens that forms the principal optical element of the telescope. This lens or mirror is the heart of the telescope, and its diameter, or **aperture,** is the measure of telescope power. Thus, for example, we speak of the 200-in. Hale telescope on Palomar Mountain or the 10-m Keck telescopes on Mauna Kea.

All the light from a single source that falls on the entire area of the lens or mirror is brought to a focus at a single point. When you look at a planet without a telescope, your retina is detecting only the amount of energy that passes through the few square millimeters of area corresponding to the size of the lens in your eye. But gazing through the eyepiece of a 2-m reflecting telescope, you have access to all of the light that falls on the 3 million mm^2 of the primary mirror (Fig. 3.6). The eyepiece, in turn, is constructed to magnify further the image of the planet formed by the telescope. Using eyepieces composed of different lenses, telescopes can achieve different magnifications.

A good pair of binoculars is a small but sophisticated pair of telescopes. Typically each has a

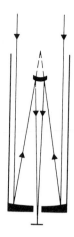

Figure 3.6 The Cassegrain focus of a modern reflecting telescope. Light from a planet or star is reflected by the large primary mirror to a smaller secondary mirror that sends it back through a hole in the primary. Instruments to examine the light may then be mounted on the back of the telescope.

main lens 35 mm in diameter, which collects several hundred times more light than the eye alone. The magnification might be 6—in which case, we call this a pair of 6 × 35 binoculars. An amateur astronomer's telescope, such as are sold widely in camera stores, typically has a main mirror diameter of 10–20 cm and magnifying powers from 20 to 100.

Binoculars can be held in the hand, but large astronomical telescopes require permanent mountings (Fig. 3.7). The motion of the telescope is computer controlled to permit precise pointing and tracking. The telescope, its computer drive system, and all the auxiliary instrumentation are housed in an observatory. This is a building with a dome that has an opening in it through which the telescope can be pointed to view the heavens. (Contrary to cartoon depictions, the telescope does not poke out through this opening.) The dome rotates to give the telescope complete access to the sky.

With appropriate modifications, these same considerations apply to radio telescopes. Once again, a large collecting area is used to gather the incident electromagnetic radiation. In this case, the collectors may be an array of antennas or a

parabolic dish, shaped to bring the radiation to a focus. The long wavelengths of radio photons lead to a requirement for very large aperture antennas or even collections of antennas linked by computer to allow studies of just a small area in the sky (Fig. 3.8). Since these metal antennas do not need protection from the weather, radio telescopes do not have domes.

SEEKING THE BEST OBSERVATORY SITES

To take full advantage of a telescope's radiation-collecting ability, it is necessary to observe from a favorable location. If you have ever looked at a

Figure 3.7 The 4-m Mayall telescope of Kitt Peak National Observatory. The telescope tube is mounted in a yoke and is in a vertical position in this picture.

distant scene near the surface of the Earth on a hot day, you have noticed that motions in the air distorted the appearance of what you were trying to see. In extreme cases, layers of air at different temperatures can produce mirages. A good observatory site should minimize these distortions as well as provide as many cloudless nights as possible.

When you look at a star or a planet in the night sky, you are looking through the Earth's entire atmosphere, so the light that reaches your eyes has passed through many different layers of air, at various temperatures and moving at various speeds in various directions. The effect of all this is to make the stars appear to twinkle, when actually they are shining with a very steady light. A telescope magnifies these effects because it is collecting light over a much greater area than the eye, thus allowing even more random properties of the atmosphere to affect the image.

These atmospheric distortions can be improved by placing the telescope at a high-altitude location, where the air is thinner and steadier. Such a mountaintop location has several additional advantages. Mountains are usually far from cities, so the sky is darker and clearer, free from dust, smog, and the light from street lights, cars,

and buildings. Furthermore, a sufficiently high altitude takes the observer above most of the water vapor in the Earth's atmosphere, thereby opening wider the infrared atmospheric windows. Some sites, particularly isolated mountains surrounded by ocean, are better than others because the local topography and wind patterns reduce turbulence and yield sharper images. Recently astronomers have invented ways to provide additional image stabilization using what are called active optics. With computer-controlled active optics, ground-based telescopes can achieve image quality as good as that of the Hubble Space Telescope.

The best sites for astronomical observations are Mauna Kea in Hawaii and several high desert peaks in the Chilean Andes. Mauna Kea, an isolated mountain that rises 4.2 km above sea level, has more large telescopes than any other location (Fig. 3.9). These include 8-m national telescopes of Japan and the United States, as well as twin 10-m telescopes of the Keck Observatory, jointly operated by the University of California and the California Institute of Technology. A location in Chile called Cerro Paranal is the home of the largest multiple-telescope instrument in the world, consisting of four 8-m telescopes, operated by a consortium of European countries.

Figure 3.8   The very large array (VLA) of radio telescopes of the National Radio Astronomy Observatory near Socorro, New Mexico. Each of the 27 telescope antennas has a diameter of 25 m. They can be moved on railroad tracks laid out in the shape of a Y to produce an effective aperture of nearly 35 km. Even though this aperture is not filled by the antennas, they are linked by a computer so the whole array functions as a single instrument.

Figure 3.9 The finest observatory site on Earth is the summit of Mauna Kea, a 4.2-km extinct volcano in Hawaii. Among the large telescopes are twin 10-m Keck telescopes and two 8-m instruments, one (called Subaru, or Pleiades) built by Japan and the other built by the U.S. National Optical Astronomy Observatories.

Even the best mountaintop is not ideal, however. The next step is to use a telescope in an airplane in order to reach still higher altitudes. This has proved particularly effective for infrared studies because the absorptions by Earth's water vapor are reduced dramatically, and NASA is developing a 2.5-m airborn infrared telescope that flies above 13 km in altitude. To do still better, one must use balloons and rockets and ultimately send a telescope into orbit about the Earth, totally outside our planet's atmosphere. Then the entire electromagnetic spectrum becomes accessible.

Orbiting Observatories

In the discussion of good sites for observatories, we have concentrated on observations made in the visible and infrared regions of the spectrum.

What about the ultraviolet? The ozone layer is so high that even an airplane doesn't get above it, and only instruments in rockets or satellites will do. Nor is this the only advantage of placing a telescope above the Earth. Without the distorting effects of the atmosphere, a space telescope can see fainter sources and distinguish finer detail than its counterpart on Earth, as well as have access to the entire electromagnetic spectrum.

While this approach has obvious advantages, putting it into practice is not so easy. Launching a large optical instrument into orbit with the ability to find and track faint sources, to detect and analyze their radiation, and then to send the information to Earth—all these are major challenges. Nevertheless, these challenges have been met, although the resulting space observatories are many times more expensive than those built on the ground.

PERSPECTIVE

DO THE PLANETS INFLUENCE OUR LIVES?

If you are an astronomer, the positions of the Moon and planets may indeed have an influence on you. When Mars is close to the Earth, you may find yourself at an observatory far from home, studying the light from this planet. Other astronomers avoid observing when the Moon is near full phase because its brilliance obscures the light from faint galaxies. But these perturbations affect only a few hundred individuals on the Earth. What about everybody else?

You all know that many nonastronomers believe that the positions of the Sun, Moon, and planets have an influence on them. The term *lunacy* is derived from a supposed tendency for irrational actions near the time of full moon. Even more pervasive is a belief that the positions of the Sun, Moon and planets at the time of birth (not, curiously, at the time of conception) have an effect on our nature and can be used to predict our future life. Under the heading of "astrology" or "horoscope," most newspapers tell us each day what to expect based on the month in which we were born. Astrology in America is a big business.

Astrology is as old as astronomy. Many ancient cultures regarded the planets as expressions of divine forces, and they studied planetary motions in the hopes of understanding their gods. The fact that the motions of the planets in the sky are complex and were often unpredictable to early observers added to the mystery and seemed consistent with the idea that the gods were sending us messages hidden in the wanderings of the planets.

The astrology that spread throughout Europe and eventually to the New World is derived from the ancient religion of Mesopotamia (modern-day Iraq). The priests believed that the motions of the seven wandering celestial objects or planets (Sun, Moon, Mercury, Venus, Mars, Jupiter, and Saturn) could be used to predict the fates of kings and nations. They invented the concept of a horoscope, which is a plot of the positions of the planets at the moment of a person's birth, as an indication of the future. To "cast a horoscope" is to write down this celestial configuration for a person based on the day and time of birth—a demanding task in those days, although it can be done today by a computer in a fraction of a second. The ancient Greeks believed that this sort of predictive system (called natal astrology) applied to everyone, not just the king. Claudius Ptolemy, who recorded so much of Greek astronomy, performed the same function for astrology with his *Tetrabiblos*, which is the "bible" of astrology even today.

Renaissance astronomers overthrew Ptolemy's astronomical theories and started us on the path of discovery toward modern science. However, today's astrologers still follow the precepts of Ptolemy. This is not a field that has shown progress. In fact, astrology as most people understand it today has moved backwards because most modern astrologers use the same birth signs as did Ptolemy, neglecting that they are now one sign behind. If you think you are an Ares, you are really a Pisces, and soon you will be an Aquarian.

Actually, few people today really believe that the time of our birth can be used to predict our future lives. We no longer think that the planets are divine beacons set in the sky just above our heads. There is no known force or influence of the planets that might have an effect on us. Yet majorities in public opinion polls still indicate some level of "belief in astrology", and many people make judgments about others, including whom they will marry or hire, based on birth sign.

➡

One would expect that astrology could be tested in the same way we gather data to test scientific hypotheses. We have the techniques, for example, to determine whether environmental exposure to toxic chemicals contributes to cancer, or whether certain drugs are efficacious in improving our health. We can apply similar techniques to test the accuracy of astrological predictions. One simple example is to use birth sign (easily determined from a person's birth date) to search for similarities in personalities. For example, we can determine whether some birth signs are correlated with choice of military career, or with financial or political success, or with excellence in sports. When we know the exact time of birth, we can calculate a full horoscope and carry out similar statistical studies. Many such tests have been done, and none has yielded any positive results. There is no evidence that astrology has any predictive value concerning either the type of person we are or the future that awaits us.

Astrology was born in the same culture as astronomy, and in the ancient world the two were virtually the same thing. People who studied the planets and their motions were the same ones who established calendars and knew how to cast horoscopes. The two studies began to diverge in the Greco-Roman world of two millennia ago, and they have parted company dramatically during the past 500 years. But astrology is still with us—one of the oldest religions practiced in the world today, predating Christianity, Buddhism, or Islam. You probably have friends who believe in it, and some of them may even now confuse it with the science of astronomy.

The most famous instrument in orbit is the Hubble Space Telescope (HST), named after the astronomer Edwin Hubble, who discovered the expansion of the universe. HST, with an aperture of 2.4 m, was launched in 1990 and is expected to operate until about 2010. The primary objective of the HST is to achieve very high spatial resolution—better than was then possible with any ground-based telescope. Unfortunately, however, the main mirror of the telescope was incorrectly shaped, so that its resolution was worse than that of ground-based telescopes. This tragic error was corrected when astronauts installed new instruments in December 1993, and since that time HST has performed flawlessly (Fig. 3.10).

Other orbiting observatories are designed to explore parts of the electromagnetic spectrum that do not reach the ground. These have included the Compton Gamma Ray Observatory, the Chandra X-Ray Facility, and the SIRTF infrared telescope.

3.6 EXPLORING THE PLANETARY SYSTEM

Even the largest telescopes on the ground or in orbit are limited by the immense scale of the solar system. To see the craters on the moons of Jupiter, to explore the rings of Saturn, to search for hydrocarbon seas on Titan, to sift the sands of Mars for signs of life—all these feats are beyond our reach at any terrestrial observatory. To accomplish objectives like these, we have no choice: We must go there. Direct planetary exploration, begun in 1959, has by now resulted in spacecraft visits to every planet except Pluto. Humans have walked on the Moon and brought nearly half a ton of lunar rocks and soil back to Earth. Instrumented spacecraft have landed on the surfaces of Mars, Venus, and the asteroid Eros. We have probed deep into Jupiter's atmosphere. Other spacecraft have flown past satellites, comets, and asteroids. We have come to know every planet in our solar system except Pluto as a familiar, three-dimensional world.

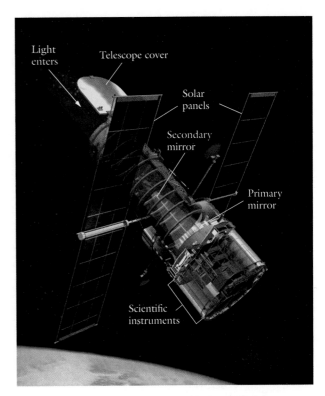

Figure 3.10 Orbiting the Earth at an altitude of 500 km, the Hubble Space Telescope (HST) was captured for repairs by the space shuttle *Endeavor* on December 9, 1993. Perched atop a foot restraint on the shuttle's remote manipulator system arm, astronauts F. Story Musgrave (top) and Jeffrey A. Hoffmann complete the last of five space walks required to refurbish the telescope. The west coast of Australia is in the background.

The results of these missions form most of the substance of this book. A list of some of the most important successful spacecraft is given inside the back cover. We will also discuss some of the individual missions when we consider the discoveries they made. Here we simply want to indicate some of the advantages of studying the planets closeup.

We have already mentioned the most obvious advantage: Getting closer allows a much more detailed inspection of the target object. Equipped with telescopic lenses, the cameras on the spacecraft can record fine detail. Spectrometers can study small areas of a planet to see whether the composition of one lava flow or type of cloud resembles that of another. Spacecraft can also view planets from angles we can never achieve from

Earth. The far side of the Moon was the first such realm to be explored in this way, in 1959 by the Soviet spacecraft Luna 3. Another example is shown in Figure 3.11, where we see a view of Saturn we can never achieve from Earth. This is a farewell image recorded by Voyager 2 in 1981.

INTERPLANETARY FLIGHT

The spacecraft that leave our planet completely are moving like bullets shot from a gun. They are given a terrific boost to escape from the Earth's gravitational field, but then they coast most of the way to their targets. Small thruster rockets are fired occasionally to provide course corrections, nudging the spaceships into more accurate trajectories. If one of these robot vehicles is to go into orbit around the

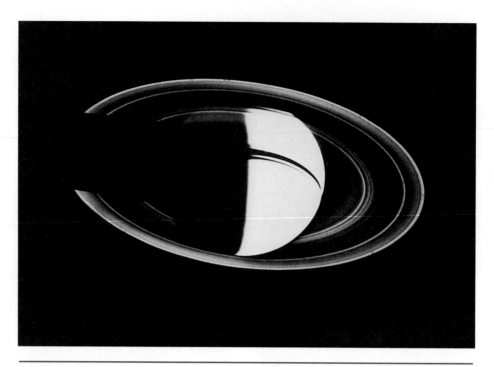

Figure 3.11   Looking back at Saturn from Voyager 2. The view from behind and below the planet shows the unlit side of the rings, as the spacecraft sets forth on its journey to Uranus.

Moon or another planet, it must reduce its speed and change its direction as it gets close. Otherwise, it will fly past in a trajectory that becomes curved in response to the pull of the planet's gravity, or it will simply crash into the target.

An important technique that is used to change the velocity of a spacecraft without using up precious fuel is to perform a close flyby of one planet in order to reach another one. This kind of cosmic billiard shot (without the impact!) helped Mariner 10 to travel from Venus to Mercury, sent Pioneer 11 completely across the solar system from Jupiter to Saturn, and permitted Voyager 2 to accomplish a grand tour of all four giant planets (Fig. 3.12). These gravity-assist trajectories work by using the target planet's moving gravitational field (as the planet itself moves in its orbit about the Sun) to accelerate the spacecraft. This acceleration can be a change both in direction and in speed. Gravity-assist has proved to be a very effective technique.

The Voyager 1 and 2 and the Pioneer 10 and 11 spacecraft achieved escape velocity with respect to the Sun, not just with respect to Earth. They are on trajectories that are taking them out of the solar system, into the vast realms of space among the stars. But while we may feel proud to be starting on this adventure, we should understand just how inadequate our current technology is for the task. The nearest star, Alpha Centauri, is 4.3 light years away. This is the distance light can travel in 4.3 years, which is about 25 trillion miles. If these speedy spacecraft were directed toward that star (and none of them is), it would take them about 100,000 years to get there.

HOW A SPACECRAFT WORKS

The trip to a planet can require just a few months, if the target is Venus, or as long as a decade if our objective is in the outer solar system. The spacecraft come in all different shapes and sizes, depending on the functions they are to perform, how far they are traveling, the capability of the launch vehicle, and such mundane factors as how much money is available to build them. The

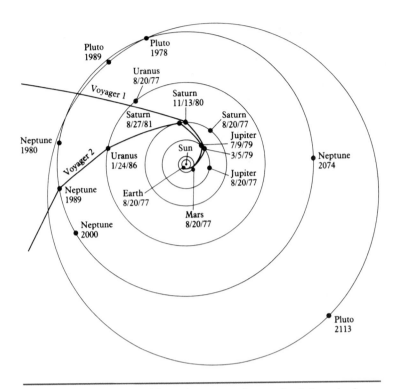

Figure 3.12 The two Voyager spacecraft were both launched in 1977 but followed divergent paths in their journeys past the planets. Both spacecraft are now exiting the solar system.

Viking landers that have been resting on the surface of Mars since 1976 are about the size of a golf cart and weigh about a ton (Fig. 3.13). They were powered by nuclear energy. In contrast, the Sojourner rover that landed on Mars in 1995 is the size of a microwave oven and had to make do with power from small solar cells. The Galileo and Cassini spacecraft to the outer solar system are nearly as big as a subcompact car.

These spacecraft are crammed with sophisticated instrumentation, the electronics and computers required to make them run, antennas for communication with Earth, and panels of solar cells or nuclear generators to provide the power to make everything work. The amount of power required is amazingly small, since everything has been carefully miniaturized. A 100-watt light-bulb consumes more energy than some of these spacecraft.

What happens when one of these remarkable devices reaches its intended target depends on its

Figure 3.13 The science test model of the Viking landers that reached the surface of Mars in 1976. Viking Project Manager Jim Martin provides useful scale. This lander is currently on display at the Smithsonian Air and Space Museum in Washington, D.C.

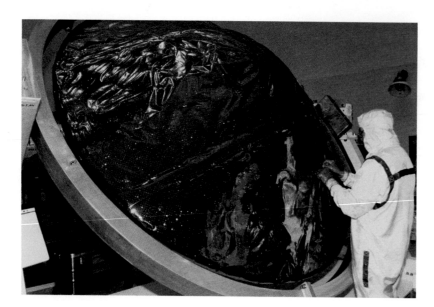

Figure 3.14 The Cassini Huygens probe, which will be deployed into the atmosphere of Saturn's moon Titan in 2005.

sophistication. Usually it measures the intensity of the magnetic field and the density of charged electrons and protons in the interplanetary medium all along its trajectory. When it reaches its destination, these measurements are continued and the magnetosphere of the planet is mapped out in detail. These kinds of instruments require no particular pointing. Indeed, it is helpful for some of them to be mounted on a spacecraft that spins slowly so they can sample the full 360° angle around the trajectory.

To record images or to make spectroscopic observations requires a stabilized platform attached to a spacecraft that does not spin. Such a spacecraft uses star sensors and small rockets (or thrusters) to maintain a fixed orientation as it proceeds along its track. The large radio antenna that it carries to communicate with Earth remains pointed toward our planet, while the platform carrying the cameras and spectrometers is turned toward the target planets and satellites in accord with a sequence of commands stored in the onboard computer.

ORBITERS, PROBES, AND LANDERS

If orbit around another planet is desired, an additional rocket engine must be carried to slow down the spacecraft just enough to permit the planet's gravitational field to capture it. In this case, the spacecraft is called an orbiter. Examples include the U.S. Magellan and the Soviet Veneras 15 and 16. These three Venus orbiters were all equipped with radar antennas and transmitters in order to penetrate the cloudy veil of Venus and gradually produce topographic maps of the planet's surface after many orbits.

An atmospheric probe or a lander might be carried on the orbiter, from which it is released to coast down to the planet itself. The former USSR mounted a very successful series of Venus probe missions in the 1970s and early 1980s, bringing us much new knowledge about that planet's atmosphere and surface. Many of these probes could properly be called landers, since they survived on the surface, sending back pictures and compositional analyses. The Galileo spacecraft deployed a probe into the atmosphere of Jupiter in 1995, and the Cassini mission will send a probe into the atmosphere of Saturn's satellite Titan in 2005 (Fig. 3.14). The Near-Earth Asteroid Rendezvous (NEAR) spacecraft orbited the asteroid Eros for a year before it touched down early in 2001 and became a lander.

SUMMARY

As a first approximation to studying what planets are composed of, we can look at four types of matter: gas, ice, rock, and metal. For the planets, a simple measurement of density gives important clues concerning composition, allowing us to distinguish a rocky from an icy world or to deduce the overall composition of the giant planets.

In the outer solar system, hydrogen dominates planetary composition, which results in a reducing chemistry; closer to the Sun oxygen dominates and conditions are oxidizing. The inner planets are relatively small, dense objects composed predominantly of rock and metal. Any primitive material they once contained has been strongly modified to produce the familiar igneous, sedimentary, and metamorphic rocks. In contrast, the outer giants have low densities and are made primarily of hydrogen and helium, the two lightest and most abundant elements in the universe. These giant planets captured most of their hydrogen and helium from the primordial solar nebula. They are surrounded by satellites that form systems reminiscent of the solar system itself.

The distinction in composition between inner and outer solar system reflects a difference in the degree to which the matter in the planets has been processed. The primitive outer planets (and their satellites) are rich in gases and ices, reflecting more closely the original hydrogen-rich composition of the nebula. The oxidized inner planets are differentiated bodies and are much more evolved. Some have outgassed thin atmospheres that are oxidizing in chemistry.

The ability of a planet to retain an atmosphere depends on its mass and temperature. Large, cold planets do better than small, warm ones, and heavier gases are easier to retain than light ones. The gases in planetary atmospheres were captured from the original nebula or outgassed from the material making up the planets themselves (or both).

To learn about the rocks and gases of another planet, we must study the electromagnetic radiation the planet sends us—either reflected sunlight or thermal emission. Using a spectrometer to analyze the radiation, we can determine the nature of the reflecting or emitting surface and the composition, density, and temperature of any surrounding atmosphere.

We help ourselves enormously in such studies by using powerful telescopes to collect the radiation we wish to investigate. These telescopes must be located in observatories on remote mountaintops if they are to realize their full potential. To do still better, we must leave the Earth and its obscuring atmosphere and launch our telescopes into space.

Most of what we are learning about the planets today is the result of visits by instrumented spacecraft. With spacecraft cameras we can see details we could never hope to glimpse from Earth, even using a telescope in orbit. Spacecraft also offer the opportunity to send probes and landers to a planet to sample its environment directly. Ultimately, such missions can even bring samples back from some distant world to allow detailed studies in our laboratories.

KEY TERMS

albedo	oxidizing
aperture	ozone
atmospheric window	primitive
basalt	reducing
differentiation	remote sensing
exosphere	sedimentary
igneous	silicates
metamorphic	solar nebula
mineral	volatiles

REVIEW QUESTIONS

1. Why is density such an important quantity to measure for a planet? List the densities of some common materials, including gold and lead. Figuring out how to determine whether an item is made of pure gold has been of interest for many centuries. Find out how the ancient Greek scientist Archimedes solved this problem based on a determination of density, and relate this to the problem of determining the composition of distant planets.

2. Distinguish between rocks and minerals. What are some common examples of each? On a cold planet (such as the moons of the outer planets) ices of various kinds are present; are these rocks or minerals?

3. Why are there no primitive rocks on the Earth? Where in the solar system are primitive rocks likely to be found? Why is the search for them important?

4. What is differentiation? Explain why a large planet is more likely to differentiate than a small one. What other properties, in addition to size, are likely to be important in determining whether a planet differentiates?

5. Distinguish among gas, ice, rock, and metal, giving examples of each. Which of these dominate for the various members of the planetary system?

6. Explain the origin and loss of planetary atmospheres. Is it clear why some objects have atmospheres while others do not?

7. What are the differences between direct measurements and remote sensing? Explain how each might be used to study a planetary surface or a planetary atmosphere.

8. What are the various advantages and disadvantages of studying the planets by ground-based telescopes, by space telescopes, by flyby spacecraft, by orbiters, or by probes/landers?

QUANTITATIVE EXERCISES

1. Suppose you wanted to send a spacecraft to Pluto by the minimum-energy trajectory, with perihelion at the Earth and aphelion at the orbit of Pluto. Calculate the one-way flight times for Pluto (a) near its perihelion, which it reached in 1989, and (b) near its aphelion, which it will reach early in the twenty-second century.

2. What would be the mass of the Earth if it retained its present diameter but was composed entirely of (a) water ice, (b) silicate rock, or (c) iron metal?

ADDITIONAL READING

Burrows, W. E. 1990. *Exploring Space: Voyages in the Solar System and Beyond.* New York: Random House. History of the exploration of the solar system by spacecraft, with emphasis on the often-complex politics behind the missions.

Field, G. B., and E. J. Chaisson. 1985. *The Invisible Universe.* Boston: Birkhäuser. Description of how astronomers use radiation from all parts of the electromagnetic spectrum, including justifications for observatories in space.

Poynter, M., and A. L. Lane. 1981. *Voyager: The Story of a Space Mission.* New York: Atheneum. Highly readable history of one of the most successful of all space missions, written for a high school audience.

Sagan, C. 1994. *Pale Blue Dot: A Vision of the Human Future in Space.* New York: Random House. Essays on the exploration of the solar system from both scientific and humanistic perspectives.

4

METEORITES: REMNANTS OF CREATION

The Pillars of Creation. This beautiful image of the Eagle Nebula was obtained by the Hubble Space Telescope in 1995. Such dense aggregations of gas and dust are the birthplaces of stars and planets.

Some of the most fundamental questions about the planets concern their origin. When was the solar system created, and how? Are the processes that formed the planets unique, or might they be a common feature of the birth and evolution of stars, implying that multitudes of planetary systems might exist throughout the Galaxy? And are these other planetary systems likely to be able to support life?

Unfortunately, the planets tend to be mute on questions of origin because they (and their larger satellites) are geologically and chemically active. They have changed considerably since their formation. Just as a study of an adult population of humans provides limited and ambiguous data on birth and childhood development, so mature planets retain only limited and ambiguous memories of the conditions of their formation and early history. Much more helpful would be examples of smaller, unmodified objects—planetary building blocks—that have changed little since the formation of the solar system. An important development in modern planetary science has been the recognition that some of these remnants of creation have survived as the comets and asteroids. Even more important, fragments of material from comets and asteroids are available for laboratory study on Earth in the form of the meteorites.

Most impacting space debris (hundreds of tons every day) is broken apart and consumed by atmospheric friction before reaching the Earth's surface. Plunging into the atmosphere at speeds of tens of kilometers per second, the debris burns away at high altitudes in fiery trails of glowing gas. This death-flash of a small piece of rock or ice is called a **meteor** or, more familiarly, a shooting star or falling star. If the incoming material survives its brief passage through the air and lands on the ground, we call it a **meteorite.** The story revealed by studies of meteorites is the subject of this chapter.

4.1 THE SOLAR NEBULA

The planetary system began with the primordial solar nebula, the cloud of gas and dust out of which the solar system formed about 4.5 billion years ago. Astronomers can identify many similar condensing clouds of interstellar material in our Galaxy today, at locations where other stars and perhaps other planetary systems are now being born. Before we begin to study the individual objects in our planetary system, we need to sketch some current ideas on the origin of our solar system.

FUNDAMENTAL PROPERTIES OF THE SOLAR SYSTEM

It is easy to give a name to the solar nebula, but characterizing it is harder. In order to deduce the properties of this long-gone disk of dust and gas, we must work our way backward from conditions in the solar system today. As we noted in Section 2.3, more than 99% of the material in the solar system is in the Sun, which is composed almost entirely of the two lightest gases, hydrogen and helium. In this respect the Sun is like the other stars and the interstellar material between the stars. Surely, therefore, the solar nebula also had approximately this same cosmic composition.

The largest planets, Jupiter and Saturn, have almost the same composition as the Sun. The smaller bodies, however, are depleted in hydrogen, helium, and other light gases. In addition, the inner planets were probably formed without ices or other volatile materials. Thus, it appears that some processes involved in planetary formation selectively concentrated the less volatile compounds, permitting lighter materials to escape. Such chemical sorting is called **fractionation.** Although the Sun and giant planets have roughly the same composition as the original solar nebula, the processes that gave rise to smaller solid bodies were highly selective in the building materials they employed.

Another basic property of the solar system is its rotation. All the planets have orbits that are approximately circular and lie roughly in the same plane, and all revolve in the same direction around the Sun. Further, the Sun itself rotates in this same direction, and its equator lies essentially in the same plane as the orbits of the planets. All these bodies apparently formed from a solar nebula that was rotating. Moreover, the material that formed the planets must have been confined to a disk in order for their orbits to settle so closely to the same plane.

Two fundamental properties of the solar system must therefore be related to the solar nebula: (1) its rotation and disk shape, and (2) its chemical composition, with the Sun and Jupiter composed of relatively unmodified cosmic material, while the smaller members of the system are made up primarily of rarer rocky, icy, and metallic substances.

FORMATION AND CONDENSATION OF THE NEBULA

In order for the solar system to form, thin clouds of interstellar gas and dust must have become concentrated enough to collapse under their own weight. Astronomers see many similar gravitationally contracting clouds at sites of active star formation today (Fig. 4.1), but we still do not fully understand what triggers the collapse. Anyway, it happens, or we would not be here to discuss the process. Let us skip over the start of the collapse and begin our story at the stage of a contracting,

Figure 4.1 The Great Nebula in Orion is a giant cloud of gas and dust 1500 light years away. Deep within this huge complex—partially illuminated by light from embedded and nearby stars—new stars, planets, comets, and asteroids are forming.

self-gravitating solar nebula with a composition representative of our corner of the Galaxy. The mass of this collapsing cloud was probably only 10%–20% greater than the mass of the Sun today.

The following outline of the origin of the solar system provides a framework for the chapters that follow. These ideas are undoubtedly incomplete and even simplistic. As we learn more about the planetary system, we will repeatedly come back to the processes of formation. In Chapter 17, we will recapitulate the formation of the solar system in greater detail, using information presented throughout this book.

As the solar nebula collapsed, its central parts were heated by the infalling material. As we described in Section 2.4, gravitational contraction generates heat, as the energy of falling material is transformed into heat energy. At the same time, the shrinking nebula began to spin faster, and its outer parts flattened into a disk.

The formation of a rapidly spinning disk from a collapsing cloud is an example of the principle called conservation of **angular momentum.** This familiar physical law is based on the fact that the angular momentum of a body depends on three quantities: mass, rotation, and size. In order for angular momentum to be conserved, or held constant, if one of these quantities decreases, another must increase proportionately. In our example, the size of the nebula decreases while the mass remains the same, so the law tells us that the rotation rate must increase. Just as a turning ballet dancer on point increases the speed of her spin by drawing in her arms, the nebula speeds up as it contracts. Thus, the contracting solar nebula develops a hot central core surrounded by a rapidly spinning disk. The core will become the Sun when its internal temperature and density are high enough to sustain thermonuclear reactions, while the disk will give rise to the planets. The outer parts of the nebula continue to fall into the disk, which increases in both mass and temperature.

Let us focus our attention on the disk (Fig. 4.2). It is heated by both infalling material and the radiation from the hot core in the center. As the initial rapid infall of nebular material subsides,

Figure 4.2 As the cloud of gas and dust that formed the solar system contracted, it rotated more rapidly, leading to the formation of a flattened disk, which in turn broke up into condensations that became the planets and their satellites. The Sun, accumulating most of the mass of the system, formed at the center.

temperature differences develop in the disk. The outer parts cool while the inner parts are heated by the proto-Sun, just as the temperatures of the inner planets are higher than those in the outer solar system today. At each point, solid grains begin to condense like hailstones in a storm cloud, but they are composed of only those chemical substances that have freezing temperatures below the local temperature of the nebula. Table 4.1 lists the composition of the condensate expected in the outer regions of the solar nebula, relative to the hydrogen and helium gas. These are the main materials available to form a giant planet such as Jupiter or Saturn.

In the warmer inner parts of the solar nebula, not all of these materials were able to form solid grains. For instance, water ice could condense at a distance of 5 AU from the proto-Sun but not at 2 AU. Similarly, many common silicate rocks could form at 1 AU but not at the higher temperatures closer to the center. Inside of about 0.2 AU no solids formed and the solar nebula remained a hot gas. In this way, a sequence of chemically distinct grains formed at different distances from the center of the nebula. This temperature-related chemical condensation sequence provides one possible explanation for some of the major differences we see today in the compositions of the planets and other members of the planetary system.

ACCRETION AND FRAGMENTATION

We now have a solar nebula in which solid grains have formed, chemically sorted according to distance from the proto-Sun. The next step in the formation of the planetary system requires that these grains come together to form larger aggregates. Initially, the grains stuck together as they bumped into one another, quickly building up to objects a few tens of kilometers in diameter, which astronomers call **planetesimals** (little planets). As they grew, the planetesimals began to attract each other gravitationally, beginning the process of **accretion,** in which the particles grew by their mutual gravitational attraction. Ultimately the innumerable tiny

Table 4.1 Expected abundances in the outer solar nebula

Material	Percent (by mass)
Hydrogen (gas)	77.0
Helium (gas)	22.0
Water (ice grains)	0.6
Methane (ice grains)	0.4
Ammonia (ice grains)	0.1
Rock and metal (solid grains)	0.3

grains were swept up into bodies that reached hundreds and then thousands of kilometers in diameter—the **protoplanets.**

The two largest protoplanets eventually grew so massive that they were able to attract and hold the uncondensed gas in their part of the nebula, becoming the giants Jupiter and Saturn. Some of the loose gas was also attracted to Uranus and Neptune. Elsewhere in the nebula, however, only the condensed solids were available as building materials for accretion.

As the planetary objects became fewer in number and their individual sizes larger, they exerted stronger gravitational forces upon each other and their smaller neighbors, which were jostled about like floating Ping-Pong balls stirred with a stick. As a result, the speeds at which one body struck another grew higher, and the violence of the impacts increased sharply. When a small body struck a large one, it gouged out a crater or, if the body was not solid, splashed liquid fragments in all directions. When two bodies of similar size crashed, the results were even more catastrophic, ending in mutual disruption.

These processes, by which impacts break down objects rather than build them up, are called **fragmentation.** Thus, we have two competing processes: *accretion*, which makes planets grow, and *fragmentation*, which breaks them down. As the system evolved, fragmentation became more important, ultimately dominating over accretion for all but the largest bodies. Thus, the big got bigger, but most of the smaller protoplanets were reduced to fragments—some of which survive today in the form of asteroids and comets.

FINAL STAGES

In the end, three processes eliminated the gas and dust of the solar nebula. One was condensation and accretion, which led to the formation of protoplanets. A second process, which accounted for a much greater loss of the original material, was the increasing activity of the young Sun at the center. Early in their lifetimes, stars go through a stage of mass loss in which they expel material at high speed from their surfaces. These streams of outflowing solar material—a kind of super solar wind—effectively swept away most of the gas in the solar nebula, as well as any grains that had not yet accreted into the larger bodies. Third, gravitational interactions with the newly formed planets eliminated most of the remaining solid fragments. Either these fragments struck the solid surfaces of the planets or they came close enough to be expelled gravitationally.

We have provided this sketch of solar system formation as an introduction to the meteorites. Much of what we know about the processes of condensation, accretion, and fragmentation in the solar nebula has been learned from these remnants of that early epoch of our system's history. The meteorites are fragments from comets and asteroids, which are themselves the rare survivors of the vast numbers of building blocks that formed from the solar nebula. When you contemplate a meteorite, don't think of it as just a piece of ugly dark rock that fell from the sky. Think of it as a survivor from the time when the planets formed, providing us a unique perspective on our origins.

4.2 CLASSIFICATION OF METEORITES

Meteorites are defined as those extraterrestrial fragments that collide with the Earth and survive to reach the surface. Before the nineteenth century, however, the idea that extraterrestrial materials were reaching the surface of the Earth was scoffed at by educated persons, who placed stories of falling stones in the same category with tales of fairies and dragons. U.S. President Thomas Jefferson, himself a distinguished amateur scientist, is reported to have reacted to information about an 1807 meteorite fall in Connecticut by commenting that he could more easily believe that Yankee professors would lie than that stones would fall from the sky. Such events were so infrequent and unpredictable, and so rarely observed by reliable witnesses, that it was easy to dismiss them (Fig. 4.3).

At the end of the eighteenth century, however, the special compositions of some meteorites were becoming recognized, and a case was made

Von dem donnerstein gefallē im xcij. lar: vor Ensisheim

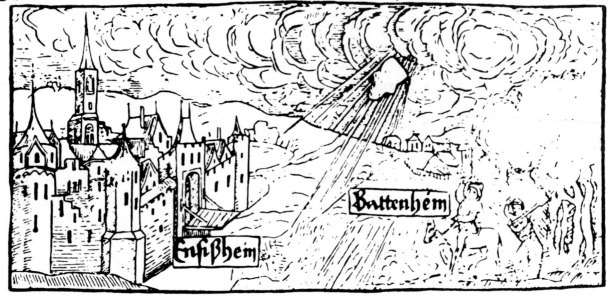

Figure 4.3 An old woodcut showing the fall of the meteorite near Ensisheim, France, in 1492. The German caption reads: "of the thunderstone (that) fell in xcii (92) year outside of Ensisheim."

that they came from beyond the Earth. For most scientists of the time, the extraterrestrial origin was proved in 1803 when a fall of stones from the sky was reported in the village of l'Aigle, France. The French Academy of Science sent a team of reputable scientists to investigate, interview the witnesses, and collect the fallen stones; further investigation confirmed that these stones were unlike any ordinary rocks. Thus, the authenticity of meteorites was established among European scientists, although apparently it took a few more years for this information to reach Jefferson.

IRONS, STONES, AND STONY-IRONS

Meteoritic materials are constantly reaching the Earth, and several falls are observed and recovered every year. The fragments that reach the ground are usually stones or metallic masses of only a kilogram or two, small enough to be held comfortably in your hand. Larger falls are produced when a mass of hundreds or thousands of kilograms strikes the atmosphere, often breaking up to scatter meteorites over many kilometers. Even larger projectiles strike at intervals of thousands of years, some of them exploding in the atmosphere, while others produce impact craters when they crash into the surface. We will return to the process of impact cratering in Chapter 7, when we consider the heavily scarred surface of the Moon.

Our meteorite collections contain a surprising variety, which suggests the existence of many different **parent bodies**—the term we use for the asteroids or other objects from which the meteorites originated. The traditional descriptions of these meteorites are based on their appearance and bulk composition: irons, stones, and the rarer stony-irons.

Irons are composed of nearly pure metallic nickel-iron and are readily recognized from their high density of more than 7 g/cm^3. Their extraterrestrial origin is obvious when we recall that iron and most other metals normally occur on Earth in the form of oxides rather than in the pure metallic state. In fact, iron meteorites were one of the first sources of iron, and their use helped to stimulate the transition from Bronze Age to Iron Age cultures. The second group, the stones, more closely resembles terrestrial rocks; these are not generally recognized as being of extraterrestrial origin unless their fall was witnessed. The third group, the stony-irons, contains a mixture of stone and metallic iron, as the name implies.

PRIMITIVE AND DIFFERENTIATED METEORITES

A more useful categorization of the meteorites is based on the history of their parent bodies. If the chemistry of a meteorite indicates that it is representative of the original materials out of which the solar system was made, little altered by the subsequent chemical evolution of its parent body, we refer to it as a **primitive meteorite.** All primitive meteorites are stones. These meteorites are in fact our only example of primitive material on Earth. Because many of the meteorites that reach the Earth are primitive, we can be confident that material still exists in the solar system that has remained relatively unchanged since before the planets were formed (Fig. 4.4).

Although the surface of a meteorite is heated to incandescence during its brief plunge through the atmosphere, the heat pulse penetrates no more than a few centimeters into the interior, most of which remains cool and undisturbed. Even the outer layers have normally cooled by the time the meteorite strikes the ground, as shown by objects that have fallen on ice or snow without melting it. Thus, primitive meteoritic material remains largely unaltered by its violent arrival at Earth. Incidentally, this fact also provides a quick way to check on press stories of possible meteorite falls. If witnesses report that the falling stones started fires, there is probably something wrong. Meteorites don't start fires.

Meteorites that have experienced major chemical change since their formation are called **differentiated meteorites** (or igneous meteorites). Like the igneous terrestrial rocks, they solidified out of a molten state. Differentiated meteorites appear to be fragments of differen-

Figure 4.4 A meteorite sawed in half to show the appearance of its interior. Shiny flakes of metal are mixed with silicate minerals. This is a primitive meteorite that condensed from the cooling solar nebula.

tiated parent bodies that experienced major episodes of heating along with loss of volatile materials. All of the iron and stony-iron meteorites, and many of the stones as well, are examples of differentiated meteorites.

While many meteorites are lumps of material of fairly uniform composition, others (both primitive and differentiated) show evidence of having been repeatedly broken, mixed, and welded by impact processes on their parent bodies. These fragmented and recemented rocks are called **breccias.** Some meteorite breccias are especially interesting because each contains fragments from a variety of regions on the crust of the parent body, thus providing a sampling over a wide area.

FALLS AND FINDS

Meteorites can also be distinguished by the way they are identified on Earth. The most obvious meteorites are ones that are seen to fall. A meteoric plunge through the atmosphere may terminate in an aerial explosion that scatters fragments over many square kilometers. Meteorites located in this way are termed **falls.**

Table 4.2   Abundances of major meteorite types
(as percent of total)

	Finds	Falls	Antarctic
Primitive stones	52	87	85
Differentiated stones	1	9	12
Irons	42	3	2
Stony-irons	5	1	1

Figure 4.5   The fifth meteorite found at Allan Hills (Antarctica) in 1981 (ALHA 81005). This basaltic breccia is a sample of the lunar surface delivered to Earth without human intervention.

A second group of meteorites is called **finds.** These are objects whose falls are not witnessed but which are later recognized to be of extraterrestrial origin. Because stony meteorites look superficially like ordinary rocks, they are rarely recognized unless their fall is seen or they land in an unusual location, such as a snowfield. Most finds therefore are the much more obvious irons and stony-irons.

Most of the meteorites exhibited in museums are irons, in part because they are more easily identified. Table 4.2 shows the frequencies of the different types of meteorites in three population groups: finds, falls, and the Antarctic meteorites discussed below. It is clear from this table that the primitive stones are the most common type of meteorite reaching the Earth.

ANTARCTIC METEORITES

The Antarctic is today the most important source of meteorites. In some parts of the Antarctic continent, the slow movement and subsequent evaporation of ice transport and concentrate any meteorites that fall over areas of tens of thousands of square kilometers (Fig. 4.5). In these "blue ice" regions, thousands of meteorites have been identified and collected, representing accumulation over hundreds of thousands of years.

The Antarctic meteorites do not suffer from the selection effects that cause stony meteorites, including nearly all of the most primitive samples, to be overlooked among meteorite finds. The fact that essentially every meteorite that has fallen within the area of ice flow can be spotted and collected is one reason the Antarctic meteorites constitute such a valuable addition to our inventory of extraterrestrial materials.

The first Antarctic meteorites were discovered in 1969 by a team of Japanese scientists. Since then expeditions have been sent out nearly every southern summer to search by helicopter and on the ground for the stones, many of which are little bigger than pebbles. Because they have remained frozen since their fall, these meteorites have suffered relatively little weathering and atmospheric contamination. In the United States, many of the Antarctic meteorites are stored under carefully controlled conditions at the NASA Johnson Space Center in Houston, along with the lunar samples brought back by the Apollo and Luna missions. From there, these samples are distributed to scientists for analysis in their laboratories. By the year 2002, more than 20,000 meteorites had been collected in the Antarctic. The total number of meteorites in our collections prior to the Antarctic discoveries was less than 1000 (where fragments from the same fall are not counted separately).

NOMENCLATURE

Each individual meteorite is given a name, usually for a town or other geographic feature near its point of recovery. Thus, for instance, we have the

Table 4.3 Characteristics of the main meteorite types

Type	Composition
Primitive meteorites (chondrites)	
Carbonaceous	silicates, carbon compounds, water
Other primitive stones	silicates, iron
Differentiated meteorites (achondrites)	
Differentiated stones	igneous silicates
Stony-irons	igneous silicates, iron, nickel
Irons	iron, nickel

Allende meteorite, a large collection of stones that fell in 1969 near Pueblito de Allende in northern Mexico. In the case of the Antarctic meteorites, where thousands have been found in just a few locations such as the Yamato Mountains and Allan Hills, a number is also used for identification, such as ALHA 81005, a first-discovered lunar meteorite. The first two digits give the year of the find (in this case 1981), and the last three represent a running index.

Table 4.3 summarizes the types and compositions of meteorites, proceeding from primitive to modified.

ORBITS AND ORIGINS

To determine where the meteorites are coming from, we would very much like to know their orbits before they struck the Earth. Of course, once a meteorite has fallen, it is too late to go back and reconstruct its path before its orbit intersected that of the Earth. Since the meteorites we deal with are all far too small to have been sighted telescopically before impact, the only solution is to make accurate enough measurements of the path of the meteorite through the atmosphere to reconstruct its pre-impact orbit.

One way to collect this information is to set up our cameras, photograph the sky continuously every clear night, and hope for luck. During about 20 years of such searches, only three meteorites (all of them primitive) were photographed in flight and subsequently recovered: Pribram, in Czechoslovakia (1959); Lost City, in the United States (1970); and Innisfree, in Canada (1977). These meteorites proved to be on eccentric orbits that carried them in from the main asteroid belt to cross the orbit of the Earth (Fig. 4.6). Although they did not identify specific parent bodies, these three observations first suggested that the asteroids might have been the source of these meteorites.

HITS AND MISSES

At this point, you may wonder whether any of these interplanetary wanderers ever impact human lives, literally or figuratively. The answer is yes, in both categories. There is only one confirmed report of a meteorite striking a human being (Annie Hodges of Sylacauga, Alabama, bruised by a ricochet on November 30, 1954), but there are several reports of animals being killed by direct hits. For example, one of the meteorites that is now known to have originated on the planet Mars is said to have killed a dog when it fell to Earth in Nakhla, Egypt, on June 24, 1911.

Impacts on human structures are much more common. On November 8, 1982, a 3-kg meteorite crashed through the roof of the Donahue home in Wethersfield, Connecticut, continued on through the living room ceiling, bounced off a carpeted wooden floor, and eventually came to rest under the dining room table. Defying all odds, this was the second meteorite to hit a home in Wethersfield in 12 years!

Another recent meteorite was seen first as a brilliant meteor, lighting up the evening sky over several eastern states before it smashed into a Chevrolet in Peekskill, New York (Fig. 4.7). Scientists were able to analyze videotapes fortuitously recorded by people under the meteor's path who had aimed their cameras at this totally unexpected phenomenon. As a result, it was possible to add a fourth orbit to the three deduced from the specific research described in the previous section (see Fig. 4.6). As with the other three, it is clear that the Peekskill meteorite originated in the asteroid belt.

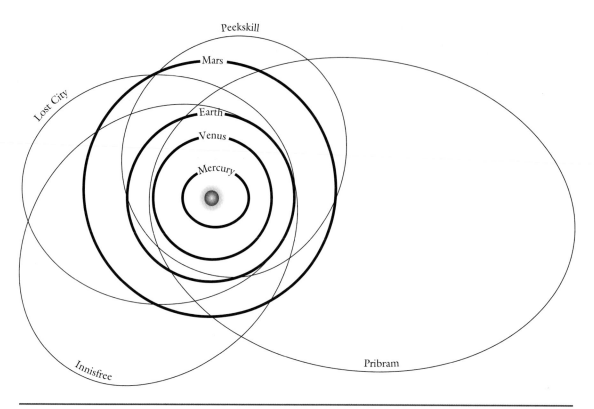

Figure 4.6 The reconstructed orbits of four meteorites—Peekskill, Lost City, Innisfree, and Pribram—whose tracks through our atmosphere were photographed. All four orbits extend into the asteroid belt.

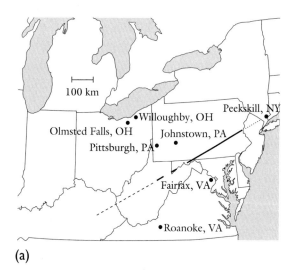

(a)

Figure 4.7 One meteorite that affected human lives. **(a)** The path of the Peekskill meteor, projected on the ground, as it flashed downward through the Earth's atmosphere at 7:45 P.M. on October 9, 1992. **(b)** A Chevrolet that suffered a most unusual collision! The meteorite that did the damage rests on the battered trunk.

(b)

4.3 AGES OF METEORITES AND OTHER ROCKS

We have asserted in this book that the solar system, including the Earth, was formed about 4.5 billion years ago. How was this figure determined? The ability to date the formation of rocks and planets is one of the triumphs of modern science, with implications for many fields ranging from astrophysics to paleontology. Rock dating, for example, provides the absolute timescale for biological evolution on Earth. It permits us to determine the time intervals between different eras originally assigned only relative ages on the basis of geological context and fossil remains.

NATURAL RADIOACTIVITY

The dating of rocks depends on the properties of radioactive decay, discovered early in the twentieth century. Radioactivity is the natural process whereby some isotopes of certain elements spontaneously change into other isotopes. This is a one-way process, accompanied by the release of energy in the form of gamma rays, electrons, or alpha particles (the nuclei of helium atoms).

Today much of our concern about atomic radiation focuses on radioactivity from nuclear reactors or nuclear bombs. Such artificially created radioactive materials can pollute the environment and, in the aftermath of a nuclear war, could result in the death of much of the life on our planet. In contrast, the radioactivity that we will be discussing here is natural, the product of minute quantities of uranium, thorium, potassium, and other radioactive elements that are distributed throughout most rocks and minerals.

As a radioactive atom decays to produce a new, nonradioactive isotope, the concentration of the original, or parent, material decreases and that of the new, or daughter, material increases. The relative concentrations of parent and daughter are the basic information from which an age can be derived.

THE RATE OF RADIOACTIVE DECAY

In order to interpret the parent/daughter ratios, we must know the rate at which radioactive decay is taking place. Fortunately, this rate is a fixed property of the parent material and is independent of external conditions, such as temperature, pressure, or even the chemistry of the minerals in which the radioactive material is located. Although there is no way to predict when any particular individual atom will undergo this change, we know that in a fixed interval of time some specific fraction of a collection of similar atoms will decay.

Usually the decay rate is expressed in terms of a radioactive **half-life,** defined as the time for one-half of the atoms to decay into their daughter product. For instance, if a rock initially contained 64 μg of a radioactive isotope with a half-life of 1 million years, we would find that only 32 μg remained at the end of the first million years, 16 μg at the end of 2 million years, 8 μg at the end of 3 million years, 1 μg at the end of 6 million years, and so on. The daughter product would increase as the parent declined.

To determine the ages of meteorites and other ancient rocks we require parent elements with half-lives in the range of hundreds of millions to tens of billions of years, commensurate with the ages of the rocks we are studying. The most useful reactions, together with their half-lives, are given in Table 4.4.

Table 4.4 Radioactive decay reactions used to date meteorites and planets

Parent	Daughter	Half-life (billions of years)
Samarium (Sm-147)	Neodymium (Nd-143)	106
Rubidium (Rb-87)	Strontium (Sr-87)	48.8
Thorium (Th-232)	Lead (Pb-208)	14.0
Uranium (U-238)	Lead (Pb-206)	4.47
Potatssium (K-40)	Argon (Ar-40)	1.31

DIFFERENT ROCK AGES

Suppose that you make a careful measurement of the concentrations of one of the parent/daughter pairs in Table 4.4 and calculate an age from the given half-life. Just what does such an age mean? This is an important question, especially since different parent/daughter pairs may not yield the same calculated age. Essentially, the age measured is the time interval over which the daughter product has been able to accumulate undisturbed from the decaying parent. Because in a liquid the parent and daughter tend to become mixed with other materials, the measured age is the time since the material became solid, referred to as the **solidification age.**

In the case of the potassium-argon pair, the daughter product is a gas, and the age measured is the **gas-retention age.** Thus, if we calculated a meteorite age as 18 million years by the K/Ar method and as 4.3 billion years by the Rb/Sr method, we would probably conclude that the rock solidified from the liquid 4.3 billion years ago but that a subsequent shock (perhaps the breakup of the parent body) allowed much of the argon gas to escape 18 million years ago, in effect resetting the clock that measures the gas-retention age.

In order to realize the full power of the radioactive age measurement techniques, we must have a way to determine the original ratio of parent to daughter isotopes in the rock at the time of solidification. This ratio is calculated using other nonradioactive isotopes of the same elements. These sister isotopes behave in exactly the same way as the radioactive isotopes in any chemical processes that alter the composition of the rock melt, thereby removing this source of uncertainty. Additional checks can be applied by making independent age measurements for different mineral grains from a single sample.

Our confidence in the measured ages of meteorites and lunar and terrestrial rocks is the result not only of the high precision achieved in the measurement of the concentrations of parent and daughter products, but also of the number of checks that can be carried out to establish the consistency of the results. Usually it is possible to apply three or more separate techniques, based on different radioactive elements with very different half-lives, and also to apply the measurements to a variety of mineral grains of differing chemistry. Generally, these results agree to within less than 1% in the derived age. Where differences do arise, they can usually be attributed to subsequent events, such as the example cited earlier where the gas-retention age was shorter than the solidification age as a consequence of impact resetting of the gas-retention clock.

Essentially all primitive meteorites have solidification ages between 4.5 and 4.6 billion years, with an average of about 4.55 to 4.56 billion years. This is the number that defines the "age of the solar system"—although of course the individual planets may have formed slightly after the solidification of these "building blocks." In this book we round this age off as 4.5 billion years, but it could just as easily be rounded to 4.6 billion years, which is a value also frequently cited for the age of the solar system.

4.4 PRIMITIVE METEORITES

Primitive meteorites have chemical compositions that have remained relatively unchanged since they formed in the cooling solar nebula about 4.5 billion years ago. Except for a shortage of gaseous and other volatile constituents such as hydrogen, helium, argon, carbon, and oxygen, the composition of the primitive meteorites is thought to be the same as that of the Sun. In fact, these meteorites are used to define the presumed solar abundances for some rare elements that cannot be observed directly in the Sun.

COMPOSITION AND HISTORY

The primitive meteorites are also called **chondrites,** named for the small, round **chondrules** that they contain. These chondrules (typically about 1 mm in diameter) appear to be frozen droplets that condensed from molten material (Fig. 4.8). Because not all chondrites contain chondrules, we prefer the more general term *primitive meteorite.* In appearance, the primitive

Figure 4.8 Chondrules are small congealed drops of rock. This image is an enlargement of a thin section of a meteorite showing chondrules (the largest a few millimeters across) embedded in a dark matrix.

meteorites are light to dark gray rocks, often with a darker crust produced during their fiery descent through the atmosphere. Many are breccias. Their densities are about 3 g/cm³, similar to that of many crustal rocks on the Earth.

Chondrites often show evidence of limited alteration by heating or the effects of liquid water that was present during some early epoch of their evolution. Some scientists call these metamorphic meteorites. Such altered meteorites are less primitive, and scientists must make due allowance for their past history in order to relate them properly to the solar nebula.

Many of the primitive meteorites contain grains of metallic iron as a major constituent, 10%–30% by weight (see Fig. 4.4). The quantity of iron leads to a classification of the primitive meteorites by their metal content. Thus, scientists speak of H chondrites for those with high iron content, L chondrites for those with low iron, and LL for very low iron. The most abundant elements after iron are silicon, oxygen, magnesium, and sulfur.

It is interesting to note the uniform age of the primitive meteorites at 4.55 to 4.56 billion years. Apparently they all formed together, and none has intruded from beyond the solar system. This fact shows just how isolated the planetary system is within the Milky Way Galaxy, even though our

system has completed some 20 revolutions around the center of the Galaxy (20 "galactic years"). Needless to say, scientists hope that someday we might find a meteorite from beyond the solar system. Any information that could be derived concerning the age and composition of such a rock would be fascinating to compare with what we know about our own planetary system.

CARBONACEOUS METEORITES

The most primitive meteorites are a special group called the **carbonaceous meteorites.** These are relatively rich in carbon (a few percent by weight), as their name implies, and also in volatile compounds such as water, which combines chemically with other minerals to form clays. They contain very little metallic iron. Carbonaceous meteorites are dark gray to black in color and are physically very weak; a fragment can generally be crushed between your fingers (Fig. 4.9). Because of their relatively high proportion of water, they are less dense than other meteorites, typically about 2.5 g/cm³. From their composition we conclude that they were formed in a cooler region of the solar nebula than were the other primitive meteorites.

One of the most interesting carbonaceous meteorites fell in the Yukon Territory of Canada in January 2000. The meteorite was extremely

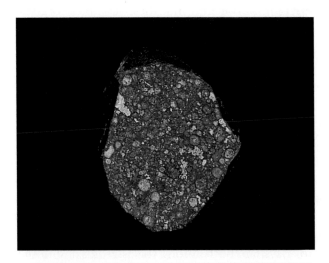

Figure 4.9 A carbonaceous meteorite. The part of the stone facing us has been broken off and shows unaltered meteoritic material.

fragile, crumbling to dust at a touch. Pieces were collected because they landed gently in snow on a frozen lake (Tagish Lake). Jim Brook, a local resident, found the first meteorite fragments while driving home on the ice a day after the bright fireball had been seen in the sky. He wrote, "I was watching closely for meteorites and suspected their identity as soon as I saw them, although I had been fooled several times by wolf droppings. It was obvious what they were as soon as I picked one up, because rocks aren't found on the ice, and I could see the outer melted crust. I was very happy and excited." Ultimately more than 500 pieces of the Tagish Lake meteorite were recovered before they were buried in subsequent snowfalls.

One of the special properties of the carbonaceous meteorites is the evidence that some of them have been altered by liquid water, which has dissolved certain minerals and redeposited them in veins. Apparently there existed parent bodies within which water could assume a liquid form, at least for some stage of early solar system history.

Many carbonaceous meteorites also contain up to 3% of complex carbon compounds. Compounds in which this element is joined with hydrogen and other elements are called **organic compounds** because scientists once thought that only living organisms could produce them. On Earth, natural carbon minerals are rare except for calcium carbonate, and these complex organic compounds (which make up coal and petroleum, among other things) are indeed produced as a part of the life process. The organic compounds in meteorites, however, are not an indication of life in the primordial solar nebula. Meteoritic carbon compounds may have played a role, however, in supplying the chemicals required for the origin of life on Earth.

Many chemical reactions could have taken place in the hydrogen-rich environment of the solar nebula to produce organic compounds. This kind of matter is actually rather common in the solar system, both among the asteroids and in the satellite families of the outer planets. Most of the carbon compounds in the primitive meteorites are complex, tarlike minerals that defy exact characterization. They also include substances that we recognize as being of fundamental importance for life: components of proteins and nucleic acids. These were first identified with certainty in the Murchison meteorite, a carbonaceous meteorite that fell in Australia in 1969, a few months after the Allende meteorite fell in Mexico. Murchison, Allende, and Tagish Lake are among the best studied of all the meteorites.

Murchison yielded 74 separate amino acids, including 55 that are rare on Earth. The most remarkable thing about these and other amino acids in meteorites is that they include equal numbers of left-handed and right-handed forms. Amino acids can have either of these kinds of symmetry, but in practice life on Earth has evolved using only the left-handed versions of these compounds to make its proteins. The fact that both symmetries are present in the meteorites demonstrates that no contamination has taken place since they arrived on Earth, and indicates that these organic compounds formed in space without the intervention of living things. This conclusion is supported by the fact that the ratio of the two stable isotopes of carbon, C-12/C-13, is different in the meteorite from its value in terrestrial organic materials.

INTERPLANETARY DUST

Probably the most primitive material available to us arrives at Earth in the form of interplanetary dust particles (IDPs), which are collected at high altitudes by specially instrumented airplanes (Fig. 4.10). These dust particles, typically no larger than the width of a human hair, are not properly meteorites at all. Most of them have a cometary origin, and we will discuss them further in Chapter 6. Laboratory analysis shows that they are rich in organic carbon, and many of them also contain unusual amounts of deuterium (heavy hydrogen) and a heavy isotope of nitrogen, both interpreted as signatures of interstellar rather than interplanetary material. Perhaps some of the interplanetary dust particles are survivors of the original interstellar dust from the solar nebula. Unfortunately, the quantities of material are too small to permit us to date the formation of these strange particles.

Because the carbonaceous meteorites and interplanetary dust particles are the most primitive

Figure 4.10 These meteoric particles (often called "Brownlee particles" after their discoverer, Donald Brownlee of the University of Washington) were captured in the Earth's stratosphere by a U-2 aircraft equipped with a special dust-sampling device. A steady rain of particles like these (here magnified 300,000 times) is constantly falling on the Earth. We are thus continuously breathing in "comet dust" along with everything else our air contains.

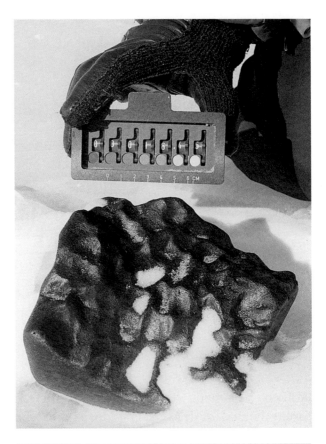

Figure 4.11 An iron meteorite on the Antarctic ice, about to be added to our meteorite collections. The pattern of "thumbprints" was caused when the thin skin of melted iron produced by the friction of the meteorite's passage through the atmosphere is pushed back away from the direction of motion.

samples of solar system material available for laboratory study, they reveal the most about the earliest stages of the planetary system. In the search for a key to unlock the secrets of the past, these cosmic fossils are proving to be our most valuable artifacts.

4.5 DIFFERENTIATED METEORITES

The differentiated or igneous meteorites are composed of materials that have been melted after their condensation from the solar nebula, at the time that their parent bodies underwent heating and differentiation. Often they are also referred to as the achondrites, a term that signifies the absence of the chondrules that are characteristic of most primitive meteorites.

IRONS AND STONY-IRONS

The most obviously differentiated meteorites are the irons, which consist of almost pure metallic nickel-iron, with trace quantities of sulfur, carbon, and metals such as platinum (Fig. 4.11). The nickel content is usually about 10% by weight. Iron meteorites make up only 2% of the falls and of the Antarctic meteorites, which probably represent an unbiased sample of meteoritic material. These meteorites can be quite beautiful when cut and polished, and subsequent etching further reveals a unique crystalline pattern, created by slow cooling of the nickel-iron melt over millions of years (Fig. 4.12).

Presumably the iron meteorites are fragments of the metal cores of their differentiated parent bodies. The cores of the Earth and other intact planets will remain forever beyond our direct investigation, but these meteorites provide a unique glimpse into the very heart of a former differentiated planetary body. Detailed chemical analysis

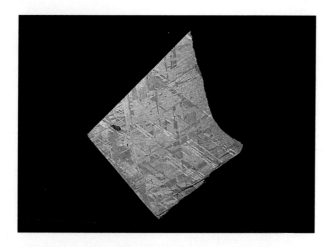

Figure 4.12   A slice of an iron meteorite that has been polished and then etched with dilute nitric acid to show a crisscross crystallization pattern, called a "Widmanstätten" pattern.

Figure 4.13   An example of a eucrite. The eucrites, which are composed of basalt, are fragments from the asteroid Vesta. They look very similar to terrestrial lava rocks.

indicates that there were a number of different parent bodies, at least several dozen, for the known iron meteorites.

Related to the iron meteorites are the stony-irons, composed of a mixture of nickel-iron and silicate minerals. The most beautiful of these meteorites, called pallasites, consist of large crystals of olivine, a green semiprecious stone, in a setting of shiny meteoritic iron. The pallasites are thought to be fragments from the interface between the metal core and silicate upper layers of their parent bodies. Fewer than 1% of the meteorites that reach the Earth are stony-irons.

The age of the iron cannot be measured by radioactive techniques, but age measurements can be carried out for the silicate fraction in stony-irons or for small silicate inclusions found in some iron meteorites. These ages lie between 4.4 and 4.5 billion years, suggesting that the processes of heating, differentiation, and cooling of their parent bodies all took place very early in solar system history.

BASALTIC METEORITES

One group of differentiated stony meteorites appears to be derived from the crusts of their parent bodies. These rocks were formed by crystalliza-tion from a cooling body of basaltic lava, which in turn represents the lower-density material that rose to the surface during differentiation. Basalt is the most common form of lava on the Earth, and we will see in Chapter 7 that basalt is also the material that makes up the dark lunar "seas." Most of the basaltic meteorites are breccias, indicating their long residence near the surface of their parent bodies, where they were fragmented and mixed together by impact cratering.

The best-studied group of basaltic meteorites are the eucrites, of which more than 30 examples are known, having identical oxygen isotopic ratios and closely related compositions (Fig. 4.13). It is generally believed that the eucrites are derived from a single parent body, the asteroid Vesta, and that their compositions are representative of a series of different lava flows.

Another group of about two dozen basaltic meteorites are known to come from Mars. Like the eucrites, these are chemically related lavas, derived in this case from a source region in the upper layers of a single parent body (Fig. 4.14). These are the

Figure 4.14 An example of one of the martian meteorites, only slightly altered by the impact that sent it to us and by the heating it experienced upon entry into the Earth's atmosphere.

youngest meteorites, with relatively recent solidification ages of about 1 billion years—consistent with their origin on a planet that has remained geologically active, as Mars has. In Chapter 11, we will discuss how we know that these meteorites come from Mars, and we will see what they can tell us about that planet.

Finally, there are more than a dozen lunar meteorites, which are similar to the Moon rocks collected in the Apollo program; these will be discussed in Chapter 7. Unlike the Apollo samples, a few of these lunar meteorites are for sale—at a cost of several hundred dollars for a tiny piece.

4.6   Meteorite Parent Bodies

The lunar and martian meteorites are exceedingly rare. For the more common meteorite types, both comets and asteroids have been suggested as parent bodies. An exploded planet was also once hypothesized. Before we pursue these ideas further, it is useful to summarize what we know about the parent bodies from our study of the meteorites.

Sources of Primitive Meteorites

Primitive meteorites must have originated in or on relatively small bodies that formed directly from dust condensing out of the cooling solar nebula. Large parent bodies are not possible because they would have retained too much internal heat generated by natural radioactivity; the thicker layers of material act as a blanket to keep the radioactive heat confined instead of allowing it to escape to space. Substantial heating, of course, would have altered the material from its primitive state.

Calculations show that in order to have stayed cool, the parent bodies must have been no more than a few hundred kilometers in diameter. Some of these must have included organic chemicals and liquid water. The chemical and isotopic variety among the primitive meteorites indicates that many different parent bodies are represented in our meteorite collections.

Sources of Differentiated Meteorites

The differentiated meteorites are fragments of differentiated parent bodies. From detailed chemical analysis, it is clear that several dozen distinct parent bodies were involved. We can set some limits on their size from the iron meteorites, which retain in their crystal patterns indications of the rate at which they cooled (see Fig. 4.12). These cooling times of the order of a million years indicate objects no more than about 100 km in diameter. Thus, the differentiated parent bodies were at least as small as the parent bodies of the primitive meteorites, and some characteristic other than size was responsible for the fact that one group of objects differentiated while the other remained in a primitive state. No one knows why one set of parent bodies differentiated and the other did not.

From the existence of basaltic meteorites, we conclude that some parent bodies experienced surface volcanism, and in the case of the eucrite parent body, the chemistry of these lava flows is well defined. The lunar meteorites have ages typical of the time of extensive lunar volcanism, between 3.3 and 3.9 billion years ago, whereas the Mars meteorites have ages from less than 1 billion to more than 4 billion years, indicative of a long history of volcanic activity on Mars.

ASTEROIDS OR AN EXPLODED PLANET?

As we will see in Chapter 5, most of the evidence points toward the asteroids as parent bodies, although comets are possible as parents for some of the primitive meteorites and interplanetary dust. What is clearly excluded, however, is the old idea of an exploded planet as the source of both the asteroids and the meteorites.

It is easy to see why the exploded planet hypothesis fails. More than 90% of the meteorites show evidence of parent bodies no larger than 200 km, and several lines of evidence suggest dozens of distinct parent objects. In addition, the prevalence of breccias points to the impact-stirred surfaces of airless precursors as a common environment for many meteorites. None of these observations would be true if the meteorites were derived from a single planetary parent body as has sometimes been proposed. In addition, the breakup or explosion of such a planet seems to be an intrinsically unlikely (and perhaps impossible) event.

4.7 QUANTITATIVE SUPPLEMENT:
RADIOACTIVE AGE DATING

Rocks are dated—that is, their solidification ages are determined—by measuring the amount of a radioactive parent isotope that has changed into a daughter isotope since a particular crystal or grain of material formed. This process is relatively straightforward if the only source of the daughter isotope is decay of a single parent. In this case, we can assume that the rock solidified with little of this daughter present, and the current measured ratio of daughter to parent in a mineral grain will establish the age. In practice, however, the situation is a little more complicated.

One common method of dating the formation of the Earth involves the decay of uranium and thorium to produce the lead isotopes Pb-206

and Pb-208, respectively (see Table 4.4). But not all of the Pb-206 and Pb-208 present on the Earth today resulted from decay of thorium and uranium. To determine the original quantity of these two isotopes, we look at a third lead isotope, Pb-204, which is not produced by radioactive decay. One can calculate what the primordial ratio of Pb-206 and Pb-208 must have been to Pb-204, and use the measured amount of Pb-204 to subtract the original quantity of the other two isotopes before calculating the age.

One of the simplest ways to use the presence of other, nonradiogenic isotopes to ensure that correct ages are being calculated is to graph the results, as we now describe. Consider the decay of Rb-87 to Sr-87, which is one of the most useful reactions for determining the ages of meteorites. After a time t, the amount of Sr-87 is the sum of the initial quantity present (Sr-87*) plus that produced by decay of Rb-87:

$$Sr\text{-}87 = Sr\text{-}87^* + Rb\text{-}87^*(e^{kt} - 1)$$

where k is a constant related to the half-life of the reaction. How do we determine Sr-87*? We note that since all isotopes of the same element have the same chemical properties, the amount of Sr-87 initially bound into the minerals of the rock should be exactly proportional to the amount of other strontium isotopes, such as Sr-86.

For each individual crystal or grain in a rock, we plot the measured ratio of the two strontium isotopes Sr-87/Sr-86 against the ratio Rb-87/Sr-86. Such a plot is illustrated in Figure 4.15, which shows data from the Guareña meteorite. Grains with a great deal of initial Rb-87 will be toward the lower left. Now consider what happens as the rock ages. Rb-87 is converted to Sr-87, causing the Rb-rich grains to shift down and to the right on the plot. The older the rock, the more the line drawn through the data points rotates in a clockwise direction, until eventually it will be nearly horizontal. Such a line is called an *isochron*, and its slope is a measure of the age of the rock, independent of any assumptions about the original quantity of Sr-87 present. In the example illustrated in Figure 4.15, the Rb-Sr isochron indicates an age of $(4.46 \pm 0.08) \times 10^9$ years.

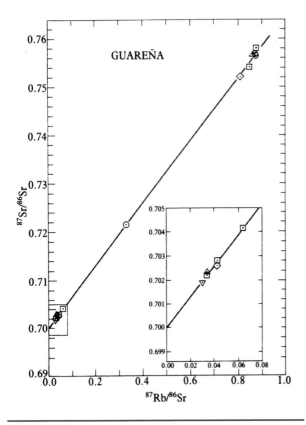

Figure 4.15 The Rb-Sr isochron for the Guareña H6 chondrite. The derived age is 4.46 ± 0.08 billion years.

the carbonaceous meteorites, which show a rich history of organic chemistry in the early solar system.

Differentiated meteorites are samples from differentiated parent bodies that, like the Earth and other planets, have experienced chemical changes due to melting. The evidence indicates, however, that the parent bodies of most of these meteorites were not full-fledged planets, but small bodies that were heated and differentiated very early in solar system history and subsequently broken up by impacts. The iron meteorites appear to be fragments from the metallic cores of such bodies, and the stony-irons come from the boundary between the core and the silicate upper layers. A few differentiated meteorites have come to us from the crusts of objects that experienced surface volcanic activity. These basaltic meteorites include samples from the Moon, Mars, and the asteroid Vesta.

Today more than twenty thousand individual meteorites are in scientific collections around the world, and hundreds of scientists are working on problems of their analysis. In some laboratories, instruments of tremendous sophistication can determine the detailed chemical and isotopic composition of even a single crystal or grain almost too small to be seen with the naked eye. In the exotic surroundings of these superclean laboratories, some of the most important results of modern planetary science are produced.

SUMMARY

The meteorites are rocky and metallic fragments that have reached the Earth and survived their plunge through the atmosphere. They provide information on conditions at the birth of the solar system, when solid material condensed out of a contracting and cooling solar nebula of interstellar gas and dust. Generally they are fragments of parent bodies that formed very early in the history of the planetary system.

Most of the meteorites that fall on Earth are primitive in composition, meaning that they and their parent bodies have not suffered significant changes since their birth. In this respect they differ from the rocks of Earth and Moon, both of which are the product of a history of planetary heating and geological activity. The most important primitive meteorites for understanding the solar nebula are

KEY TERMS

accretion	gas-retention age
angular momentum	meteor
breccia	meteorite
carbonaceous meteorite	organic compound
chondrite	parent body
chondrule	planetesimals
differentiated meteorite	primitive meteorite
falls	protoplanet
finds	radioactive half-life
fractionation	solidification age
fragmentation	

REVIEW QUESTIONS

1. Distinguish between meteors and meteorites. What do any of these objects have to do with the weather? (The word *meteor* is closely related to *meteorology*.)

2. What is meant by primitive objects? Which objects are the most primitive? What kinds of information do they provide on the origin of the solar system?

3. Describe the competing processes of accretion and fragmentation in the solar nebula.

4. What are meteorite falls and finds? Why are the relative numbers of irons and stones so different for falls and finds?

5. How do we measure the age of a rock, and what exactly does the measured age mean?

6. Why do you think there are differentiated meteorite parent bodies? What might these bodies have been like? Can you explain how they might have been heated enough to cause differentiation?

7. Could the meteorites have originated in an explosion of a planet between the orbits of Mars and Jupiter? Give the arguments pro and con.

QUANTITATIVE EXERCISES

1. Consider the relative proportions of meteorites in Table 4.2. If the observed falls and the Antarctic meteorites represent the true distribution of meteorite classes in space, and if the iron meteorites are the remnants of the cores of differentiated parent bodies, what percentage of the total would you expect to be from the upper layers of these same parent bodies? What can you conclude about the relative fractions of primitive and differentiated parent bodies?

2. Suppose that the eucrite parent body was a differentiated asteroid with diameter 500 km that is now completely disrupted and reduced to fragments. Further suppose that the basaltic crust of this parent body had a thickness of 10 km and the layer below it had a thickness of 200 km. What relative amounts of basaltic and upper layer material would be expected when this body fragmented?

3. Consider the radioactive decay of Rb-87, Th-232, and U-238 (see Table 4.4). After 5 billion years, what fraction of each of these isotopes remains?

4. Look up some real data for meteorites or lunar samples as measured in modern laboratories to see how well the isochron ages agree for different dating methods on the same samples.

ADDITIONAL READING

Burke, J. G. 1986. *Cosmic Debris: Meteorites in History.* San Francisco: University of California Press. Interesting discussion of the role of meteorites in history and our changing attitudes toward these remarkable rocks.

Dodd, R. T. 1986. *Thunderstones and Shooting Stars: The Meaning of Meteorites.* Cambridge, MA: Harvard University Press. Excellent popular-level introduction to meteorites by a leading researcher in the field.

Hutchison, R. 1983. *The Search for Our Beginnings.* New York: Oxford University Press. Well-written popular account, with emphasis on the message of the meteorites concerning the origin and early evolution of the solar system.

Norton, O. R. 1995. *Rocks from Space: Meteorites and Meteorite Hunters.* Missoula, MT: Mountain Press. An account of meteorites primarily from the perspective of those who find them.

5

ASTEROIDS: BUILDING BLOCKS OF THE INNER PLANETS

Asteroid Ida with its tiny moon Dactyl, as seen by the Galileo spacecraft camera.

In order to interpret the scientific message of the meteorites, we need to understand where they come from. These interplanetary wanderers appear to be fragments of larger parent bodies that have been broken up and eroded by impacts in space. The two classes of objects in our planetary system that are their potential parent bodies are the comets and the asteroids, the remnants of the population of small bodies that date back to the birth of the solar system.

As we saw in Section 4.2, the orbital evidence from four meteorites connects them to a probable source in the asteroid belt. The rocky composition of most meteorites also seems a better match for the asteroids than for the volatile-rich comets (although a comet that has lost its volatiles might be indistinguishable from an asteroid). In this chapter, we discuss the asteroids, deferring discussion of the comets until the next chapter.

Some astronomy textbooks refer to the comets, asteroids, meteors, and meteorites as debris. We prefer to think of them as objects from which we may be able to deduce events in the distant past. We study them in much the way an archaeologist sifts through the ruins of past civilizations, in the hope that we too can locate a Rosetta stone that will unlock secrets of the birth of the planetary system.

5.1 DISCOVERY OF THE ASTEROIDS

When you look at a diagram of the solar system, you probably notice the wide apparent gap between the orbits of Mars and Jupiter. Eighteenth-century astronomers hypothesized that there might be an undiscovered planet in this gap. Today we know that this gap is filled by millions of asteroids, too small to be seen without a telescope.

THE FIRST ASTEROIDS

On New Year's Day, 1801, Guiseppe Piazzi at Palermo discovered the first asteroid, which he named Ceres for the Roman patron goddess of Sicily. Since this faint object was in the middle of the gap at an orbital distance of 2.8 AU, it was at first hailed as the missing planet. Shortly thereafter, German astronomers discovered three more asteroids—Pallas, Juno, and Vesta—also orbiting between Mars and Jupiter. These were even smaller than Ceres, although Vesta is slightly brighter due to its more reflective surface. Even when combined, however, the masses of these four objects came nowhere near that of the Moon, let alone that of a real planet. Thus, the gap remained a mystery, and some suggested that the asteroids were the remnant of an exploded planet that had once occupied this space. Only much later did studies of meteorites discredit the hypothesis of an exploded planet (see Section 4.6).

The next asteroid was not discovered until 1845, but from then on they were found regularly by visual observers who scanned the sky looking for them. By 1890 the total number had risen to 300. At that time photographic patrols began, and the number of known objects rapidly increased, reaching 1000 (named Piazzia) in 1923, 3000 in 1984, and 5000 in 1990. While not confined entirely to the space between Mars and Jupiter, the majority of asteroids do occupy this part of the solar system. We define the **main belt asteroids** as asteroids with average distance from the Sun between 2.2 and 3.3 AU (Fig. 5.1).

Today new asteroids are relatively easy to find, and their numbers have grown rapidly. It required 197 years to find the first 10,000, but 3 years later this number had doubled to more than 20,000. The responsibility for cataloguing asteroids and approving new discoveries is assigned to the International Astronomical Union Minor Planet Center in Cambridge, Massachusetts. Asteroids are designated by both a number, given in order of discovery, and a name, usually selected by the discoverer (e.g., 4 Vesta, 433 Eros). Initially these were the names of Greek and Roman goddesses, such as Ceres and Vesta, and later were expanded to include female names of any kind. Recently the requirement of female gender has been relaxed, and asteroids today are named for a bewildering variety of persons and places, famous or obscure. For instance, asteroid 2410 is named Morrison, for one of the authors of this book, and asteroid 2309 is named Mr. Spock.

NUMBERS

Although thousands of small asteroids remain undiscovered, we can draw some general conclusions from present data. Ceres is the largest, with a diameter just smaller than 1000 km. The next largest asteroids are about half this size (Fig. 5.2). The total mass of all the asteroids amounts to only 1/2000 of the mass of the Earth (less than 1/20 the mass of the Moon), which indicates that the gap in the planetary system between Mars and Jupiter is real, a topic we will return to in Chapter 17.

Our census of the larger asteroids is by now fairly complete. Probably we have discovered all the main belt asteroids 25 km or more in diameter, and discovery should be at least 50% complete for diameters down to 10 km. Our knowledge is better for the closer asteroids in the inner part of the asteroid belt, and most of the larger undiscovered bodies are probably beyond 3 AU from the Sun.

There are many more small asteroids than large ones. An estimate of the relative numbers of

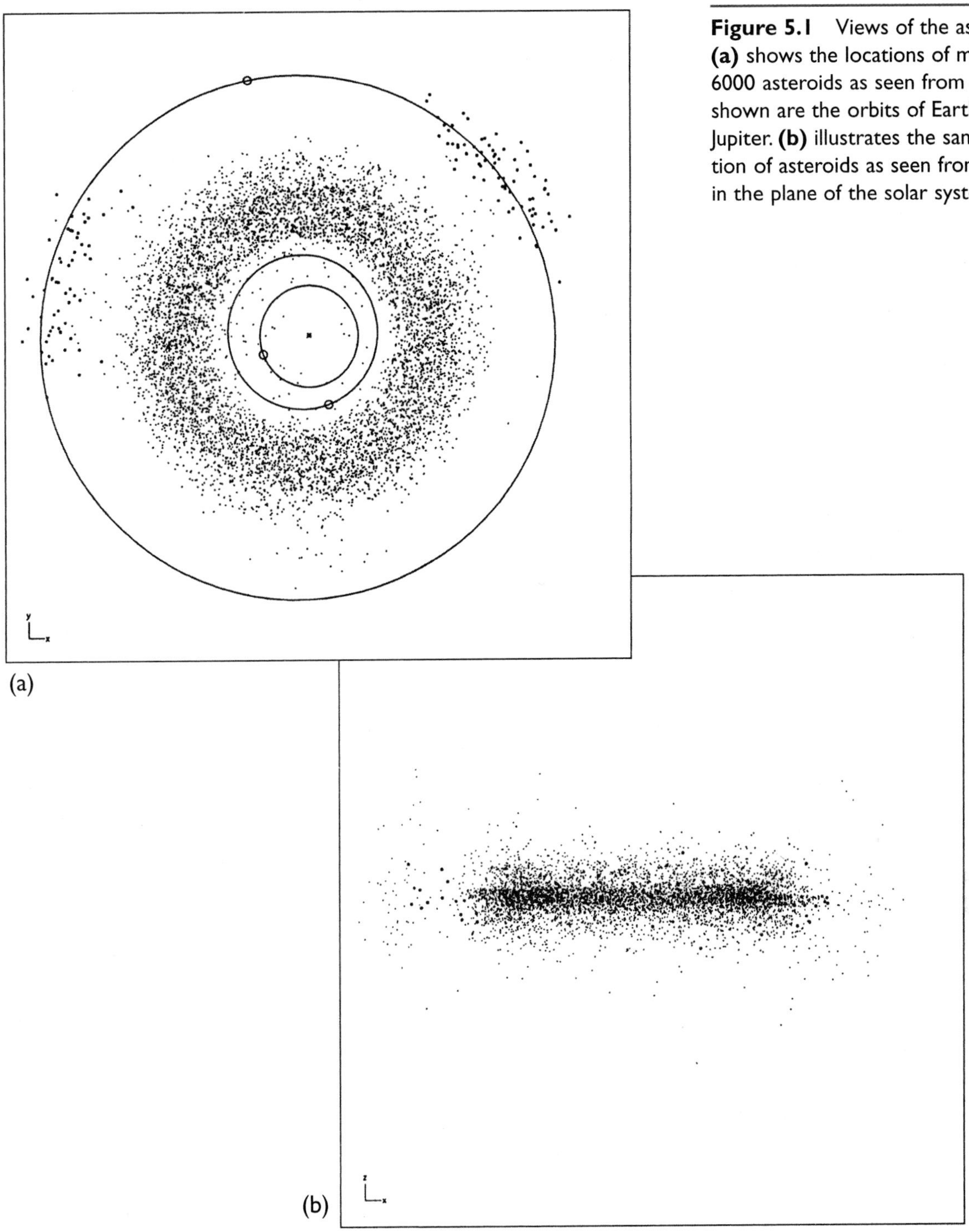

(a)

(b)

Figure 5.1 Views of the asteroid belt: **(a)** shows the locations of more than 6000 asteroids as seen from above. Also shown are the orbits of Earth, Mars, and Jupiter. **(b)** illustrates the same distribution of asteroids as seen from a position in the plane of the solar system.

objects of each size is interesting as a characterization of the asteroid population, and it will also prove important when we look at the size distribution of lunar craters produced by asteroid and comet impacts.

As a rule, many processes in nature, including fragmentation, produce approximately equal masses of material in each size range. Consider asteroids with diameters of 100 km, 10 km, and 1 km. Each 100-km asteroid has 1000 ($10 \times 10 \times 10$) times the mass of a 10-km asteroid, which in turn has 1000 times the mass of a 1-km object. Therefore, if the total mass is to be the same in each size range, there must be 1000 times more 10-km objects than

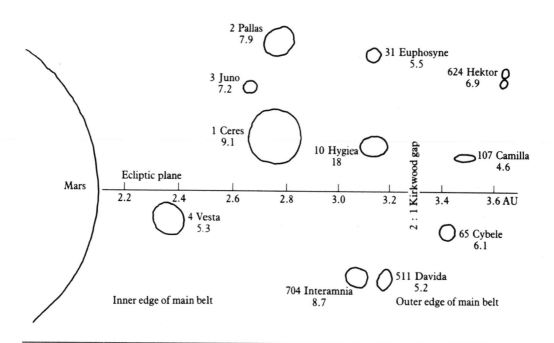

Figure 5.2 This scale drawing shows the relative sizes of some of the larger asteroids compared with the size of the planet Mars. The numbers next to the names are the rotation periods in hours. The horizontal scale gives the average distance from the Sun in astronomical units (AU).

there are 100-km ones, and 1 million (1000 × 1000) more at 1 km than at 100 km. In other words, the number of objects of a given diameter (D) is inversely proportional to the cube of their diameter ($1/D^3$).

Actual measurements of the asteroids indicate that the numbers do not rise quite this fast with declining size. Instead of the number increasing as the inverse cube of the diameter, it increases more nearly as the inverse square ($1/D^2$), resulting in a distribution with most of the mass in the larger objects (Fig. 5.3). This is why we are relatively certain of the total mass of the asteroids, even without having counted all of the small ones.

Think a moment about a size distribution in which number is proportional to $1/D^2$. What would we say, for example, is the average size of the objects? The great majority are very small, near the lower limit of size, but most of the mass is in the larger objects. The concept of average size is not very meaningful, at least not in the way we usually think about the term. As we shall see later, this type of size distribution applies to lunar craters and to the particles in planetary rings as well as to asteroids.

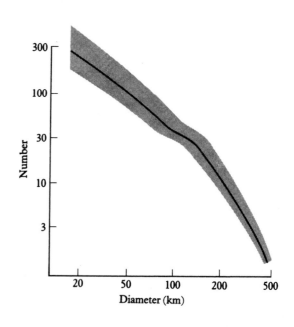

Figure 5.3 The number of asteroids increases rapidly with decreasing size.

PHYSICAL STUDIES

As seen through most telescopes, an individual asteroid is an unresolved starlike point. (The word *asteroid* means "starlike.") Before about 1975, little was known about the physical nature of asteroids, and research was confined to discovery and charting of orbits and the determination of rotation rates from observations of periodic brightness variations. In the past 25 years, however, the application of new observing techniques and larger telescopes has revealed a great deal about the physical and chemical nature of the asteroids.

Only the largest asteroids can be resolved and their sizes measured directly with the Hubble Space Telescope or with modern adaptive optics on ground-based telescopes. One of the first challenges was to measure the size and reflectivity of the majority of asteroids, which are too small to appear as a disk in the telescope. How can we make such measurements? The most accurate technique involves timing the rare passage of an asteroid in front of a bright star. Since we know exactly how fast the asteroid is moving against the stellar background, a measurement of how long the star is obscured yields an accurate size for the asteroid. If timings of the same event made from different locations on Earth are combined, the profile of the asteroid can also be derived. An example of the results for Juno is shown in Figure 5.4.

More generally, we would like to have a way to determine whether an asteroid of a given

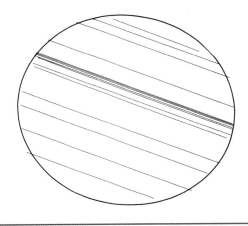

Figure 5.4 The approximate shape of the asteroid Juno as determined from its passage in front of a star. This phenomenon was observed from several different locations on Earth and hence from several different viewing angles. Each straight line in the figure represents the apparent path of the star as viewed from each observing location on Earth as the asteroid passed in front of it.

brightness is large and dark or small but highly reflective. One technique that can be applied to large numbers of asteroids involves measurement of the heat radiated as well as the solar light reflected. A dark object absorbs most of the incident sunlight and reemits it in the infrared as heat; it will appear relatively bright in the infrared and faint in the visible. A highly reflective object will be bright in the visible but will be a weaker source of heat (Fig. 5.5). Application of techniques based on this principle has resulted in the

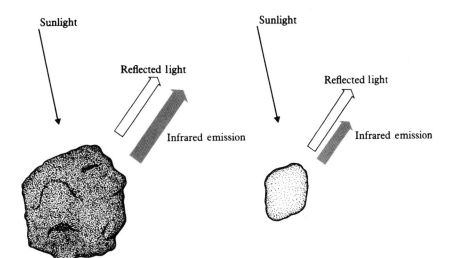

Figure 5.5 Comparison of a large, dark asteroid with a small, bright (highly reflective) asteroid. In this example, both reflect the same amount of sunlight, so both appear equally bright in visible light. The large, dark asteroid emits much more infrared thermal radiation, owing to its larger size and hotter surface. Thus, measurement of reflected and emitted radiation allows us to determine both size and reflectivity (albedo).

determination of diameters and reflectivities (albedos) for more than 1000 asteroids.

A second important type of information about asteroids is obtained by measuring the spectra of sunlight reflected from their surfaces. Different minerals have different colors. Variations with wavelength in the reflectance of the surface material can indicate the composition of the asteroid. While not as rigorously diagnostic as the sharp spectral lines produced when light passes through a gas, the broad absorption features in the spectrum of an asteroid are often sufficient to identify the major minerals present. This identification is aided by the similarities between the spectra of asteroids and of many meteorites, permitting a laboratory comparison between individual meteorites and asteroids (Fig. 5.6).

The use of spectral and reflectance data has yielded preliminary determinations of composition for more than 1500 asteroids. In most cases, the minerals inferred to be present on asteroid surfaces are similar to those in the stony meteorites. Exact identifications are difficult, however. Our current state of knowledge is like the verbal description of an unknown criminal suspect. We can categorize in terms analogous to height, weight, age, gender, and hair color, but we cannot specify the unique properties that identify an individual. Asteroid research therefore tends toward broad statistical studies rather than detailed investigation of particular objects—that is, until we have an opportunity to get close in and personal with a spacecraft (as we will describe in Section 5.4).

5.2 MAIN BELT ASTEROIDS

The great majority of the asteroids are located in the main asteroid belt, at average distances from the Sun between 2.2 and 3.3 AU, corresponding to orbital periods of 3.3 to 6.0 years. From the observed distribution of sizes we can estimate that there are more than 100,000 asteroids down to a diameter of 1 km.

Although 100,000 sounds like a lot of objects, space in the asteroid belt is still empty. The main belt asteroids occupy a very large volume, roughly doughnut-shaped, about 100 million km thick and

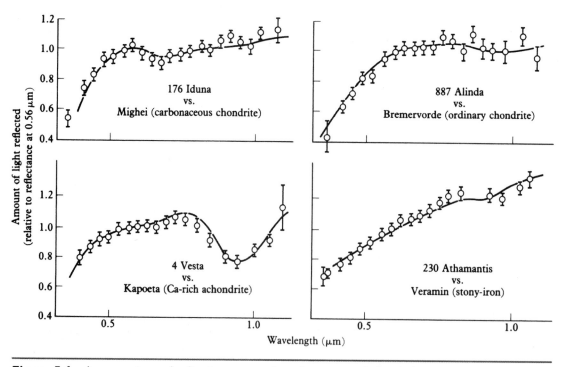

Figure 5.6 A comparison of reflection spectra from four asteroids (circles) with laboratory-determined spectra of meteorites (solid lines).

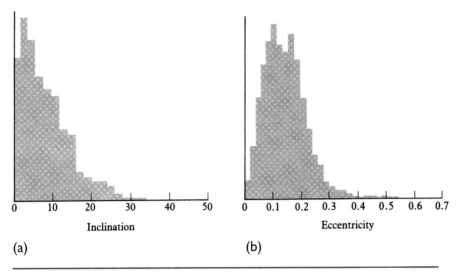

Figure 5.7   (a) The number of asteroid orbits with various inclinations.
(b) The number of asteroid orbits with various eccentricities.

nearly 200 million km across. Typically they are separated from each other by millions of kilometers. They pose no danger to passing spacecraft; in fact, it was difficult to locate asteroids near enough to the trajectory of the Jupiter-bound Galileo spacecraft to allow close asteroid flybys.

ORBITS AND RESONANCES

The orbits of the main belt asteroids are for the most part stable, with eccentricities less than 0.3 and inclinations less than 20° (Fig. 5.7). In the past, when there were more asteroids in this region of space, collisions may have been common, but by now the population has thinned to the point where

each individual asteroid can expect to survive for billions of years between collisions. Still, with 100,000 objects, a major collision somewhere in the belt is expected every few tens of thousands of years. Such collisions, as well as lesser cratering events, presumably yield some of the fragments that eventually reach the Earth as meteorites.

The orbits of asteroids within the main belt are not evenly distributed. As shown in Figure 5.8, some orbital periods are preferred, while others are nearly unpopulated. These unpopular sections of the belt are **resonance gaps,** also known as Kirkwood gaps for the nineteenth-century American astronomer Daniel Kirkwood, who discovered them.

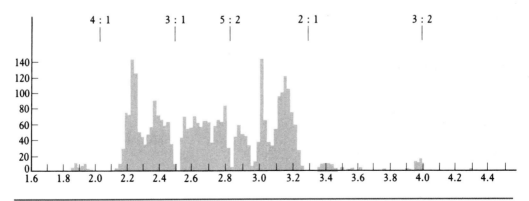

Figure 5.8   The number of asteroid orbits with various average distances from the Sun. The positions of orbital resonances with Jupiter are shown (2:1 = 1/2 Jupiter's orbital period, etc.).

The gaps in the asteroid belt occur at orbital periods that correspond to resonances between these periods and the orbital period of Jupiter. A **resonance** is defined as follows: If an asteroid should find itself in a resonant orbit, repeated small gravitational effects of Jupiter would add together to alter the orbit, in much the same way that repeated small pushes on a swing, applied at the proper time, lead to a large oscillation. A resonance takes place when the orbital period of one body is an exact fraction of the period of another. In this case, the underpopulated asteroid orbits correspond to periods that are 1/2, 1/3, 1/4, and so on, that of the period of Jupiter.

Consider an asteroid with a 6-year period; it would revolve about the Sun exactly twice for each revolution of Jupiter. Its average distance from the Sun would be 3.3 AU. It is closest to the giant planet once each 12 years, at exactly the same place in its orbit. The cumulative effects of Jupiter's gravity, always applied at the same place, will alter the orbit of the asteroid; in contrast, an object on a nearby orbit that is not resonant with Jupiter experiences gravitational nudges all around its orbit with no net effect. In this way Jupiter will perturb and ultimately eliminate asteroids in resonant orbits. Indeed, the period corresponding to 3.3 AU is one of the observed resonance gaps, the one that marks the outer edge of the main asteroid belt. Similar processes involving the satellites of Saturn play an important role in creating resonance gaps in Saturn's rings (see Chapter 16).

The resonance gaps represent missing orbital periods rather than physical gaps in the belt. Since most asteroids have orbits of significant eccentricity, their in-and-out motions effectively fill the space within the belt. In contrast, the resonance gaps in the rings of Saturn show up as physical gaps also, since the Saturn ring particles are in essentially coplanar circular orbits.

ASTEROID FAMILIES

Asteroid orbits display other patterns in addition to the resonance gaps. A **family of asteroids** is defined as a group of objects with similar orbits, suggesting a common origin. About half of the known main belt asteroids are members of families, with nearly 10% belonging to just three: the Koronos, Eos, and Themis families (named after their largest member). Although not clustered together in space at the present, the members of an asteroid family were all at the same place at some undetermined time in the past.

Members of the same asteroid family tend to have similar reflectivities and spectra. Apparently the family members are fragments of broken asteroids, shattered in some ancient collision and still following similar orbital paths. According to some estimates, almost all the asteroids smaller than about 200 km in diameter were probably disrupted in earlier times, when the population of asteroids was larger. The families we see today may be remnants of the most recent of these inter-asteroid collisions.

THE LARGEST ASTEROIDS

The largest asteroid in the main belt is Ceres, the first to be discovered. Its diameter is nearly 1000 km, and its mass of 1/5000 that of the Earth amounts to nearly half the total mass of all the asteroids. Two other asteroids are about 500 km in diameter: Pallas and Vesta. The 20 largest asteroids (all larger than 220 km) are listed in Table 5.1. Note that several of the larger asteroids, such as Interamnia, Davida, and Patientia, were not among the early discoveries. These objects are in the outer part of the belt and have low reflectivities, making them faint in spite of their large sizes.

COMPOSITIONAL CLASSES

In order to characterize the asteroids further, we use the classifications based on measurements of reflectivity and spectra discussed in the previous section. As soon as the data on a substantial sample of asteroids were examined, it became apparent that most of them fell into one of two classes based on their albedo or reflectivity: They were either very dark (reflecting only 3%–5% of incident sunlight) or moderately bright (15%–25% reflectivity). A similar distinction exists in their spectra: The dark asteroids are fairly neutral reflectors with no major absorption bands in the visible range to reveal their composition, although some of them show spectral evidence for H_2O in the infrared. Most of the lighter asteroids are reddish and show the spectral signatures of common silicate minerals such as olivine and pyroxene. Since the dark gray asteroids had spectra similar to the carbonaceous meteorites, they were called **C-type asteroids.** The lighter class was named the **S-type asteroids,** indicating a silicate or stony composition. A third major group appears to be metallic (like the iron meteorites) and is called the **M-type asteroids.**

Table 5.1 includes the assigned compositional classes for the asteroids larger than 220 km. Among the largest asteroids, Ceres and Pallas are in the C class, but Vesta fits into none of the three classes we have just defined. The special case of Vesta is discussed later.

Given the uncertainties inherent in deducing composition from such limited data, few scientists are willing to state positively that C-type asteroids are made of exactly the same thing as carbonaceous meteorites, but the connection seems probable. In any case, the extreme darkness of their surfaces and the presence of H_2O indicate that C-type asteroids are primitive and carbon-rich. The interpretation of the rarer M type as metallic is even clearer, thanks to radar studies of the large M-type asteroid Psyche; the high radar reflectivity of this object is a clear signature of the presence of metal on its surface. Least certain was the identification of the S-type asteroids with the primitive (chondritic) stony meteorites, until the NEAR-Shoemaker spacecraft landed on the asteroid Eros and verified this hypothesis—as we will discuss later.

Let's look at the distribution in space of the C, S, and M types. At the inner edge of the belt, near 2.2 AU, the S-type asteroids predominate.

Table 5.1   The 20 largest asteroids

Asteroid Number	Name	Year of Discovery	Semimajor Axis (AU)	Diameter (km)	Compositional Class
1	Ceres	1801	2.77	940	C
2	Pallas	1802	2.77	540	C
4	Vesta	1807	2.36	510	*
10	Hygeia	1849	3.14	410	C
704	Interamnia	1910	3.06	310	C
511	Davida	1903	3.18	310	C
65	Cybele	1861	3.43	280	C
52	Europa	1868	3.10	280	C
87	Sylvia	1866	3.48	275	C
3	Juno	1804	2.67	265	S
16	Psyche	1852	2.92	265	M
451	Patientia	1899	3.07	260	C
31	Euphrosyne	1854	3.15	250	C
15	Eunomia	1851	2.64	245	S
324	Bamberga	1892	2.68	235	C
107	Camilla	1868	3.49	230	C
532	Herculina	1904	2.77	230	S
48	Doris	1857	3.11	225	C
29	Amphitrite	1854	2.55	225	S
19	Fortuna	1852	2.44	220	C

As we move outward, the fraction of C-type objects increases steadily, and in the belt as a whole these dark, carbonaceous objects make up 75% of the population, compared to 15% S and 10% other types. Beyond the main belt, all the asteroids appear to be carbonaceous. The M types are located near the middle of the belt, but they constitute only a few percent of the known asteroids.

In principle, an excellent way to characterize an asteroid is by measuring its density. As we saw in Section 3.1, the density of a planet provides a good indication of its bulk composition, distinguishing between metal, solid rock, and volatile-rich primitive material. A handful of asteroids have accurately measured densities, primarily based on masses that are calculated from the motions of satellites using Kepler's laws. Unfortunately, however, the density seems to reflect the interior structure of the asteroid as well as its composition. Some asteroids are loosely bound rubble piles rather than compact, solid rocks. Even a metallic asteroid can have a surprisingly low density, presumably resulting from the presence of many voids in the interior. Therefore, the interpretation of density is ambiguous. We will discuss some specific cases below when we turn to spacecraft investigations.

Clearly, most asteroids are composed of relatively primitive material. If the asteroids are still near the locations where they formed, we can use the distribution of asteroid types to map out the composition of the solar nebula. Since carbonaceous meteorites formed at lower temperatures than the other primitive stones, the concentration of C-type asteroids in the outer belt is consistent with their formation far from the Sun, where the nebular temperatures were lower. It is also possible, however, that the asteroids formed elsewhere and were herded into their present positions by the gravity of Jupiter and the other planets. In that case, the C-type asteroids could have formed far beyond Jupiter and subsequently diffused inward to their present positions in the outer part of the asteroid belt. Similarly, the S-type asteroids near the inner edge of the belt could either have formed where we see them today or have been gravitationally scattered to their present locations

from still closer to the Sun. The solar nebula temperatures that we would deduce from the application of these two alternative models are quite different. So far, however, we have not been able to settle on which model for the origin of the asteroids is to be preferred.

VESTA AND ITS OFFSPRING

Vesta has an unusual spectrum, indicative of igneous rock (see Fig. 5.6). Only a handful of very small asteroids share this basaltic surface composition. But recall that we introduced the eucrites as a group of about 30 basaltic meteorites (see Section 4.5). When the spectra of the eucrites were measured in the laboratory, it was found that they matched the spectrum of Vesta closely, though not exactly. A more careful comparison showed that there are regions on the surface of Vesta that have the compositions of the main subgroups of the eucrites. As the asteroid rotates, these regions—essentially large lava flows of slightly different composition—are viewed successively. Is this spectral similarity perhaps fortuitous? Might the eucrites be fragments from the breakup of another large, differentiated asteroid similar to Vesta, rather than samples of Vesta itself? This seems like a reasonable hypothesis, and to evaluate it we must apply a careful chain of logic.

The lava rock of the eucrites represents the crust of the parent body. From the chemistry of the eucrites it is possible to predict the composition of the much thicker mantle of the parent. Yet an examination of our meteorite collections reveals no meteorites with this predicted mantle composition. Since more mantle debris than crust debris would be produced by the total breakup of this hypothetical parent, it follows that no such breakup took place. Therefore, the "eucrite parent body" is still largely intact, and the eucrites themselves were chipped from its crust by impact cratering.

If we accept this argument for an intact eucrite parent body and combine it with the fact that no other large asteroid exists with the same basaltic surface as Vesta, we are drawn to the conclusion that Vesta really is the parent body of the eucrites. This conclusion is reinforced when we consider that only a rather large asteroid could have been heated sufficiently to generate a basalt surface. It is possible that the immediate parents are some of the small "vestoids"—asteroids that themselves seem to be chips of Vesta. But Vesta is probably the ultimate parent body of the eucrites.

If the connection between Vesta and the eucrites is correct, we can apply the meteoritic evidence to this asteroid. Vesta then becomes the fourth solar system object for which we have samples (the others are Earth, Moon, and Mars). Thus, we can date the Vesta lava flows at 4.5 billion years ago, the solidification age of the samples, and fix the impact that released the meteorites at about 3 billion years ago from the gas-retention age. Whatever triggered the volcanic activity of this asteroid was short-lived, as we might expect for such a small object.

5.3 ASTEROIDS FAR AND NEAR

A few asteroids—perhaps 1% of the number in the main belt—have orbits that cross the orbits of Mars or the Earth. Most of these wanderers are fragments of main belt asteroids spun into more eccentric orbits as a result of collisions in the belt. There are also asteroids outside the main belt. Beyond 3.3 AU, the distribution of asteroids changes, with most of the objects concentrated at just a few orbital periods with wide gaps in between. Many asteroids are found near the orbit of Jupiter, and a few travel deep into the outer solar system.

TROJAN ASTEROIDS

A particularly interesting group of dark, distant asteroids is orbitally associated with Jupiter. Although the gravitational attraction of this giant planet generally has the effect of making nearby asteroid orbits unstable, exceptions exist for objects with the same orbital period as Jupiter but leading or trailing it by 60° (Fig. 5.9). These two stable regions are called the leading and trailing Lagrangian points, named for the French mathematician (Joseph Louis, Comte Lagrange) who demonstrated their existence in 1772. While he was examining mathematically the possible motions of three mutually gravitating bodies, Lagrange found two regions where a small object could occupy a stable orbit within the gravitational fields of two larger objects. If the larger objects are Jupiter and the Sun, a small object in one of the Lagrangian points occupies one corner of an equilateral triangle, with the Sun and Jupiter at the other two points.

The regions of stability around the two Lagrangian points are quite large, and each contains several hundred known asteroids. The first of these Lagrangian asteroids was named Hektor when it

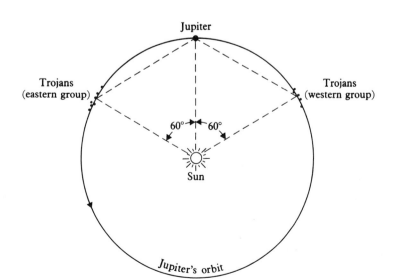

Figure 5.9 The Trojan asteroids are located 60° ahead of, and behind, Jupiter in that planet's orbit. Mars and Neptune have also captured asteroids in their Trojan points, and we also find small satellites in the Saturn system that occupy similar positions with respect to Saturn (in the role of the Sun) and a larger satellite (in the role of Jupiter).

was discovered in 1907. All of them are named for the heroes of Homer's *Iliad* who fought in the Trojan War, and collectively they are known as the **Trojan asteroids.** Their spectra are distinctive, suggesting that they represent a group of special, primitive objects that have been trapped in this region of space since the birth of Jupiter. If we could detect the fainter members of these Trojan clouds, we might find that the Trojan asteroids are nearly as numerous as those in the main asteroid belt.

As the most massive planet, Jupiter has the most stable Lagrangian orbits. Once located in one of the Trojan clouds, an asteroid should remain there for the lifetime of the solar system. Other planets can support their own Lagrangian orbits, but they tend to be less stable and would not be expected to contain so many objects as the jovian Trojans. Several small asteroids have been discovered in the Lagrangian regions for Mars and Neptune, but none for Earth or the other outer planets. Within the Saturn satellite system there are also tiny Lagrangian satellites, as we will see in Chapter 16.

Although they are dark and carbonaceous, the detailed compositions of these more distant asteroids appear to be different from those of the C-type asteroids of the main belt. Their colors are redder, and they do not look the same as any known carbonaceous meteorite. Because these objects do not seem to be represented in our meteorite collections, scientists hesitate to commit themselves concerning their composition. It is generally thought, however, that they are primitive objects, and that a fragment from one of them would be classed as a carbonaceous meteorite, though of a different kind from those already encountered.

Asteroids Beyond Jupiter

A few asteroids are known to venture beyond Jupiter. The first to be discovered was Hidalgo, a dark object occupying a cometlike orbit of high inclination and eccentricity that carries it from the inner edge of the main belt out almost as far as Saturn. Second is Chiron, discovered in 1977; its orbit carries it from 8.5 AU, near Saturn, out to 19 AU, near Uranus. A still more distant object discovered in 1992 is called Pholus, with an orbit that takes it as far as 33 AU from the Sun, beyond the orbit of Neptune, and more of these faint objects continue to be found. The names Chiron and Pholus are taken from the two good centaurs of Greek mythology. Pholus is the reddest object in the solar system, indicating a very strange (but yet unknown) surface composition. Figure 5.10 illustrates the orbits of Chiron and Pholus, together with those of Comet Halley and the first object discovered in the Kuiper comet belt (both discussed in Chapter 6).

Should distant objects with highly eccentric orbits be classed as asteroids or comets? Their orbits suggest that they could be comets, but the true distinction between a comet and an asteroid is based on the nature of the volatile compounds the body contains. For objects in the inner solar system, this distinction in volatility translates directly into a difference in telescopic appearance: If the object develops a visible atmosphere, it is a comet; if it appears starlike, it is an asteroid. However, at the great distances of these objects from the Sun, no cometary atmosphere would be expected, even if the objects were composed in large part of water ice. If they contain frozen nitrogen or carbon monoxide, however, enough material might evaporate from their surfaces to be detected even at the orbit of Neptune. We suspect that if these objects came closer to the Sun, all three would develop atmospheres and be reclassified as comets.

This speculation became a reality in 1988, when astronomers discovered that Chiron had brightened by more than a factor of two from the previous year, presumably as the result of the formation of a tenuous atmosphere of gas and dust. The following year this atmosphere was photographed, confirming the transition of Chiron from asteroid to comet. By 1988 Chiron's orbital motion had brought it within 10 AU of the Sun, and its higher temperature was apparently sufficient to begin the release of volatiles from the surface.

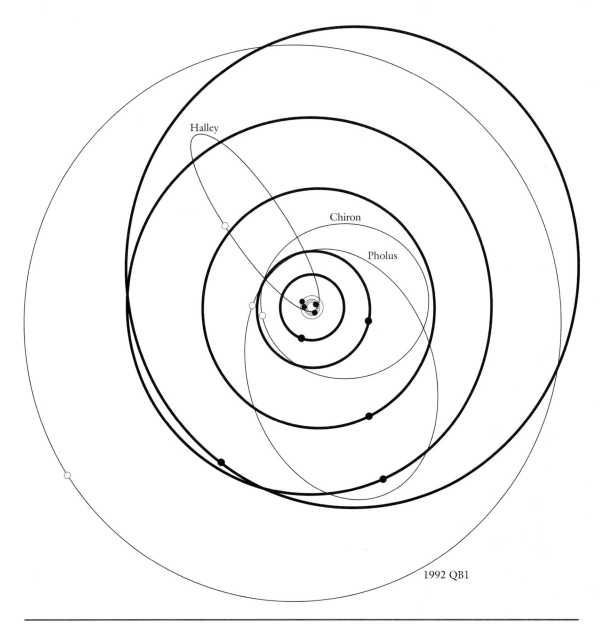

Figure 5.10 View of the solar system from above, showing the orbits of Comet Halley and three of the newly discovered objects in the outer solar system: Chiron (initially designated an asteroid but subsequently seen to display cometary activity); Pholus, the reddest known object in the solar system; and the first object found in the Kuiper belt (discussed in Chapter 6).

NEAR-EARTH ASTEROIDS

Although only about 1% of the asteroids stray inside the asteroid belt and approach the Earth, we have a special interest in those that do. The Earth-approaching asteroids are potentially useful to humanity as sources of raw materials in space, and they also pose a threat to us if they should collide with the Earth (as many eventually will; Fig. 5.11).

The first asteroid found on an orbit that crossed the Earth's orbit (coming within less than 1 AU of the Sun) was Apollo, discovered in 1948, and thus the Earth-crossing objects are often called the Apollo asteroids. Here we use the

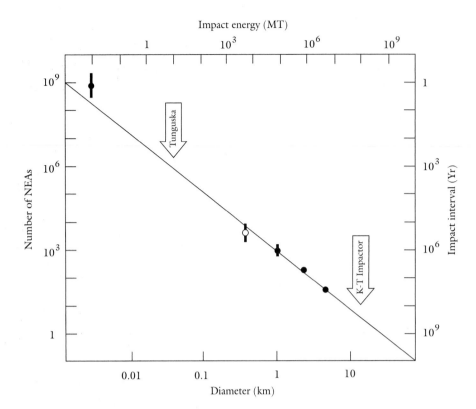

Figure 5.11   The frequency of asteroid impacts on the Earth, showing how the frequency increases as the size of the objects decreases. Also indicated is the impact energy (in megatons of TNT) for asteroids of various diameters. Objects smaller than about 40 m in diameter (5 MT energy) disintegrate in the upper atmosphere and do no damage. Shown for reference are the 1908 Tunguska event (15 MT airburst, average interval of several hundred years) and the K-T impact of 65 million years ago (100 million MT; average interval about 100 million years).

somewhat broader term **near-Earth asteroid (NEA)** to include potential Earth-crossers, objects whose orbits do not now intersect that of the Earth but could do so under the shifting gravitational nudges of the other planets.

As a result of searches dedicated to providing the Earth advanced warning of any impact (see Perspective), about 2000 NEAs have been discovered. The largest is Eros, a highly elongated asteroid about 30 km long, but most of the known Earth-approachers are much smaller. The asteroid that has the smallest perihelion distance is Icarus, which comes inside the orbit of Mercury. It is estimated that approximately 1000 Earth-approaching asteroids exist down to 1 km in diameter, and about a quarter of a million down to 100 m in diameter.

The NEAs that have measured reflectivities and spectra are a diverse lot. Most are S-type objects, but some are C-type and others have spectra that are uncommon among main belt asteroids. Some of these may be dead comets.

The orbits of near-Earth asteroids are unstable, as a result of the constantly varying gravitational influence of Earth, Mars, and Venus. Any object that finds itself caught in such an orbit will experience one of two fates, each about equally probable. The object may be gravitationally

ejected as the result of nearly missing a planet, or it may terminate its existence dramatically in a crater-forming impact. Most of the craters on the Moon and on the terrestrial planets are caused by collisions with NEAs.

Calculations indicate that 10–20 percent of the asteroids whose orbits currently cross the Earth's orbit will eventually hit our planet. An Earth-approaching asteroid will either strike a planet or be ejected after an average interval of about a hundred million (10^8) years. A hundred million years may seem like a long time to us, but it is short in comparison with the 4.5-billion-year history of the planetary system. Any asteroids that began as Earth-approachers at the time the solar system was formed would long since have been ejected or crashed. Therefore, the NEAs present today must have come from somewhere else, and there has to be a continuing source (or sources) for these objects, such as collisions among main belt asteroids, several of which take place every million years.

The occasional impact on our planet of near-Earth asteroids or comets is an important aspect of our solar system environment. Although these impacts have negligible effect on the planet as a whole and are far too small to influence the

Table 5.2 Spacecraft encounters with asteroids

Asteroid	Class	Date	Dimensions (km)	Density (g/cm^3)	Resolution (m)
Gaspra	S	1991	18 × 11 × 9	—	50
Ida	S	1993	60 × 25 × 19	2.6	25
Mathilde	C	1997	66 × 48 × 46	1.3	160
Eros	S	2000	31 × 13 × 13	2.67	0.1

rotation or orbit of the Earth, they can have severe environmental effects. Both the initial ejection of molten rock from such an impact and the subsequent presence of fine dust in the Earth's atmosphere can first heat and then cool the surface, with disastrous consequences. Only recently have scientists recognized that impacts have had a major influence on the extinction and evolution of life, a topic we will return to in Chapter 12.

5.4 ASTEROIDS CLOSE UP

On its long flight path to Jupiter, the Galileo spacecraft made close flybys of two S-type main belt asteroids, Gaspra and Ida. The NEAR-Shoemaker spacecraft made a similar flyby of the C-type main belt asteroid Mathilde before orbiting and eventually landing on Eros (S-type). Table 5.2 summarizes the four spacecraft encounters. In addition to the asteroids studied by spacecraft, we have good radar images of several NEAs that have passed especially close to the Earth.

All of the spacecraft photos of asteroids show the shallow circular depressions called craters. The origin and structure of craters will be described in detail in Chapter 7. For the present, simply note that craters are produced by high-speed collisions with smaller debris, and that a large number of craters generally indicate a long exposure to such cosmic impacts.

GASPRA AND IDA

Gaspra is one of many members of the Hungaria asteroid family, while Ida is a member of the Koronos family. The Galileo flybys revealed both Gaspra and Ida to be highly irregular in shape and heavily cratered, with only slight differences in color or reflectivity across their surfaces (Fig. 5.12). Gaspra has fewer craters, indicating a relatively young age (where age is the time since the last global-scale impact). In contrast, Ida is saturated with craters, and it appears to have a broken-up surface layer that is tens of meters thick (similar to that of the Moon).

Figure 5.12 Images printed to the same scale of the first three main belt asteroids studied with spacecraft: Mathilde, Gaspra, and Ida.

PERSPECTIVE

DEFENDING THE EARTH AGAINST ASTEROIDS

The craters of the Moon provide evidence of a long history of bombardment by asteroids and comets. If the Earth were as geologically dead as the Moon, with little volcanism or erosion, our planet would also be covered with craters. As it is, only the largest and youngest of these craters survive on Earth. The atmosphere protects us only from small impacts—those by objects smaller than a house. For an incoming asteroid larger than this, the atmosphere offers no effective resistance.

People became aware of the continuing hazard of impacts in 1980, after scientists linked the extinction of the dinosaurs to the impact of a large asteroid 65 million years ago (a topic we shall discuss in Chapter 12). In 1991, the U.S. Congress asked NASA to evaluate the current impact hazard and suggest ways to deal with the problem. The House Science Committee wrote that "the detection rate must be increased substantially, and the means to destroy or alter the orbits of asteroids when they do threaten collisions should be defined and agreed upon internationally. The chances of the Earth being struck by a large asteroid are extremely small, but because the consequences of such a collision are extremely large, the Committee believe it is only prudent to assess the nature of the threat and prepare to deal with it."

The NASA studies (chaired by author Morrison) included a quantitative evaluation of the risk. We concluded that the greatest hazard was from near-Earth asteroids (NEAs) larger than 1 km in diameter, and that this risk was substantial—similar to that of other natural hazards such as earthquakes or severe storms. The biggest danger is not from the blast itself, but from the global pall of stratospheric dust that would follow. This dust would darken the sky, causing temperatures to drop and killing most crops. We also concluded that if such a collision were predicted far enough in advance, we had the technology (in principle, at least) to deflect or destroy the threatening asteroid. Impacts represent not only the worst imaginable natural catastrophe, but also the only one that we might be able to effectively defend ourselves against. This led to the recommendation that we begin a concerted effort to find all of the NEAs larger than 1 km in diameter (there are estimated to be about 1000) and determine their orbits in order to identify any that threaten collision. This search for NEAs is called the Spaceguard Survey. The name was taken from a science fiction novel by Arthur C. Clarke, who coined the term to describe an early warning system built in the twenty-first century to protect the Earth from impacts.

When the Spaceguard Survey was first proposed, astronomers had discovered fewer than 10% of the estimated 1000 NEAs larger than 1 km in diameter. Most of the discoveries came from photographic patrols carried out with a handful of small telescopes. The photographs were scanned by hand to find the rare image of an asteroid moving against the background of millions of star images. The result was approximately one new 1-km NEA discovery per month. The total effort of the citizens of our planet directed toward protecting us from a catastrophe that could kill more than a billion people was less than the staff of an average McDonald's restaurant. To accelerate the search, astronomers invented automated telescopes equipped with electronic detectors to image the sky. Computers compared the images to identify the moving asteroids and determine whether they were NEAs

➥

↪ or members of the main belt. The U.S. Air Force contributed telescopes and technical equipment, and some officers thought that the protection of the Earth from space rocks should be part of the mission of a new "air and space" defense force. By the late 1990s, the survey was dominated by two small (1-m aperture) Air Force telescopes in New Mexico, operated with NASA funds. The discovery rate rose from one per month to more than ten per month, pointing toward cataloguing 90% of the NEAs larger than 1 km by 2008.

So far no NEA has been found on a collision course with our planet. Probably none will hit within the next century. But already people are thinking about how to respond if an impact is predicted. Given decades of warning, we could send a series of spacecraft to the asteroid, first to study it scientifically ("know your enemy") and then to change its orbit (probably with nuclear explosives). We are the first species on Earth that understands the cosmic impact hazard and the first that is prepared to defend our planet against such a catastrophe.

The most unexpected discovery was the presence of a small satellite (named Dactyl) in orbit around Ida, which it circles in 23 hours at an altitude of about 100 km. The presence of the satellite allowed Ida's mass and density to be measured. The density is 2.6 g/cm^3, similar to that of primitive rocks. Partly on this basis, it appears that these two S-type asteroids are probably coherent and composed of materials similar to the ordinary chondrite primitive meteorites. However, the spectral mismatch between these objects and known chondrites in our meteorite collections continued to baffle us following these two flybys.

MATHILDE

The first spacecraft dedicated to asteroid studies was the Near-Earth Asteroid Rendezvous (or NEAR) mission, launched in 1996. The NEAR mission was subsequently named for pioneering planetary scientist Gene Shoemaker, and it is therefore known by the somewhat awkward name NEAR-Shoemaker. On the way to its primary target Eros (discussed below), the spacecraft flew past Mathilde, a 60-km main belt asteroid and the first C-type object visited by spacecraft.

As shown in Figure 5.12, Mathilde has a very different shape from Gaspra and Ida, as well as the dark color characteristic of carbonaceous asteroids. It has the largest craters known for any object in the planetary system, relative to its own size. If it were a hard, rocky object, any of the impacts that produced these giant craters would have fragmented the asteroid, perhaps blowing it apart but at the least shaking it so hard that any previously existing craters would be erased. The only way that Mathilde can support several giant craters simultaneously (as opposed to just the one resulting from its most recent collision) is for the asteroid to be a soft, porous, dusty object—the self-gravitating remnant of the first of these big impacts, which reduced it to a pile of rubble. This hypothesis is supported by the low density of Mathilde, only 1.3 g/cm^3, which suggests a loosely packed interior with 50% porosity.

EROS

NEAR-Shoemaker missed its original rendezvous date with Eros in December 1998 due to a malfunction, but the spacecraft was able to recover after one more trip around the Sun and finally arrive in February 2000. It achieved an initial high orbit and then gradually lowered its altitude over

Figure 5.13   Near-Earth asteroid Eros as imaged by the NEAR-Shoemaker spacecraft, which orbited the asteroid for 12 months in 2000 and 2001.

the next year, studying Eros with a variety of instruments. The spacecraft obtained thousands of images and more than 10 million laser altimetry measurements, making Eros one of the best-mapped objects in the solar system (Fig. 5.13).

After 1 year in orbit, NEAR-Shoemaker began a staged descent to the surface, taking pictures of ever-increasing resolution. It landed on February 12, 2001, with an impact velocity of 1.6 m/s. Fortunately, the spacecraft was not damaged, even though it had not been designed for such a maneuver. Using its low-gain antenna, it continued to radio data from the surface for more than a week, providing the best measurements of elemental composition.

The quantitative measurement of radioactivity from K, Th, and U as well as gamma-ray lines of Fe, O, Si, and Mg demonstrated that Eros has a primitive composition. Since Eros is a normal class-S asteroid, this *in situ* result finally settled questions that had remained open for decades concerning the nature (primitive or differentiated) of the S-type asteroids. The density of Eros (2.67 g/cm³) is also generally consistent with this meteorite identification, although it still implies a substantial bulk porosity of about 25%.

Long ridges seen in some of the images demonstrate that Eros is a consolidated and coherent body with global-scale cracks. As suspected for several other asteroids, Eros is a solid collisional fragment of a larger parent body (but not a rubble pile like Mathilde), with an interior that has been heavily fractured. The surface is cratered, but there is a surprising deficiency of small craters, combined with an excess of boulders up to 100 m in size (Fig. 5.14). There are actually more boulders than craters in the tens-of-meters sizes. On crater walls, dark material has flowed downslope, exposing underlying bright material (Fig. 5.15). The effects of exposure to space (space weathering) are evident in the different spectral reflectivities of soils of differing ages. This explains why scientists have had such a difficult time over the years comparing the colors and reflectivities of freshly broken meteorites with those of the natural surfaces of asteroids, which have been exposed for hundreds of millions of years to charged particles from the Sun as well as the innumerable impacts of interplanetary dust grains.

Figure 5.14   High-resolution views of Eros taken by the NEAR-Shoemaker spacecraft as it descended toward its landing on the surface. Notice the large number of bumps or boulders, which greatly outnumber the craters in the two closest photos.

Figure 5.15 Remarkable flat areas on Eros that appear to be filled with fine loose dust. These are informally called "ponds" although no water has ever been associated with them.

RADAR STUDIES

One of the most powerful tools for the investigation of asteroids is **radar.** The principles of radar are essentially the same whether applied to a distant planet or a nearby airplane on its approach to a foggy airport. An intense pulse of radio radiation is emitted by a transmitter in the direction of the target, and a very, very much weaker echo reflected from the target is picked up by a sensitive radio receiver (Fig. 5.16). There are two major planetary radar facilities, both of which were upgraded in the late 1990s. NASA operates the Goldstone (California) planetary radar facility as part of the Deep Space Net, while the 1000-foot Arecibo dish in Puerto Rico is operated by the National Astronomical and Ionospheric Center with National Science Foundation and NASA support. The two facilities are complementary: Arecibo has greater sensitivity but Goldstone has greater sky coverage. For the asteroids that come closest to the Earth, radar allows astronomers to construct geologically detailed

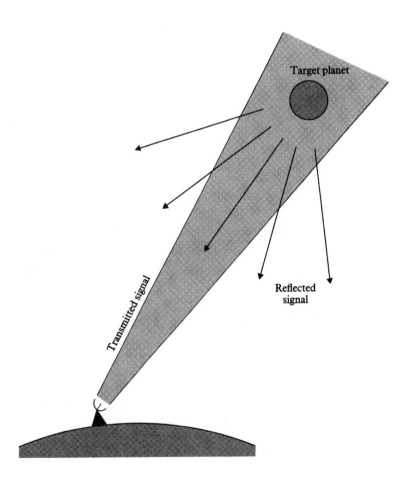

Figure 5.16 A radar transmitter sends a powerful signal into an Earth-based antenna (lower left). This antenna focuses the microwave signal into a narrow beam and directs it toward the target. The target in turn reflects the signal, and a small fraction of this reflected radiation reaches the terrestrial receiver (usually a large radio telescope, often the same instrument as the transmitting antenna).

Target planet

Transmitted signal

Reflected signal

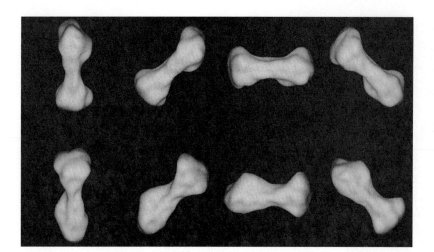

Figure 5.17 Radar images of the metallic main belt asteroid Kleopatra. The "dog-bone" shape is presumably the result of past collisions, in which fragments settled back into this strange shape after a collision had disrupted the asteroid.

three-dimensional images that sometimes rival the resolution of spacecraft cameras.

By 2001, radar had detected more than 120 asteroids, with sizes as small as 30 m. These include large objects in the main belt as well as more than 80 of the smaller NEAs. One of the early radar contributions was to search for direct evidence of metallic surfaces for a few asteroids from their high microwave reflectivity. Observations of M-type asteroids Psyche and Kleopatra provide the best evidence linking the M class with metallic composition. However, these two asteroids have provided numerous surprises. In spite of its apparently metallic surface, Psyche has a density of only about 2 g/cm³, suggesting that its interior has extremely high porosity if composed of metal. Kleopatra is even stranger, with a remarkable "dog-bone" shape that suggests reaccretion of material following a catastrophic impact (Fig. 5.17). There is also evidence of a low-density surface of unconsolidated rubble on Kleopatra—again not what we would have expected by comparison with the lumps of iron-nickel in our meteorite collections.

The best resolution imaging has been achieved for Toutatis, a 5-km-long NEA (Fig. 5.18). It was discovered by French astronomers in 1988 and named for a popular French cartoon character. In December 1992, Toutatis passed within 3 million km of the Earth—less than ten times the distance to the Moon. The asteroid is elongated and lumpy, which suggests that two or more preexisting frag-

ments may have collided at low speed and stuck together. These observations also reveal that Toutatis has a wobbling spin, consisting of two apparent simultaneous rotations around different axes, with periods of 5.4 and 7.3 days. Radar images of a smaller NEA called Castalia are even stranger (Fig. 5.19). Castalia consists of two objects of similar size, barely in contact as they spin about each other. Other NEAs have been found to have satellites not much larger than an office building.

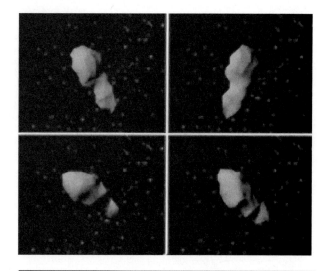

Figure 5.18 The near-Earth asteroid Toutatis in views reconstructed from radar images by Steven Ostro and Scott Hudson. The asteroid is 5 km long and is shown at a resolution of about 100 m. The narrow neck suggests that two originally separate asteroids collided at low velocity to produce this composite object.

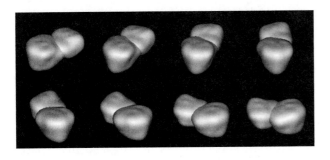

Figure 5.19 Looking like a double dinner roll, the asteroid Castalia is shown here as modeled by a computer from radar observations of the object during a close (5.6 million km) approach to the Earth. The asteroid is 1.8 km across at the widest point and appears as a "contact binary."

THE SATELLITES OF MARS

The moons of Mars are captured asteroids. After Galileo Galilei's discovery of the four large satellites of Jupiter, scientists began to speculate about moons orbiting other planets. Since Venus had no satellite, Earth had one, and Jupiter had four, some concluded that Mars, the planet between Earth and Jupiter, should have two (an example of numerology, not science). Johannes Kepler was one of those who suggested the existence of two martian satellites, and these hypothetical objects were widely publicized when Jonathan Swift included them in his popular satire *Gulliver's Travels*, published in 1726. Swift wrote that the astronomers of his mythical Laputia had discovered

> *two satellites, which revolve about Mars, where-of the innermost is distant from the center of the primary planet exactly three of the diameters, and the outermost five; the former revolves in the space of ten hours, and the latter in twenty-one and a half; so that the squares of their periodical times are very near in the same proportion with the cubes of their distance from the center of Mars. . . .*

Note that Swift is explaining Kepler's third law to his readers! It was to be another 150 years before these fictional moons were actually discovered, however.

The martian satellites were among the last important planetary discoveries made by the human eye looking through a telescope. After many nights of careful searching, the American Asaph Hall spotted both satellites during the favorable martian opposition of 1877. They were named Phobos (fear) and Deimos (panic), for the horses that pulled the chariot of Ares (Mars) in Greek mythology.

Phobos, the larger of the two satellites (about 21 km in diameter), is closest to Mars; its distance (from the center of the planet) is 9378 km and its orbital period is 7 hr 39 min. Deimos, which is about half the size of Phobos, is 23,459 km from Mars and has an orbital period of 30 hr 18 min (both of them surprisingly similar to Swift's fantasy). Because these satellites seem to have so little in common with Mars, and because scientists would not expect the processes that formed Mars to generate two small satellites as well, these moons are assumed to have been captured from elsewhere.

To be captured, a passing asteroid must be slowed near the planet, losing energy and falling into orbit—the energetic equivalent of firing a retrorocket on a spacecraft. Probably the capture of Phobos and Deimos took place while Mars still had an extended early atmosphere, and friction with this gas provided the capture mechanism. We can imagine many asteroids approaching Mars while the atmosphere was denser and being slowed so much that they crashed into the surface. Later, the atmosphere was too thin to affect passing asteroids. Thus, Phobos and Deimos represent just two of many potential captured bodies, the two that happened to come along at the right moment in the evolution of the early atmosphere of Mars.

Both martian satellites were photographed by the U.S. Viking spacecraft in 1977. They are dark, cratered, irregular objects about the same size as Gaspra. Their colors are similar to those of C-type asteroids and of carbonaceous meteorites. Their densities (based on masses derived from close fly-bys of the Viking spacecraft) are only 2.0 g/cm³, lower than that of rock and suggesting that the interior is porous.

Figure 5.20   Seen in this close-up view, the surface of Phobos is marked with a series of linear grooves that appear to be surface fractures caused by the impact that formed the largest crater on this satellite.

Close-up views of Phobos show that the surface is covered by parallel grooves or valleys, typically several hundred meters across and several tens of meters deep (Fig. 5.20). Apparently the grooves are fractures related to the impact that formed Stickney, the largest crater on Phobos (10 km in diameter). Indeed, this crater is so large relative to the size of the satellite that its formation must very nearly have broken Phobos apart. Like the other asteroids that have been visited by spacecraft, these two small moons are battered remnants that show the scars of a difficult history, dominated by collisions that have very nearly destroyed them.

5.5 QUANTITATIVE SUPPLEMENT:
ASTEROID COLLISIONS

Although there are about 100,000 asteroids larger than 1 km diameter in the main belt, they are widely spaced and collisions are rare. It is quite easy to make some approximate calculations to estimate collision frequencies.

First let us estimate the average spacing between asteroids. The main asteroid belt can be considered a volume stretching around the Sun from 2.2 to 3.3 AU and about 0.2 AU thick. Its volume (circumference × width × height) is thus about 4 AU^3, or about 10^{25} km^3. Dividing this total volume by the number of asteroids gives 10^{20} km^3 per asteroid. The average spacing is approximately the cube root of this, or 5×10^6 km. Thus, it should come as no surprise that spacecraft crossing the asteroid belt rarely come within a million kilometers of any asteroids unless they are specifically targeted to do so.

We can now calculate the frequency of collisions. Observations have shown that the typical asteroid has a speed relative to its neighbors of about 4 km/s, resulting from their different orbital eccentricities and inclinations. Take an asteroid with diameter 10 km as average. Its cross section (given by πR^2) is about 100 km^2, and it thus sweeps out a volume in space relative to its neighbors equal to its cross section times its speed, or 400 km^3/s. Since there are 3×10^7 seconds in a year, the volume swept out each year is equal to about 10^{10} km^3.

To estimate the frequency of collisions, we compare the volume swept out by each asteroid in a year with the volume of 10^{20} km^3 associated with each asteroid, which we calculated earlier. The ratio is 10 billion years, older than the solar system. This tells us that most asteroids do not collide with others, even over billions of years, if there are 10^5 asteroids larger than 1 km in diameter. However, there is some current evidence (based in part on the heavy cratering observed on asteroid Gaspra) that there are as many as 10^6 asteroids of this size. In that case, you can see that collisions would be ten times more frequent and nearly every 1-km asteroid would have suffered several such catastrophes.

Finally, we can estimate how often a collision of the sort just described takes place in the asteroid belt today. This is simply the probability of an impact for one asteroid in 1 year (10^{-10}) times the number of main belt asteroids (10^5). The answer is one collision every 100,000 years. This frequency is sufficient to generate a great deal of dust and meteoritic fragments, and such collisions are probably responsible for the prominent asteroid dust bands discovered by the IRAS (Infrared Astronomy Satellite).

Summary

Asteroids are interesting in their own right as members of the planetary system and also as possible parent bodies for the meteorites. Because they are much smaller than the planets, many (but not all) of the asteroids have remained relatively unmodified since the origin of the solar system.

Most of the asteroids are found in the main asteroid belt, between the orbits of Mars and Jupiter. In this location, and in the Trojan Lagrangian clouds that lead and trail Jupiter in its orbit, are the only stable orbits where small objects could have survived for the lifetime of the solar system. Even though their orbits are stable, however, there have been many collisions among asteroids over the history of the solar system, and the asteroids we see today are a fragmented remnant of the original population of small bodies.

The largest asteroid (Ceres) is only 1000 km in diameter, and the total mass of the asteroids is only about 1/20 the mass of the Moon. Asteroids are classified by analogy with the meteorites as C, S, and M types. Indeed, the asteroids appear to be similar in composition to many of the meteorites. The asteroids, like the meteorites, include both primitive bodies (which are in the majority) and differentiated objects. The best-known differentiated asteroid is Vesta, which has a basaltic surface and is probably the parent body of the eucrite meteorites.

Four asteroids have been observed at close range from spacecraft: the main belt S-type asteroids Gaspra and Ida, the C-type asteroid Mathilde, and the near-Earth S-type asteroid Eros. We have also studied two small satellites of Mars, Phobos and Deimos, which are probably captured asteroids. All are irregular, cratered, and apparently the product of catastrophic fragmentation of their parent objects. The NEAR-Shoemaker spacecraft landed on Eros and verified that it has a primitive composition like the chondrite meteorites. Radar imaging of near-Earth asteroids has revealed both asteroid satellites and binary asteroids, with the two fragments rotating in contact with each other. Many asteroids have low densities, indicating that they are heavily fractured or, in a few cases, are rubble piles of loosely consolidated fragments.

Near-Earth asteroids are mostly derived from the main belt, but probably some may also be extinct comets. They have relatively short lifetimes (100 million years), and nearly one-quarter of those with Earth-crossing orbits will eventually strike our planet. The largest impacts are capable of causing global environmental catastrophe and thus influencing the evolution of life on Earth.

Key Terms

C-type asteroid	radar
family of asteroids	resonance
M-type asteroid	resonance gap
main belt asteroids	S-type asteroid
near-Earth asteroid (NEA)	Trojan asteroids

Review Questions

1. Describe the geography of the main asteroid belt and the kinds of objects that populate it.

2. Summarize the kinds of basic information that can be obtained for an asteroid from telescopic studies. Compare this information with that available for various planets and satellites in the planetary system.

3. What is meant by a size-frequency distribution in which the number of objects is proportional to $1/D^2$? Compare this with the size-frequency distribution of people in a shopping mall crowd, or of trees in a forest.

4. Distinguish between main belt asteroids and near-Earth asteroids in terms of their orbits and their physical and chemical properties.

5. Summarize the evidence that meteorites come from asteroids, and indicate which kinds of asteroids are the probable parent bodies of the different kinds of meteorites.

6. Explain the phenomenon of resonance gaps in the asteroid belt. Where else in the planetary system would you expect similar resonance effects to be seen?

7. Compare Ida and Mathilde in size and shape. They seem to have very different internal structures. Explain.

8. Compare the near-Earth asteroid Eros with the main belt asteroids Gaspra and Ida. Describe their probable history and future fate.

QUANTITATIVE EXERCISES

1. Use Kepler's laws to find the asteroid semimajor axes that correspond to resonances with Jupiter—that is, to periods that are 1/2, 2/5, 1/3, and 1/4 that of Jupiter.

2. Ceres has a diameter of 940 km. How does its volume compare with that of the Earth and Moon? How many objects the size of Ceres would you expect to find if the asteroids had originated from the breakup of a planet the size of the Earth, if all the results of this explosion were the size of Ceres?

3. Use reasoning similar to that in Section 5.5 to estimate the probability that a spacecraft crossing the asteroid belt will collide with an asteroid 1 km in diameter or larger. Use the fact that there are about 10^{20} km^3 per asteroid, and assume that the spacecraft has an area (cross section) of 10 m^2 and that it spends 6 months crossing the belt at an average speed of 8 km/s.

4. Use the same reasoning as in Exercise 3 to estimate how close the spacecraft will pass an asteroid of 1 km in diameter during its 6-month trip through the belt.

ADDITIONAL READING

Chapman, C. R. 1982. *Planets of Rock and Ice*, Chapter 3. New York: Scribner's. Well-written discussion by a leading asteroid researcher (somewhat out of date).

Kowal, C. T. 1988. *Asteroids: Their Nature and Utilization*. New York: John Wiley & Sons. Discussion that includes the potential of asteroids as resources for future development in space.

Steel, D. 2000. *Target Earth: The Search for Rogue Asteroids and Doomsday Comets That Threaten Our Planet*. Pleasantville, NY: Reader's Digest. Well-illustrated and up-to-date account of the danger of asteroid impacts.

6

COMETS: MESSENGERS FROM THE COLD, DARK PAST

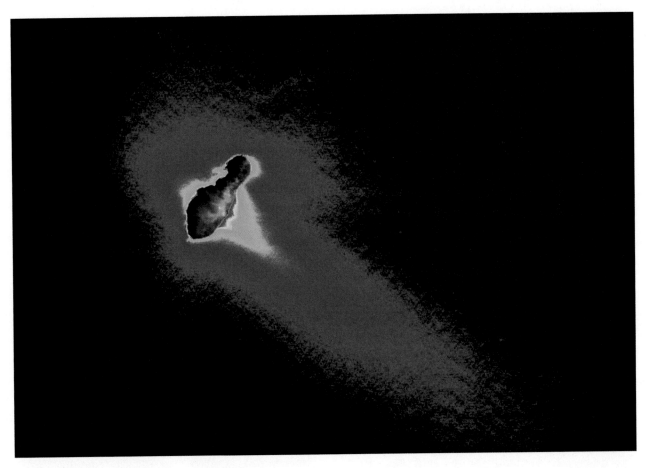

The nucleus and atmosphere of Comet Borrelly in a composite image obtained by the NASA DS-1 spacecraft.

Comets are small primitive objects like the asteroids, but their composition is fundamentally different, indicating a different origin. Comets contain a substantial quantity of water ice and other frozen volatiles. When a comet approaches the Sun, these ices evaporate to form a tenuous, transient atmosphere. It is this extensive but short-lived cloud of gas and dust that makes the comet readily visible and forms its characteristic long tail. (*Komitis* in Greek means "longhaired," "feathered," or "a comet.") In order to have preserved these volatiles in the frozen state, the comets must have formed at low temperature and must have spent most of their existence in the cold, dark outer reaches of the Sun's gravitational field.

The comets and asteroids provide complementary information about the formation of the solar system. As we saw in Chapter 5, asteroids are remnants of material that formed in the inner part of the solar nebula and became the building blocks of the inner planets. In contrast, the comets represent a much larger reservoir of material from colder parts of the solar nebula, more closely related to the outer planets and their systems of satellites and rings.

6.1 COMETS THROUGH HISTORY

Comets have been known since antiquity. Approximately one comet visible to the unaided eye appears each year, and a moderately bright comet comes along an average of once per decade. Written records of these long-tailed wanderers go back to at least 1140 B.C.E. in the Middle East and nearly as far in China.

Perhaps because their appearances were unpredictable and their forms were variable, comets have frequently been regarded with apprehension (Fig. 6.1). In the past, the apparition of a comet was sometimes believed to herald some remarkable event on Earth, usually unfavorable. Thus Shakespeare wrote in *Julius Caesar*, "When beggars die, there are no comets seen / The heavens themselves blaze forth the death of princes." Milton characterized Satan as a comet that "from its horrid hair / Shakes pestilence and war." No one knows how comets acquired such a bad reputation, but particularly during the medieval period in Europe, nothing good was associated with them. In Asia, however, comets were treated more like other celestial phenomena, associated with good as well as evil omens.

The scientific study of comets may be traced back 2000 years to the Greeks and Romans. Because of their belief that the celestial realm was unchanging, most of the classical philosophers thought that comets were located in the atmosphere, and efforts were made to use them to predict the weather. As late as the sixteenth century, the scientists of Renaissance Europe were arguing over whether comets were astronomical or meteorological phenomena.

APPEARANCE OF A COMET

These days people who read about the great attention afforded comets throughout history are often disappointed when they actually see a comet. Most of these icy messengers appear as dim, fuzzy patches of light smaller than the Moon in apparent size and a great deal fainter (Fig. 6.2). Nor is the comet distinguished by its motion, which is imperceptible unless you watch for many hours. In photos, comets may look as if they are moving rapidly, and this is the way they are often depicted in TV shows, leading to a common confusion between comets and meteors (shooting stars). In real life, comets are faint and essentially stationary. To

Figure 6.1   Representation of Comet Halley on the Bayeux Tapestry commemorating the conquest of England in 1066. For King Harold of England (shown seated) the comet was a bad omen, but presumably William the Conqueror felt differently about it. The Latin words say "They marvel at the star."

Figure 6.2   Comet Halley as seen from Mauna Kea in the spring of 1986. Like most bright comets, it appeared as a small nebulous patch of light with a tail. Note the silhouette of one of the observatories. This photograph is a good approximation of a naked-eye view.

see one at all, you must escape the smog and light pollution of our cities, but even under a dark sky you will probably need binoculars for a good view, unless the comet is unusually bright.

When we see a comet, we are observing primarily the extended atmosphere that is its trademark. The small solid body from which this atmosphere is released is called the **nucleus.** This is the real heart of the comet, even though it is too small to be seen. The atmosphere that surrounds the nucleus is called the head or **coma** of the comet. The long streamers of less substantial gas and dust sweeping away from the Sun are called the **tail.** All comets have a nucleus and coma, and all except the faintest have tails.

Comets seem less important to us today than they did to our ancestors, primarily because most of us no longer feel that celestial events have any effect on our lives (unless something smashes into the Earth!). Furthermore, we no longer know the sky the way our ancestors did, so a comet has to be very bright to attract our attention. Nevertheless, a bright comet in a dark, star-filled sky is a spectacular sight, as the millions who witnessed Hyakutake and Hale-Bopp in 1996 and 1997 can attest (Fig. 6.3). Despite our fond hopes for the growth of rationality, a tiny cult of people in southern California actually committed suicide on that occasion, in the belief that Comet Hale-Bopp was a spaceship, which their souls would join for a great cosmic voyage.

TYCHO AND HALLEY

A turning point in cometary studies came with the Great Comet of 1577, which was extensively investigated by Tycho Brahe, the greatest of the pretelescopic astronomical observers. He carefully measured the position of the comet against the background of stars, finding that it did not shift back and forth with the change in viewing angle afforded by the Earth's rotation, as would a nearby object. (This is an example of parallax, discussed in Section 2.1.) Thus, Tycho concluded that this comet was not in the atmosphere of the Earth but was well beyond even the Moon.

Figure 6.3 Comet Hale-Bopp. This photo shows the straight pale blue plasma tail as well as the brighter, curved dust tail of this remarkably bright comet.

At about the same time that Tycho was studying the comet of 1577, others demonstrated that comet tails always point away from the Sun, rather than back along the direction of motion. After Johannes Kepler's elaboration of the heliocentric motion of the planets, it was quickly realized that comets also were bound to the Sun, although their orbits were so eccentric that they were essentially parabolic rather than nearly circular ellipses.

One astronomer whose name has been immortalized for his work on comets was Edmund Halley, a contemporary of Newton in late seventeenth-century England (Fig. 6.4). It was Halley who first realized that cometary orbits could be closed ellipses, with a given comet reappearing at regular intervals. He reached this conclusion from a study of the Great Comets of 1531, 1607, and 1682, all of which had similar orbits. Halley concluded that these were successive appearances of the same comet, which he predicted would return in 1758.

Although Halley did not live to see his calculations verified, the comet returned just as he had expected, and it was quickly hailed as Halley's Comet,

Figure 6.4   British astronomer Edmund Halley, who successfully predicted the return in 1758 of the comet that now bears his name.

a name it retains today. Comet Halley, with its 76-year orbit, is the brightest of the periodic comets. It appeared twice during the twentieth century, in 1910 and in 1986. Not until 1820 was a second periodic comet identified: the much fainter Comet Encke, with a period of only 3.3 years.

Understanding the physical nature of comets has proved even more difficult than pinning down their elusive motions. Until Tycho's work on the Great Comet of 1577, comets were believed to be "exhalations" in the atmosphere of the Earth. Tycho's determination that this comet was beyond the Moon implied that the visible head of the comet was very large, comparable in size to the Earth. Soon astronomers were measuring cometary tails millions of kilometers in length, and public fear of plagues from comets was augmented by fear of collision with them.

COMETS IN THE NINETEENTH AND TWENTIETH CENTURIES

The fact that stars could be seen easily through the cometary atmosphere showed that in spite of their large size, comets are mostly rather tenuous objects. Isaac Newton was among the first to suggest that most of what was seen was thin gas expelled from a solid nucleus as it was heated by the Sun. Nineteenth-century astronomers also correctly attributed the streaming of the tail away from the Sun to the force of solar radiation on the insubstantial atmosphere of the comet.

During the second half of the nineteenth century, public interest in comets throughout the United States and parts of Europe provided a substantial boost to astronomy and led to the endowment of a number of observatories. In 1910 Comet Halley was particularly well placed for viewing, and the Earth even passed through the tail of the comet on May 20 of that year. A few months earlier, an even brighter new comet had appeared, making 1910 a memorable year for comet watchers. However, the 1986 return of Halley was a disappointment in comparison with 1910. The comet was very poorly placed; when it was closest to the Sun and therefore brightest, it was located in its orbit on the opposite side of the Sun from the Earth.

The primary interest in the 1986 return of Comet Halley did not depend on ground-based observations. The predicted arrival of this famous comet allowed the USSR, the European Space Agency, and Japan to send instrumented spacecraft to fly into the comet a few weeks after perihelion. Later in this chapter we will discuss the results from these missions.

In spite of the difficulty of seeing comets in our modern world, a devoted group of comet hunters continues to discover new ones each year, as well as to greet the return of periodic comets. This interest is encouraged by the International Astronomical Union policy of naming comets for their discoverers, a custom that began about a century ago. Comet Hale-Bopp, for example, is named for two American amateur astronomers. Up to three independent discoverers can be recognized, which leads to some awkward names, such as Comet Honda-Mrkos-Pajdusakova or Comet IRAS-Araki-Alcock, named for the IRAS satellite as well as astronomers Araki and Alcock. Just before Comet Hale-Bopp got close enough to be visible to the unaided eye, the other bright recent comet, Comet Hyakutake, made a sudden appearance in late 1996. There was no time to prepare spacecraft to investigate these unexpected visitors, but extensive observations from telescopes on Earth and in orbit greatly increased our knowledge about comets, as we shall see.

ORBITS OF COMETS

Comets are classified on the basis of their orbits as **long-period comets** (greater than 200 years), intermediate-period or Halley-class comets (30–200 years), and **short-period comets** (less than 30 years). Since the intermediate-period comets (of which only a handful exist) apparently have the same origin as the long-period comets, we will combine them in a single class. Most long-period comets are on extremely eccentric orbits, falling in toward the Sun from great distances and returning again to the depths of space. Their ellipses are so elongated that they can be approximated very well by a parabola with the Sun at one focus, with corresponding orbital periods of a million years or more. Of the nearly 1500 comets discovered through 2001, 90% are long-period comets. Most of the rest are short-period comets with aphelia near the orbit of Jupiter; they are sometimes called Jupiter-family comets. Table 6.1 lists some famous comets.

Table 6.1 Some famous comets

Name	Period	Significance
Great Comet of 1577	long	Found by Tycho to be farther away than the Moon
Great Comet of 1843	long	Brightest ever (visible in daylight)
Daylight Comet of 1910	long	Brightest of twentieth century
Hyakutake	long	Bright because of close approach to Earth (1997)
Hale-Bopp	long	Brightest recent comet (1997)
Swift-Tuttle	133.0 yr	Parent comet of Perseid meteors
Halley	76.0 yr	Studied by multiple flyby spacecraft (1986)
Chiron	51.0 yr	Largest nucleus (formerly designated as an asteroid)
Biela	6.8 yr	Broke apart and never returned
Borrelly	6.8 yr	Flyby by Deep Space 1 spacecraft (2000)
Wirtanen	6.7 yr	Target of Rosetta mission (2012)
Giacobini-Zinner	6.5 yr	First spacecraft encounter (ICE in 1985)
Wild 2	6.4 yr	Target of Stardust sample return mission (2004)
Tempel 1	5.7 yr	Target of Deep Impact mission (2005)
Encke	3.3 yr	Shortest period
Shoemaker-Levy 9	—	Broke up (1992) and crashed into Jupiter (1994)

Because the paths of comets cut across those of the planets, comet orbits are inherently unstable. Like the Earth-approaching asteroids, they must be coming from elsewhere. As we will discuss later in the section on cometary evolution, scientists have identified two sources. One is a very large diffuse cloud of comets at the outer fringe of the solar system, containing perhaps a trillion (10^{12}) comet nuclei, known as the Oort comet cloud. It is the source of the long-period comets. The short-period comets are thought to originate in a smaller disk of comets beyond the orbit of Neptune called the Kuiper belt. The comets we see represent the tiny fraction that leak into the inner solar system from these two immense reservoirs.

6.2 THE COMET'S ATMOSPHERE

The visible part of a comet is its *atmosphere*, a term we use to include both the coma and the tail. Since most astronomical studies of comets deal with the atmosphere, we discuss it first before turning to the relatively tiny nucleus that is the source of the atmospheric gas and dust.

THE COMA

The brightest part of a comet is the inner coma, which consists of gas and dust recently ejected from the nucleus (Fig. 6.5). Sometimes there appears to be a starlike condensation of bright material at the center, fooling observers into thinking they are seeing the nucleus itself, when actually it is only the brightest inner part of the coma, hundreds to thousands of kilometers in diameter. Sometimes this inner coma is symmetric about the nucleus, but more often it is brightest in the direction of the Sun, and frequently it displays structure in the form of fans of denser material apparently streaming away from gas jets on the nuclear surface. We can see these jets very clearly from the pictures of the nucleus of short-period Comet Borrelly, recorded by the spacecraft Deep Space 1 (Fig. 6.6).

Figure 6.5   The head of Comet Halley as photographed on May 8, 1910.

Figure 6.6   Photo of Comet Borrelly, recorded by the Deep Space 1 spacecraft in 2001. This comet is smaller and more fully outgassed than Halley, yet it shows distinct jets of escaping gas and dust.

The gases released from the nucleus are quickly broken down by ultraviolet sunlight to create such molecular fragments as OH, CH, and NH. Most of these reactions are straightforward and easy to calculate, but others, such as the source(s) of the carbon molecules C_2 and C_3, remain mysterious. A great deal of complex chemistry takes place in the inner atmosphere of a comet, within minutes of the release of gas from the surface, and not all of it is understood.

Near the surface of an active comet, the gas density is only about a millionth as great as the density of the Earth's atmosphere, and this density falls off rapidly with distance as the gas streams away from its source. At distances as far as 10,000 km, the gas flows smoothly; beyond this range, the individual molecules cease interacting with each other and with the embedded dust. Surrounding the rest of the atmosphere is a cloud of glowing hydrogen atoms that can extend more than a million kilometers from the nucleus, briefly making the comet larger than the Sun. Beyond a hundred thousand kilometers, however, most of the gas molecules become ionized, and this plasma is swept away by the blast of the solar wind.

COMPOSITION OF THE GAS

Spectroscopic studies of comets generally refer to the inner coma, which is the source of most of the light from a comet. Astronomers measure the composition of this gas and then try to work backward to identify the parent materials in the nucleus that gave rise to the molecular fragments found in the coma. The central role of water ice in comet chemistry has been confirmed by observations of the spectra of various comets. When water vapor is released from the nucleus, it is quickly ionized to become H_2O^+, and it is then broken down by sunlight into its constituent parts: hydrogen (H) and hydroxyl (OH). Emission by OH molecules is a common feature in spectra of bright comets. In recent years, both H and OH have been detected from observatories in space, while emission from H_2O^+ plasma has been detected optically and measured by radio astronomers. Moreover, in 1986 neutral water itself was detected in the atmosphere of Comet Halley and was subsequently observed in Hale-Bopp, Hyakutake, and other comets.

In addition to water, emissions have been seen from carbon-containing gases such as C_2, C_3, CO, CO_2, CH_4 (methane), and CH_3OH (methanol). These emissions indicate the presence of additional ices, consisting of both hydrocarbons (compounds containing hydrogen and carbon) and oxides of carbon. It is unusual to find both reduced and oxidized compounds of the same element on the same object, since simple chemical reactions tend to convert molecules from one extreme to the other, just as methane released by bacteria on Earth is quickly oxidized to CO_2. The preservation of these compounds on comets confirms the low-temperature history of these objects: Chemical reactions are strongly inhibited at temperatures below 40 K.

Nitrogen compounds, including N_2, HCN, NH_3, and CH_3CN, have also been detected in cometary spectra, as well as several sulfur compounds. HCN is hydrogen cyanide, a gas that is lethal to humans; the deduction that HCN was in the atmosphere of Comet Halley frightened people in 1910 when they were told that the Earth would pass through the comet's tail. Of course,

Table 6.2 Atoms and compounds observed in comets

Metals	Other Atoms	Diatomics	Triatomics	Polyatomics
Na	H	CH	H_2O	NH_3
K	C	CO	CO_2	CH_4
Ca	N	CN	NH_2	C_2H_2
V	O	C_2	HCN	C_2H_6
Cr	S	CS	C_3	H_2CO
Mn		OH	HCO	HCOOH
Fe		NH	HNC	$HCOOCH_3$
Co		S_2	H_2S	CH_3OH
Ni		CN	SO_2	CH_3CN
Cu		SO	OCS	HC_3N
			CS_2	HNCO
				H_2CS
				NH_2CHO

the density of the tail is so low that there was absolutely no danger. A summary of constituents detected in cometary atmospheres (as of 2002) is given in Table 6.2.

COMET TAILS

The tenuous gases that make up the comet's atmosphere are not gravitationally bound to the nucleus. As they expand, these gases are ionized and caught up by the solar wind to assume the characteristic wispy form of the cometary tail. This tail is called a **plasma tail.** We see plasma tails primarily by the blue glow of ionized carbon monoxide (CO^+) stimulated to **fluorescence** by absorption of ultraviolet light from the sun. (Fluorescence is the emission of light that was absorbed at wavelengths different from those at which it is emitted. The most common example is the fluorescent light, in which a substance coated on the inside of a glass tube emits visible light after absorbing ultraviolet light from an electric discharge within the tube.) Other ionized gases such as water vapor, carbon dioxide, and molecular nitrogen have also been detected. Plasma tails are straight, point away from the Sun, and usually are made up of individ-

ual streamers or rays only a few thousand kilometers across. An active comet such as Hale-Bopp can develop a plasma tail 100 million km long, approaching the distance between the Sun and the Earth (see Fig. 6.3).

Within a plasma tail the actual density of material is incredibly low, typically only a few hundred molecules per cubic centimeter. Thus, a giant tail tens of millions of kilometers long contains no more molecules than a modern supertanker—perhaps half a million tons of mass. For most comets, the mass of material in the tail is far smaller yet.

Since the plasma tail is driven by the solar wind, it provides a kind of celestial wind vane to record the speed and direction of flow. By tracking individual kinks and twists as they move downstream, astronomers can measure properties of the solar wind. Typical solar wind velocities determined from comets are about 400 km/s, sufficient to traverse the entire length of the comet tail in less than a day. It is not surprising, therefore, that plasma tails can alter their form from hour to hour and that the overall appearance of a comet can change dramatically from one night to the next (Fig. 6.7).

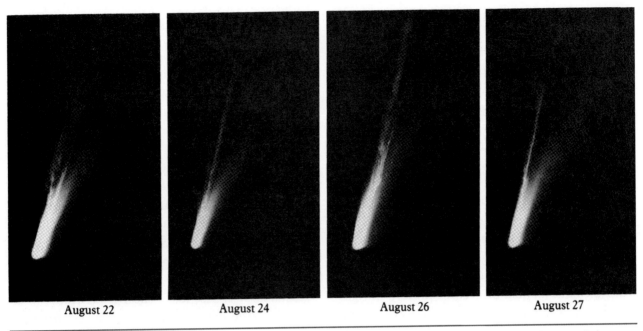

| August 22 | August 24 | August 26 | August 27 |

Figure 6.7   Comet Mrkos photographed during six nights in 1957. Note the large changes in the plasma tail that occur from night to night, while the curved dust tail remains relatively unchanged.

Most comets have two tails: the plasma tail, which we have been discussing, consisting of charged molecules caught in the solar wind, and a second tail composed of dust grains, appropriately called the **dust tail**. Although generally shorter than plasma tails (usually less than 10 million km long), dust tails can be as bright or even brighter. They are readily distinguished by their color (yellow-white, from reflected sunlight) and from the fact that they are curved rather than straight.

Dust tails consist primarily of small grains only a few micrometers in size, indicating that much of the solid material in the nucleus must be of the same consistency, similar to the finest cake flour. Once decoupled from the expanding and thinning gas, the dust grains follow their own orbits, moving under the joint influence of solar gravity and of pressure generated by solar radiation. Because they move relatively slowly, the dust grains tend to mark the location of the comet at the time of their release. Thus, the dust tail traces the path of the comet across the sky. Measurements of the material in dust tails suggest a mass that is roughly comparable to the gas in the

plasma tail, indicating that similar quantities of ice and dust are present in the nucleus.

6.3 THE COMET'S NUCLEUS

The most important part of a comet is its nucleus; the rest is insubstantial show. We know a nucleus exists because the transient gases of the comet's atmosphere must come from somewhere. Telescopic observers cannot see the nucleus directly, however, since it is relatively tiny and is lost in the glare of the head of an active comet. Only recently has radar penetrated the atmospheres of several comets to provide unambiguous detection of their nuclei, while the spacecraft that flew past Comet Halley in 1986 and Comet Borrelly in 2001 were able to photograph the nuclei of those comets at close range.

THE DIRTY-SNOWBALL MODEL

Before we recognized that comets contain a great deal of water ice, which evaporates under solar heating to generate the atmosphere, some scientists thought that the nucleus consisted of a loose agglomeration of grains and boulders similar in composition to the primitive meteorites. Because this meteoritic material contains limited quantities of trapped gas, it was necessary to hypothesize a great number of particles to account for the observed gas production near the Sun.

In 1950, however, Fred L. Whipple of Harvard University made a fundamental step forward when he developed a model for the nucleus as a single, relatively small "dirty snowball," made up of roughly equal quantities of silicates and ice (Fig. 6.8). Whipple suggested that the rock (or dust) and ice were intimately mixed, as might be expected for a primitive body that formed directly from the solar nebula or even from dust and gas clouds in interstellar space. However, several comets have produced more dust than gas, leading to their characterization as "icy dirt balls" rather than "dirty snowballs." These variations from comet to comet suggest both different places of origin and different evolutionary histories, as we shall see.

Figure 6.8 Fred Whipple (left), the originator of the dirty-snowball model for comet nuclei, listening to Roald Sagdeyev (right), the director of the VEGA mission to Comet Halley and of the USSR Institute for Space Research, as the data from the 1986 comet flyby were received at Earth. The VEGA pictures provided the first visual evidence that Whipple's model—conceived 36 years earlier—was indeed correct.

COMPOSITION OF THE NUCLEUS

At first very little was known about the composition of the nucleus, although Whipple concluded that water (H_2O) ice was likely to be the major constituent. This hypothesis was supported by observations that most cometary activity turns on as the approaching comet reaches about 3 AU from the Sun. At this distance we can calculate that the surface temperature, produced by the steadily increasing intensity of sunlight, should reach the point (about 210 K) at which water ice begins rapid evaporation. Other ices, with different evaporation rates, would begin activity at different temperatures and hence at different distances from the Sun. A comet like Hale-Bopp, which developed a large coma at more than 5 AU from the Sun, must contain other more volatile materials than H_2O, such as CO and N_2.

We have some direct measurements of the gases evaporating from the nucleus of Comet Halley, obtained by the VEGA and Giotto probes within a few thousand kilometers of the nucleus itself. In addition, there are spectroscopic measurements of the atmospheres of many comets (see Table 6.2). It is clear that water ice is the predominant volatile component, but that a few percent of the ice also consists of frozen carbon monoxide (CO), carbon dioxide (CO_2), and methyl alcohol (CH_3OH), plus traces of hydrocarbons and other compounds. One possible model for the ice chemistry of the nucleus is illustrated in Table 6.3.

In addition to the volatiles listed in Table 6.3, the cometary nucleus contains substantial quantities of dark carbonaceous and silicate dust. The 1986 spacecraft measurements of Comet Halley showed that, for this comet at least, carbon and hydrocarbon dust predominates over silicates. While no dust measurements were made of Comet Borrelly, the images showed that the surface of the nucleus was extremely dark, reflecting only 3% of the incident sunlight. The darkest spots on Comet Borrelly have an albedo of less than 1%, the lowest ever measured (darker than a lump of coal or the toner in a photocopy machine). Thus, this comet is surely covered by carbon-rich material as opposed

Table 6.3   Volatile abundances in comets

Molecule	Abundance (percent by mass)
H_2O	65–80
CO	5–20
CO_2	2–10
CH_3OH	2–10
CH_4	<1
NH_3	<1
HCN	<1
H_2S	<1
Hydrocarbons	<1

Adapted from a 1993 summary by M. Mumma, P. Weissman, and A. Stern. (*Protostars and Planets III*, ed. E. Levy and J. Lunine, University of Arizona Press, 1993.)

to silicate dust or ice. Therefore, the evaporating ices that produce the jets from these nuclei must originate below their surfaces. As the ices evaporate, they also release the dust particles, which stream into space carried along by the flow of gas. This dust contributes to the cometary tail and is the primary source of the meteors that flash through our skies when these particles encounter the Earth.

It is possible that some of the carbonaceous and silicate material in comets predates the origin of the solar system, having originated in the interstellar dust before the solar nebula formed. Many of the known or suspected volatiles in comets have been identified by radio astronomers in interstellar "molecular clouds" throughout our Galaxy. Thus, some of the frozen cometary volatiles may be unaltered materials from such interstellar sources. If this idea is correct, it implies that comets may be vehicles for carrying molecules created out among the stars to the surfaces of satellites and planets throughout the solar system. This includes the possibility of bringing water and organic material to the early Earth.

PHYSICAL NATURE OF THE NUCLEUS

How large is the nucleus of a comet? For most comets only a rough estimate can be made. Once the atmosphere of a comet forms, the tiny nucleus is lost in the midst of the bright dust and gas cloud that forms the coma, and its size cannot be measured. The only cometary nuclei that have been photographed from close-up are those of Halley and Borrelly.

Another way to penetrate through the obscuring gas and dust is to use radar. The first radar signal was bounced from the nucleus of Comet Encke in 1980, and in 1983 radar data were used to derive a diameter of 5–10 km for the faint comet IRAS-Araki-Alcock. Radar observations not only can determine the diameter of the nucleus but also have been used to detect clouds of large particles (boulders?) surrounding the nucleus. Whether any of this material is represented in our meteorite collections remains an open question.

In 1986, the European Giotto craft used information transmitted from the pathfinder VEGA spacecraft (which preceded it) to target its flyby within 500 km of the nucleus of Comet Halley. Thus, the Giotto cameras pierced the dust fog and imaged the nucleus itself, an irregular dark mass about $16 \times 8 \times 8$ km in size (Fig. 6.9). The reflectivity of this nucleus was only 3%–4%, similar to that of the darkest asteroids.

The flyby of Comet Borrelly in September 2001 was at a distance of 2000 km. This short-period comet is much less active than Comet Halley, so that the Deep Space 1 spacecraft cameras could operate well even at this greater distance. The best photo of Comet Borrelly has a resolution of 50 m, substantially higher than that of the Halley nucleus (Fig. 6.10). Borrelly is 8 km long and 3–4 km wide, with a shape something like a bowling pin (compare its shape with that of the slightly smaller asteroid Toutatis in Fig. 5.18). The terrain appears rough and jumbled. Unfortunately, we don't have a mass for either Borrelly or Halley, so we cannot calculate their densities.

The very low reflectivities of Halley and Borrelly, indicative of primitive, carbonaceous material,

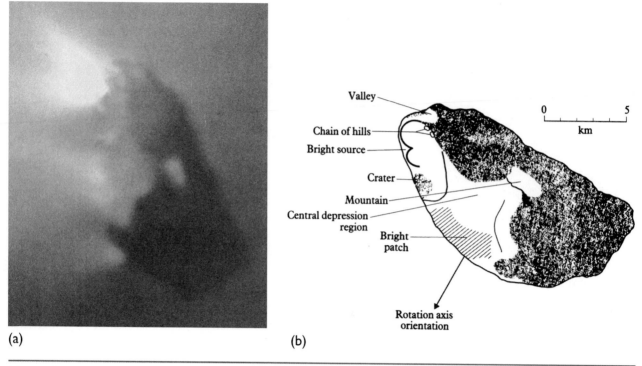

(a)   (b)

Figure 6.9 **(a)** The nucleus of Comet Halley is about twice as large as the island of Manhattan. It is composed primarily of water ice, but frozen CO_2, CO, NH_3, and other gases are present, all covered by a layer of dark carbon-rich material like a giant ice cream bar. This view was produced from 60 different images obtained by the Giotto spacecraft in March 1986. The smallest details are only 60 m across. Jets of dust and gas emanate from at least three bright regions along the left (sunward) side of the nucleus. **(b)** Schematic diagram illustrating the main features visible on the Halley nucleus.

confirm the pre-Halley conclusion by NASA astronomer Dale Cruikshank and a few other observers that most comets are covered by dark material similar to that found on some distant asteroids. Cruikshank originally encountered great skepticism when he reported that these icy objects were as black as coal, but now we see that the nucleus is a very dirty snowball indeed. Presumably the loss of ice and other volatiles from the upper layers of the nucleus allows a crust of dark dust to accumulate over the entire surface. A similar effect darkens the toe of a melting glacier on Earth by concentrating nonvolatile material as the ice evaporates.

The three comets with measured sizes—Halley, Borrelly, and IRAS-Araki-Alcock—all have diameters near 10 km, in spite of a difference in brightness (hence activity level) of a factor of a hundred, which indicates that comets with nuclei of about the same size can look very different, depending on the outflow of gas and dust from the nucleus. There are few smaller comets—say, 1–2 km in size—unlike the asteroids, which increase rapidly in number toward smaller sizes. At the opposite extreme, a few spectacular comets from past centuries are estimated to have been much larger. For example, a comet in 1727 was visible without a telescope even though it came no closer than the orbit of Jupiter, where most comets are exceedingly faint. It must have had a diameter on the order of 100 km. The Centaur object Chiron, which now seems firmly established as a comet, has a measured diameter of about 200 km. The corresponding range in masses is from about 10^{11} tons for Borrelly up to more than 10^{15} tons for Chiron.

COMETARY ACTIVITY

The activity of a comet refers to its outgassing rate. As a typical comet nucleus approaches the Sun, the rapid evaporation of its ices begins at a

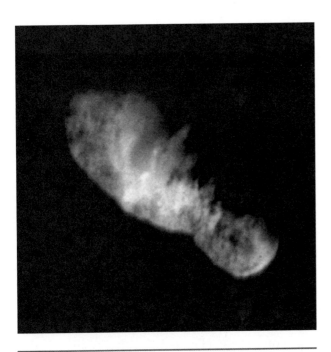

Figure 6.10   Spacecraft view of the nucleus of Comet Borrelly. Its length is 8 km.

surface temperature of a little above 200 K, out in the main asteroid belt. By the time it crosses the orbit of Mars, the comet develops a full-scale atmosphere and tail. Nearer the Sun, the solar energy evaporates more and more ice as well as heating the surface. Eventually, most of the energy goes into evaporation, although the darker parts of the cometary crust can become quite warm.

Measurements of the total brightness of comets have yielded estimates of the mass of gas and dust lost during one trip through the inner solar system. Typically this is 10 million to 100 million tons for an active comet, corresponding to less than 0.1% of the total cometary mass. Clearly, at this rate of loss the comet will not last forever, but it will exhaust its store of ices after a few thousand passes through the inner solar system. If the solid or dirty component is carried away with the evaporating ices, the comet will simply shrink down to nothing. On the other hand, if a residual core of solid material remains after the ices are gone, this core would be indistinguishable from a dark near-Earth asteroid. We do not know which of these scenarios for the death of a comet is correct. Nor are they mutually exclusive; some comets may end one way, some the other.

The colors of some of the near-Earth asteroids are virtually identical to those of old comet nuclei, suggesting that extinct comets indeed contribute to this population of objects that have the potential of striking the Earth.

NONGRAVITATIONAL FORCES

While a comet remains active, the escaping gases can have an effect on its orbit. Recall the way a rocket engine operates, generating thrust from the ejection of mass according to Newton's third law of motion. The same thing occurs with a comet. Evaporation of ice takes place primarily in jets on the warmest part of the comet, which corresponds to the "afternoon" side facing the Sun. As dust and gas stream away from the nucleus at a speed of hundreds of meters per second, the reaction creates a force in the opposite direction, away from the Sun (Fig. 6.11). Acting continuously over a period of weeks, such a nongravitational force can have a measurable effect on the orbit, and the influence of such forces must be taken into account when predicting the future paths of comets.

One of the immediate successes of Whipple's dirty-snowball model for the nucleus was its ability to explain nongravitational forces by the rocket effect. A nucleus consisting of a loose collection of gravel and boulders would not behave in this way.

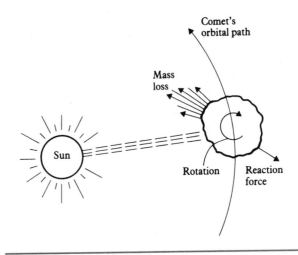

Figure 6.11   A schematic illustration of the rocket effect exerted by gases escaping in jets from a comet nucleus.

In his original 1950 paper on the cometary nucleus, Whipple showed how the rocket effect could explain the observed changes in the orbit of Comet Encke, which had been tracked by astronomers for more than a century. Subsequent analysis has also yielded the spin rates of comets, which are typically a few hours. Comet Halley, as measured in 1986, has an unusually long rotation period of three days.

Although the dirty-snowball model provides a satisfactory understanding of many cometary phenomena, it is incomplete and leaves unanswered important questions concerning details of the activity. What would it really be like to ride along on a comet as it approaches the Sun? The spacecraft data seem to indicate a highly dynamic environment, with huge jets of material like continuous geysers dominating the activity near perihelion. Perhaps eruptions burst out from different spots on the nucleus, as the dark surface absorbs sunlight and heats the ices below. Gases trapped in the ice could cause the activity of distant comets. Mysteriously, Comet Halley brightened suddenly in 1991, when it had already retreated 14 AU into the cold, dark regions of the outer solar systems; we do not know what triggered this event. Finally, we do not know the ultimate fate of comets: Do they become asteroids or just fade away?

THE COMET CRASH OF 1994

The fate of one comet is not in question: In 1994 a multiple comet called Shoemaker-Levy 9 crashed into Jupiter, ending its life in a spectacular display of celestial fireworks. The presence of craters on the surfaces of planets and satellites in the outer solar system is evidence that comets sometimes strike planets, but never before had such a cosmic collision been witnessed by astronomers.

This comet, called S-L 9 for short, was discovered in 1993 by Gene and Carolyn Shoemaker and their amateur colleague David Levy as part of a regular telescopic patrol carried out with a small telescope at Palomar Observatory. At the time of its discovery, S-L 9 was no ordinary comet, for it had about 20 separate nuclei spread out like pearls on a string (Fig. 6.12). A backward calculation of the orbit indicated that the comet had passed just 35,000 km above the clouds of Jupiter in the summer of 1992, close enough for its nucleus to have been disrupted by the gravity of the giant planet. Detailed calculations of this disruption process indicated that the original nucleus was between 2 and 3 km in diameter and consisted of many loosely bound fragments—not the orbiting gravel bank of pre-1950 models, but lacking any significant structural strength. These calculations also provided the first actual estimate

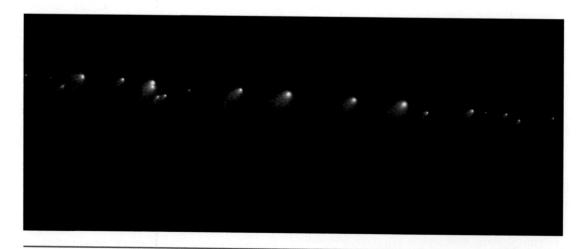

Figure 6.12   Comet Shoemaker-Levy 9 (S-L 9) as it appeared prior to its collision with Jupiter in July 1994. There are more than 20 fragments forming a "string of pearls" with the largest fragments near the center.

of the density of a comet, yielding a value near 1.0 g/cm³.

When the positions of S-L 9 were calculated into the future, even more surprises were in store. It turned out that the comet was not technically in orbit about the Sun at all, but that it had been captured into orbit about Jupiter. In effect, Jupiter had acquired about 20 new small satellites. However, the calculation showed that this jovian orbit was not stable, and all of the fragments would smash into the planet on their next close approach, in July 1994.

The impacts all took place on the back side of Jupiter, just over the horizon as seen from the Earth. The only direct view of the impact sites was obtained from the Galileo spacecraft, then about 200 million km from Jupiter. However, telescopes all over the Earth (and in orbit above the Earth) were trained toward Jupiter during the week when these unprecedented impacts took place, and the results were not disappointing. The Hubble Space Telescope and other instruments were able to see the impact explosions as they rose above the edge of Jupiter, and each of the events left huge dark smudges in the planet's atmosphere that were visible even with small amateur telescopes.

Each comet nucleus smashed into the jovian atmosphere with a speed of 60 km/s, creating a brilliant meteor and violently disintegrating as it plunged into the deeper layers of the atmosphere. The largest nucleus was probably about 1 km in diameter, and some were not more than a few hundred meters across, yet they exploded with energies of millions of megatons—each one hundreds of times greater than all the nuclear weapons on Earth put together. Each explosion generated a hot fireball that erupted upward to an elevation of about 3500 km, easily seen protruding over the jovian horizon (Fig. 6.13). The material from the fireball spread out to distances of more than 10,000 km from the point of impact as it collapsed and heated the jovian upper atmosphere. As the rotation of Jupiter carried each impact site into view, astronomers saw a huge dark cloud of cometary material, which remained visible in the jovian atmosphere for weeks (Fig. 6.14). In retrospect, astronomers realized that the impact geometry was nearly ideal, permitting

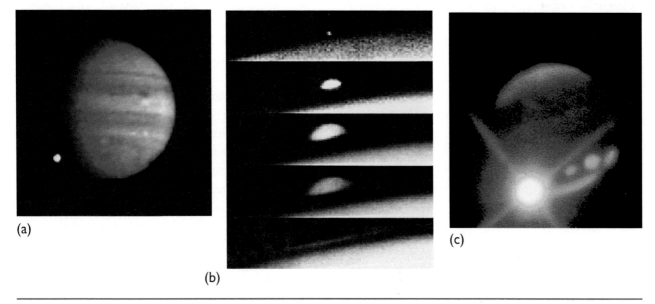

(a)

(b)

(c)

Figure 6.13   Impact of Comet S-L 9 with Jupiter. **(a)** The Galileo spacecraft, on its way to Jupiter at the time, got a direct view of the impacts on Jupiter's night side. **(b)** In under 20 minutes, the Hubble Space Telescope observed an impact plume rise thousands of kilometers above Jupiter's clouds and then collapse back down. **(c)** This bright infrared flare was formed when the debris from the impact fell back into the atmosphere of Jupiter, about 20 minutes after the impact.

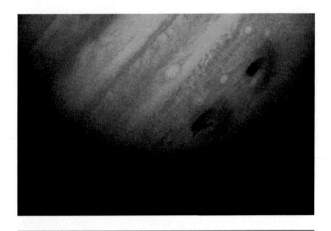

Figure 6.14 A Hubble Space Telescope image of Jupiter showing the effects of the impacts of two fragments of S-L 9 on Jupiter's atmosphere. The large circular feature with a dark spot inside it and a crescent-shaped feature to the right were produced by fragment "G." The image was recorded less than 2 hours after impact. The dark spot above and to the right was produced a day earlier by fragment "D."

them to view the erupting fireballs in profile against dark space.

In addition to the information revealed about the comet and the jovian atmosphere, these events raised an obvious concern about similar impacts on Earth. Many people asked, What if the same thing happened to our planet? We will return to this interesting question in Chapter 9.

6.4 COMET DUST

The dirty-snowball model for the comet nucleus predicts that comets will release large quantities of dust or larger solid material into the inner solar system. The production of solid material is further supported by the curving dust tails of comets. What is the ultimate fate of this dust?

METEORS AND METEOR SHOWERS

Comet dust fills the inner solar system. Most of it either falls into the Sun or is swept outward by the solar wind, but a tiny fraction of it strikes the Earth, burning up in the atmosphere to produce the meteors.

Meteors or shooting stars can be seen on any clear night, usually at a rate of several per hour. The typical meteor is no larger than a pea, but this is sufficient to generate a much larger cloud of glowing gas high in our planet's atmosphere which can be seen as far away as 200 km. Perhaps as many as 25 million meteors bright enough to be seen strike the Earth every day, amounting to hundreds of tons of cosmic material added to our planet.

Many meteors are produced by bits of material in random orbits. These can come from any direction at any time, and they are called sporadic meteors. Sometimes, however, the Earth encounters a stream of particles moving together along similar orbits around the Sun, and we see a **meteor shower.** It is the shower meteors that are most directly linked to the comets.

To illustrate the connection between cometary dust and meteors more specifically, we can look to one of the famous comets of the nineteenth century called Comet Biela, discovered in 1826 on a 6.8-year orbit. In 1846 this comet split in two, and upon its next return in 1852 both components were again present, separated by 2 million km. Neither part of Comet Biela was ever seen again; between 1852 and 1866, it simply ceased to exist. Nevertheless, astronomers watched with interest when the Earth passed through the orbit of Comet Biela in 1872, and they were not disappointed. Instead of the comet, they saw a wonderful meteor shower, with thousands of meteors visible from any spot on Earth during the night the Earth crossed the comet's orbit. The comet had been transformed into a stream of meteoric particles.

A similar shower is associated with the Leonid meteors that appear each November. Moving in the orbit of this stream is a particularly dense clump that the Earth encounters every 33 years. The most recent meeting with this clump occurred over the 3-year interval 1999–2001, which produced more than a shower—it approached a meteor storm, with more than 1000 per hour in some locations. One of the impressive aspects of such a shower/storm is that sometimes several meteors can appear almost simultaneously; we saw as many as four at the same time during the peak hour in 2001.

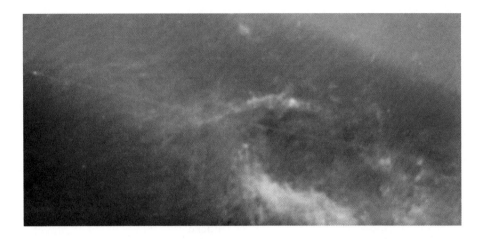

Figure 6.15 The IRAS infrared image showing the first cometary dust trail ever detected directly, the Tempel 2 dust trail. The dust trail is the thin blue line. False colors are used to accentuate the faint dust trail.

In 1983, the IRAS observatory succeeded in imaging some of the meteor streams associated with cometary orbits by recording the thermal infrared radiation emitted by these orbiting dust particles (Fig. 6.15). The most dependable annual meteor showers are listed in Table 6.4. In addition, however, an unpredicted shower can occur any time the Earth encounters a clump of meteoritic dust in space.

THE NATURE OF METEORIC MATERIAL

Most of the meteors that strike our atmosphere are associated with showers and therefore with meteor streams in space. Many of the nonshower, or sporadic, meteors are probably also the remnants of dispersed meteor streams. And most known meteor streams in turn are associated with comets.

If most meteors are really cometary dust, we naturally ask whether meteorites might not also be from comets. However, meteorites are not associated with meteor showers, and even on the rare occasions when the sky is filled with "falling stars," no actual meteorites fall to Earth. The shower meteors are not simply small meteorites that do not reach the surface; they are fundamentally different material.

Something of the nature of the cometary meteors can be inferred from observations made during their flights through the atmosphere. Their densities, calculated from the rate at which they slow due to atmospheric friction, are less than 1 g/cm^3, in contrast to the densities of meteorites, which are 3–7 g/cm^3.

As noted in Chapter 4, tiny meteoritic particles have been collected from the upper atmosphere. These fragments of cosmic dust are strange, fluffy bits of chemically primitive matter. Careful analysis of these tiny particles has demonstrated that many of them contain organic matter that must have been assembled in the interstellar cloud that formed the solar system. This conclusion is based on studies of the nitrogen and hydrogen isotopes in this material, which are found to have very different proportions from normal solar system material. So here is another piece of evidence suggesting that comets indeed preserve at least some pristine interstellar material.

6.5 ORIGIN AND EVOLUTION OF COMETS

The comets we see are temporary residents of the inner solar system. Many are traveling on nearly parabolic orbits and will not return for millions of years. Even the short-period comets have limited lifetimes. After a few thousand orbits—less than 10,000 years for many of them—their volatile ices

Table 6.4 Meteor showers

Name	Date of Maximum	Associated Comet	Comet's Period (yr)
Quadrantid	January 3	unknown	—
Lyrid	April 21	Thatcher	415
Eta Aquarid	May 4	Halley	76
Delta Aquarid	July 30	unknown	—
Perseid	August 11	Swift-Tuttle	133
Draconid	October 9	Giacobini-Zinner	7
Orionid	October 20	Halley	76
Taurid	October 31	Encke	3
Andromedid	November 14	Biela	7
Leonid	November 16	Tempel-Tuttle	33
Geminid	December 13	Phaethon	1.4

will be exhausted and they will no longer be comets, although the dead nucleus might remain as an Earth-approaching asteroid. Comets are also dynamically unstable. Like all objects with planet-crossing orbits, they run the risk of either striking a planet or being gravitationally expelled from the solar system. Their average residency in the inner solar system is no more than a few million years. Therefore, it is important to determine the sources of these comets.

We would not have to look for a continuous source of new comets if the comets were young, having all formed in some catastrophe that took place within the last few million years. Although such a hypothesis cannot be absolutely disproved, there is no real evidence to support it, and it contradicts our understanding of other solar system phenomena such as asteroids and meteorites. An explanation that does not require a unique event, but rather can be understood in terms of a general model for the origin and evolution of the solar system, would be much more reasonable.

Astronomers have concluded that there are two distinct source reservoirs for the comets, with different locations and origins. These are the Oort comet cloud and the Kuiper belt.

THE OORT COMET CLOUD

The first satisfactory theory for the origin of the comets was proposed by the Dutch astronomer Jan Oort in 1950, the same year that Whipple published his model for comet nuclei. Oort noted that in all cases where the orbits of long-period, nearly parabolic comets had been carefully determined, the orbits indicated an aphelion at a distance of about 50,000 AU, a thousand times more distant than Pluto. Very few comets seemed to come from greater distances, and none showed evidence of originating outside the solar system in interstellar space. Oort therefore suggested the existence of a comet cloud associated with the Sun but very far beyond the known planets. Since the orbits of new comets are not confined to the ecliptic but show a full range of inclinations, Oort concluded that this cloud was roughly spherical. This distant reservoir of comet nuclei is now called the **Oort comet cloud.**

At any given time virtually all the comets in the Oort cloud are too far away for detection, and their ices will be preserved indefinitely in the cold, dark void of outer space. Even if the orbits of these comets bring them in as close as Neptune, they remain frozen and are unlikely to be discovered unless they are very large (100 km). However, occasionally some of these comets can be perturbed by the slight gravitational tugs of nearby stars to bring their perihelia into the inner solar system. Only then will they develop the coma and tail we associate with comets.

To account for the several new comets discovered each year, Oort calculated that the comet cloud contained about 100 billion (10^{11}) comets. More recent calculations have suggested that the true number is ten times greater. If there are 1 trillion comets in the Oort cloud, and ten of these are lost each year by falling too near the Sun, the total number of comets lost to the cloud is still only about 5% of the original population over the history of the solar system.

THE KUIPER BELT

The idea of a large, spherical Oort cloud accounts for most of the new comets, which can descend upon us from any direction in space at any time. However, most short-period comets share the sense of revolution of the planets and do not have particularly large orbital inclinations. Detailed calculations show that these must originate in a region that is shaped like a flattened disk and not like a large sphere centered on the Sun. This disk-shaped source region should lie just beyond the orbit of Pluto and consist of myriads of potential comet nuclei, objects known generically as **icy planetesimals.**

The existence of such a region can also be anticipated by trying to answer the simple question: What determined the original outer edge of the planetary system? One might imagine that the formation of Neptune and Pluto probably did not consume all of the solid material in the outer reaches of the solar nebula. Some 50 years ago, G. P. Kuiper postulated the existence of a belt of icy planetesimals extending outward from the orbits of these two planets, a kind of cosmic junkyard of orbiting spare parts. At the time, the mass and diameter of Pluto were poorly constrained. Kuiper assumed that Pluto was sufficiently massive to have scattered these objects out into the Oort cloud, leaving trans-Neptunian space essentially empty.

Subsequent efforts by astronomers to find some evidence of trans-Neptunian planets or planetesimals were totally unsuccessful, apparently supporting this picture. Finally in 1992, David Jewitt and Jane Luu discovered the first small member of a collection of objects that became known as the **Kuiper belt,** as a result of a systematic search with the 2.2-m telescope of the University of Hawaii on Mauna Kea (Fig. 6.16). This was the first addition of a new component to the solar system since the discovery of Pluto itself in 1930. Subsequently, Jewitt, Luu, and other astronomers have found more than 500 Kuiper belt objects (KBOs, sometimes also called TNOs or trans-Neptunian objects), and more are added to the list each year. Contrary to Kuiper's initial idea, this region of space is far from empty, as Pluto is much too small to scatter these objects into the Oort cloud. In fact, Pluto is now recognized as simply the largest member of the Kuiper belt. Other objects like Pluto must have existed in the early solar system and ultimately ended up inside the giant planet Neptune. Triton, Neptune's largest satellite, is probably another example of a survivor from this early epoch.

Figure 6.16 David Jewitt and Jane Luu, who discovered the Kuiper belt in 1992 and continue to investigate its structure and the properties of its members.

With these additional discoveries, it has become possible to determine that the outer edge of the main Kuiper belt is at about 50 AU, although some members in highly eccentric orbits reach distances of 1000 AU or more. It turns out that Neptune exerts a gravitational influence on the Kuiper belt resembling that of Jupiter on the main belt of asteroids. Most KBOs have orbits that cluster around one of three resonances with the orbit of Neptune (Fig. 6.17). Pluto itself is in a 3:2 resonance, and several other KBOs (dubbed "Plutinos") have orbits with this same characteristic.

CHANGING COMET ORBITS

Just as asteroids suffer collisions and some are perturbed out of the main belt into orbits that cross those of Mars and Earth, the same things occur to icy planetesimals in the Kuiper belt. KBO collisions create dust, which was observed by the Voyager spacecraft that first ventured into this region of space (see Chapter 13). The resulting dust ring may be analogous to similar rings observed around nearby stars such as Beta Pictoris (see Chapter 17). Collisions plus gravita-

tional perturbations send some KBOs into orbits that take them closer to the Sun. The original population of this belt was therefore some 100 times greater than the number of objects we find there today.

The new orbits of the inward-moving icy planetesimals can be strongly disturbed by encounters with the other giant planets. Some encounters lead to catastrophic collisions, as happened on Jupiter in 1994 (as we have described), while others force the wandering KBOs into much larger orbits, perhaps joining the Oort cloud or leaving the system entirely. In yet another fraction of these encounters, there is a kind of "hand-off" from one giant planet to another, until the KBO is trapped by Jupiter into a short-period orbit around the Sun and becomes a member of the Jupiter family.

The intricacies of these celestial odysseys have not yet been worked out in detail, but we can see some of the results. The Centaurs appear to be the intermediate case of inward-moving KBOs that are presently in unstable orbits. The short-period comets represent the final stage of the journey, the capture by Jupiter into orbits whose

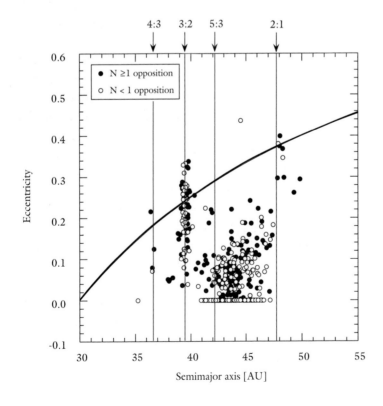

Figure 6.17 This plot of eccentricity versus semimajor axis of Kuiper Belt Objects (KBOs) shows the clustering at the resonances with Neptune's orbit as described in the text. Pluto is a member of the highly populated group in the 3:2 resonance, indicated at the top of the chart.

close proximity to the Sun soon leads to the exhaustion of all the volatiles originally trapped in these icy objects when they formed at the outer edge of our planetary system.

We have hints from Earth-based observations that there are some subtle compositional differences between the KBOs (as manifested in the short-period comets) and their brethren from the Oort cloud. Given the differences of 10–30 AU in the places of their formation in the solar nebula, such variations in composition are not surprising. Because the KBOs were formed in the coldest region of the original solar nebula, they must be the most primitive of the surviving icy planetesimals. Thus, spacecraft measurements of both long- and short-period comets may allow us to deduce some of the original conditions in the outer solar nebula and perhaps even in the interstellar cloud from which it came.

THE FATE OF COMETS

Although a comet may peacefully circle the Sun in deep space for billions of years, once it is diverted into the inner solar system, its life expectancy is limited. There is some chance that it will not survive even its first plunge toward the Sun; many comets have been seen to strike the Sun or to pass so near its surface that they are destroyed by the heat in a single pass. Other comets die even before reaching the Sun; in 1906 and 1913, for example, new comets that had been expected to put on a good show simply faded to invisibility as they neared perihelion.

Comets may also break apart for reasons not always understood. Shoemaker-Levy 9 was torn apart by the gravity of Jupiter. But what happened to Comet West, which apparently split three times within a few days early in March 1976? The four components drifted apart, extending out more than 10,000 km by the end of the month. The smallest fragment survived only a few days, but the other three components of the nucleus still retained their individual identities as the activity level declined with increasing distance from the Sun. We do not know why this breakup occurred.

More likely than hitting the Sun or breaking apart, a new comet may pass near enough to some planet—usually Jupiter—to suffer a change in orbit. This change can either increase its energy, ejecting it completely from the solar system, or cause it to lose energy, making it a short-period comet. The reason most short-period comets have aphelia near the orbit of Jupiter is that they were captured by the gravitation of this giant planet. But the comet has now gone out of the frying pan and into the fire, for it still runs a significant risk of impact with one of the terrestrial planets. Meanwhile, the heat of the sun is also consuming its icy substance at a rapid rate.

ORIGIN OF THE OORT CLOUD AND THE KUIPER BELT

The idea of the Oort cloud and the Kuiper belt provides a framework for understanding the continuing supply of comets to the inner solar system, but it begs the question of the ultimate origin of the comets. How did the icy planetesimals get into the Oort cloud or the Kuiper belt in the first place? Might the Oort cloud comets have been formed in place, tens of thousands of AU from the Sun? We have already mentioned this possibility, but at such distances in the solar nebula it seems difficult to imagine solids condensing. More likely, these comets were formed in the realm of the planets, at the same time as the other members of the planetary system. In this scenario, the comets condensed in the colder, outer region of the solar nebula. The presence of a variety of ices in comets suggests formation temperatures in the range of 30–100 K, corresponding to the region of space now occupied by the giant planets and the inner edge of the Kuiper belt. Perhaps all the Kuiper belt comets (with perihelia equal to or greater than Neptune's orbit) formed in approximately the locations where we find them today. However, closer to the Sun, many of the icy planetesimals must have been ejected into the Oort cloud from regions where they interacted gravitationally with the giant planets, while others were gravitationally dispersed throughout the solar system, crashing into the planets or the Sun, or expelled from the system entirely.

We have stressed the primitive nature of comets in this chapter, how they have preserved both hydrogen- and oxygen-rich compounds in their ices, and how some of these may be treasures preserved intact from the interstellar cloud that gave birth to our planetary system. But it is also possible that comets contain some materials that were processed at the higher temperatures of the inner solar nebula and then transported to the outer regions where they mixed with more primitive matter falling in from the interstellar cloud to make the comets. Thus, these icy remnants of the early solar system may not be as primitive as we think.

The same ambiguity clouds our efforts to determine how important comets have been in shaping our destiny here on Earth. As soon as Edmund Halley had successfully calculated the orbit of the comet that bears his name, he realized that it was possible for comets to strike the Earth. Ever since that time, scientists have wondered whether these icy messengers could have delivered the water in the Earth's oceans, the nitrogen and other gases in our atmosphere, and the organic compounds that might have "jump-started" the origin of life. Studies of Comets Halley, Hale-Bopp, and Hyakutake have demonstrated that the proportions of light and heavy isotopes of hydrogen in the water ice of these three comets, all coming from the Oort cloud, are twice as great as the proportions of these isotopes found in ocean water on Earth. This shows that the Earth's oceans were not formed by impacts that melted Oort cloud comets. We don't yet have isotope measurements for Kuiper belt comets, nor do we have noble gas measurements in any comets. Thus the jury is still out on the question of how much comets have contributed to inner planet volatiles. As we will see in Chapter 13, a completely different line of evidence suggests that icy planetesimals related to comets played a major role in building the giant planets. In Chapter 12, we shall find intriguing hints of cometary delivery of water to Mars.

To find out what is hidden inside these icy sarcophagi and explore the comet-planet connection, we need missions to comets that will provide more discriminating data. Fortunately, several such missions are under way. Stardust will fly past Comet Wild 2 in January 2004 and will return samples of comet dust to Earth in 2006, providing a definitive calibration of data obtained from dust captured at high altitudes on Earth. Deep Impact plans to smash a 500-kg ball of copper into Comet Tempel 1 in 2005, allowing a look at what's inside the nucleus. Rosetta, a mission developed by the European Space Agency, will reach Comet Wirtanen in 2012, stay with it for several months as the comet passes through perihelion, and deploy a lander on the nucleus. All of these missions will include instruments to examine the composition of the comet nuclei. It is safe to say that our knowledge about comets is about to undergo a major revolution!

6.6 QUANTITATIVE SUPPLEMENT:
ALBEDOS AND TEMPERATURES

In this text, we use the term *reflectivity* to indicate the fraction of incident sunlight that is reflected from a planetary surface. Generally, what we mean is the *normal reflectivity*—that is, the fraction of the light reflected straight back from a beam incident normal (perpendicular) to the surface. But there are several other ways in which reflectivity can be defined, depending on the application.

The normal reflectivity is a property of a particular area of the surface. Unless we have close-up photos from a spacecraft, however, we may not be able to distinguish individual surface elements. Suppose, instead, that we are interested in the global properties of a planet or asteroid. We define the *albedo* (more properly, the *Bond albedo*) as the ratio of total reflected light to total incident light. The Bond albedo, which is generally numerically smaller than the normal reflectivity of the surface, is a measure of the global reflection and absorption of sunlight.

Since the Bond albedo (called A) tells us how much solar energy is absorbed, we can use it to calculate an effective temperature for a planet. Remember that the Stefan-Boltzmann law expresses the radiation emitted by a perfect radiator at temperature T. A planet is not a perfect radiator, nor is its surface all at the same temperature, but we can define effective temperature as the value of T that satisfies the energy balance when the absorbed sunlight is equated to the emitted thermal power:

$$(1 - A)\left(\frac{P_{\text{Earth}}}{D^2}\right)(\pi R^2) = \sigma T^4 \times 4\pi R^2$$

Here P_{Earth} is the sunlight per unit area at the distance of the Earth from the Sun (the solar constant), D is the distance of the planet from the Sun (in AU), and R is the radius of the planet. Note that the area (πR^2) on the left side is the cross section of the planet, while that on the right side $(4\pi R^2)$ is the total area of a sphere; we assume that the planet radiates from its entire surface, not just the part facing the Sun. Simplifying this equation, we have

$$T^4 = (1 - A)\left(\frac{P_{\text{Earth}}}{4\sigma D^2}\right)$$

Another kind of albedo is defined to describe the apparent brightness of an object. This is the *geometric albedo* (called p), and it is the ratio of the global brightness, viewed from the direction of the Sun, to that of a hypothetical, white, diffusely reflecting sphere of the same size at the same distance. Numerically, the geometric albedo is approximately the same as the normal reflectivity. Note that some objects can be more reflective than the hypothetical white reference sphere, so that geometric albedos greater than 1.0 are possible; the geometric albedo of Saturn's satellite Enceladus is about 1.1. The Bond albedo can never be greater than 1.0, however.

Finally, there is a relationship between the two albedos for any object, called the *phase integral, q*. The phase integral, which depends on the scattering properties of the surface, is defined by the relation

$$q = \frac{A}{p}$$

For the Moon and many other objects in the solar system, the value of the phase integral has been measured to be near 2/3; in other words, for typical surfaces of planets, the Bond albedo is equal to about two-thirds of the geometric albedo.

SUMMARY

The comets are the most primitive members of the planetary system. Composed in large part of water ice and other frozen volatiles, they were formed at temperatures below 100 K and have spent most of the past 4 billion years in the deep freeze beyond Pluto. Only when they are diverted into the inner solar system do we see these small icy bodies and identify them as comets.

As it is heated by sunlight, a comet develops a thin but extensive atmosphere. Spectral analysis shows that the cometary atmosphere contains a complex mixture of water vapor, carbon monoxide, carbon dioxide, and methanol, as well as many variable hydrocarbons. Once ionized, this atmosphere streams away from the Sun to form the comet's plasma tail, which can extend for millions of kilometers. Copious dust is also released, consisting of both silicate and carbonaceous particles.

The comets can best be understood in terms of Whipple's dirty-snowball model, in which the small solid nucleus is composed of approximately equal quantities of ice (primarily water ice) and silicate and carbonaceous material (largely in the form of small dust grains). Rapid evaporation of the ice under the influence of solar heating leads to the formation of the atmosphere. Preferential evaporation from the Sun-facing hemisphere of the nucleus also gives rise to jets, which produce the nongravitational forces that alter the orbits of many comets. The gas released from the nucleus dissipates into space, while the dust contributes to meteor showers when the Earth intersects the orbit of a comet. The surface of the nucleus is darkened by the carbonaceous dust left behind by the evaporation process, as shown directly in the spacecraft photos of the dark nuclei of Comets Halley and Borrelly.

Comets are on unstable orbits and must originate elsewhere. Two source regions have been identified: the Oort cloud, a roughly spherical halo of icy bodies that extends to about 50,000 AU from the Sun, and an inner, flattened Kuiper belt beyond the orbit of Pluto. The objects in these regions (as many as a trillion of them) are thought to be leftover building blocks from the outer solar system, perhaps the same materials that made up the cores of the giant planets. Pluto is the largest of these objects that still has an independent existence. Although many scientists think that comets must have delivered some fraction of the volatiles we now find on the inner planets, we do not yet know how big this fraction is or what volatiles it contained. We also do not know whether comets contain unaltered interstellar material. Several space missions are under development to seek answers to these questions.

KEY TERMS

coma

comet nucleus

dust tail

fluorescence

icy planetesimal

Kuiper belt

long-period comet

meteor shower

Oort comet cloud

plasma tail

short-period comet

REVIEW QUESTIONS

1. Distinguish among comets, main belt asteroids, and near-Earth asteroids in terms of their orbits and their physical and chemical properties. What relationships exist among these three classes of small bodies?

2. Describe the parts of a comet and their appearance as seen from the Earth. Can you think of any reason comets have traditionally been associated with disasters, such as wars and plagues?

3. Describe the dirty-snowball model for the nucleus of a comet. Compare this model with the observations. Explain in particular how it accounts for the nongravitational forces that disturb the orbits of comets.

4. Imagine yourself riding on a comet. Describe what you would see as the comet approached the Sun. Do you think that an instrumented lander on a comet nucleus would survive for long? What kind of measurements might such a lander make?

5. Describe how the atmosphere and tail of a comet are formed. How do the phenomena we see in the head and tail relate to the properties of the underlying nucleus?

6. What are the relationships between comets and meteors? If shower meteors come from comets, isn't it likely that the meteorites come from comets, too? Explain.

7. Describe the Oort comet cloud. While a comet is in this cloud, how might it change over time? What is the orbit of a comet in the cloud like? Under what circumstances will a comet in the Oort cloud be able to enter the inner solar system?

8. What is the difference between the Oort cloud and the Kuiper belt? Which one supplies the short-period comets? Do you think the comets in these two reservoirs have different compositions? Explain.

9. Surely the bright "new" comets like Hale-Bopp have more to tell us about the primitive material we think comets contain than the tired old comets like Borrelly that have been "baked out" by the Sun. Yet all the cometary missions that have been carried out or are planned for the future go to these short-period comets, not the big, bright ones coming in from the Oort cloud. Can you think of a reason involving spacecraft trajectories that accounts for this strange discrimination?

10. Compare the history of a near-Earth asteroid with that of a short-period comet. Is there any possibility that these two classes of small objects could have some members in common? Explain.

QUANTITATIVE EXERCISES

1. What is the eccentricity of a comet with perihelion at the Earth's orbit (1 AU) and aphelion at Jupiter's orbit (5 AU)? What are its semimajor axis and period of revolution?

2. If the comet in Exercise 1 has a Bond albedo of 0.03, what is its effective temperature at perihelion and aphelion? (For comparison, the effective temperature of a black sphere [$A = 0$] at 1 AU is 270 K.)

3. Saturn's satellite Enceladus has a geometric albedo of 1.1 and a Bond albedo of 0.8. Find its phase integral and effective temperature.

4. The equilibrium temperature of a black surface facing the Sun at 1 AU from the Sun is 270 K. What is the effective temperature of a spherical object, also black ($A = 0$), at 2 AU? What is its effective temperature at 2 AU if its Bond albedo is 0.5? What are the effective temperatures of both objects at 1 AU?

5. What is the effective temperature of a comet at the inner edge of the Kuiper belt? In the Oort cloud?

ADDITIONAL READING

Brandt, J. C., and R. D. Chapman. 1992. *Rendezvous in Space: The Science of Comets*. New York: Freeman. A semitechnical work by two comet experts, this is one of the few books that includes analysis of the observations of Comet Halley in 1986, including the spacecraft missions to the comet.

Sagan, C., and A. Druyan. 1985. *Comet*. New York: Simon & Schuster. Beautifully written popular summary of our knowledge of comets on the eve of the 1986 apparition of Comet Halley, with discussions of cultural and historical comet lore.

Whipple, F. L. 1985. *The Mystery of Comets*. Washington, DC: Smithsonian Institution Press. Popular pre-Halley summary of our knowledge of comets by the leading cometary scientist of our time.

Yeomans, D. K. 1991. *Comets: A Chronological History of Observation, Science, Myth, and Folklore*. New York: John Wiley & Sons. Fascinating historical information on bright comets through history and changing cultural attitudes toward these space visitors.

7

THE MOON: OUR ANCIENT NEIGHBOR

Astronaut on the lunar surface.

The objects in our planetary system are remarkably varied. As a general rule among planets, larger size leads to geological complexity because internal heat can be retained by larger objects to power an active surface geology. It is therefore reasonable to expect that smaller planets will be less active, and hence probably simpler to understand, than larger ones. This is why we began in Chapters 5 and 6 with the smallest bodies, and it is why we now take up the study of the Moon before we tackle the larger and more complex planets.

A relatively simple geologic history is not the only reason for beginning a detailed study of the individual planets with the Moon. The lack of a lunar atmosphere also eases our task considerably. There are no obscuring clouds or vapor, no wind or precipitation to erode the surface, and no oxygen or water to weather the rocks or alter their chemistry. Change on the Moon is a slow process; a million years from now, the footprints of Apollo astronauts in the lunar soil will still be fresh. Unlike the Earth, where the surface geology is often hidden under vegetation, and rock and soil chemistry are altered by local conditions, on the Moon what you see is what you get.

The Moon also provides an appropriate introduction to the other planets and satellites because of the dominant influence of impacts in shaping its surface, producing mountains and abundant craters. Such impacts provide a common thread weaving together the histories of different worlds. We have seen that the asteroids are fragments from collisions, with smaller impact craters superposed. On the Moon, with its much larger gravity, craters behave somewhat differently. By studying impacts on geologically inactive worlds like the Moon, we will gain the understanding needed to interpret more complex planets where internally driven processes compete with cratering in sculpting the surface.

7.1 THE FACE OF THE MOON

Although large in comparison with the comets and asteroids discussed in Chapters 5 and 6, the Moon is a small world. With a diameter of 3476 km, its total surface area is about the same as that of Asia. A flight from Los Angeles to New York traverses a greater distance than the lunar diameter. The mass of the Moon is less than 2% that of the Earth, although still substantially greater than that of all of the asteroids combined.

Chemically the Moon differs from the primitive objects such as comets and C-type and S-type asteroids. The Moon is not primitive; it has been heated and differentiated, and it lacks water, ice, carbon, and organic compounds. From its relatively low density of 3.3 g/cm^3, we conclude that the Moon also lacks a metallic core like that of the Earth. Note that the Moon is large enough to compress its interior and squeeze out most void spaces, so that (unlike asteroids and comets with their considerable interior porosity) its density is representative of the rocks themselves.

RESOLUTION OF THE SURFACE

The Moon is the only planetary body that can be distinguished with the naked eye as a globe, and even without a telescope we can see that its surface is not uniform. Many cultures have associated names and myths with the markings familiarly known as the "Man in the Moon," but it was not until Galileo Galilei turned his first small telescopes on the Moon in 1610 that it became clear that the surface of our satellite was rugged and mountainous like that of the Earth. Galileo's observations provided the foundation for considering the planets as other worlds and thus indirectly gave birth to the science of planetary geology.

A sharp-eyed observer can see features on the Moon as small as 1/15 of its apparent diameter, or approximately 200 km across. At this **resolution,** the larger light and dark markings can be distinguished, but topographic features, such as mountains and craters, are undetectable. You can demonstrate this conclusion to your own satisfaction by sketching the Moon at different phases and using these sketches to produce a naked-eye map. This exercise is especially instructive when you note that the best Earth-based telescopic views of other planets, such as Venus and Mars, have about the same resolution as naked-eye maps of the Moon.

A telescope, with its higher resolution, is able to reveal lunar surface topography. Galileo's early telescope was barely sufficient to distinguish the largest craters and mountains. You can do as well today with a good pair of 7–8-power binoculars, which perform better in many ways than Galileo's best 30-power telescope. Today's astronomical telescopes are far more powerful. As we noted in Section 3.5, telescopes on the surface of the Earth have historically been limited in resolution by the atmosphere. For the Moon, this limiting resolution is about 1 km.

The angle at which sunlight illuminates a surface plays an important part in determining our ability to distinguish topographic detail. When the Sun is low, features cast shadows, whereas at moderate angles of illumination, slopes and contours are revealed by shadings that vary as the surface is tilted toward or away from the Sun. When the sunlight streams down from directly behind the observer, details of topography become indistinguishable. Thus, an image of the full moon emphasizes differences in reflectivity between light and dark areas but suppresses topography (Fig. 7.1). Craters and mountains are best studied near first and last quarter, where we see the border between night and day. In Figure 7.2 the Sun is low

Figure 7.1 Seen at full phase, the Moon shows little topographic detail even through a telescope. The dominant features visible under these lighting conditions are the dark volcanic maria and the bright rays associated with young impact craters. The bright young crater Tycho (lower left) is the source of many of these rays.

Figure 7.2 The nature of the lunar surface is better revealed at quarter phase, when the sunlight strikes obliquely, highlighting topographic features such as craters and mountains. Image is rotated 90° with respect to Fig. 7.1.

in the lunar sky, and every topographic detail is sharply etched.

LUNAR HIGHLANDS AND MARIA

When we look at the Moon through binoculars or a small telescope, we are immediately aware of two different kinds of surface terrains. The predominant surface type is relatively light (reflectivity about 15%) and extremely rugged, with craters of all sizes piled one upon the other. Since these brighter, heavily cratered regions also generally lie at higher elevations, they are called the lunar **highlands.** The second surface type is darker (reflectivity about 8%) and smoother, with relatively few large craters. These regions, which make up the features of the Man in the Moon, are called **maria** (singular: **mare**). *Mare* is the Latin word for "sea"; when the term was first applied to the Moon in the seventeenth century, these darker regions were thought to be water oceans. (Galileo himself was careful not to overinterpret what he saw. In his *Siderial Messenger*, written just a few months after he made his first astronomical observations, he stated only that if the Earth were seen from the Moon, its seas would probably look similar to the dark lunar maria revealed by his telescope.) Figure 7.3 contrasts the appearance of highlands and maria.

Most of the maria are found on the side of the Moon that faces the Earth. The opposite, or far-side, had never been seen until 1959 when the Soviet Luna 3 radioed back to Earth the first rudimentary photos. To the surprise of almost everyone, no major maria appeared on the lunar farside, which was almost all highland terrain. Any explanation for the existence of highlands and maria must address this basic asymmetry in mare distribution. Subsequent mapping of the entire Moon revealed that only 17% of the surface area consists of mare material.

Maria and highlands differ in many ways. Most obvious is the distinction in color and reflectivity, implying that their chemical compositions are not alike. Second, a difference in cratering suggests different geologic histories. Finally, there is a difference in elevation, related to the largest scale forces that have molded the lunar surface. The

Figure 7.3   The contrasting nature of the lunar highlands and maria is clearly shown in these Apollo orbital photos of the two types of terrain. The crater density is 10–20 times greater on the highlands (left) than on the mare (right).

maria and highlands are representative of different chemical and geological regimes; they provide windows into two different periods of solar system history.

LUNAR CRATERS

The prevalence of the characteristic circular features called craters is one of the most striking aspects of the lunar surface. The word *crater*, derived from the Greek for "cup" or "bowl," refers to the shape of the feature. Note that a crater is a depression; the popular application of the term to a volcanic mountain, such as Vesuvius in Italy or Haleakala in Hawaii, is incorrect. A volcano frequently has a crater at its summit, but the mountain itself is not a crater.

Even binoculars or a small telescope can reveal the larger lunar craters, which are hundreds of kilometers in diameter. At the best Earth-based resolu-

Figure 7.11 The *Apollo 15* rover. These vehicles permitted the astronauts to investigate the lunar geology over a much wider area than had been accessible in early missions.

veyor 3 spacecraft and retrieved parts of it for return to Earth. (34 kg of samples collected.)

Apollo 13: April 1970. Mission aborted after explosion of oxygen tank in command module. Astronauts returned safely after harrowing flight around the Moon in their disabled spacecraft.

Apollo 14: January 1971. Landing on the Fra Mauro ejecta from the Imbrium Basin. First use of lunar rickshaw to haul equipment and samples; astronauts traversed 5 km on foot. (43 kg of samples collected.)

Apollo 15: July 1971. Landing at edge of the Imbrium Basin at foot of Apennine Mountains. First in new phase of missions with extended scientific capability. Carried improved command module instruments and launched subsatellite into lunar orbit. First use of lunar rover vehicle permitted 24-km traverse, including visit to Hadley Rille, an ancient lava channel. First measurement of lunar heat flow (Fig. 7.11). (77 kg of samples collected.)

Apollo 16: April 1972. Only landing in a highland site, near crater Descartes. Additional detailed science from the surface and orbit. (95 kg of samples collected.)

Apollo 17: December 1972. Final Apollo landing, in Taurus-Littrow Valley on margin of Mare Serenitatis. Included only scientist-astronaut to visit moon, geologist Harrison Schmitt. (111 kg of samples collected.)

During the development of the Apollo program, the United States competed with the USSR to be the first to land humans on the Moon. The Soviet effort finally faltered when their large rocket, comparable to the U.S. Saturn 5, failed several crucial tests in 1968 and 1969. However, the Soviets pursued a vigorous robotic exploration program during the early 1970s. Three robot sample return missions, Lunas 16, 20, and 24, brought back 300 g of material, including one core sample, and in 1970 and 1973, two mobile vehicles called Lunakhods were successfully operated on the lunar surface.

PERSPECTIVE

WERE THE APOLLO LANDINGS FAKED?

More than half the people in the world have been born since the Apollo landings. Already this extraordinary feat of technology and human courage has faded into legend. Within a single decade the Apollo program progressed from a dream to reality, only to be abandoned at the peak of its success.

Most people today honor the achievement of the Apollo program and the astronauts who went to the Moon. The movie *Apollo 13* helped remind us of their accomplishment. But there are also those who claim that the entire lunar program was a hoax. The Fox TV network ran shows in 2001 that supported this position and even accused NASA of murdering several astronauts to protect the secret. In some fundamentalist Islamic schools, it is firmly asserted that no human has been (or ever will go) to the Moon. Even in the United States, opinion polls have indicated that between 6% and 20% of the people today question the reality of the Moon landings.

Many of the reasons given for doubting the authenticity of Apollo are pretty silly. One argument is that if the photos were really taken on the airless Moon, it would not be possible to see anything in the shadows—overlooking light scattered into shadowed regions from the Earth and the lunar lander. Another statement is that the motion of the flags indicates that the films were made in a studio rather than on the Moon—neglecting that the flags moved only when bumped by one of the astronauts. Or the claim is made that if the photos were real, we could see stars in the sky—ignoring that with a brightly lit foreground, both the human eye and camera lenses close down so that faint sources such as stars can't be seen.

The 2001 Fox TV show went much further, suggesting that NASA had to kill a number of people, including the three astronauts who died in the 1967 *Apollo 1* fire, in order to

➡️

Figure 7.12 The giant Saturn 5 rockets of the Apollo program were the largest rockets ever built. Unfortunately, Apollo was terminated before its completion, and several Saturn 5s, such as this one on public display in Houston, were left to rust on Earth rather than being sent to the Moon.

cover up their plot. Such accusations are hardly worth answering. The whole "Moon hoax" hypothesis depends on literally tens of thousands of people being in on the deception and keeping it quiet—everyone involved in the Apollo launches and in tracking the spacecraft, the mission controllers in Houston, the manufacturers of key hardware, the Apollo science teams, and of course the two dozen astronauts themselves who went to the Moon. And what about the additional thousands who would have been involved in setting up and filming an elaborate stage set? They too would have to be part of this plot. Finally, there is the problem of the race with the USSR to send humans to the Moon. Were the Soviets also involved in this remarkable conspiracy?

On the positive side, let's think about the Apollo legacy. Thousands of large-format photos of the Moon were taken from the orbiting command modules—pictures that could be faked only by people today with an intimate knowledge of lunar geology and the vast computational power that permits animated films such as *Jurassic Park*. There are the ALSEP radio transmitters and laser retroreflectors placed on the Moon as part of the Apollo program, all of which were regularly tracked from Earth for many years after the astronauts came home. And of course there are the 382 kg of lunar samples, which have been analyzed by thousands of scientists around the world—samples that are clearly of extraterrestrial origin and could not be faked even with today's technology.

Who do these accusations of a hoax come from? What has happened to people who treat the Apollo astronauts not as heroes but as part of a vast government conspiracy of lies and murder? Perhaps they cannot believe in any great accomplishments of the past, rather like those who insist the pyramids of Egypt were built by aliens because they cannot imagine that humans had the technology or the will to produce such magnificent creations.

It is ironic that 30 years after Apollo, neither the United States nor any other nation has a capability for human lunar exploration. Tourists gawk at the giant Saturn rockets on display at Cape Canaveral and Houston (Fig. 7.12). Leftover Apollo spacecraft, built at costs of hundreds of millions of dollars, take the place of honor in museums instead of resting where they were intended to, on the surface of the Moon. No forecaster of the future or writer of science fiction had ever predicted that humans, having once attained the Moon, would so quickly abandon it. The scientific legacy of Apollo continues, but what of the exploration potential? Where has that legacy gone?

7.3 IMPACT CRATERING

Craters are the dominant geological features on the Moon, readily visible to generations of telescopic observers. Yet their impact origin was not widely recognized until about 50 years ago. It is interesting to examine why this fundamental principle of planetary science remained hidden for so long, and to see what finally provided convincing evidence in favor of an impact origin for the lunar craters.

VOLCANIC OR IMPACT ORIGIN?

Throughout the nineteenth century, most geologists thought that the lunar craters were volcanic in origin. The argument was really fairly simple.

No impact craters had been recognized on Earth, and the largest projectiles known to strike our planet were meteorites. In those days, before the discovery of near-Earth asteroids, there was no evidence for large objects in the inner solar system that could strike either Earth or Moon. Volcanoes, on the other hand, were well known to terrestrial geologists, and volcanoes do have craters. Therefore, the craters of the Moon, by analogy with the Earth, must also be volcanic.

The first detailed arguments against the volcanic crater hypothesis were presented in the 1890s by Grove Gilbert, then director of the U.S. Geological Survey. Gilbert was among the few geologists who were interested in the Moon, and he was a strong proponent of quantitative measurements rather than subjective descriptions of what the Moon looked like through a telescope.

Assembling data on the sizes, shapes, and distribution of lunar craters and of terrestrial volcanoes, Gilbert pointed out many differences between the two. Most significant was the fact that lunar craters do not appear at the summits of mountains. Gilbert was among the first to emphasize that the floors of lunar craters actually lie well below the level of the surrounding plains (Fig. 7.13). Using these and similar arguments, he developed a case for the dissimilarity between lunar and terrestrial craters. With the apparent similarity demolished, the argument by analogy crumbled.

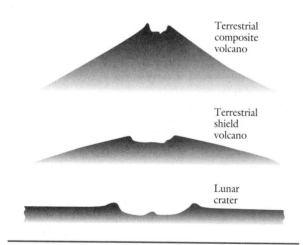

Figure 7.13 Comparison of the profiles of typical lunar impact craters and terrestrial volcanic craters.

Through elimination of alternatives, impacts emerged as the most probable cause of lunar craters, and Gilbert presented as strong a case for this hypothesis as was possible in the 1890s. He did, however, perpetuate a fundamental misconception about the impact-cratering process. In his mind, the formation of an impact crater on the Moon was similar to the way a small crater is formed when one throws stones into mud or sand. Such sandbox craters are round only if the stone strikes the surface from above; an oblique impact produces an elongated crater. Since virtually all lunar craters are circular, Gilbert had to hypothesize peculiar circumstances in which the asteroids or comets would fall on the Moon from nearly overhead. These efforts were not very convincing, undermining his arguments for an impact origin of lunar craters.

THE PROCESS OF IMPACT CRATERING

Gilbert's fundamental problem lay in his failure to understand that the high-speed impact on the lunar surface generates an explosion similar in many ways to that of a bomb. (Actually, there are significant differences between impact craters and explosion craters that are apparent to the trained geologist, but we will not probe that deeply into the crater-forming process.) Once the explosion is under way, it obliterates any evidence of the direction from which the projectile struck the surface. To cite a modern analogy, shell and bomb craters are always circular, independent of the direction from which the bombardment occurred.

Why does the impact of a piece of cosmic debris produce an explosion? Not because fragments of asteroids or comets are inherently explosive. Rather, it is because of the tremendous energy acquired by the projectile as it falls to the surface. For the Moon, the minimum speed of impact (which is equal to the escape velocity) is 2.4 km/s, and for Earth, with its stronger gravitational pull, the minimum impact speed is 11 km/s. To these values must be added the original orbital speed of the projectile, relative to the target. Such speeds endow the impactor with energy greater than that of an equivalent mass of TNT, so that it explodes upon impact, regardless of its own composition or the nature of the target.

Imagine a small asteroid striking the Moon at a speed of several kilometers per second. Its energy is so great that it penetrates two or three times its own diameter below the surface before it stops. The force of the blow shatters the surface and generates seismic waves—moonquakes—that rapidly spread throughout the Moon. Meanwhile, most of the energy goes into heating the projectile and its immediate surroundings. The material forms a pocket of superheated gas, and the expansion of this hot, high-pressure gas contributes to excavation of the crater.

Figure 7.14 illustrates the stages of crater formation. At the speeds associated with lunar impacts, the energy released is sufficient to form a crater with a diameter about ten times larger than the projectile. The material removed from the crater, which consists of the original projectile plus several hundred times its own mass in excavated rock, is thrown upward and outward. Part of it falls back into the hole, partially filling it, while the rest spreads over the surrounding area.

THE EJECTED MATERIAL

The ejected material consists primarily of broken and shattered rock fragments mixed with gas and liquid droplets generated by the heat of the explosion. The bulk of it falls within one crater diameter of the rim, where it produces a rough, hilly deposit known as an **ejecta blanket.** There may also be a surge of back-falling debris that flows outward from the explosion center. On the Earth, similar hot, fluid surges are a characteristic (and very dangerous) aspect of some volcanic explosions, such as that of Mount Saint Helens (Washington, 1980). Impact craters formed on Mars, where the subsurface is (or was) saturated with ice, have a unique form of ejecta blanket, as we will see in Chapter 11.

Large fragments thrown out by an explosion rain down on the surrounding lunar surface, where they generate their own small craters, called **secondary craters.** Sometimes a group of these fragments strikes the surface together to produce a chain of secondary craters.

A third type of crater ejecta consists of high-speed streams of material that are spewed out of

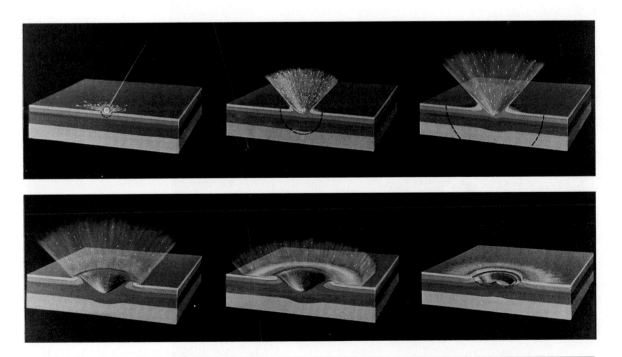

Figure 7.14 This diagram illustrates the formation of a large lunar crater from the explosive impact of an asteroid or comet.

large impacts like the splash created when a careless diver plunges into a pool. Such streamers can arc for hundreds or even thousands of kilometers across the lunar surface. Where they strike the ground, they produce many small secondary craters and generally stir up the surface, lightening its color. These long streamers or rays are well illustrated in Figure 7.15, which shows an oblique Apollo view of the Imbrium Basin with the 91-km crater Copernicus in the distance. Several rays stretch toward the observer across the dark mare surface. Near the foreground, we can see the association between the light ray material and clusters of irregular secondary craters. Since the impact speed of secondaries is rather low (less than 1 km/s), they do not generate true explosions. Therefore, these craters can be irregular or elongated, just like the pits produced by throwing stones into a sandbox.

EVOLUTION OF THE CRATER

Meanwhile, back at the crater, the lunar crust is trying to adjust in the aftermath of the impact. For a small crater, up to perhaps 10 km in diameter, little adjustment is necessary, and the crater retains approximately its original bowl shape. Larger craters, however, cannot be supported for long once the explosion is past. Under the force of gravity, sections of the crater wall collapse and slide downward, partially filling the center and creating a series of steplike terraces along the walls, thus reducing the depth/diameter ratio. Near the center of the crater, where a great weight of overlying material has been removed, the crust may rebound to create a central peak or group of peaks (Fig. 7.16).

Large craters tend to have flat floors, often flooded by later lava flows. Let us return to Tycho, where you may imagine yourself landing in the

Figure 7.15 This spectacular view of the Imbrium Basin with the crater Copernicus on the horizon illustrates the ejecta patterns from a large lunar impact. Extending toward the viewer are many rays and patterns of secondary craters formed by the Copernicus impact about a billion years ago.

Figure 7.16 King crater on the lunar farside. This 75-km-diameter crater is relatively fresh and shows all the features expected in a large impact crater, including central peak, flat floor, collapsed terraces on the inside walls, and a surrounding ejecta blanket.

interior. Standing on the floor, you would not have the sensation of being in a bowl-shaped depression. Instead, you would find yourself on a rather level rocky plain, with the distant crater rim visible as a low line of mountains serrating the horizon. The central peak would appear as a huge pile of rubble.

CRATER DENSITIES AND SURFACE AGES

The number of impact craters on a planetary surface is a measure of the age of that surface. On an active planet like the Earth, erosion and other geologic processes rapidly degrade and destroy craters. However, there is little degradation on the Moon, and the number of craters increases with time. We define the **crater-retention age** as the time over which the surface has been sufficiently stable to preserve craters once they are formed. On the Moon, practically the only events that can destroy craters are later impacts or lava

flooding during periods of large-scale volcanism. In the lunar maria, the crater-retention ages generally represent the time elapsed since the most recent major eruptions.

When we see differently cratered regions (for instance, in Fig. 7.3), we naturally interpret the differences as the result of different crater-retention ages. After all, we know the impacting projectiles came from outside, and there is no reason to think that they preferentially struck in some regions—the highlands, for instance—while sparing others. The situation resembles a city street in the midst of a long, windless snowstorm. As you walk along, you will find the sidewalk in front of some houses deeply covered with snow, while in other places the depth is less, with a few areas of sidewalk nearly clear. Do you conclude that different amounts of snow have fallen in front of the Joneses' house than at the Smiths' next door? No; you attribute the differing depth to the time that has passed since that section of walk was shoveled. The less snow, the shorter the "snow-retention age." It is just the same with craters.

To determine a crater-retention age, we must first count the number of craters that can be seen on a particular terrain being studied. The number of craters on a given area is called the **crater density**. This density has nothing to do with the material density used throughout this book to characterize the bulk composition of an object. It is simply the expression of the number of craters of a given size on a well-defined area of the surface, usually taken as 1 million km^2, about the size of the state of Texas on Earth or of Mare Imbrium on the Moon.

CRATER SIZE DISTRIBUTIONS

Look at the photographs of the Moon throughout this chapter. You will notice that there are always many more small craters than large ones. The reason is obvious: As we saw in Chapter 5, small meteoroids are much more abundant than large ones. Fragments in space, from asteroids down to pebbles, tend toward a size distribution in which there are more than a hundred 10-km meteoroids for each one that is 100 km in diameter, more than a hundred 1-km objects for each 10-km one, and so forth. Since each lunar crater has a diameter approximately ten times that of the impactor, the craters will display the same sort of size distribution, scaled up by a factor of ten. Indeed, scientists have turned the argument around. By counting lunar craters in different size ranges, they have determined the size distribution of the projectiles (which are mostly near-Earth asteroids; see Section 5.3).

Figure 7.17(a) illustrates schematically the crater size distribution observed on the Moon for craters larger than 3 km in diameter. There are many more small craters than large ones, as we expect. In Figure 7.17(b), exactly these same craters are randomly distributed over a surface, illustrating the typical appearance of a planetary surface where there is no erosion and all craters are retained once they form.

The measure of crater density that is commonly used is a cumulative one; that is, we count all craters larger than a given minimum size, frequently 10 km. For example, the 10-km cumulative crater density on the average lunar mare is 50 per million square kilometers, while that on the

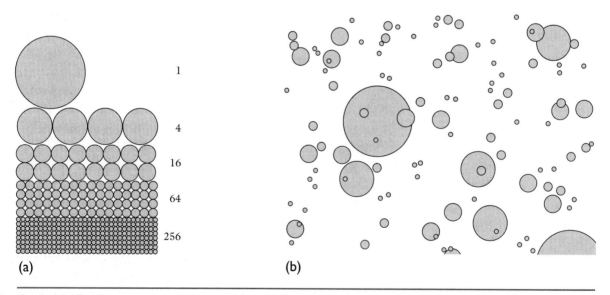

(a)   (b)

Figure 7.17 **(a)** The relative number of objects that could strike the Moon, according to a law in which the number of objects increases in inverse proportion to the square of the object's radius. **(b)** A random distribution of craters made by the population of objects shown in (a).

Table 7.1 Cumulative crater densities on the lunar surface

Lunar Region	Crater Density (10-km craters per 1 million km^2)	Age (billion years)
Highlands	1000	> 4.0
Fra Mauro (*Apollo 14*)	130	3.9
Apennine (*Apollo 15*)	95	3.9
Tranquillitatis (*Apollo 11*)	50	3.7
Fecunditatis (Luna 16)	30	3.4
Putredinis (*Apollo 15*)	25	3.3
Procellarum (*Apollo 12*)	20	3.3
Copernicus Crater	10	0.9
Tycho Crater	2	0.2

lunar highlands is 1000 per million square kilometers. Cumulative crater densities for a number of areas of the Moon are listed in Table 7.1.

How much of the lunar surface is actually covered by craters? Here is an approximate calculation. For the lunar crater size distribution, the average size crater is about twice as large as the lower limit counted. In our case, where we are considering craters down to 10 km in diameter, the average diameter is about 20 km, so the area of each crater is about 400 km^2. A quick multiplication shows that on the maria, where the crater density is 50 per 1 million km^2, only about 2% ($50 \times 400/1,000,000$) of the surface is cratered. You can verify these calculations by examining the photos of highland and mare terrain in this chapter and counting the craters yourself.

ABSOLUTE AND RELATIVE AGES

The greater the crater density on a planet, the older the surface. But how much older? Only for the Moon, where scientists have measured both the crater densities and the associated radioactive ages from returned samples, can this question be answered with any precision. On Earth, there are too few craters, whereas on other planets, there is no absolute age scale defined by returned samples.

All of the lunar maria have roughly similar crater densities, and all therefore appear to have comparable ages. This fact was recognized before the Apollo missions, but without returned samples there was a lively debate among lunar scientists concerning the absolute age of the maria and, hence, the period since the cessation of major lunar volcanism. If one assumes that the rate of impacts has remained constant throughout solar system history, then the fact that the mare crater densities are about 1/20 of the highland densities suggests that the maria are only 1/20 as old as the highlands. If the highlands are as old as the Earth, 4.5 billion years, the maria would be only about 200 million years old.

Is this conclusion consistent with what we know about the current impact rates from comets, near-Earth asteroids, and other meteoroids? In the 1960s, many scientists thought not. Starting from the known numbers of objects in near-Earth space, they calculated that it would require closer to 4 billion years, rather than 200 million, to accumulate the observed mare crater density. In other words, the flooding of the maria by lava flows was restricted to the early period of lunar history. They therefore argued for an inactive Moon, on which the volcanic fires cooled billions of years ago.

This explanation poses a problem, however. If the maria and the highlands are both billions of years old, how can the crater density in the highlands be 20 times greater than on the maria? In order for these parts of the Moon to have accumulated so many craters, a much higher early impact rate would have been required, between the formation of the highland crust and the period of lunar volcanism represented by the maria.

When the first lunar samples were returned to Earth in 1969, the most eagerly awaited scientific result was the radioactive age determination for mare materials. The solidification age, for a variety of Mare Tranquillitatis rocks and soil samples, was measured to be 3.7 billion years. Subsequent measurements on samples from other mare sites confirm that all the mare solidification ages are between 3.2 and 3.9 billion years. Highland rocks, in contrast, are older than 4 billion years, whereas the estimated age of the entire Moon (since differentiation) is about 4.5 billion years.

7.4 EARLY LUNAR HISTORY

Early in its history (in a time period that has been erased on Earth by subsequent geologic events), the Moon was subject to many impacts, some of them of gargantuan size. Compared with the rate of impacts we see today in the solar system, this was a period of catastrophic changes. **Catastrophism** is a term used in geology to refer to sudden or violent events that substantially modify the landscape. Often it is contrasted with **uniformitarianism,** which seeks explanations of the world around us in terms of very slow processes acting over vast time spans. Large impacts and massive volcanic eruptions are examples of catastrophic events, whereas the gradual accumulation of ocean sediments is the product of uniformitarian processes. Although proponents of the two concepts have been in conflict in the past, we now recognize that both types of geological processes are important. Occasional catastrophes are a natural part of the long span of both geological and biological evolution.

A UNIFYING CONCEPT OF CRATERING ON THE INNER PLANETS

The combination of lunar crater counts, dated lunar samples, and measurements of the numbers of projectiles in near-Earth space today has given rise to the current picture of early planetary history. Judging from the lunar evidence, the impact rates on all of the inner planets have not changed greatly during the past 3.8 billion years. Presumably the impacting bodies over this timespan have been the comets and near-Earth asteroids that we recognize today. If similar numbers of projectiles have struck Mercury, Venus, or Mars, then the lunar chronology can be applied to crater densities on these planets as well.

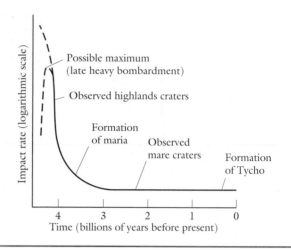

Figure 7.18 Schematic illustration of the variation over history in the flux of projectiles on the Moon. The rate of formation of craters was much higher during the first few hundred million years of lunar history, but has remained nearly constant for the past 3.8 billion years.

According to this theory, however, quite different conditions prevailed during the first 700 million years of solar system history. Before about 3.8 billion years ago, the rate of impacts must have been much higher; otherwise the high crater densities on the highlands cannot be understood. From lunar research, it appears that the impact rate was a thousand times greater at 4.0 billion years ago than at 3.8 billion years ago (Fig. 7.18). Earlier it may have been much higher yet, but the evidence concerning lunar history before 4.0 billion years ago is limited.

This period around 3.9 billion years ago is variously called the late heavy bombardment or the terminal bombardment of the Moon. It is presumed to have involved a different source of impacting objects from those present today. Because the impact rates were very high at this time, many scientists suspect that they were the result of a unique event, such as the disruption of several large asteroids, which scattered fragments through the inner solar system. Alternatively, the terminal bombardment may represent the final stages of planetary accretion. Either way, this period of heavy cratering has largely obliterated direct evidence of the first half-billion years of lunar history.

Figure 7.19 Eugene Shoemaker of the U.S. Geological Survey and the Lowell Observatory played a leading role in developing a theory of the role of impacts from comets and asteroids over the history of the solar system. He is shown here with his wife, Carolyn, who has discovered more comets than anyone else in history.

The idea that the inner solar system experienced a late heavy bombardment 3.9 billion years ago, followed by nearly constant impact rates extending to the present, constitutes one of the paradigms, or generally accepted theories, of modern planetary science. One of the many scientists who developed this paradigm was Eugene Shoemaker, who is sometimes called the father of planetary geology (Fig. 7.19). Shoemaker studied impact craters all his life and wrote his doctoral dissertation on the geology of Arizona's Meteor Crater. In the 1960s, he was one of the Apollo geologists who guessed the wrong lunar chronology, predicting relatively young mare ages. After Apollo, Shoemaker began a major observational and theoretical program to understand the orbits and size distributions of asteroids and comets in Earth-Moon space today. His work demonstrated that the number of small Earth-crossing asteroids is consistent with the lunar impact rates deduced from mare crater counts. Shoemaker later extended this cratering paradigm to the satellites in the outer solar system. As we mentioned in Chapter 5, the NEAR mission to the asteroid Eros was named for Shoemaker after he died during a field trip to map impact craters in Australia. A capsule containing his ashes was also sent to the Moon on the Lunar Prospector mission, so that a memento of Shoemaker now rests on the world he so much wanted to visit in person.

Application of the cratering paradigm to other planets, particularly Mercury and Mars, has yielded important insights concerning solar system history. But all unproven hypotheses must be treated with some caution. We know the lunar chronology, but its application elsewhere carries a certain risk, at least so long as we do not know the origin of the late heavy bombardment. We will try to keep this reservation in mind when we come to interpret crater counts on other planets.

THE LUNAR HIGHLANDS

On the Moon, most of the catastrophic geological events took place early in its history. The lunar highlands are the oldest surviving part of the lunar crust (Fig. 7.20). The degree of highland cratering far exceeds that on the maria. Since the highland craters are packed virtually shoulder to shoulder, we have no way of telling how many

Figure 7.20 This view of the heavily cratered lunar highlands was taken from the *Apollo 17* command module.

impacts took place in the past. In these areas, the crater density has apparently reached a state of saturation, in which new impacts do not create additional craters; they simply replace old craters with new ones.

An indication of the magnitude of highland cratering is provided by the behavior of moonquakes studied by the Apollo ALSEP instruments. Seismic waves provide an excellent way to probe interior structure, as we will discuss further in Chapter 9. The reverberation of seismic waves in the highlands indicates that the crust has been shattered to a depth of 25 km, and that the fragmented surface layer of rubble has an average depth of hundreds of meters. In contrast, the fragmented soil on the maria is only about 10 m deep.

As might be expected from this history, the rocks of the lunar highlands are breccias, composed of recemented pieces broken and scattered by earlier impacts (Fig. 7.21). Many highland breccias are extremely complex, indicating three and even four generations of shattering and reforming of the rocks. Often the highland breccias contain frozen droplets of impact-melted mate-

Figure 7.21 Many lunar rocks are breccias, like this sample from the highlands. These rocks, consisting of individual fragments cemented together, display a long history of impacts. Most lunar breccias formed during the high-bombardment period of the first half-billion years of lunar history.

rial. By comparison, the meteoritic breccias discussed in Section 4.2 are relatively simple, consistent with their formation on smaller bodies.

HIGHLAND SAMPLES

The oldest dated highland samples are not whole rocks but tiny individual fragments within breccias. Most of these show solidification ages near 4.2 billion years, although a few go back as far as 4.4 billion years. The apparent age cutoff of 4.2 billion years does not mean that the surface remained molten up to that time. More likely, the radioactive clock was repeatedly reset for highland samples by heating and intense shock from multiple impact events.

Dating of Apollo samples indicates that the period of lunar catastrophism extended up to nearly 3.8 billion years ago, when the rain of impacts subsided to more nearly its present low rate. As lunar scientist Stuart Ross Taylor observed, "The meteoritic bombardment has drawn a curtain across the landscape through which we peer dimly to discern the earlier history. The [original] structure of the lunar crust is obliterated, and there is no vestige of a beginning."

One way to try to pull this curtain aside is to look at the bulk composition of the highland lunar crust. This material, like all lunar samples, is severely depleted in **volatile elements**—that is, in those elements that can be evaporated or vaporized at relatively low temperatures. Note that this is a slightly different use of the term *volatile*, to mean elements rather than ices as used in Chapters 4 and 5. Ices are extremely volatile, but geologists also consider any element that evaporates at the temperature of lava, about 1000°C, to be volatile. On the Moon water is absent; the lunar minerals include no clays or other familiar terrestrial forms that incorporate chemically bound water. Other common volatiles that are depleted include nitrogen, carbon, sulfur, chlorine, and potassium. The lack of both metals and volatiles on the Moon is an important clue to its origin, as we will discuss in the next chapter.

The primary material of the highland crust is a class of igneous silicate rocks called **anorthosites.** The lunar anorthosites consist primarily of mineral oxides of silicon, aluminum, calcium, and magnesium. Anorthosites are rocks that might form out of an originally molten Moon, and their presence argues for an early differentiation, before the lunar crust solidified. Battered though these fragments of the crust may be, they still provide some information on this earliest stage of lunar history.

During the hundreds of millions of years of heavy bombardment that followed the formation of the original crust, many episodes of remelting triggered by impacts probably occurred. We can imagine large projectiles breaking through the crust to release floods of still liquid rock from the interior. At some point after the crust had stabilized and thickened, more conventional volcanic activity also began. Although now nearly obscured by subsequent events, indications exist of the formation of the first basalts as early as 4.2 billion years ago.

LUNAR BASINS

The surviving lunar impact basins were formed during the final stages of the heavy bombardment. We use the term *basin* to refer to any impact feature larger than about 300 km in diameter. The impacting meteoroids that blasted out these features must have been 30–100 km in diameter. Possibly many basins were formed from a cluster of asteroidal impacts that occurred during the terminal bombardment, in the relatively brief interval from about 4.1 to 3.9 billion years ago. Today lunar scientists have identified about 30 ancient basins, including many on the lunar farside. The largest known basin is located near the south pole and has a diameter of roughly 2200 km.

The youngest of the great basins are Imbrium and Orientale (see Figs. 7.15 and 7.5), each about the size of the state of Texas. The Imbrium event can be precisely dated using samples from *Apollo 14* and *15*. The time of the impact was just over 3.9 billion years ago, and the extensive ejecta, which overlie older features on much of the nearside of the Moon, provide a reference marker for lunar stratigraphy. The ejecta from Orientale are found on top of Imbrium features, indicating that the Orientale event occurred later. Although no Apollo landing took place on Orien-

tale ejecta, indirect evidence dates this last great impact between 3.8 and 3.9 billion years before the present.

Both Imbrium and Orientale are mountain-ringed circular features; today they differ in appearance primarily because the Imbrium Basin was later flooded by lavas to produce Mare Imbrium, while the Orientale Basin escaped major volcanic modification. The outer mountain ring of Imbrium, consisting of the lunar Carpathians, Apennines, and Caucasus, rises as high as 9 km above the present mare material, similar to the height of the Hawaiian volcano Mauna Loa above the ocean floor. The diameter of this outer ring is 1200 km. The Fra Mauro formation, which is the ejecta blanket from Imbrium, extends outward nearly a thousand kilometers and varies in thickness from more than a kilometer down to a fraction of a meter (Fig. 7.22). The Orientale Basin is defined by a 900-km outer mountain ring, the Cordilleras, and a 600-km inner ring, the Rook

Figure 7.22 The Fra Mauro formation, shown in this view of the *Apollo 14* landing site, is the hummocky material extending through the center of the frame. This material was ejected from the impact that formed the Imbrium Basin, located about 600 km north of the location shown here.

Figure 7.23 One of the best views of mountains on the Moon was obtained by the *Apollo 15* crew, who landed near the base of the Apennine Mountains. Mount Hadley, shown here, was formed by uplift associated with the impact that created the Imbrium Basin.

Mountains. The floor of this unflooded basin consists of rough and hilly contours and patches of impact-melted rock.

The process of formation of basin-rim mountains is not entirely understood, but it apparently involved a combination of uplift, produced by the blast itself, and later subsidence along concentric cracks as the lunar surface adjusted following the impact. A cross section of the Apennine mountains is shown in Figure 7.23, which reveals layers in the uplifted basin rim. Other isolated mountains on the Moon are probably rootless piles of ejecta.

A Perspective

Compared with conditions today, the era of heavy bombardment was a time of great violence, with impact rates tens of thousands of times greater than those experienced now. Thinking back to that time, we tend to picture a sky filled with projectiles or a surface pockmarked with fresh craters. A bit of arithmetic will quickly convince us otherwise, however. Today a 1-km crater is formed over 1 million km² about once per million years. When the impact rate was 10,000 times greater, such a crater was still formed once per century. An individual witness is aware of events only to a radius of about 100 km, or about 4% of 1 million km². Thus, a hypothetical observer standing on the Moon or Earth during the period of heavy bombardment would have witnessed the formation of a 1-km crater only once in 2500 years! Events that seem catastrophic on a planetary timescale are still rare from our human perspective.

7.5 Lunar Volcanism

Even before the end of the heavy bombardment, volcanic vents were erupting on the lunar surface. However, the major period of lunar volcanism apparently did not begin until shortly after the formation of the Imbrium and Orientale Basins, about 3.8 billion years ago. During the following half-billion years, repeated outpourings gradually filled the nearside basins with dark basalt to create the familiar pattern of lunar maria.

The Lunar Maria

The story of lunar volcanism is essentially the story of the maria. Although there may have been isolated examples of volcanic activity elsewhere, they were of negligible significance compared with the great outpourings of mare lava. Fortunately, several landing sites on the flat mare surfaces were selected by the safety-conscious Apollo planners, thereby providing us with many observations and large quantities of returned mare rocks and soil.

The mare basalts resemble their terrestrial counterparts and are believed to have originated in a similar way, from subsurface melting and eruption of lava. However, lunar basalts have a higher iron content, including metallic iron, and are free of water and its effects. Since all lunar rocks are igneous and formed in an environment lacking both water and atmospheric oxygen, they have simpler compositions than terrestrial rocks. Only about 100 separate minerals have been identified on the Moon, in contrast to more than 2000 on the Earth. There is no clay or other water-bearing rock and only a tiny amount of carbonaceous material.

The oldest lunar basalts, obtained from the *Apollo 14* and *Apollo 17* sites, have solidification ages of 3.8 to 3.9 billion years. The flows on the margin of Mare Imbrium sampled by *Apollo 15* are dated at 3.3 billion years, which presumably fixes the late stages of volcanism in this basin. *Apollo 12*, which landed on Oceanus Procellarum, yielded the youngest mare basalt, with a solidification age of just under 3.2 billion years.

THE MARIA-FORMING ERUPTIONS

Unlike most of the volcanic activity with which we are familiar on Earth, the lunar eruptions did not normally create volcanic mountains. Instead, the eruption of large volumes of fluid lava from long fissures yielded thin, flat flows of enormous extent. On Earth, similar flood basalt eruptions are responsible for the lava plains of eastern Washington State and southwestern India. Some geologists suspect that the largest flood basalt eruptions on Earth may have been implicated in mass extinctions of species, a subject we will return to in Chapter 12.

Most of the lunar lava flows have been buried by subsequent eruptions, but the final large out-

pourings can still be identified. On Mare Imbrium, the last major eruptions originated from a fissure about 20 km long and spread more than a thousand kilometers from the vent, covering a total area of 0.2 million km^2, or about the size of the state of Utah. Along the flow fronts the thickness is typically 30–50 m. A hundred flows of this thickness are required to account for the total depth of the mare, which is estimated at about 5 km. As much as a million years might have elapsed between individual flows.

After the eruption of the mare material, some settling and subsidence took place, perhaps as a result of shrinkage as the lava cooled. In places, "high-water marks" show that lava once reached up to 100 m higher than the current level. Many cracks have formed around the margins of the maria, often concentric with the center of the basin. In other areas, the lavas seem to have been compressed to form wrinkle ridges resembling the wrinkles formed in a rug by pushing it against a wall (Fig. 7.24).

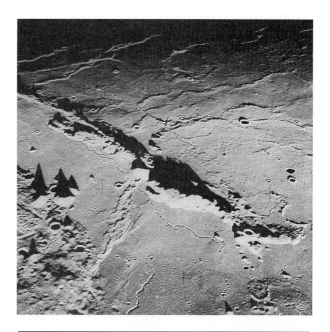

Figure 7.24 The Aristarchus Plateau, seen in this *Apollo 15* view, shows the many wrinkle ridges that characterize the lava plains that make up the lunar maria.

VOLCANIC VALLEYS AND MOUNTAINS

Among the most remarkable geologic features of the maria are the curving valleys called sinuous rilles. Before Apollo, many scientists thought these broad, meandering channels, which resemble terrestrial rivers, had been carved by running water. We now know that the Moon is, and always has been, dry. Since the sinuous rilles are found in the maria, it seems likely that they too represent some kind of volcanic phenomenon. Subsequently, the Magellan spacecraft revealed even larger lava channels on Venus, a subject we will discuss in Chapter 10.

Apollo 15 was targeted to visit one of the largest, Hadley Rille, near the edge of Mare Imbrium (Fig. 7.25). When the astronauts drove their rover to the edge of the channel, they found a curving valley 1.2 km wide and 370 m deep, littered with fallen boulders, some as large as a small house. On the far wall they could make out an exposed 50-m-thick layer of solidified lava. Hadley Rille seems to be a channel scoured by an ancient lava river.

Although volcanic mountains are extremely rare on the Moon, they may not be completely absent. A group of domes in Oceanus Procellarum, the Marius Hills, are thought by some geologists to be true volcanoes. The Marius Hills were to have been visited by a later Apollo flight, but when the program was canceled in 1972, this mission was among the casualties.

SOURCES OF LUNAR ERUPTIONS

Why are the maria confined primarily to the side of the Moon facing the Earth? The main reason seems to be that the average surface elevation is lower on the Earth-facing hemisphere. The subterranean pressures that forced lava to the surface appear not to have been great enough to produce major eruptions in the highlands, including most of the farside. In this respect the maria do indeed resemble seas, filling low-lying areas just as the oceans of Earth occupy the basins between the continents.

Substantial information on the source regions of the lunar eruptions has been derived from a detailed study of the composition of the lunar

(a)

(b)

Figure 7.25 One of the largest of the ancient lava rivers on the Moon is Hadley Rille, visited by the *Apollo 15* astronauts: **(a)** the curving valley, more than 1 km wide, is seen from orbit; **(b)** the view from the surface.

basalts. Liquified rock in the interior of a planet is called **magma.** The mare lava samples are derived from magma that has been three times chemically separated or fractionated from the original material of the solar nebula. First there was the loss of water and other volatiles that is characteristic of the entire Moon. Second was differentiation, in which the lunar interior separated into core, mantle, and crust. Finally, there was a process of "partial melting," in which the more easily melted minerals of the mantle were liquified to create large magma reservoirs hundreds of kilometers below the surface.

The major lunar volcanic activity ceased a little more than 3 billion years ago. This means that if we saw the Moon 3 billion years ago it would look much as it does today (absent young craters such as Tycho and Copernicus), whereas the continents of the Earth's crust would have been unrecognizable even a few hundred million years ago.

Figure 7.26   This astronaut's bootprint in the lunar soil illustrates the strength and cohesiveness of the fine upper layer of the lunar regolith.

7.6 THE SURFACE OF THE MOON

When the first Apollo astronauts stepped onto the lunar surface, they found themselves in a stark but beautiful world. The airless sky was deep black, the surrounding plains a dark brownish gray. The force of gravity was just one-sixth that at the surface of the Earth, making their bulky spacesuits seem relatively light. With no haze, distant details stood out as sharply as those in the foreground. The mare plain itself was flat and covered with scattered irregular rocks of all sizes and shapes. Some distance away the low profiles of small craters could just be identified, but at first sight there was little evidence to establish that the astronauts stood on a heavily cratered planet.

THE LUNAR REGOLITH

Once they began to move about, the astronauts quickly became aware that the surface of the Moon was layered with a fine dark dust. Every step raised clouds of this material, which soon coated their spacesuits and eventually found its way into all of their equipment. In spite of this omnipresent dry dust, they found the lunar soil to be firm underfoot. Their boots sank only a few centimeters into the soil, and the bootprints they made were crisp and sharp-edged, as if they had been walking in damp dirt or crunchy snow (Fig. 7.26).

Although later Apollo flights landed in rougher and more scenic locales than that first site in Mare Tranquillitatis, the basic character of the immediate surroundings was the same everywhere. The entire Moon appears to be covered with fragmented rock and dust, which represents the ejecta from impact craters, far and near. This continuous debris blanket has been named the lunar **regolith.** The finer component is generally referred to as *lunar soil,* whereas the term *regolith* includes the full range from fine dust to house-sized blocks.

The depth of the regolith varies with the age of the surface. On the maria, it is typically 10 m thick, whereas in the highlands, the regolith may be as thick as hundreds of meters. As shown by core samples, the regolith is composed of overlapping layers of ejecta, each typically a few centimeters thick. One *Apollo 17* core, which penetrated nearly 3 m of regolith, revealed 42 distinct layers, each representing ejecta from a single impact.

In the post–heavy-bombardment era, the lunar regolith accumulates at an average rate of about 2 mm per million years, or 2 m per billion years. In comparison with terrestrial erosion and deposition processes, this is slow change indeed. Along with the buildup by addition of ejecta layers, the upper part of the regolith is frequently disturbed by the impacts of very small meteoroids, which stir or "garden" the top few millimeters of material and help to maintain the loose fine dust.

Studies of the processes that form the regolith indicate that most of the ejecta at any one site is derived from relatively nearby impacts. If this were not the case, the sharp boundaries between mare and highland materials would have been blurred by horizontal mixing, and we would not see the Man in the Moon so clearly. Only about 5% of the material comes from as far away as 100 km, and only 0.5% from 1000 km. These calculations apply to the normal cratering process, not to ejecta from the major basins, which have blanketed substantial fractions of the lunar surface to depths of many meters.

In addition to original lunar material, the regolith includes meteoritic fragments from billions of years of cosmic bombardment. Typically, between 1% and 2% is meteoritic, apparently about the same composition as the primitive meteorites collected on Earth. Most of the small amount of carbon in the lunar soil is derived from impacting carbonaceous meteorites.

When examined under a microscope, all lunar soils are found to contain large quantities of glass in the form of spherules up to a millimeter in size (Fig. 7.27). Glass is simply silicates that have been melted and then rapidly cooled so that no crystals can form. These glasses, which are characteristic of the lunar soil, are the product of melting of lunar rock during meteorite impact. Since its atmosphere protects the Earth from these micrometeorites, our soil does not contain similar glass spherules, except at a few sites associated with terrestrial impact craters. The tiny spherules can themselves be pitted by even smaller micrometeorites when they are exposed on the lunar surface (Fig. 7.28). Small impacts also liberate tiny grains of pure metal from the soil, which darken the surface.

Figure 7.27 Viewed under a microscope, the lunar soil is composed of a variety of small rock fragments mixed with glass spheres produced by impact melting. Scale markers in the lower right are 1 mm apart.

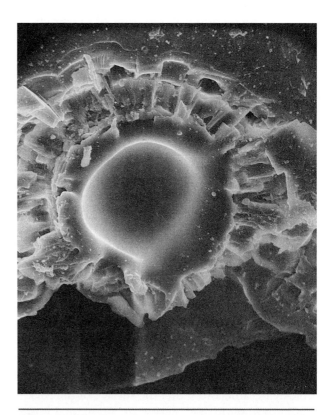

Figure 7.28 This photograph of a droplet of lunar glass, made with a scanning electron microscope, shows a beautiful microcrater produced by the high-speed impact of a tiny fragment of interplanetary dust.

PERSPECTIVE

BEING THERE AND GOING BACK

"That's one small step for a man, one giant leap for mankind."
—Neil Armstrong

What is it like to walk on the Moon? Long before the astronauts arrived there was plenty of speculation on this topic. The lunar gravity if about 1/6 that of the Earth, so people naturally thought that leaping about would be easy. Here is an astronaut's account:

BUZZ ALDRIN: I took off jogging to test my maneuverability. The exercise gave me an odd sensation and looked even more odd when I later saw the films of it. With bulky suits on, we seemed to be moving in slow motion . . . My Earth weight, with the back pack and heavy suit, was 360 pounds. On the Moon, I weighed only 60 pounds.

Jack Schmitt found that getting into the lunar roving vehicle required a certain art.

JACK SCHMITT: You stand facing forward by the side of the vehicle, jump upward about two feet with a simultaneous sideways push, kick your feet out ahead, and wait as you slowly settle into the seat, ideally the correct one.

Those big pressure suits and backpacks were of course essential to survive on the Moon's airless surface. The absence of an atmosphere enhanced the contrast of sky and ground, although the color of the ground itself is remarkably uniform.

BUZZ ALDRIN: At one point I remarked that the surface was "Beautiful, beautiful. Magnificent desolation." I was struck by the contrast between the starkness of the shadows and the desert-like barrenness of the rest of the surface. It ranged from dusty gray to light tan . . . We could also look around and see the Earth, which seemed small—a beckoning oasis shining far away in the sky.

Even for the men who once walked there, the Moon now seems very far away.

STU ROOSA: I look at the Moon all the time and I say these things: "I was there!" Sometimes I joke, "Was that another life? Another lifetime away?" And it really was. It was another life.

There is certainly much more to be learned from the Moon. While there are no approved human missions, there are many plans, some of them very ambitious. Whether or not any of this ever happens will depend on political events on Earth that are impossible to predict. Sending humans to the Moon is an enormously expensive enterprise, and there are many pressing social needs that have a more immediate claim on our resources. It is not at all obvious what set of stimuli comparable to the ones President Kennedy responded to will motivate the nations of the world to unite in the effort to make a return trip to that distant looking glass. But one day, surely, they will return. And not just as an end in itself, but as the first step toward the wider exploration of the planetary system.

DAVE SCOTT: Arriving at the lunar module [for the trip home] I experience a sense of impending loss. Soon I will leave the Moon, probably forever And, in a peculiar way, I have come to feel a strange affection for this peaceful, changeless companion of the Earth.

SURFACE CONDITIONS AND EROSION

In the absence of moderating air or oceans, the surface of the Moon experiences much greater temperature ranges than the Earth does. The contrast between day and night is further magnified by the fact that day and night are each two weeks long. At the near-equatorial Apollo sites, the maximum surface temperature is about 110°C, higher than the boiling point of water. During the long lunar night, the surface temperature drops to −170°C, only about 100 K. It is a tribute to the design of the Surveyor spacecraft and the ALSEP instruments that they were able to operate over this huge temperature range. The astronauts themselves were never subjected to such extremes, since they always landed in the lunar morning when the temperatures were similar to those on the Earth.

The last three Apollo flights visited mountainous areas: the Apennines, the Descartes highlands, and the Taurus Mountains. Notable in each of these landscapes were the gentle, rounded contours of the lunar mountains and hills. The illustrations in science fiction stories had depicted the mountains of the Moon as extremely steep and spiky, but the reality was otherwise. A partial explanation for the soft contours of lunar mountains can be found in the ejecta that blankets all lunar features, but the primary reason there are no sharp peaks or steep cliffs on the Moon is that there is no water or ice erosion to sculpt them. Water and ice erosion on Earth do not simply wear down mountains; they also cut the valleys and shape the peaks. In the absence of such natural forces, the mountains on any planet will be as gentle as those of the Moon.

7.7 QUANTITATIVE SUPPLEMENT:
IMPACT ENERGIES

In order to determine the size of a crater formed from the impact of an asteroid or comet, we need to calculate the energy released by such an impact. This energy depends on the mass (m) and speed (v) of the projectile, as expressed by the standard equation for kinetic energy:

$$E = \frac{1}{2}mv^2$$

The speed is made up of two components: the speed of the projectile in space before it approaches the planet, and the additional speed it gains as it falls to the surface accelerated by the planet's gravitation. The minimum impact speed is the escape velocity of the planet, to which the initial relative speed must be added. For large planets, the speed gained during the fall generally exceeds the initial speed.

The escape velocity of the Earth is 11 km/s, or 1.1×10^4 m/s. Thus, we calculate the minimum impact energy per kilogram of the projectile to be 6×10^7 joules. This is about ten times larger than the chemical energy of TNT, which is 4×10^6 joules/kg, explaining why we have emphasized that the crater-forming event is in many ways similar to a bomb explosion.

Since the escape velocity of the Moon is only 2.4 km/s, impacts of much lower energy are possible. Typically, a projectile strikes the Moon with only about half the speed with which it would strike the Earth, resulting in an energy only about one-fourth as great. The mass (and volume) of material ejected in the impact is roughly proportional to the energy, and the linear dimensions (diameter and depth) of the crater are proportional to the cube root of the volume. On the basis of this calculation, we expect that a given impacting body makes a lunar crater only two-thirds as large as a terrestrial crater (about 10 and 15 times the projectile diameter, respectively).

Just how large are these impacting energies? A 10-km-diameter asteroid with a density of 2.5 g/cm³ has a mass of about 1.3×10^{15} kg. Striking the Earth with a speed of 20 km/s, it would have an energy of about 2.6×10^{23} joules. We have already noted that 1 kg of TNT is equivalent to 4×10^6 joules, so the energy of impact can also be expressed as 65×10^{15} kg of TNT = 65,000,000 megatons of TNT = 65 gigatons of TNT. For comparison, the fission bombs dropped on Hiroshima and Nagasaki had a yield of 20 kilotons, and the largest nuclear device ever tested was a Soviet bomb with a yield of 53 megatons.

Summary

The Moon is the best-studied planetary body after the Earth, as the result of the Apollo expeditions and their treasure of returned lunar samples. Unlike the Earth, however, our satellite is no longer geologically active, nor does it possess an atmosphere. Therefore, the Moon is a much easier place to understand, although still complex enough to leave us with a number of fundamental questions unanswered.

The Moon is a differentiated planet, with a highland crust of anorthosite that solidified at least 4.4 billion years ago, much older than any rocks that still are present on the Earth. During its first half-billion years, the Moon (and probably the other inner planets as well) was subjected to an intense bombardment, which saturated its surface with craters and created a deep regolith of broken rock. The major basins, including Orientale and Imbrium, were formed near the end of this period. About 4 billion years ago, a final burst of impacts occurred, after which the bombardment dropped to approximately its current rate. The highland crust preserves a record of this early period, while the much less cratered lunar maria represent younger surfaces.

The maria, which cover 17% of the surface, are composed of very fluid basaltic lava flows that erupted between 3.9 and 3.2 billion years ago from magma reservoirs deep beneath the lunar crust. They generally occupy low-lying areas on the nearside of the Moon, primarily within impact basins. There is no evidence of any major activity during the past 3 billion years, a period when the lunar surface has been modified only by impact cratering, including the creation of a fine, glass-rich soil from the eroding effects of many small impacts.

The dominant landform of the lunar surface is the impact crater, ranging in size from basins with diameters of 2000 km down to microscopic pits. We discuss the craters in some detail because they are a common phenomenon on other planets; their characteristic shapes (including central peaks, ejecta blankets, and secondary craters) are the result of violent explosions caused by the sudden impacts of fast-moving projectiles. The density of craters is determined by the rate at which asteroids and comets strike the surface and by the age of the surface. Crater counting provides a good measure of relative surface ages, but it must be used with caution when trying to estimate absolute ages.

The Moon differs greatly from the Earth. In spite of their proximity in space, the Moon's bulk chemical composition is depleted in both metals and volatiles compared with the Earth. This is a major mystery, which we will return to in Chapter 8. In addition, the Moon's small mass, less than 2% that of the Earth, has led to much lower levels of geological activity. With its relatively cool, solid interior, the Moon has been unable to form the continents, mountain ranges, volcanoes, and other characteristic geological features of our own planet. Also because of its small mass, the Moon has no atmosphere, which creates a very different surface environment of extreme temperatures, with constant bombardment by micrometeorites and cosmic rays. Although unlike the Earth, the Moon is rather similar to one other planet, Mercury. In the next chapter, we will compare the Moon and Mercury and probe further into the origin and evolution of both.

Key Terms

anorthosite	magma
catastrophism	mare / maria
crater density	regolith
crater-retention age	resolution
ejecta blanket	secondary crater
highlands	stratigraphy
impact basin	uniformitarianism

Review Questions

1. Describe the appearance of the Moon, noting what kinds of features can be seen at different levels of resolution.

2. Impact cratering is an important process in planetary geology. Describe how craters are formed and note what features are characteristic of impact craters. How can you tell an impact crater from a volcanic crater? Why did it take so long for scientists to recognize that the lunar craters are of impact origin?

3. Determining the age of a planet's surface is another basic part of planetary geology. Discuss what is meant by stratigraphy, with illustrations from the history of formation and flooding in the Imbrium Basin. Discuss the concepts of crater density and crater-retention age. How are these relative ages translated into absolute ages for the lunar surface?

4. Describe the standard paradigm for the cratering history of the inner solar system. Note how our detailed knowledge of the Moon is used to define the stages in this history. What role is played by the asteroids and comets, and is this theory consistent with what we learned about asteroids and comets in Chapters 5 and 6?

5. Discuss volcanism on the Moon. When and where did the magmas form? Why do the lava flows occupy the nearside basins almost exclusively? Why are there so few volcanic mountains? Why did the volcanic activity cease billions of years ago?

6. Discuss the soil on the Moon. What erosional processes are at work? How do large impacts create and modify the surface layers? What are the roles of micrometeorites? What forms the lunar glass?

QUANTITATIVE EXERCISES

1. Compare the impact energy of a 1000-ton primary impact on the Moon (at 15 km/s) with the energy of a secondary impact of the same mass but at a speed of 1.5 km/s.

2. According to the crater scaling law discussed in this chapter, the diameter of a stony asteroid striking the Moon is about one-tenth the diameter of the resulting crater. Calculate the diameter and mass of the projectile that formed Tycho. What was the energy of this impact in megatons if the impacting speed was 15 km/s?

3. Consider a meteoroid striking the surface at exactly the escape velocity. If it makes a 1-km crater on the Earth, what size crater would it make on the asteroid Ceres, which has an escape velocity

given by $\sqrt{2\,GM/R}$, where G is the gravitational constant (6.67×10^{-11} newton m^2 kg^{-1}) and M and R are the mass and radius, respectively?

4. It can be estimated that the gravitational binding energy on the Moon is given approximately by GM^2/R, where G is the gravitational constant (6.67×10^{-11} newton m^2 kg^{-2}) and M and R are the Moon's mass and radius, respectively. An impact that releases more than this amount of energy is capable of breaking the Moon apart. Calculate the diameter of an impacting asteroid that would generate this amount of energy. Also calculate the size of the resulting crater, using the simple arguments for crater size scaling given in this chapter, and compare this size with the diameter of the Moon itself. This will give you an idea of the largest possible crater or basin that could be made on the Moon without disrupting the object entirely.

ADDITIONAL READING

Chapman, C. R. 1982. *Planets of Rock and Ice: From Mercury to the Moons of Saturn*, Chapters 2, 3, 5, and 8. New York: Scribner's. Popular introduction to planetary geology by a leading planetary scientist.

Cooper, H.S.F. 1970. *Moon Rocks*. New York: Dial Press. Journalist's account of the acquisition and study of lunar samples, originally published in *The New Yorker*.

Cortwright, E. M., editor. 1975. *Apollo Expeditions to the Moon (NASA SP-350)*. Washington, DC: U.S. Government Printing Office. Beautifully illustrated NASA publication describing the Apollo missions and their scientific harvest of information.

Masursky, H., and others, editors. 1978. *Apollo over the Moon: A View from Orbit (NASA SP-362)*. Washington, DC: U.S. Government Printing Office. Beautifully illustrated book of the best Apollo orbital photos, with informative text on lunar geology.

Taylor, S. R. 1975. *Lunar Science: A Post-Apollo View*. New York: Pergamon Press. Authoritative and well-written monograph on lunar geology and geochemistry, aimed at the science student (rather technical and now somewhat dated).

8

MOON AND MERCURY: STRANGE RELATIVES

Mercury was traditionally known as the messenger of the gods. Fleet of foot, he is usually depicted with wings on his heels and a winged helmet, as in this bronze statue by Francesco Righetti (1749–1819).

Our next step toward larger-size planets takes us from the Moon to Mercury. In this chapter, we discuss Mercury and compare this planet with our satellite. The five terrestrial bodies—Mercury, Venus, Earth, Moon, and Mars—are an appropriate group for comparison. The larger three, Venus, Earth, and Mars, all have substantial atmospheres and long histories of geological activity. In contrast, Mercury and the Moon invite comparison to each other as smaller, airless, less active worlds.

Although the Moon and Mercury are similar in many ways, they have one outstanding difference: their bulk composition. The Moon is a rocky planet, depleted in metals. In contrast, Mercury is a planet made mostly of metal, as is evident from its high density. Compositionally, Mercury is like the core of the Earth, and the Moon is like Earth's mantle. Even though the two objects look alike, we must remember that the resemblance is only skin deep: In their interiors they are very different.

Although the Moon and Mercury are simpler planets to understand than Venus, Earth, or Mars, we should not delude ourselves into thinking we have solved all of their mysteries. Both are complex worlds, and even for the Moon our available information is infinitesimal in comparison with what we know about the Earth. In the case of Mercury, we have never even seen half the surface! It is sobering to remind ourselves that, with all the missions that have studied the Moon and hundreds of kilograms of returned samples, we are still not certain how and where our satellite originated. How much more uncertain may be the conclusions drawn about other, more distant worlds?

8.1 AN ELUSIVE PLANET

Because of its proximity to the Sun, Mercury is a difficult planet for astronomers to study. There is a story, perhaps apocryphal, that Copernicus said he had never seen the planet. Yet Mercury was well known to observers in antiquity, although not all of them recognized that this twilight "star" was the same object when seen in the morning before sunrise and in the evening after sunset.

APPEARANCE OF MERCURY IN THE SKY

If you observe the motions of Mercury or Venus for several months, you can see that these two planets are closely tied to the Sun, never moving very far from it in the sky. Because Mercury and Venus have smaller orbits than that of the Earth, they are sometimes called inferior planets. The maximum angular distance of Mercury from the Sun is about 30°, compared with 45° for Venus.

Figure 8.1 shows the apparent position of Mercury as evening star over a 2-month period as you might draw it each evening about half an hour after sunset. Mercury is actually quite bright, about as bright as the brightest stars, but it is harder to see because of its proximity to the twilight horizon. Mercury never strays far from the Sun, but Venus, with its larger orbit, can set as long as 3 hours after the Sun, which allows it to be seen against the dark sky.

GEOCENTRIC AND HELIOCENTRIC PERSPECTIVES

Do the planets circle the Earth or the Sun? This fundamental question was debated in the Western world for two millennia and not settled until the time of Kepler and Galileo. During most of that time, the geocentric (Earth-centered) ideas prevailed. Yet the apparent motions of Mercury and Venus, which oscillate from one side of the Sun to the other, draw us naturally to the idea of a Sun-centered system. Figure 8.1 shows how Mercury appears to move away from the Sun until it reaches a maximum distance (eastern elongation). It then appears to reverse its motion and gradually swings back to disappear in the Sun's brilliance. A few weeks later the planet reappears in the morning sky, moves out to its greatest western elongation, and then reverses. The entire pattern repeats every 4 months. Venus does the same thing, but with a period of about 20 months.

In the geocentric cosmology, Mercury and Venus each circled the Earth, but they were attached to an imaginary line that connected the Earth with the Sun (Fig. 8.2a). In this theory their apparent motion was not due to their path around the Earth but was the result of motion on a second circle, called an epicycle, which rotated about a hub on the Sun-Earth line. According to this theory, these planets were always illuminated from behind by the Sun, and they should therefore always display a crescent phase.

Figure 8.2(b) illustrates the heliocentric theory for the motion of Mercury. The description is now simplified; if Mercury really circles the Sun,

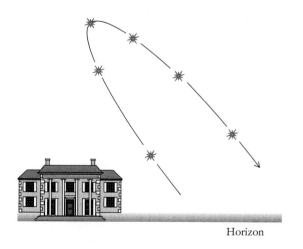

Horizon

Figure 8.1   Appearance of Mercury as an "evening star," showing its apparent position at weekly intervals, as seen in the western sky a few minutes after sunset.

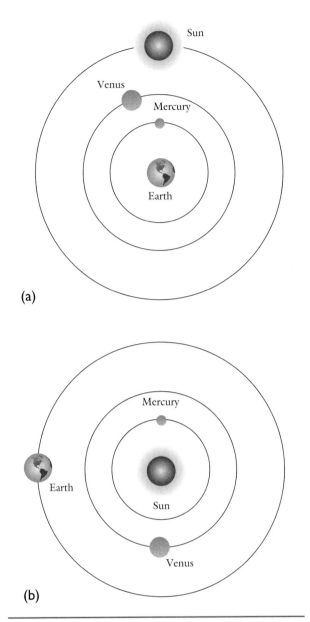

(a)

(b)

Figure 8.2   Motion of Mercury as described in the geocentric and heliocentric systems. **(a)** Geocentric view: Mercury is constrained to move along only a part of its supposed orbit around the Earth, staying always near the Sun in the sky. **(b)** Heliocentric view: Mercury orbits the Sun just like all the other planets. (Compare Fig. 1.16.)

then its close attachment to the Sun is immediately understood without a requirement for epicycles or other artificial contrivances. But note also that the phases of the planet are different. It is now crescent only during the half of its orbit when it lies between the Sun and Earth. For the other half of its orbit, Mercury is on the far side of the

Sun as seen from the Earth and is therefore more fully illuminated. The same sequence of phases applies to Venus as well. Galileo's telescopic observations of the phases of Venus provided strong early support for the Copernican theory, as we discussed in Chapter 1.

8.2   TIDES AND SPINS

Not all planets rotate rapidly like the Earth. Both Mercury and the Moon have been slowed down by tidal forces.

THE DOPPLER EFFECT

Mercury is a disappointment to any telescopic viewer, with its small size and the difficulties of observing it near the Sun. It is a lucky (and persistent) astronomer who has seen any markings on its surface. For this reason, Mercury was a neglected object until the development of radar astronomy in the early 1960s. In Chapter 5 we mentioned using radar as a tool to study asteroids, and in Chapter 6 we reported how radar provided evidence of boulders near the nucleus of one comet. The most powerful planetary radar system in the world uses the 1000-ft antenna suspended in a bowl-shaped valley near the town of Arecibo in Puerto Rico as both a transmitter and a receiver (Fig. 8.3).

The radar echo from Mercury is typically only 1 millionth of a trillionth (10^{-18}) as strong as the transmitted pulse, and the round-trip time (at the speed of light) is more than 10 minutes. By transmitting a radio pulse of carefully controlled duration and frequency, a radar system can reveal much more about a target than would be achieved by simply illuminating it with ordinary light. The time required for the radio pulse to cover the two-way path to the target and back provides a measure of the distance to the target. In addition, the motion of the target toward or away from the observer can be measured using the Doppler effect.

The **Doppler effect** describes the dependence of radio wavelength or frequency on the relative motion of the source and receiver. We experience the Doppler effect most commonly in the realm of

Figure 8.3 The 300-m (1000-ft) Arecibo radio telescope in Puerto Rico is the most powerful radar instrument on our planet. Radar first determined the true rotation period of Mercury.

sound, which is propagated, like light and radio, in the form of waves. A sound source moving toward a listener has a higher pitch (shorter wavelength), whereas motion away results in lower pitch (longer wavelength) (Fig. 8.4). We hear these effects daily in the sounds of police and fire truck sirens or in the familiar "whoosh" as trains or cars pass by us: Their pitch suddenly drops as the relative motion of the vehicles changes from approaching us to moving away.

When either sound waves or radar waves are reflected from a moving target, the Doppler shift in frequency is proportional to the target's relative speed toward or away from the transmitter/receiver system. In the case of planetary radar systems, the frequency of the transmitted pulse is controlled precisely, so that even tiny changes in the frequency of the echo can be detected and analyzed.

The first results from planetary radar were the determination of the radar reflectivities of the

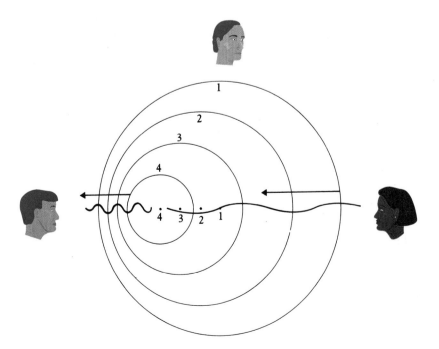

Figure 8.4 The Doppler effect. Suppose a source of sound or light is moving from position 1 to position 4. Motion toward the observer at the left crowds the waves of sound or light together, so the wavelength is shorter and the frequency is higher. Meanwhile, the motion away from the observer at the right spreads the waves, increasing the wavelength and decreasing the frequency. The observer at the top sees no change in wavelength or frequency, since the source moves neither toward nor away from him.

Moon, Venus, and Mercury, and precise measurements of their distances. Careful tracking of the motion of Venus, in particular, was critical in defining the distance scale of the solar system and thereby making possible the accurate navigation of spacecraft on complex interplanetary trajectories. The planetary radar observations of the 1960s were an essential prelude to the Mariners, Pioneers, and Voyagers of the 1970s. They also provided new data on the planets.

ROTATION OF MERCURY

Radar made the first measurements of the rotation rates of Venus and Mercury. The determination of a planet's rate of rotation by radar uses the Doppler effect in a clever way. Since the incident radar beam illuminates an entire planetary hemisphere, the rotation of the planet as well as its orbital motion alters the frequency of the reflected signal (Fig. 8.5). As seen from the Earth, the effect of rotation is to make one side of the planet approach us while the other recedes, relative to the overall motion. The echo from the approaching side of the planet is shifted toward higher frequency, and the echo from the receding side is lowered in frequency. As a result, the single frequency of the transmitted pulse is broadened in the returned echo, which includes reflections

from a full hemisphere. The faster the rotation, the broader the frequency range in the echo.

When the Doppler radar technique was first applied to Mercury in 1965, the frequency width of the echoes did not conform to astronomers' expectations. Ever since nineteenth-century maps had been made showing faint markings on Mercury, it was believed that Mercury always kept the same face toward the Sun, just as the Moon does with respect to the Earth. Astronomers thought that Mercury, unlike the Moon, really had a dark side, a hemisphere never illuminated by the Sun. Mercury was therefore thought to be simultaneously the hottest and coldest place in the planetary system. (We now know it is neither.) All maps of the visible markings on the planet were of the supposed Sun-facing side only.

The radar results did not yield the expected 88-Earth-day rotation period, equal to the period of revolution. Instead, the indicated rotation period was approximately 59 days. Subsequent data have demonstrated that the rotation period of Mercury is almost exactly two-thirds of its 88-day period of revolution, or 58.65 days.

It is no coincidence that the rotation period of Mercury is an exact fraction of the orbital period. Before we discuss the reason for this curious spin-orbit coupling, let's examine the consequences it has for Mercury itself. In the process,

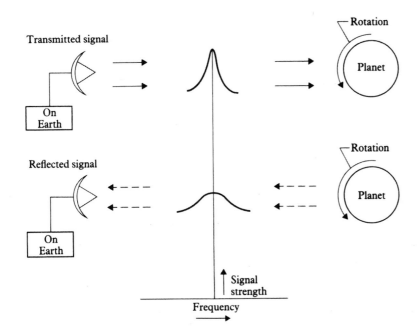

Figure 8.5 The broadening of a radar signal by planetary rotation. The side of the planet approaching the radar shifts the signal toward higher frequencies while the side moving away is shifted to lower frequencies. The net effect is to broaden the frequency of the transmitted signal.

we must consider just what we mean by the terms *day* and *year* as applied to another planet. The *year* on a planet is its period of revolution about the Sun, the time necessary to complete one orbit. The term *day* is more ambiguous, at times referring to the rotation period—that is, the time for one complete spin relative to the fixed stars. Here we shall adopt the other, more familiar meaning of *day* as the time interval between successive sunrises. This is also called the **solar day**.

A year on Mercury lasts approximately 88 Earth days, during which time the planet rotates one and a half times on its axis. Since the orbital motion and the spin are in the same direction, however, one partially cancels the other, and the Sun appears to move only halfway through its diurnal cycle in this span of time (Fig. 8.6). The solar day on Mercury is equal to two orbits and three complete rotations, or about 176 Earth days. Thus, on Mercury the solar day is longer than the year.

The situation for Mercury is further complicated by the large eccentricity of its orbit, which takes it from 0.308 AU away from the Sun at perihelion to an aphelion distance of 0.467 AU. The corresponding range in apparent diameter of the Sun is from 1.6° to 1.1°, compared with an apparent size of the Sun as seen from the Earth of about 0.5°. Each solar day on this planet includes two perihelion and two aphelion passages.

STRANGE DAYS AND SEASONS

Let us imagine the appearance of the Sun to a race of hypothetical Mercurians, beginning with a group living at a longitude where the Sun is overhead at perihelion. There are two such hot longitudes, on opposite sides of the planet. At sunrise and sunset, the Sun will be at its most distant. The Sun, therefore, rises small and increases in size as it approaches the zenith. At the same time, the combination of rotation and orbital motion causes the apparent motion of the Sun to slow, until it hangs nearly motionless overhead while the planet races through perihelion. Speeding up as it shrinks, the Sun then dips toward its setting point. The stars will meanwhile be moving through the sky about three times as fast as the Sun. A star that rises with the Sun will set before noon and will rise again before sunset.

Now consider an observer situated at a longitude 90° away. For her, the Sun will be small and rapidly moving at noon but large and nearly stationary at sunrise and sunset. Will such a Mercurian wonder about the favored lands 4000 km to the east or west where the Sun stands still at noon, or will she be grateful that the Sun lingers at rising and setting when its light is most desired? What interesting conversations two Mercurians separated by 90° in longitude would have were they to compare their cosmologies!

In addition to their different cosmological perspectives, the two longitudes that face the Sun at perihelion have a very different thermal history from the longitudes that face the Sun at aphelion. The perihelion longitudes are called the hot longitudes because they receive extra heating as a result of both the proximity of the Sun at noon and the way it lingers overhead for many Earth days. In contrast, the maximum surface temperatures at the cool longitudes 90° away are more than 100°C lower, although still a sizzling 550 K.

Both the Moon and Mercury have rotation periods closely linked to their orbital motion. For

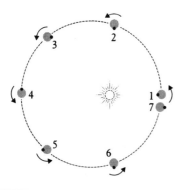

Figure 8.6 Mercury rotates on its axis once every 58.65 days and orbits the Sun every 88 days. Hence the planet rotates three times (3 × 58.65 = 176 days) for every two revolutions about the Sun (2 × 88 = 176 days). Starting at position 1, the planet completes one full rotation when it reaches position 5. At 7, the point on the surface that originally faced the Sun now points directly away from it. After the second orbit, this point will again face the Sun.

the Moon, the two periods are equal and the same side always faces the Earth. Mercury has a unique 2:3 resonance between the two periods. Many outer-planet satellites also experience spin-orbit coupling. How were these linkages forged, and what are the consequences for the evolution of planets and satellites?

The Nature of Tides

The most effective agents for altering the spins of planets are tidal forces—the forces that generate tides. A tide is a distortion in the shape of one body induced by the gravitational pull of another nearby object. The tides with which we are most familiar, of course, are those generated in the Earth's oceans by the attraction of the Moon and Sun.

If a planet were completely rigid, the gravitational attractions of other bodies would act on it as if its mass were all concentrated in a point at its center, the so-called center of mass. However, there must also be a differential gravitational force that results from the facts that planets are not point masses, and that the part of one body facing toward another is more strongly attracted than the part that is turned away. This tendency to pull harder on nearer parts and thereby to distort the shape of one planet when it is near another is a tidal force. It is easy to see that the part of the Earth that faces the Moon is more strongly

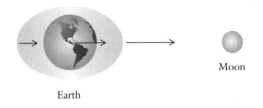

Figure 8.7 Tidal forces. The gravitational pull of the Moon is strongest on the side of the Earth facing the Moon and weakest on the opposite side. The arrows (exaggerated) illustrate these tidal forces.

attracted than is the center of mass. Note, however, that there is an equally effective differential force acting on the opposite side of the Earth and directed away from the Moon, since this part of our planet's surface experiences less attraction than does the center of mass.

The tidal force of the Moon on the Earth sets up horizontal forces that cause the water of the oceans to move toward two areas, one approximately facing the Moon and one on the opposite hemisphere. The result is two tidal bulges in the ocean, oriented toward and away from the Moon (Fig. 8.7). As the Earth rotates, each point on the surface passes through these bulges, experiencing two high tides and two low tides each day. Similar but smaller tidal bulges result from the pull of the Sun, and the highest tides occur twice a month when the two bulges are aligned (Fig. 8.8).

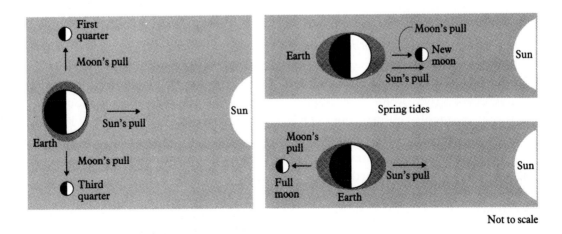

Figure 8.8 When the Sun and the Moon pull on the Earth at right angles to each other (left panel), the tides are smaller than when they pull together or opposite each other (right panel).

Tidal forces are strongly dependent on the separation of the objects; mathematically they are proportional to the mass of the perturbing body and to the inverse cube of its distance (M/D^3), as opposed to an inverse square law for gravity (see Section 1.4). Thus, tides affect most strongly satellites in close orbits around their primaries, and tides raised by the Sun are much more important for Mercury than for any more distant planet.

TIDAL FRICTION AND ORBITAL CHANGE

If the Earth did not rotate with respect to the Moon, the lunar-induced tidal bulges would align directly with the Moon. On a rotating Earth without friction between the oceans and the land, the bulges would maintain their lunar alignment while the solid Earth rotated beneath. This is approximately the case, and an observer on the shore watches the tide rise and fall twice a day. Friction occurs, however, as the restless seas wash against the shores of the continents, inducing a lag in the tidal bulges. This lag, which varies from one part of Earth to another, explains why high tide does not take place when the Moon is overhead (or underfoot) but usually follows by several hours.

The constant friction between the ocean tides and the land has several important consequences. First, it slows the rotation of the Earth. About 2 billion horsepower is dissipated by ocean tides, which lengthens the day by about 500 millionths (5×10^{-8}) of a second per day. Part of the energy lost by the spinning Earth results in a very small frictional heating of the ocean and our planet's surface. Part also goes to alter the orbit of the Moon.

As the Earth slows its spin, the Moon is gradually forced away into a larger orbit—an example of the conservation of angular momentum. The total angular momentum of the Earth-Moon system includes the spin of both objects and the revolution of the Moon around the Earth. If the Earth's rotation is slowed, losing angular momentum, then some other part of the system must gain equivalent angular momentum. The Moon absorbs this additional momentum by moving outward.

Long ago, the Moon was much closer to the Earth and the length of the month much shorter. Calculations suggest that at one time the Moon may have circled the Earth in as little as a week, while at the same time the solar day on the Earth would have been as short as six hours. Barring unforeseen events, the coupled evolution of the lunar orbit and the rotation of the Earth will continue until the Earth has slowed down to the point where it keeps the same side always facing the Moon.

Total eclipses of the Sun are possible on Earth because the Moon is just large enough in apparent size to cover the face of the Sun (see Section 1.1). We now see that this situation applies only briefly on a cosmic timescale. In the past, the Moon was nearer and solar eclipses were more common, but in the future, the Moon will be too far away to block the Sun completely, and there will be no total eclipses of the Sun.

Everything we have said about the effects of the lunar tides on the Earth is equally applicable to the tidal effects of the Earth on the Moon. To be sure, the Moon has no oceans, but a significant tidal bulge can be formed even in a solid planet by gravitational distortion of the rock. Such body tides can also dissipate energy. As the Moon rotates with respect to the bulge, it is subjected to forces that compress and bend its rocky crust and mantle. Some energy is released through tidally induced moonquakes, some by heating of the rocks. The seismometers deployed as part of the ALSEP experiments were able to record these moonquakes. Just as in the case of ocean tides, this release of energy slows the Moon's, or any planet's, spin until the same side faces its companion.

In the distant past, the effects of energy dissipation on the Moon from Earth-induced tides slowed the rotation of our satellite until it reached a rotation period precisely equal to its period of revolution. This **synchronous rotation** is a highly stable situation because it minimizes frictional energy loss. Once synchronous rotation is reached, it will be maintained; if the orbital period should change slightly, the rotational period adjusts itself accordingly. Most of the satellites in the solar system are in synchronous rotation about their planets.

MERCURY AND THE EFFECTS OF SOLAR TIDES

If synchronous rotation is favored and Mercury is the planet most strongly affected by solar tides, why does it not keep the same face toward the Sun? How did it get into its present peculiar rotational state? The answer lies in its large orbital eccentricity, which trapped this planet into a 2:3 (rather than 1:1) synchronous rotation.

Imagine that Mercury started its existence with a relatively rapid rotation and that early in its history it was slowed by solar tides. At the same time, its eccentric orbit carried it first closer to the Sun and then farther away. Since tidal forces are so sensitive to distance (varying as the inverse cube), most of the tidal dissipation took place near perihelion. As Kepler's second law tells us, however, the orbital velocity of the planet near perihelion is much greater than the average (see Section 1.3), and in Mercury's case, the orbital speed at perihelion matches a spin period close to two-thirds of the orbital period. That is, near perihelion, when the tidal forces are strongest, Mercury finds itself turning on its axis in apparent synchronism with its orbit for a rotation period near 60 days. This is the same effect that causes the Sun to appear to stand still in the mercurian sky near perihelion.

Once Mercury's tidal evolution carried it into the 2:3 synchronous rotation period, it was trapped. Given the orbital eccentricity, this spin minimizes tidal dissipation near perihelion, where it is most effective. Therefore, the planet remains in a stable rotational state, with two fixed tidal bulges that face the Sun at alternate perihelion passages.

What would happen if we magically intervened and placed Mercury into a 1:1 synchronous rotational state? It could not speed up to return to the 2:3 state, since friction can only slow a body's spin and not increase it. In spite of keeping the same face toward the Sun on the average, however, Mercury would still experience significant tidal heating. Each time it neared perihelion, the increased force of solar gravity would swell its tidal bulge, while the changing direction to the Sun would twist and distort it. Jupiter's satellite Io

finds itself in precisely this situation today, where the resulting tidal heating causes a high level of volcanic activity (see Chapter 15). Past tidal heating has also been implicated in the thermal histories of many bodies in the planetary system, as we will see in later chapters.

8.3 THE FACE OF MERCURY

Almost everything we know about the geology of Mercury was learned from a single spacecraft, Mariner 10, which made three flybys of the planet in 1973 and 1974. For a few months this small planet held center stage in the planetary exploration program, before settling back again into relative obscurity. Currently, however, two new orbiter missions to Mercury are being planned—one from NASA and one from the European Space Agency (ESA).

THE VIEW FROM MARINER 10

Mariner 10 provided resolution and coverage of Mercury very similar to that of the Moon as seen by telescopes from the Earth: One hemisphere was photographed at a resolution of approximately 1 km, while the other hemisphere was not seen at all. The spacecraft cameras revealed a planet that looked remarkably similar to the Moon (Fig. 8.9). Although an abundance of impact craters had been expected for this small, airless world, the degree of similarity of the mercurian surface to the lunar highlands was surprising. Most geologists had anticipated greater evidence of internal geological activity, since Mercury is larger than the Moon, but the Mariner photographs revealed only one lava-flooded impact basin, suggesting that mercurian volcanism was less than on the Moon or that it had been expressed in different ways.

Although the composition of the surface material on Mercury has not been measured directly, the colors and reflectivity of this planet are similar to those of the lunar highlands; the surface therefore likely consists of igneous silicate rock. Basaltic lavas such as those of the lunar maria are possible, but the generally higher reflectivity of the surface is

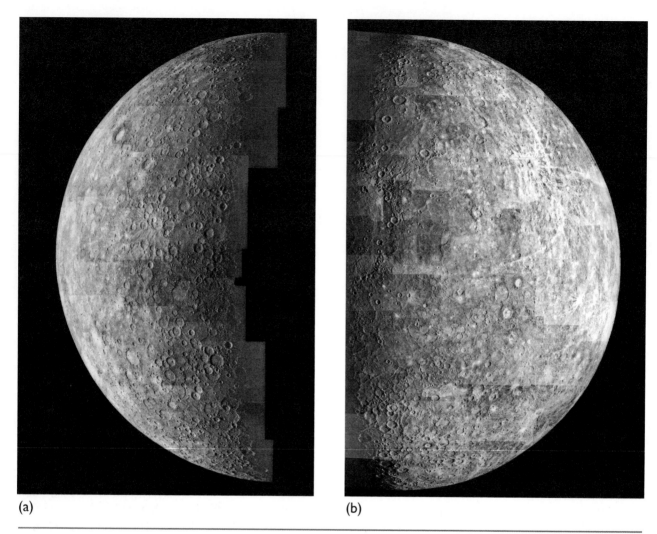

(a)　　　　　　　　　　　　　　　　　　　　(b)

Figure 8.9 Two mosaics produced by the Mariner 10 spacecraft in 1974 show the visible hemisphere of Mercury as the spacecraft approached **(a)** and as it receded **(b)** from the planet. Mariner 10 was able to examine about 45% of Mercury's surface. The remaining 55% remains unexplored.

suggestive of compositional differences, perhaps associated with a higher metal content in the mercurian soil. There is no way to know for certain whether the crust of Mercury is depleted in volatiles like that of the Moon, or whether it shares any of the other chemical peculiarities of our satellite. Nor is there any way to measure the age of the surface, beyond noting that the heavy cratering probably implies great antiquity. In the absence of returned samples, interpreting the composition and history of another world is very difficult!

One intriguing possibility is to look for possible mercurian samples among our large collection of meteorites, which have already yielded dozens of rocks from Mars and the Moon. In 2001, a mercurian source was proposed for a basaltic meteorite called NWA 011, found in the Moroccan Sahara in December 1999, but not confirmed.

On a small scale, we know that Mercury has a regolith and that its surface soil has a texture resembling that of the Moon. These similarities can be deduced from the similar way the two objects reflect sunlight and emit thermal radio energy. With our post-Apollo knowledge of the Moon as a benchmark, we can interpret these astronomical studies of Mercury with greater confidence. Surface temperatures have also been measured, ranging from a high of about 675 K at noon at the hot longitudes (high enough to melt zinc) to a low of about 90 K over most of the night hemisphere (low enough to freeze carbon dioxide). The lowest temperatures on Mercury are nearly equal to those on the Moon, since the higher daytime temperatures are balanced by a longer cooling period during the 88-Earth-day mercurian night.

POLAR ICE ON MERCURY AND THE MOON

One of the most surprising discoveries about Mercury in the years since Mariner 10 is the presence of polar caps, as revealed by radar. A **polar cap** is a permanent deposit of water or other frozen volatiles in the cool polar regions of a planet. The Earth and Mars have polar caps, but no one expected such a phenomenon on Mercury, which is much closer to the Sun. Yet both poles of Mercury show enhanced radar reflectivity of the sort expected from surface or near-subsurface deposits of water ice.

Since the spin axis of Mercury is almost exactly perpendicular to the plane of its orbit, very little sunlight falls on the poles and they are therefore cool. Strictly in terms of temperature, ice caps at the poles are possible. But where did the ice come from? With no atmosphere on Mercury, the water cannot have been transported to the poles from lower latitudes. Does this primordial water date from the formation of Mercury? And if Mercury has polar caps, why doesn't the Moon?

The first hint of ice on the Moon came from a radar experiment carried out with the Clementine lunar orbiter, which measured a slight enhancement in reflectivity near the poles (much smaller than the radar signal on Mercury). The NASA Lunar Prospector mission in 1998 measured neutrons over both lunar poles that indicated the presence of hydrogen, probably in the form of ice. The quantities were not large—perhaps equivalent to a shallow pond of a few square kilometers—but even this is enough to tantalize people who hope to set up permanent human settlements on our otherwise stone-dry satellite.

The lunar ice cap (if it is real) is probably the result of comet impacts on the Moon. When a comet strikes, much of its ice is vaporized to form an extremely thin, short-lived lunar atmosphere of water vapor. Permanently shadowed regions near the two poles are the coldest places on the lunar surface, so they can act as a "cold trap" to collect some of this water vapor. As long as the poles stay cold, this little deposit of ice will grow each time a comet hits the Moon. The same process is probably responsible for the polar caps on Mercury.

NOMENCLATURE FOR SURFACE FEATURES

Most of the craters on the Moon are named for scientists or philosophers of the ancient world. Mars, which was the next planet to be mapped extensively, follows a similar scheme, although many of the scientists commemorated on Mars are from the nineteenth and early twentieth centuries. When Mariner 10 arrived at Mercury, however, astronomers decided to adopt a different scheme, naming the newly discovered craters for great artists, writers, composers, architects, and other heroes of world culture.

Among the most prominent craters on Mercury are Bach, Shakespeare, Tolstoy, Mozart, and Göthe. Other craters bear the names of Al-Hamadhani, Balzac, Basho, Bramante, Imhotep, Ibsen, Juda Ha-Levi, Mark Twain, Phidias, Raphael, Rublev, Valmiki, Verdi, Vyasa, and Wang Meng (to name a few).

GEOLOGY OF MERCURY

The ubiquitous impact craters on Mercury bear many resemblances to their lunar counterparts. Where differences occur, they are probably due to the higher surface gravity on Mercury (0.38 of the terrestrial value, as opposed to 0.21 for the Moon). As a result, ejecta blankets are smaller, rarely extending more than one crater radius beyond the rim, and the secondary craters are similarly more confined in their distribution (Fig. 8.10).

The density of craters on Mercury varies from values near that of the lunar highlands down to about one-tenth of this value (from approximately 1000 down to 100 10-km craters per 1 million km^2). There are no large regions with crater densities as low as those found on the lunar maria. If

Figure 8.10   Close up of a crater on Mercury. Compare this terraced crater 98 km in diameter with King Crater on the Moon (see Fig. 7.16).

Figure 8.11   The cratered plains of Mercury are less heavily cratered than the lunar highlands but more heavily cratered than the lunar maria. They may be lava plains similar to the maria but somewhat older, or they may have some other origin.

Mercury experienced a period of volcanism that produced a counterpart to the lunar maria, this period must have been contemporary with the late heavy bombardment rather than occurring after the impact flux had dropped, as was the case for the Moon.

The less cratered regions of Mercury do not look like lunar maria (Fig. 8.11). They do not differ from their surroundings in brightness or color, nor are they as flat as maria. No flow fronts, sinuous rilles, or other distinctively volcanic features have been identified in the Mariner pictures. Thus geologists are left with a fundamental question: What destroyed some older craters to produce these moderately cratered plains? Was it volcanism that took place during the later stages of heavy meteoritic bombardment but did not leave the characteristic marks of the lunar maria? Or was some other process responsible, such as blanketing of the surface by impact-produced ejecta? No one knows, although a vote among geologists would probably favor some kind of volcanic process.

The largest impact basin in the half of Mercury seen by Mariner is located near one of the two hot longitudes and is appropriately named the Caloris (Latin for "heat") Basin. With a diameter of 1300 km, the Caloris Basin resembles Orientale or Imbrium on the Moon (Fig. 8.12). The most important difference, however, is in the distinctive patterned interior floor of Caloris. If this material is basalt, we ask, why does it lack the smooth surface and dark color of Mare Imbrium and other ancient lunar basalt flows? If it is not basalt, what is it? One hypothesis proposes that the floor of Caloris is made of material that melted as the direct result of the impact, rather than being filled in later as were the lunar maria.

Geologically, the most remarkable features on Mercury are compressional scarps or cliffs that have no lunar counterpart (Fig. 8.13). These scarps were produced when the planet's crust was compressed by shrinking of the interior. The total extent of such features indicates that the crust has lost approximately 4% of its area, probably as a result of a 2% decrease in the radius of the planet. Since they formed after most of the craters (as you can see in Fig. 8.13), these scarps apparently represent an internal event on Mercury that took place hundreds of millions of years after the solidification of the crust. Interior changes resulting from the slowing of the planet's spin to achieve the present 2:3 resonance have been suggested as a possible cause.

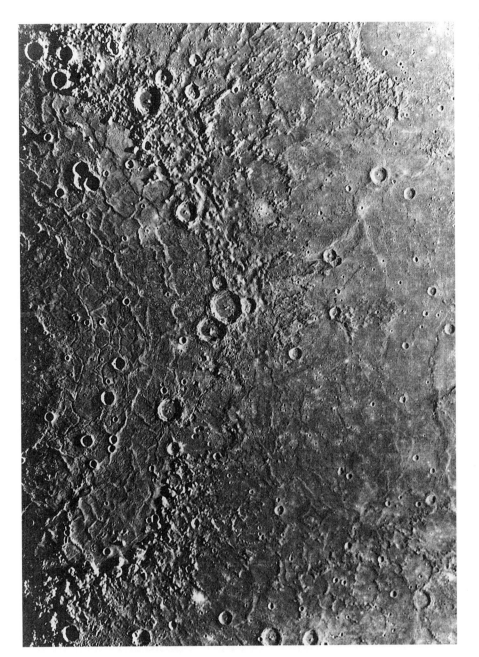

Figure 8.12 This mosaic of images obtained by Mariner 10 shows the half of the Caloris Basin that the spacecraft was able to see. It is interesting to compare this feature with the Orientale Basin on the Moon (see Fig. 7.5).

Figure 8.13 The scarp named Discovery stretches more than 500 km across Mercury's surface. In places it is 3 km high, about the same height as Mount Olympus in Greece. This scarp is younger than the craters it crosses. It probably formed when Mercury's crust contracted as the planet cooled early in its history.

8.4 INTERIORS OF THE MOON AND MERCURY

Planetary interiors are mysterious places, forever inaccessible to direct measurement. What little is known about them must be learned indirectly and is inevitably subject to misinterpretation. Yet the effort must be made, for without some understanding of the bulk of a planet lying beneath the visible surface, an intelligent discussion of planetary evolution is impossible.

THE SIGNIFICANCE OF DENSITY

The most important clue to the bulk composition of a planet is provided by its density (see Section 1.6). The Earth has a density of 5.5 g/cm^3; the Moon, 3.3 g/cm^3; and Mercury, 5.4 g/cm^3. The fact that these densities differ suggests differences in composition, but we must be careful; the average density is also affected by the size of a planet. We have already seen that very small asteroids are likely to be underdense, with interiors of high porosity. They are too small for gravity to crush and compress the interiors. The limiting size for an asteroid to squeeze out any interior porosity is about the same as the maximum size for a highly irregular shape—about 500 km in diameter (like Vesta).

At the other extreme, very large planets with strong gravity are able to compress their interior materials so much that they increase their densities *beyond* what we would expect from our experience on Earth's surface. For example, the iron in the Earth's core has a density of about 15 g/cm^3, while ordinary iron on the surface has a density of only 9 g/cm^3. Thus, a better starting point for a chemical interpretation would be an **uncompressed density**—that is, the density corrected for the effects of self-compression in a planet.

The uncompressed densities of Earth, Moon, and Mercury are, respectively, 4.5 g/cm^3, 3.3 g/cm^3, and 5.2 g/cm^3. The Moon, because of its relatively small size, has experienced little internal compression. Mercury has the highest uncompressed density of any planet, with the Earth second. Therefore, Mercury must have the largest proportion of high-density constituents, which are the metals, and the Earth has the second-highest allotment.

The uncompressed density of the Moon is only slightly greater than that of ordinary rocks, including the returned Apollo samples. The density must increase with depth; it is hard to imagine the alternative in which heavier materials could have "floated" above lighter ones when the Moon was young and molten. Therefore, the materials near the center of the Moon can be at most only slightly denser than the surface rocks, and there simply is no place to hide a substantial metallic core. One of the fundamental properties of the Moon is its lack of metals. If Earth and its satellite formed together from the solar nebula, this basic difference in composition is hard to understand.

In contrast, the only plausible model for a planet as dense as Mercury requires a lot of metal, roughly 60% by mass. If 60% of Mercury consists of an iron-nickel core similar to that of the Earth, this core must have a diameter of 3500 km and extend to within about 700 km of the surface. Mercury can then be thought of as a metal ball the size of the Moon covered with a 700-km-thick crust of dirt.

The proper way to look at this situation is to realize that Mercury is deficient in silicate rocks, rather than enriched in metals. Either temperatures in the solar nebula were so high that only metals and some high-temperature silicates could condense and form the solid grains that ultimately became the building blocks of Mercury, or else the planet formed initially with a composition more like that of the Earth and Venus but subsequently lost part of its share of silicates. We will discuss these two possibilities later in this chapter and try to choose between them.

MOONQUAKES AND HEAT FLOW

In the case of the Moon, considerable insight into interior conditions has been derived from the Apollo ALSEP experiments, as described in Section 7.2. The lunar seismometers measured the response of the whole Moon to impacts, such as those of spent Apollo rockets (Fig. 8.14), while an additional experiment on *Apollo 15* and *Apollo 17* directly measured the rate of heat flow from the interior.

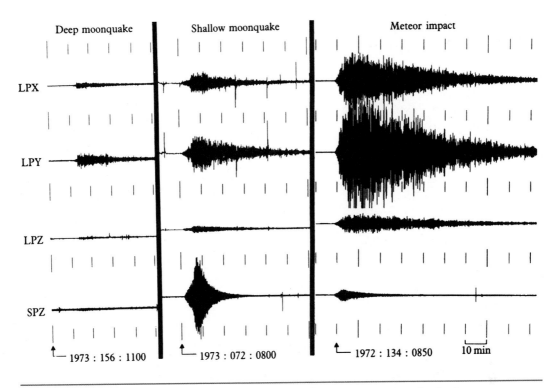

Figure 8.14 A comparison of seismograms obtained from various types of seismic events. The long "ringing" of the Moon in response to impacts is especially noteworthy.

Relative to the Earth, the Moon is a very quiet place. No moonquake measured by any of the Apollo instruments was large enough to have been felt by an astronaut standing on the Moon, and the total energy released each year by moonquakes is a hundred billion times less than the energy of earthquakes. Internally generated moonquakes arise primarily at a depth of around 1000 km. Below this depth, temperatures may be high enough to produce some melting of rocks (Fig. 8.15a).

The heat flow from the interior of the Moon also suggests that temperatures are higher at great depths. Most of this heat is being generated today by radioactive elements in the Moon. If a similar heat flow measurement were made on the surface of Mercury, it would be possible to determine whether the metal core is liquid, but in the absence of a lander no such data exist.

Another line of evidence on interior structure comes from the measurement of a planet's magnetic field. Although the exact mechanisms for generating such a field are not understood, it is thought that a strong field requires motions in a liquid, electrically conducting core. Since the Moon has no metallic core, it was no surprise when early space missions showed no global magnetic field. But no one knew what to expect from Mercury, which has a metal core but not necessarily a liquid metal core.

THE MAGNETIC FIELD OF MERCURY

One of the discoveries made by Mariner 10 was the presence at Mercury of just such a planetary magnetic field after none had been detected for the Moon. Although only about 1% as strong as the field of the Earth, the magnetic field of Mercury is similar in character and seems to indicate the existence of a liquid core. Two decades after the Mariner flybys, the issue of a liquid core remains unclear, since we still have no data on the state of the planet's interior. But the presence of a magnetic field on Mercury is undeniable, and most scientists conclude that the metal core must therefore be at least partially molten (Fig. 8.15b).

Mercury presents a paradox to challenge planetary scientists. Its interior is in many ways like that of the Earth, with a large metal core and

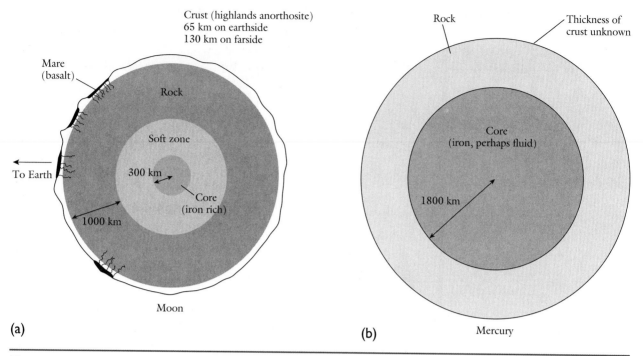

Figure 8.15   **(a)** A model of the interior of the Moon derived from seismic data. It is still uncertain whether there is a small liquid core at the Moon's center. **(b)** A model of the interior of Mercury based on the planet's high density and its requirement for a large, iron-rich core.

a planetary magnetic field. Yet its surface is like that of the Moon, with little evidence of geological activity for at least several billion years. If the interior is hot enough to generate a magnetic field, why has it not also reworked more of the surface? Or have we incorrectly estimated the age of the surface, and the cratering on Mercury was produced more recently than that on the Moon? Or, perhaps, the magnetic field does not indicate a molten core? These are questions that will be addressed by the next generation of space missions to Mercury—missions that are very likely to force much of this chapter to be rewritten.

8.5   COLLIDING PLANETS: ORIGIN OF THE MOON AND MERCURY

The geological history of a planet is written in its rocks and surface topography. This record can be read using the techniques of stratigraphy supplemented by detailed study of selected rock samples.

Because both Mercury and the Moon are relatively inactive internally and lack atmospheric weathering and erosion, the span of solar system history available for study covers a great expanse of time. Approximately 4 billion years of geologic evidence is exposed on the Moon, and the surface of Mercury probably preserves a similar record, although in the absence of dated samples we cannot be sure. Yet there is a limit beyond which we see dimly if at all. As noted before, heavy bombardment during the first half-billion years of solar system history destroyed most of the evidence from this epoch.

MODELS OF PLANETARY EVOLUTION

Direct geological evidence from the surface can be augmented by theoretical calculations of the probable *thermal evolution* of a planet—the changes in its internal temperature over time. To trace this life history, we must begin with what we know today and calculate backward and forward in time. Such calculations require the powers of large computers if they are to represent realistically the laws of physics and chemistry that govern planetary evolution.

Models also require knowledge of the behavior of materials at the unfamiliar pressures and temperatures of planetary interiors. Near the center of the Earth, for instance, the pressure is about 4×10^6 bars (that is, 4 million times the sea-level pressure of the Earth's atmosphere) and the temperatures are measured in thousands of degrees. These calculations also require certain assumptions concerning the initial state of a planet, especially assumptions about its chemical composition. If these assumptions are reasonable and the computer program works properly, an outline or *scientific model* of planetary history is produced (as we discussed for the evolution of the Sun in Section 2.5). This model can then be tested and modified to match new observations. This is the essence of the scientific method.

The output of a planetary model calculation is many columns of numbers, each representing the pressure, temperature, and chemical and physical properties at various points inside a planet. All of these values are calculated for a series of time steps, separated by perhaps a few million years. From these numbers, it is possible to trace the changing conditions at each interior point over a span of billions of years.

Lunar history is comparatively well defined by the wealth of evidence obtained by Apollo, whereas Mercury's past remains more speculative. One way to approach Mercury is to adopt the lunar-based paradigm for the cratering history of the inner solar system (see Section 7.3). If Mercury experienced impact rates similar to those of the Moon, including a terminal heavy bombardment about 3.9 billion years ago, then a consistent picture emerges. Of course, consistency does not guarantee truth, but it does suggest that planetary scientists may be moving in the right direction.

EARLY HISTORY

Before discussing the actual origin of the Moon and Mercury, let's look at the period starting 4.4 billion years ago, when we think both objects were largely molten and fully differentiated. By this time, the lunar highland crust had begun to form by freezing of the surface of a magma ocean. The oldest highland rocks provide us with samples from this period of planetary history. During this period, the heavy rain of debris still striking the lunar surface repeatedly shattered and remelted the crust, destroying any detailed evidence of its early structure. The entire surface of the Moon came to consist of overlapping ejecta from innumerable impacts. By 4.2 billion years ago, the magma ocean of the Moon had cooled and solidified, while at the same time the terminal bombardment was occurring, perhaps as the last stages of an accretion process that had continued since the formation of the Moon.

Mercury's earliest history differed from that of the Moon primarily because of its different composition. Mercury was depleted in silicate rocks rather than in metals, giving it the highest metal concentration of any planet. Differentiation must have taken place early, which led to a magma ocean (shallower than that of the Moon) and a rocky crust. Early heating and expansion of the planet may have triggered widespread volcanism near the end of the period of heavy bombardment.

FROM THE TERMINAL BOMBARDMENT TO THE PRESENT

On the Moon, the first 700 million years are called the pre-Imbrium era, which ended 3.8 billion years ago with the formation of the Imbrium Basin and widespread blanketing of the nearside of the Moon with ejecta from this impact. Shortly thereafter, the formation of the Orientale Basin marked the last of the large asteroid impacts. As the impact rate declined, large-scale volcanism began to fill the deep impact basins on the lunar nearside. The magma sources were located hundreds of kilometers below the surface, in regions being heated by internal radioactivity. Slight expansion triggered by changes in internal temperature may have opened fractures, permitting basaltic lava to rise and flood the basins. Imbrium was flooded in this way between 3.4 and 3.3 billion years ago.

On Mercury at approximately the same time, the presence of a huge iron-nickel core apparently led to a different evolutionary path. Instead of expanding, calculations show that Mercury began a slow contraction, amounting to perhaps 2% of its radius. Cracks in the crust were squeezed shut, inhibiting large-scale volcanism, and compressional forces became great enough to form the scarps we see today on the surface. Shrinkage rather than expansion of Mercury may explain the apparent paradox that this planet seems to have had less surface volcanic activity than the Moon in spite of higher interior temperatures.

Large-scale volcanic activity on the Moon ceased about 3 billion years ago, and slow cooling of the interior continued until the entire object solidified. Nothing much has happened on the surface for the last 3 billion years except for rare impacts by comets or near-Earth asteroids and continuing slow erosion by smaller impacts. Mercury is similarly winding down. Although it may still have a partially liquid iron core, the outer parts of the planet have cooled, and major geological activity ended long ago.

It is interesting to contrast these histories with that of the Earth. Except for very rare rocks in the most ancient parts of the continents, the oldest terrestrial rocks have all formed since the end of the period of lunar volcanism. Even the young (billion-year-old) lunar crater Copernicus formed before the appearance of hard-shelled life-forms in our oceans. The youngest major lunar crater, Tycho, formed only 100 million years ago, about the time the outlines of the major terrestrial continents were established. In many ways, the lunar and mercurian records are the perfect complement to that of Earth, with all the action on the smaller bodies ending at just about the time our own geologic story begins to unfold.

Moon and Earth: Similarities and Differences

The evidence for the origin of the Moon contains some serious contradictions. Study of the Apollo lunar samples revealed a fundamental similarity between the Earth and the Moon in the detailed isotopic composition of their rocks. Not only are many of the lunar minerals similar to those found on our own planet, but the relative proportions of different isotopes of oxygen and other elements are also the same. Recall from Chapter 4 that these isotopic ratios differ among the meteorites and that these differences are used to infer incomplete mixing of material in the early solar nebula. The close similarity of the Earth and Moon in this respect suggests that these two bodies formed together out of the same mix of materials condensing from the solar nebula.

In spite of the isotopic similarities, the Moon is fundamentally different from the Earth in its bulk composition. In many ways the composition of the Moon resembles that of the terrestrial mantle, which also has a density of about 3.3 g/cm^3, but significant differences remain between their bulk compositions. Water and other volatiles are severely depleted on the Moon. There is five times less potassium, while the uranium has about the same abundance. Other elements such as calcium, aluminum, and titanium are more abundant on the Moon. If the Moon and Earth formed together, we would not expect such differences in composition.

Three Theories of Lunar Origin

There are three traditional theories regarding the origin of the Moon. While none of them is now believed to be wholly correct, they illustrate the range of alternatives considered. These theories are sometimes called the *daughter theory*, the *sister theory*, and the *capture theory*. Let us look at each and confront it with the evidence cited above concerning the similarities and differences between Earth and Moon.

The daughter, or fission, theory proposes that the Moon formed from the Earth. It was first proposed in 1880 by astronomer George Darwin, the son of the famous biologist Charles Darwin, who calculated that a rapidly spinning molten Earth could split (fission) to form a double planet. If the alleged split took place after the differentiation of the Earth, the smaller Moon might be formed purely from mantle material and thus be depleted in metals. Furthermore, this idea seemed

of the Moon,
slowly moving
us flaws in the
them associ-
tion between
he deficiency
r example, is
nal mechani-
lly proposed
planetary fis-
ts.

e Earth and
a spinning
at most sci-
s of Jupiter,
oblem here
o explana-
etween the
ter theory
med with-
ot seem to
e.

The third possibility is the capture theory. The idea here is that the Moon formed elsewhere in the solar system as an independent body and was subsequently captured into orbit around the Earth. The compositional differences are understood as representing different condensation conditions in separate locales in the solar nebula, but again there are two problems. First, the isotopic ratios seem to indicate that the Earth and Moon formed from the same pool of material. Second, the Earth could not have captured the Moon into a stable orbit unless a third body intervened, and no helpful go-between has been identified. Capture by atmospheric drag, the method proposed for the capture of the martian satellites, does not work for an object as large as our Moon. Today, the capture theory has few, if any, defenders.

A Synthesis

If the Moon is not daughter, sister, or interloper, what is it? Apparently the three basic theories just described are too simple, and we must seek a more complex scenario. What is required is a mechanism that permits the Moon or its precur-

sor materials to form initially in the same part of the solar nebula as the Earth and then to undergo some fractionation process or processes that can remove most of the metals and volatile elements before solidification. The daughter theory comes closest, but we need a better way to get a piece of the terrestrial mantle into orbit where it might be able to evolve into the Moon.

The most likely answer to this dilemma is that one or more *giant impacts* during the period of planetary accretion ejected huge quantities of mantle material from the Earth, at the same time heating this material sufficiently for most of the volatiles to escape. This event must have taken place after the Earth differentiated, but as we have seen differentiation occurred quickly, probably as the Earth was forming. Impact heating and fractionation would change the chemical composition of the ejected material without altering the isotopic ratios for individual elements, which would remain the same in both the ejected material and the rest of the Earth.

Subsequently, some fraction of the ejecta could aggregate in orbit to form the Moon. Initially, the Moon was in a close orbit, but tidal forces would drive it away to its present position. Figure 8.16 is an illustration of this scenario as currently envisioned.

The idea of a giant impact appears to match what we know of the composition of the Moon, but does it make sense otherwise? How large would this impact have been? Is it reasonable that an impact could eject material into orbit rather than blasting it away from the Earth entirely? In the late 1980s, scientists started to investigate these questions. From their work, it appears that the impactor must have had a mass about one-tenth that of the Earth (about the size of Mars), and it must have struck at an oblique angle. If the mass had been much larger, the Earth would have split apart; much less and insufficient material would have been ejected. An impact of this magnitude could have ejected about 10% of the mass of the Earth into space, with perhaps 10% of the ejecta (1% of the mass of the Earth) remaining in orbit. Since the mass of the Moon is only about 1% of the current mass of the Earth, the giant impact hypothesis seems to pass this plausibility test.

Figure 8.16 The formation of the Moon apparently resulted from the impact of a Mars-sized body with the early Earth. Part of the ejecta from this oblique collision went into orbit about the Earth and subsequently aggregated to form the Moon.

ORIGIN OF MERCURY

If the Earth was struck by a Mars-size projectile during its formation, we may ask whether other planets were similarly disrupted by impacts. George Wetherill of the Carnegie Institution of Washington has carried out extensive calculations on the early evolution of the terrestrial planets. His models indicate that as many as a hundred protoplanets of lunar to martian size were probably formed in the inner solar system, many of them on highly eccentric orbits. In such chaotic circumstances, large impacts would have been a natural part of the final accumulation of the planets. Indeed, Wetherill's models not only predict the existence of giant impacts, but also suggest

that several would-be planets were probably shattered beyond repair in this cosmic demolition derby. The four remaining terrestrial planets are the survivors.

If Wetherill is correct, then giant impacts also provide an explanation for the anomalous composition of Mercury. Suppose that Mercury was originally larger, with about half the mass of the Earth or Venus. It would have differentiated to form a core and a thick silicate mantle with the same proportions of iron and silicates as the Earth. Suppose then that it was hit by a Mars-size projectile. In this case a large fraction of its mantle mass might have been ejected, leaving a thin silicate layer above a now disproportionately large core. Thus, the same theory of giant impacts early in solar system history provides us a plausible explanation for both the low metal content of the Moon and the high metal content of Mercury.

If a multitude of protoplanets formed with eccentric orbits, then each of the surviving terrestrial planets would be expected to have about the same composition, except for the fractionation effects of giant impacts. We will return to these and other ideas of planetary formation in the final chapters of this book.

8.6 QUANTITATIVE SUPPLEMENT:
RADAR AND THE DOPPLER EFFECT

Radar provides one way to determine the rotation rate of a planet, using the Doppler effect. As an illustration, let us calculate the effect of the rotation of Mercury on a radar signal transmitted at a wavelength of 10 cm. The equivalent frequency of this radar wave is given by

$$f = \frac{c}{\lambda} = 3 \times 10^9 \text{ Hz} = 3 \text{ GHz}$$

where c is the velocity of light (3×10^8 m/s).

Mercury has a diameter of 4878 km. Its circumference is 1.5×10^4 km $= 1.5 \times 10^7$ m. If it rotates in 59 days (5×10^6 s), the equatorial rotational speed is a little more than 3 m/s. Thus, one edge will be moving away from us at 3 m/s, and the other edge will approach at 3 m/s, relative to the center of mass of the planet. The total range in apparent velocity is 6 m/s, corresponding to a frequency (or wavelength) spread $(\Delta\lambda)/\lambda$ in the reflected radar signal given by the Doppler formula:

$$\frac{\Delta\lambda}{\lambda} = \frac{\Delta f}{f} = \frac{v}{c} = 2 \times 10^{-8}$$

If the transmitted wavelength is exactly 10 cm and there is no motion of the center of mass of Mercury relative to the Earth, the wavelengths of the returned signals would lie between 9.999999 and 10.000001 cm. If we measure this bandwidth in the returned signal, the rotation rate of the planet can be determined.

Note that one would rarely measure the particular apparent rotation calculated in the above example. At any given time, the apparent rotation of Mercury has two components, one due to its actual rotation and the other resulting from our changing perspective as Mercury passes by the Earth. Thus, we must know the relative orbits of Mercury and the Earth before we can use such radar observations to derive the true rotation rate.

8.7 QUANTITATIVE SUPPLEMENT:
SYNODIC PERIOD

The apparent motion of a planet in the sky does not represent its true period of revolution about the Sun. We look out from a moving platform, the Earth, and the motion we see results from a combination of the true motion of a planet with that of the Earth. Consider, for example, the case of a distant planet like Saturn, which has an orbital period of 30 years, making it nearly fixed among the stars. Yet the Earth swings around in its orbit every year. The result is that the apparent period of Saturn as it shifts from morning to evening sky follows a cycle of nearly 1 year, which is dominated by the motion of the Earth. This apparent period is called the *synodic period* of a planet; for Saturn the synodic period is 378 days.

The case of nearby planets is a little more complicated. Mercury has a synodic period of 115 days; that of Venus is 596 days; and that of Mars is 780 days. However, there is a simple

formula for calculating the synodic period of any object from its true period of revolution, as given by Kepler's laws. You take the reciprocal of the true period (in years), subtract 1, and take the reciprocal again. This is easy with a hand calculator; try it! More generally, we can find the synodic period for any two planets with the formula

$$\frac{1}{P_{inner}} - \frac{1}{P_{outer}} = \frac{1}{P_{synodic}}$$

where P_{inner} is the period of the inner of the two planets and P_{outer} corresponds to the outer of the two planets.

SUMMARY

Because Mercury is small and near the Sun and has been visited by only one spacecraft, Mariner 10, our database is limited. One important source of information, however, has been planetary radar. We saw how radar astronomy worked, and in particular noted the power of the technique of using Doppler shifts in the frequency of the reflected radio pulse to determine planetary rotation. Radar has revealed the unusual rotation period of Mercury, which is exactly two-thirds of its 88-day period of revolution around the Sun. We examined how tides raised by one body on another can slow planetary rotation and modify orbits, providing an explanation for the fact that the Moon always keeps the same face toward the Earth, as well as for the more unusual 2:3 synchronous rotation of Mercury. Tidal forces are important for many planets, satellites, and rings, as we will see later in this book.

Mercury is a small airless world heavily scarred by impact craters. Its surface looks very much like that of the Moon, although the indications of past volcanic activity are less obvious. However, Mercury is very different in its bulk properties. Its high density (the highest of any planet) indicates that much of its interior is made of iron, like the core of the Earth, and this hypothesis is further supported by the presence of a small planetary magnetic field. Roughly speaking, Mercury has an interior like that of the Earth and a surface like that of the Moon.

The lunar interior is unique. Our satellite has lost not only volatile elements but also most of its expected allotment of iron and other metals. As a result, the Moon has at most a very small iron core and no global magnetic field. The compositional differences between Earth and the Moon are great enough to frustrate simple theories in which the two objects formed together, out of the same raw materials in the solar nebula.

We looked at three simplified theories of the origin of the Moon and found difficulties with each. Apparently the correct answer is more complex, involving the ejection of terrestrial mantle material in a giant impact and subsequent aggregation of the iron- and volatile-depleted ejecta into the Moon. Paradoxically, the same giant impact theory can explain Mercury's loss of silicate mantle material. Apparently giant impacts were an important element of the early evolution of the terrestrial planets.

Key Terms

Doppler effect	synchronous rotation
polar cap	uncompressed density
solar day	

Review Questions

1. Describe how radar is used in the study of the planets. In particular, make sure you understand how the rotation of a planet broadens the returned signal by the Doppler effect, even though the signal is reflected from an entire hemisphere.

2. Discuss how tides are formed. Why are there two high tides every day on Earth? Why does high tide not generally correspond to the time when the Moon is overhead? What sort of tides are raised on the Moon by the Earth?

3. Compare the tidal and rotational histories of the Earth, the Moon, and Mercury. Can tidal theory account for their different rotational periods today? For each planet, is the present rotational state stable, or is evolution continuing? What will the final rotation period be?

4. Compare the geology of the Moon and Mercury. What do they have in common? Consider in particular the craters, impact basins, and plains (or mare) areas.

5. Describe and contrast what an astronaut would see and feel on the surfaces of the Moon and Mercury. Could you easily tell whether you were standing on Mercury or the Moon?

6. Use what you now know about Mercury to critique the standard paradigm for solar system history. How would you expect the bombardment history of Mercury to compare with that of the Earth and Moon? Is there any evidence for a late heavy bombardment on Mercury? What are the likely ages for the craters, basins, and plains on Mercury, according to the standard model?

7. Compare the interiors and histories of the Moon and Mercury. What does the geology of each tell us about the interiors? How useful are interior models? How would you go about distinguishing among several models to see which is most likely to represent correctly the thermal history of a planet?

8. Describe the three traditional scenarios for the origin of the Moon. How does each compare with the data we have on the Moon? Show how an appropriate combination of elements from these theories may be consistent with what we know about the Moon. How does the origin of Mercury relate to this theory for the Moon?

Quantitative Exercises

1. Calculate the synodic period for each of the planets from Mercury through Saturn.

2. Find the synodic periods of Mercury and Earth as viewed from Venus. How does the synodic period of Venus as seen from the Earth compare with the synodic period of Earth as seen from Venus?

3. Calculate the energy of an impact with the Earth by a projectile the mass of Mars striking at a speed of 20 km/s. Compare this energy with the gravitational binding energy of the Earth, given by GM^2/R, where G is the gravitational constant (6.67×10^{-11} newton m^2 kg^{-2}) and M and R are the Earth's mass and radius, respectively.

4. The Arecibo radar instrument is sometimes used to study Earth-approaching comets at a wavelength of 12 cm. Suppose the nucleus of such a comet has a diameter of 5 km and its period of rotation is 10 hr. What are the relative speeds of

its leading and trailing edges as seen from Arecibo? What is the total broadening of the 12-cm transmitted signal due to this rotation?

ADDITIONAL READING

Dunne, J. A., and E. Burgess. 1978. *The Voyage of Mariner 10: Mission to Venus and Mercury (NASA SP-424)*. Washington, DC: U.S. Government Printing Office. Interesting account of the Mariner 10 mission and its results.

Murray, B. C., M. C. Malin, and R. Greeley. 1981. *Earthlike Planets: The Surfaces of Mercury, Venus, Earth, and Mars*. San Francisco: Freeman. Excellent (but moderately technical) introduction to planetary geology, with good discussions of Mercury and the Moon.

Strom, R. 1987. *Mercury, the Elusive Planet*. Washington, DC: Smithsonian Institution Press. Comprehensive semipopular book on Mercury, combining results from Mariner 10 with other information about this planet.

9

THE EARTH: OUR HOME PLANET

It is hard to think of the Earth as a planet. We humans are so intimately involved with our world that we have difficulty establishing a planetary perspective. Yet the Earth is a planet, quite ordinary in terms of location and size, and it is subject to many of the same processes that shape the other worlds of the solar system. It is one of the achievements of the space age that humans have begun to appreciate the close relationships among the planets and to use information about one to gain insight into another. The other planets have much to teach about the Earth. But even more, since we know it so well, the Earth can teach us a great deal about other planets.

In order to study the Earth the way we do other planets, we must step back from the wealth of detail that surrounds us and try to address the same kinds of questions that motivate our study of other worlds. Where and how did the Earth form? What are its composition and internal structure? What sources of energy maintain its geologic activity, and how have these energy sources varied over time? Why does the Earth, alone of all the planets, have oceans of liquid water and an atmosphere rich in oxygen?

What maintains the circulation of the oceans and atmosphere, and how has each evolved with time? And perhaps most intriguing of all, what role has life played in the evolution of the planet? All of these questions are addressed in this chapter.

9.1 EARTH AS A PLANET

Let us begin our study of the Earth by imagining ourselves visitors from another solar system, making our preliminary reconnaissance of the planets. Undoubtedly our first attention would be directed to the giant planets with their spectacular systems of rings and satellites. Later, perhaps, we would check up on the smaller worlds of the inner solar system. Among these the largest would be Earth, the only inner planet with a major satellite.

THE VIEW FROM SPACE

Even from a distance, Earth would appear unique (Fig. 9.1). It is the only planet with liquid water on its surface, and seen from the correct angle, sunlight glints brightly off its oceans. Ice perma-nently blankets its south polar continent as well as Greenland and other isolated land areas in the northern regions. The polar caps show large sea-sonal changes; in winter, sea ice covers the polar seas and a transient snowcap invades the temper-ate landmasses, extending halfway to the equator. These surface changes could be seen only dimly through a shifting canopy of white clouds that re-flects nearly half the sunlight striking the Earth. A few tawny brown land areas are normally clear of clouds, while other regions, primarily near the seas, are eternally clouded. Enigmatic hazes of dust sometimes obscure even the cloud-free areas, and hints of changes might appear in sur-face color or reflectivity with the passage of the seasons.

More remarkable than its surface appearance is the chemistry of the terrestrial atmosphere. Ni-trogen, the primary gas, makes up 78% of the at-mosphere, but the real surprise to an alien explorer would be oxygen, the second most abundant gas at 21%. On Mars and Venus, oxy-gen, a highly reactive gas, is trapped in minerals in the surface rocks. Perhaps equally surprising is the

Figure 9.1 Earth is the only planet in the solar system with oceans of water, blue skies, and abundant life. On average, more than half of the planet is shrouded by water clouds at any time.

absence of more than a trace of carbon dioxide, which is the primary constituent of the atmospheres of Venus and Mars. The atmosphere and oceans of Earth represent a triple enigma: too much oxygen, too much water, and not enough carbon dioxide. All of these chemical anomalies are related to the presence on Earth of abundant and varied life, a subject we will return to in Chapter 12.

TERRESTRIAL GEOLOGY

The Earth's oceans contain enough water to cover the entire globe to a depth of 3 km, yet nearly a third of the surface of the planet rises above this watery layer. Most of this land area is concentrated in the six continents, much larger masses of rock than can be explained by individual volcanoes or upthrust mountain ranges. The highest mountains rise nearly 9 km above sea level, while in places the sea floor near the continental margins can drop as low as 11 km below sea level.

A closer examination would reveal to the space visitor many additional indications of geological activity: erupting hot springs and volcanoes; frequent earthquakes; youthful mountain ranges with sharp ridges and peaks sculpted by the forces of water and ice; and subtle variations in gravity, revealing that parts of the crust are not in equilibrium but are currently undergoing uplift or being drawn down by internal forces.

To learn something about the interior of this planet, the visitor might look at space surrounding the Earth. There, electrons and ions from the solar wind are trapped by a magnetic field produced in the planet's core. Noting the high density of the Earth, the space visitor might conclude that the Earth had differentiated and that its core consists primarily of iron, a cosmically abundant metal.

CYCLIC PROCESSES AND PLANETARY EVOLUTION

The next step in this initial investigation should focus on the processes taking place. Earth is a remarkably active planet in many ways, in dramatic contrast to the Moon and Mercury. How should we look at this activity? One possibility is that we are seeing evidence of rapid evolution and change: The Earth is in transition from one state to another in much the same way we viewed the history of the Moon and Mercury, although most of their transitions took place a long time ago. But a little thought quickly shows the fallacy of this perspective when applied to Earth. Consider the rivers flowing into the sea. Do they indicate that the sea is filling up and that soon the land will be submerged? Certainly not, because the water evaporates from the ocean as fast as it is added from the land. What we have here is a *cycle*, called the *water cycle*, in which input and outflow balance.

Many processes on Earth are cyclic. Volcanoes deposit lava on the surface, but this rock is eventually drawn down again to the interior to be recycled through new volcanic eruptions. The land is eroded into the sea, but the sea does not silt up; instead the sediment is used to create new continental crust. Marine animals trap carbon dioxide to form vast deposits of carbonate rocks, but eventually some of these rocks also are reprocessed, and their carbon dioxide is released again into the atmosphere. The Earth behaves like a giant machine, with many interlocked cycles turning at different rates.

The idea of Earth as a machine is useful for understanding many processes, but not all. True evolution does take place. Life, for instance, has clearly evolved, and along with it the atmosphere has changed greatly. There are long-term trends in geology that can be separated from the cyclic events. One objective in looking at Earth as a planet is to distinguish the long-term evolution from shorter-term cyclic processes, in order to see the many ways in which our planet's history resembles that of other planets.

9.2 THE EARTH'S INTERIOR

Few places in the planetary system are less accessible to direct study than the Earth beneath our feet. The deepest boreholes have penetrated less than 10 km, 0.1% of the distance (6378 km) to the center. Determining the structure and composition of the interior of the Earth is a difficult scientific puzzle.

Probing the Interior

Our first clue to the composition of the interior of the Earth is its relatively high density of 5.5 g/cm^3. Even when corrected for the compression in the interior, the uncompressed density of our planet is 4.3 g/cm^3, a much higher value than that of the Moon. Evidently the Earth has an invisible metallic core in addition to its obvious rocky upper layers.

A powerful technique for exploring the interior structure of the Earth is provided by earthquakes. Whenever movement of the brittle crust produces an earthquake, low-frequency vibrations are generated. These vibrations are analogous to sound, but with much longer wavelengths, typically several kilometers. The shaken Earth responds like a giant bell, and its interior structure determines the tones that will be detected at different places on the surface. By measuring the vibrations of the Earth from many locations, we can reconstruct the temperature and pressure of the deep interior. The waves generated by earthquakes are called **seismic waves,** and instruments that measure them are **seismometers.**

The compressional waves that are produced can pass through liquids, whereas the shear waves cannot. The velocities of these two types of waves depend on the densities of the rock layers through which they pass. Thus, by monitoring these waves with seismometers that are deployed at stations scattered around the Earth, we can obtain a remarkably accurate picture of the structure of our planet's interior (Fig. 9.2). This was the way that the outer liquid layer of the Earth's core was discovered.

Additional information on the interior of the Earth is provided by measurements of the chemical and physical properties of rocks. Most of the crustal rocks of the Earth have been derived, like the lunar basalts, by partial melting of subsurface layers, and they retain some information on the composition and nature of their source regions. In addition, there are places on Earth where

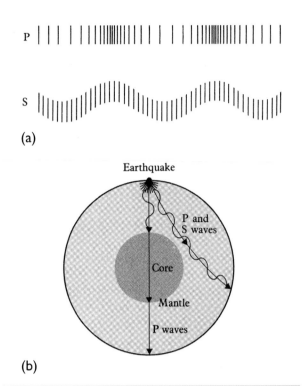

Figure 9.2 **(a)** Schematic diagram of P (pressure or compression) waves and S (shear) waves. **(b)** Paths of such waves through the Earth. S waves cannot propagate through the core, indicating that it must include a liquid component.

magma from as deep as 200 km has reached the surface carrying rocks called kimberlite. Kimberlite, named for Kimberly in South Africa, is best known as the source of most of the world's diamonds. Diamond, formed from carbon at great pressures, is an artifact from the mysterious world a few hundred kilometers beneath us.

Even the deepest rocks sample only the upper few percent of the planet, but some useful information on the deep interior is provided by the magnetic field of the Earth. Although no one understands all the details of the mechanisms that generate a global planetary magnetic field, an electrically conducting, turbulent liquid is probably required. Combining this information with that obtained from rock chemistry and from seismic studies allows us to undertake a scientific journey into the interior of our planet.

There are six major and several minor landmarks along the path to the interior. We will list

pointed south. As a matte[r]
the Earth's field has been s
past few decades, and if
tinue, we might experienc
the next 200 years. More
seeing only a short-term fl[u]

9.3 THE CHANGI[NG]
OF THE EART[H]

All around us we see the
Look at a road cut, and y
often tilted and twisted, of
posited by erosion in the
and hills are worn down,
plateaus and jagged peak
processes must be acting t[o]
Volcanoes erupt, and earth[
ments in the Earth's crus[t]
total absence of impact c[
craters are destroyed as fast

THE GEOLOGIC TIMES[CALE]

The geologic history of th[e]
rocks, which record the even
years. The strata deposited i
like the pages of a history
however, the pages have fre[
or even lost in subsequent u
record, therefore, the geolog

Table 9.1 Comparison of geo[

Time before Present (billion years)	E[o
0.03	C[
0.2	M
0.4	P[
1.0	P[
2.0	P[
3.2	A[
3.6	A[
3.8	A[
4.0	A[
4.4	A[

them here and discuss each one in more detail later. The six major divisions into which it is convenient to divide the Earth are, starting from the outside (Fig. 9.3):

1. The magnetosphere, the region of charged atomic particles that extends from the upper atmosphere, about 200 km high, out to the boundary with the solar wind and true interplanetary space, at an altitude of about 100,000 km.

2. The atmosphere, the layer of gas that extends from about 200 km in altitude down to the surface.

3. The ocean or hydrosphere, the layer of liquid and frozen water that covers about three-fourths of the surface of the Earth.

4. The crust, the solid surface of the Earth bounded on the top by the oceans and atmosphere. Its thickness is typically 10–30 km.

5. The mantle, consisting of solid but plastic rock extending from the bottom of the crust down

to a depth of 2900 km. The mantle makes up two-thirds of the total mass of the Earth.

6. The core, composed of dense metal, divided into an outer liquid core and an inner solid core.

Let's begin at the surface and work our way down; we will discuss the ocean and atmosphere later.

THE CRUST

The crust of the Earth is defined as the uppermost solid layer, made mostly of igneous rock derived from the mantle. Under the ocean basins, the crust is thin, averaging only 6 km in thickness, and composed of basaltic rocks, generally similar to the lunar and meteoritic basalts. This crust, which covers 55% of the surface, is all relatively young, having solidification ages of less than 200 million years. Although apparently analogous to the basaltic maria of the Moon, the origin of the terrestrial basalts is rather different. The ocean crust is formed from magma rising from the interior along well-defined rifts or

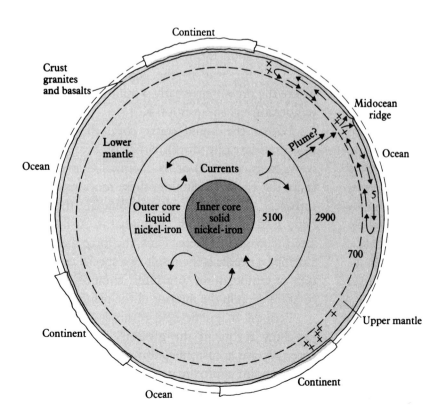

Figure 9.3 The interior of the Earth, according to current thinking. A small solid core is surrounded by a liquid region, which in turn gives way to the mantle. Rapid convection in the liquid core generates the planet's magnetic field. Slow convection in the mantle causes the motions of crustal plates, moving the continents, forming mountains, and subducting eroded material. (Highly schematic.)

spreading centers (discus:
is marked by several mid
these rifts, the ocean cru
stroyed by sinking into tl
and recycled. This proce
struction of the oceanic c
in Section 9.4.

The continental crust
lower density than the
45% of the surface and
mass of the Earth. Its p
rocks called **granites.** Gr
are igneous, but their co
and they solidified below
pressure. The thickness c
varies from about 20 km
The continental crust can
ing on top of the denser
the igneous granites, the
cludes a great deal of sedi
phic rock accumulated ove
and even billions, of years
on the Earth are found ii

The Mantle

Below the crust is the
higher-density minerals th:
its own radioactive elemer
from the hotter core belo
plastic than brittle. Within
tinct layers can be identifie

First is the rigid upper p
to a depth of about 100 km
the crust to which it is atta
the lithosphere. The bound
and upper mantle represent
tion; the boundary between
mantle underneath represen
cal properties. The lithosph
tached to the mantle beneat
it like an ice cube sliding on

Next, below the lithosp
a depth of about 700 km, i
tle that convects, transport
slow mechanical turnover o
gish convection currents pro
that moves the plates of th

Figure 9.4   The Alps, like most other high mountains on Earth, represent a balance between uplift from pressures in the crust and rapid erosion from ice and water. It is only because of ice and water that terrestrial mountains have sharp peaks.

Much of the geologic structure that we see around us is comparatively young. The jagged peaks of the Rocky Mountains or the Alps have been sculpted by ice within the past hundred thousand years (Fig. 9.4); the volcanic islands of the Hawaiian chain are at most a few million years old; and even so large a feature as the Grand Canyon in Arizona has been cut by the Colorado River within the past 10 million years. Thus, many of the major landforms around us have ages comparable to the short span of human existence on this planet.

Volcanoes

Volcanism is one of the most spectacular forms of geologic activity. On our planet, heat from within the mantle drives several types of volcanic activity. The largest but rarest types of eruptions lay down vast plains of fluid lavas very similar to the lunar maria; examples (each about 60 million years old) are the Columbia River flood basalts that cover eastern Washington State (400,000 km³ in volume) and the even larger Deccan lavas of the Indian subcontinent. More familiar, however, are the eruptions that build the distinctive mountains we call volcanoes.

Although geologists distinguish among many types of volcanoes, we will consider only two basic classes, both of which have been found on other planets as well as our own. First are the **shield volcanoes,** which are dome-shaped mountains built up by the repeated eruption of relatively fluid lavas from a single vent or fissure. Although the individual lava layers are typically only a few meters thick, the mountains, such as Mauna Loa in Hawaii (Fig. 9.5), can reach heights as great as 9 km. Typical shield volcanoes have slopes of about 10° and are topped by a large, shallow, flat-floored crater called a **caldera,** produced by subsidence rather than explosions.

At the opposite extreme are a variety of explosive eruptions that build (and often then destroy) steep-sided **composite volcanoes.** Their distinctive cones are formed by the fallback of plumes of viscous lava fragments ejected to great heights by the force of the eruption. The slopes of composite volcanoes are typically about 40°. Perhaps the best-known example is Mount Fuji in Japan (Fig. 9.6). The most violent kinds of eruptions, such as

Figure 9.5 Mauna Loa in Hawaii is a shield volcano, with gradual slopes formed from successive flows of fluid lava. This photo shows the summit eruption of 1982. Note the constellation of the Southern Cross above the mountain.

the eruption of Vesuvius in A.D. 79 that destroyed Pompeii, can eject many cubic kilometers of rock dust into the atmosphere as well as devastating the surrounding countryside.

OTHER GEOLOGIC ACTIVITY

Another easily recognized form of geologic activity is erosion by ice. In glacial regions, the movement of ice sheets can gouge deep, U-shaped valleys in times as short as a few thousand years. It is primarily glacial action that is responsible for the sharp peaks of the best-known mountains of the Earth. As we have seen on the Moon, mountains tend to be low and smooth in profile on a world without ice erosion (see Section 7.6).

Finally, **sedimentation** is a process that can alter the landscape rapidly. Shallow lakes can silt up in a few decades, and the recession of the sea due to sedimentation has left the docks of once-great ports like Ephesus, Miletus, and Pisa dry and deserted. Most sediment is deposited in the sea. The fact that the seas have not filled up is one of the properties of the Earth that requires an explanation.

By mapping the distribution of rock types and landforms, and determining ages from the fossil record, radioactivity, and rock magnetism, scientists have begun to read the book of geological history, although the records of events earlier than the last few hundred million years are blurred and some of the pages lost entirely.

Figure 9.6 Painting of Mount Fuji by the Japanese artist Hokusai. The slopes of this volcano are not as steep as the artist has painted them, but they are much steeper than those of Mauna Loa. The volcanism here is a different type than that in Hawaii.

Beyond 2 billion years ago, entire chapters are missing, and of the earliest periods of terrestrial history not a trace remains.

In this chapter, we have discussed many features of terrestrial geology, but conspicuously absent have been the craters that dominate the surfaces of the Moon and Mercury. Why does the Earth have so few impact craters relative to the Moon? Both objects occupy the same region of space, and Earth, with its greater surface area and stronger gravity, should receive many more impacts than does the Moon. The presence of an atmosphere does not explain the absence of craters, for while the air effectively filters out the smaller debris, any projectile capable of digging a large crater will punch right through the atmosphere.

WHERE HAVE ALL THE CRATERS GONE?

How did the Earth avoid being heavily cratered? The fact is, it did not avoid impact cratering. The difference between Earth and the Moon lies not in the rate at which craters are formed, but in the rate at which they are destroyed. We are deluded into thinking we are immune to large impacts because the evidence of past catastrophes has largely been removed. Remember, in this connection, that most of the craters on the Moon were formed more than 3.8 billion years ago, during the time when we have little record of geologic events on the Earth.

Only recently, largely with the aid of surveys from Earth orbit, have geologists learned to recognize the faded scars of ancient impacts on the continental crust of our planet. These are destroyed rapidly (in geologic terms) by erosion and other forms of geologic activity. Even the largest crater of the last 100 million years, located in Mexico and called Chicxulub, is completely invisible to someone standing on the surface.

Meteorites penetrate the atmosphere daily, but only about once a century does one strike that is large enough to form a small crater. In terms of significant effects on the surface of our planet, however, we must consider the much rarer impacts of objects that weigh from hundreds of thousands up to billions of tons. Calculations show that the atmosphere effectively shields us from most impacting bodies with mass less than

50,000 tons, corresponding to an energy of about 5 million tons (5 megatons) of TNT. Above this threshold, incoming projectiles will either explode in the lower atmosphere or reach the surface to form a crater.

RECENT IMPACTS

A remarkable impact event took place above the Tunguska River in Russian Siberia on June 30, 1908. Witnessed by only a few of the native population, an explosion of unprecedented magnitude occurred, registering its shock wave on atmospheric instruments around the world. A brilliant fireball in the sky briefly rivaled the Sun in brightness, followed by an explosion and fire. Over a region of a thousand square kilometers the forest was flattened, the trees stripped of leaves and branches and toppled in parallel rows radiating from the blast center (Fig. 9.7). Hundreds of caribou died, and a man in a trading post 60 km away from the explosion was thrown from his seat and knocked unconscious. Yet, in spite of all this violence, no crater was formed, and by now the Siberian forests have returned to erase the last evidence of this explosion.

Figure 9.7   A small part of the forest destroyed by the 1908 explosion of a small (60 m in diameter) asteriod in the Earth's atmosphere near Tunguska, Siberia. The energy of the airburst was about 15 megatons of TNT.

From the damage caused by the Tunguska blast, we conclude that its energy was equivalent to that of a large nuclear weapon, probably between 10 and 20 megatons. The diameter of the stony projectile was thus about 60 m, the size of a large office building. It is estimated that such an impact should occur every millennium or so somewhere on Earth. Fortunately, most of the Earth's surface is covered by ocean and much of the land area is still relatively unpopulated, so on statistical grounds we can hope that the next such impact does not take place near a populated area.

For an example of a much rarer impact from an iron asteroid, we turn to Meteor Crater in northern Arizona, near the volcanic region of the San Francisco Peaks. The time is 50,000 years ago, before there were humans living in the region to witness the impact of 100,000 tons of iron on the Earth. There was no chance that this solid mass of metal would break up in the atmosphere. Rather, the force of its impact carried it several hundred meters below the surface before it finally came to a halt, transferring its energy into an explosion that destroyed the impacting body and blasted a hundred million tons of pulverized rock into the atmosphere, leaving a crater a kilometer in diameter. Around the edges of the crater the thick layers of sediment were twisted up and folded back (Fig. 9.8). Fragments of metal from the impacting object were dispersed over

(a)

(b)

Figure 9.8   **(a)** Meteor Crater, near Winslow, Arizona, is 50,000 years old, only yesterday on the geologic timescale. A similar impact could occur anywhere on Earth at any time. Meteor Crater is about 1.3 km across and 200 m deep. A similar crater would be a tiny pit on the lunar landscape as viewed from Earth but quite a nuisance if it were formed today in downtown Chicago. **(b)** The Manicougan Lakes Crater in northern Quebec. This circular feature, 70 km in diameter, was caused by an impact in the late Triassic period, about 200 million years ago.

hundreds of square kilometers around the crater.

Today, Meteor Crater is the best known and most obvious impact crater on Earth. Tens of thousands of tourists take the short detour from Interstate 20 each year to visit it. Although erosion has partly filled the interior with sediment and has washed away much of the ejecta, the characteristic uplifted and folded-over strata around the edges clearly testify to the explosive origin of this feature, and tiny metal fragments can still be picked up in the surrounding desert. Yet even here, the recognition that this is the scar of an extraterrestrial object has come only during the past 50 years; previously, most geologists thought it was an unusual volcanic crater associated with the nearby volcanic field.

Impacts can influence the history of life as well as the geology of the surface. The impact that formed the Chicxulub crater 65 million years ago caused the extinction of most life on Earth, including all of the dinosaurs. We will discuss the role of impacts in evolution in Chapter 12.

9.4 PLATE TECTONICS: A UNIFYING HYPOTHESIS

In the 1960s, the science of geology underwent a dramatic change that altered the basic assumptions of the field. Until then, almost all geological science had been based on the idea that the major features of the Earth—the continents and ocean basins—were fixed. This was not an unreasonable hypothesis, given that most geologists thought the mantle was solid and unyielding. Within this framework, geology was interpreted in terms of periods of slow uplift to form mountains in competition with the forces of erosion gradually tearing them down again. This outlook is still reflected in older books and museum displays that explain the landscape in terms of alternating uplift and subsidence, as though the crust were part of a seesaw, always moving up and down but never sideways.

WEGENER AND CONTINENTAL DRIFT

The first serious challenge to the orthodox geological viewpoint came from Alfred Wegener, a German meteorologist and amateur geologist.

Early in the twentieth century, Wegener became convinced that the strong similarities between rocks on either side of the Atlantic could best be understood if the east coasts of North and South America had once fitted, like the pieces of a jigsaw puzzle, against the west coasts of Europe and Africa. Even casual inspection of a world map suggests that this fit is rather good. Wegener found other places where continents, now separated by thousands of kilometers, appeared once to have been joined together, and in Iceland he studied the active volcanoes, concluding that this mid-Atlantic island was located on the rift where the New World and the Old were still separating. Based on this evidence Wegener developed a hypothesis of continental drift, in which the continents slowly moved while the lower crust and ocean basins remained relatively fixed.

The hypothesis of continental drift was rejected by most geologists, however, because its proponents could suggest no plausible force to explain the migration of continents across the solid oceanic crust, to which they seemed firmly attached. In spite of the empirical evidence supporting this hypothesis, the motion of the continents proposed by Wegener was deemed physically impossible.

PLATE TECTONICS

The idea of continental drift was finally accepted when, in the 1960s, observations clearly showed that the Atlantic Ocean was widening, as a line of volcanoes, approximately halfway between the two continental masses, injected new lava along the ocean floor. These data showed for the first time that the ocean basins themselves were part of the movement, and that the continents do not drift over an unyielding crust. Instead, the crust itself is being forced apart by the injection of magma. Soon many other pieces of the puzzle fell into place. The result is a unifying geological theory called **plate tectonics.**

The term **tectonic** refers to the forces that stress a planet and the way that the crust responds to such forces. On planets with active geology, such as the Earth, Venus, and Mars, many of the mountains, valleys, and other landforms are of tectonic origin. The scarps on Mercury are also tectonic landforms. What makes terrestrial geology unique is the organization of the crust into six major and about ten smaller sections or plates that float like sheets of ice upon the plastic mantle beneath (Fig. 9.9). The plates constitute the **lithosphere,** about 100 km thick and made up of both the crust and the upper mantle. When tectonic forces are applied in the Earth's crust, these plates are capable of lateral motion—plate tectonics.

The motive power for continental drift comes from convection in the mantle. Driven by this convection, the lithospheric plates move at the rate of a few centimeters per year. Today we can measure this motion directly using the Global Positioning System (GPS). Since there is no free space on the surface of the Earth, the plates bump into each other as they move.

Four kinds of boundaries between plates are possible. First is the **rift zone** or spreading center, such as the mid-Atlantic ridge, where the plates are separating. Second is the **subduction zone,** such as the deep trenches off the east coast of Asia, where plates collide and one is forced under the other.

Third is a region of collision where neither plate slides under the other so that folded mountains are produced, such as the Himalayas. Finally, there are **faults** where one plate scrapes along parallel to another, such as the San Andreas Fault in California. We will discuss each of these cases in turn.

FORMATION AND DESTRUCTION OF OCEANIC CRUST

Most of the surface of the Earth is ocean floor. Sediment washed down from the land is initially deposited on the continental shelves, but gradually it moves out and down into the oceans proper, where it typically accumulates to a depth of 2 km or more. Much of the ocean floor consists of vast, gently rolling plains deeply buried in sediment.

There are 60,000 km of active rifts in the ocean where new crust is formed by basaltic magma rising from below. The few rifts on land, such as the great rift valley that extends through central Africa, probably represent places where a continent is being torn apart to be replaced by

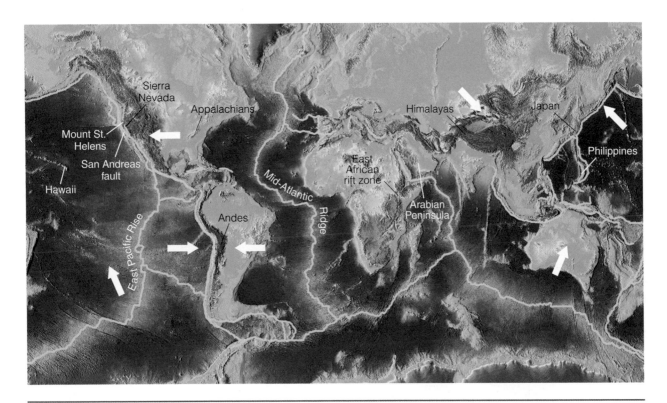

Figure 9.9   The main crustal plates on the Earth.

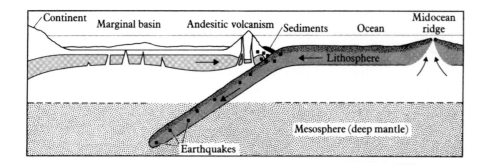

Figure 9.10 Here a slab of oceanic crust is being subducted. As the slab slides down under the granitic continental plate, earthquakes are generated and volcanoes form at the plate boundary. The vertical scale is greatly exaggerated.

new ocean. The average spreading rate is a few centimeters per year, corresponding to the creation of about 2 km² of new oceanic crust each year. Since the total area of oceanic crust is 260 million km², we can see that the lifetime of a given piece of oceanic basalt is only a little more than 100 million years (260 million km² divided by 2 km²). Otherwise, the entire planet would be covered by oceanic crust.

Oceanic crust is destroyed by subduction when convection pulls a plate into the mantle. As two oceanic plates are forced together, one generally slides under the other to produce a deep ocean trench (Fig. 9.10). The sinking slab grinds down into the mantle, generating deep earthquakes, which permit its descent to be traced. At a depth of 200–300 km, it succumbs to increasing heat and melts. Some of the melted rock can rise to the surface to form lines of volcanoes (such as the Pacific "arc of fire"), which mark the position of the subduction zone.

Volcanoes are common along both rift zones and subduction zones. In addition, volcanoes can erupt in the middle of lithospheric plates above mantle hot spots, which represent rising plumes that heat the crust from below. The Hawaiian Islands are an example of volcanoes produced by a mantle plume. Since the Pacific plate is moving with respect to the plume, the volcanic center has traced out an island chain over tens of millions of years (Fig. 9.11). It is not the plume that moves, however, but the plate.

In Section 2.5, we discussed the three ways that heat can be transported: conduction, convection, and radiation. At the temperatures of the Earth, radiation is not important, but both conduction and convection play a role in the movement of energy to the surface. First there is heat conduction through the crust, which is thought to be the dominant process on most other planets. Second is the volcanic activity associated with hot spots. But the most important way heat is released from our planet is at the rifts and subduction zones produced by tectonic activity. Without plate tectonics, the heat balance of the Earth's interior would be very different.

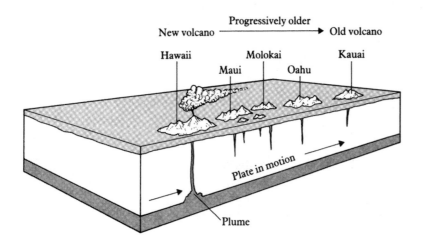

Figure 9.11 The gradual motion of the Pacific plate past a long-lasting hot spot (representing the top of a mantle plume) accounts for the formation of the chain of Hawaiian Islands.

THE CONTINENTAL CRUST

The continental crust, composed mostly of granite, has a different history from that of the oceans. Like a frothy slag floating on the top of the plates, it is rarely subducted. Continental granite is produced much more slowly than oceanic basalt, but it also lasts longer. The creation and destruction of the ocean floor are a cyclic process, but the continents evolve, slowly increasing in total area as more new granite is added by tectonism than is destroyed by subduction or worn away by erosion.

Jostled by mantle convection, the floating continental masses are sometimes pushed together and other times pulled apart. The last time nearly all of the landmass of the Earth was jumbled together was about 200 million years ago. This supercontinent has been named Pangaea. As illustrated in Figure 9.12, Pangaea has gradually been pulled apart and rearranged by continental drift. Not everything is coming apart, however. Some of the landmasses are colliding with each other, most notably the Indian subcontinent with Asia. At such points mountain ranges are folded and uplifted by the pressure of the moving plates. The Himalayan Mountains and Tibetan Plateau produced by this collision have the highest elevations on our planet.

The formation of continents appears to occur through the mountain-building process that takes place when crustal plates collide. Igneous rocks like granite as well as some metamorphic rocks, such as marble, are produced in such collisions. These rocks form the cores of the mountain ranges that are thrust upward, capped by layers of deformed sedimentary rock.

In many places one plate moves parallel to another along a fault or crack in the crust. California's San Andreas Fault, for instance, marks the break where the Pacific plate (which includes the southern coastal regions of California) slides

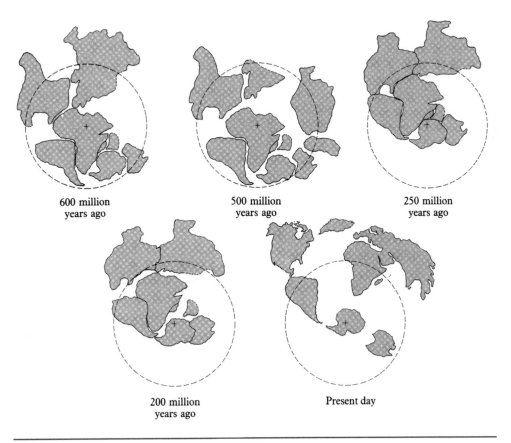

600 million
years ago

500 million
years ago

250 million
years ago

200 million
years ago

Present day

Figure 9.12   Motions of the continents with respect to the South Pole (cross) through geologic time. The dashed circle represents the position of the Earth's equator.

northward at a rate of a few centimeters per year with respect to the main part of the North American continent. At this rate, Los Angeles will be next to San Francisco in about 20 million years. Instead of sliding smoothly, however, the jagged edges of the plates tend to stick until the strain becomes so large that the restraints suddenly give way. The resulting motion is sudden and violent, generating earthquakes. In California, the interval between major plate motions is about a century, resulting in typical movements of several meters when the strain is finally released. The longer the interval between earthquakes, the larger they are likely to be. The last major earthquake in the part of the San Andreas Fault near San Francisco took place in 1906, but the southern part of California has not moved significantly since the Fort Tejon earthquake of 1857. The 145-year buildup of tension along the southern part of this fault is obviously a matter of concern for the residents of southern California.

The mostly cyclic geological processes outlined here are highly dynamic, with much of the landscape of our planet produced by forces deep in its interior. Because the Earth is much larger than either the Moon or Mercury, our planet has cooled less and these forces continue to act today. This logic would lead us to predict that Venus, with nearly the same mass and composition as the Earth, should exhibit a similar tectonic style. As we shall see in the next chapter, however, that is not the case.

9.5 Ocean and Atmosphere

Floating on top of the solid Earth are two layers of great importance to us: the hydrosphere and the atmosphere. The liquid water that covers 70% of the surface and the oxygen-rich atmosphere that shields us from solar ultraviolet light and replenishes the land with life-giving moisture are unique to our planet. Without them there would be no life. In return, both atmosphere and ocean are shaped by the evolution of the very life they make possible. In this and the following section, we examine the intricately interwoven threads of ocean, atmosphere, and life.

Composition of the Oceans

The hydrosphere of the Earth is defined to include the freshwater lakes and streams, the ice caps that blanket the polar landmasses and high mountains, and the sea ice that floats on much of the arctic and Antarctic seas. Its main component, however, is the oceans. The oceans, with a volume of 1.3 billion km^3 and a mass of more than 10^{18} tons, are the great reservoir of water on our planet.

Ocean water is not pure H_2O. As we all know, seawater is salty. In addition to common NaCl, the ocean contains a variety of other water-soluble salts, which together make up about 3.5% of the mass of the oceans. Water is an excellent solvent (which is why we wash with it), breaking down many compounds into their individual atoms. Six elements constitute more than 90% of the salts dissolved in the ocean: chlorine, sodium, magnesium, sulfur, calcium, and potassium.

The concentration of each of these elements in seawater represents a balance between inflow—primarily from weathering and erosion of continental rocks—and precipitation into ocean sediment. Calcium provides an example. About 10^8 tons of dissolved calcium enters the ocean each year, yet the total calcium content of the ocean is only a million times larger—10^{14} tons. Thus, the amount of calcium in the oceans could be accounted for by only 1 million years of input. However, this does not mean that the oceans have existed for only 1 million years, as was once thought. Rather, it means that the average residence time of a calcium atom in the seawater is 1 million years. This is how long it takes calcium to combine with carbonate molecules and become incorporated into limestone. This cycle of calcium from surface rocks to ocean to limestone sediment is completed when the ocean sediment is melted in a subduction zone and recycled back into the continental crust.

Gases as well as salts are dissolved in the ocean water. The most important of these are oxygen and carbon dioxide. The oxygen in the water makes possible the existence of advanced forms of marine animal life, such as the fish, which use gills to extract the oxygen they require from the water. However, the total oxygen content of the ocean is less than that of the atmosphere. In contrast, about 60 times more carbon dioxide is dissolved in the oceans than is present in the air. As a result, the role of the oceans is critical in determining the amount of carbon dioxide in the atmosphere and therefore in establishing the climate of the Earth.

OCEAN TEMPERATURES

Because of its huge mass and thermal capacity, the ocean serves as a great heat reservoir. The average temperature of the ocean is 4°C, only a little above the freezing point. Although the surface temperature varies from about 30°C near the equator to below freezing on the polar ice caps, the temperature a kilometer or more below the surface changes very little with location or season.

In a number of areas near plate boundaries and regions of volcanic activity, heated water containing many dissolved minerals is issuing from massive underwater vents. These remarkable thermal springs support unique forms of life. Creatures live here without sunlight, deriving both nutrients and energy from the hot, mineral-laden water. In this dark, mysterious world, giant sea worms, clams, and other exotic creatures thrive in an environment where no life was thought possible. As we will see later, there may be similar environments today on Jupiter's satellite Europa.

The surface of the ocean is heated by sunlight and stirred by the air. The wind generates waves and helps to drive the major ocean currents. Even the most violent storms, however, affect only the upper few tens of meters of the sea. Although this fact is of no comfort to the mariner caught in a wild storm, we should remember that the most fearsome hurricane or typhoon is a superficial event, of little consequence for the placid reservoir of water beneath the surface.

COMPOSITION OF THE ATMOSPHERE

Above the ocean is the atmosphere, which consists primarily of molecular nitrogen (N_2) and molecular oxygen (O_2). These two gases make up 78% and 21%, respectively. The remaining 1% of dry air is primarily argon (Ar), with carbon dioxide (CO_2)

Table 9.2 Composition of Earth's atmosphere

Gas	Percent
N_2	78.1
O_2	20.9
H_2O	3.0–0.1 (variable)
Ar	0.93
CO_2	0.036 (increasing)
Ne	0.0018
He	0.0005

the fourth most abundant gas at 0.036% (or 360 parts per million). In addition, the atmosphere contains considerable water vapor, ranging from 3% in wet, warm regions down to much less than 1% in cold, dry locations. The compositional measurements are summarized in Table 9.2.

The atmosphere exerts a pressure on the surface, which is just the weight of the gas. On the Earth, the force of gravity and the amount of gas generate a sea-level pressure of 1 bar, the same pressure that would be exerted by a 10-m-thick layer of water. Note, however, that the pressure is not a measure of the total amount of gas in an atmosphere. Pressure is the product of the amount of gas multiplied by the force of gravity. On Saturn's satellite Titan, for instance, about ten times as much gas is required to generate a pressure of 1 bar as is the case on Earth, since the surface gravity of Titan is only about a tenth of our own planet.

Nitrogen is a chemically lazy gas, little inclined to react with other substances. Even more inert is argon, one of the noble gases discussed in more detail in Chapter 12. Nitrogen and argon are also heavy enough to resist escape from the upper atmosphere. The present abundance of argon therefore reflects the total accumulation released into the atmosphere over the history of the Earth. In contrast, both oxygen and carbon dioxide are highly reactive chemically, so the abundances of both must represent an equilibrium between competing reactions that produce and destroy them. Nitrogen is also removed from the atmosphere, primarily by organisms. The primary sources and sinks of both oxygen and carbon dioxide also involve life, and their atmospheric abundances have evolved along with life itself.

Both water and carbon dioxide are underrepresented in the current atmosphere. Both are in the atmosphere only as trace gases, but consider the difference that a slightly higher temperature would make. If the oceans warmed only a little, both water vapor and dissolved CO_2 would be released into the atmosphere. If the warming continued up to 100°C, the boiling point of water, the entire ocean would vaporize and become part of the atmosphere.

How much water vapor would be produced if the ocean boiled away? We have noted that the average depth of the ocean is about 3 km, at which depth the weight of the water is 300 bars. (Remember that a 10-m thickness of water exerts a pressure of 1 bar.) The weight of this water would be the same if it were in vapor form rather than liquid; therefore, the pressure of the atmospheric water, if the oceans boiled away, would also be 300 bars. Water vapor would thus be the dominant constituent of the atmosphere. The CO_2 content would also rise, particularly if the carbonate minerals in the ocean sediment were broken down and their CO_2 released into the atmosphere (Fig. 9.13). We will return to the Earth's carbon reservoirs later.

If the Earth were hotter, it might have a massive atmosphere consisting primarily of H_2O and CO_2. If it were just a little cooler, the oceans would freeze and there would be no liquid layer on the surface (a condition, called "snowball Earth," that may have existed for a time about a billion years ago). Only within a narrow temperature range can the Earth maintain liquid water oceans and a nitrogen-dominated atmosphere.

LIFE AND ATMOSPHERIC CHEMISTRY

Liquid water is essential to the survival of life on Earth. As far as we know, life originated in the seas of Earth, and the presence of life today testifies to the continuous presence of liquid water on Earth over the past 4 billion years. Throughout this vast expanse of time, life has played an important role in the evolution of the Earth's atmosphere.

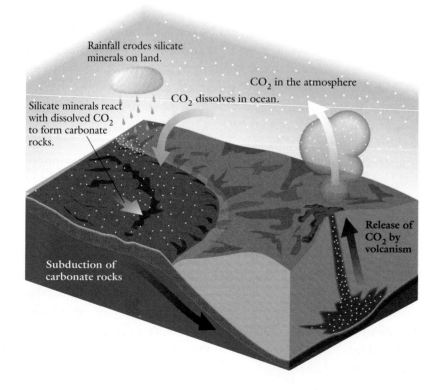

Rainfall erodes silicate minerals on land.

CO_2 in the atmosphere

CO_2 dissolves in ocean.

Silicate minerals react with dissolved CO_2 to form carbonate rocks.

Release of CO_2 by volcanism

Subduction of carbonate rocks

Figure 9.13 One of the factors that controls the amount of carbon dioxide in the atmosphere is the carbonate-silicate cycle, another example of a cyclic process on Earth. Carbon dioxide is released from volcanoes and then reacts with silicates and water to form carbonate rocks. Most of the carbon in the crust is in the form of carbonate sediments. When these sediments are subducted, part of the carbon is again released into the atmosphere from the new volcanoes. Both plate tectonics and the presence of abundant water are thus critically important for recycling carbon on our planet.

In the absence of life we would expect the Earth to have maintained a modestly oxidized atmosphere composed primarily of CO_2, N_2, and Ar, like the contemporary atmospheres of Venus and Mars. The predominant gas should be CO_2, with a surface pressure of tens of bars. Such an atmosphere can persist in chemical equilibrium with the ocean and land. Instead, we find a relatively thin and much more oxidized atmosphere that is deficient in CO_2 and contains molecular oxygen, O_2.

The transformation of the atmosphere from its initial state to its present composition is a consequence of the development of photosynthetic life. Atmospheric oxygen is manufactured from H_2O and CO_2 by green plants, and in the absence of such plants it would quickly recombine with surface minerals and be lost to the atmosphere. Studies of ancient rocks show that the transformation to an oxidizing atmosphere took place gradually between 2 and 1 billion years ago, apparently as a consequence of the proliferation of life and the burial of carbon in sediments. The removal of carbon inhibited the recombination of oxygen and carbon to form CO_2 and left us the present unique atmospheric chemistry.

The chemical interaction of life with the atmosphere is a complex saga. As far as we can tell, life could never have originated in a strongly oxidizing environment. Yet life is responsible for the gradual oxidation of the atmosphere. As the atmosphere changed, life evolved ways to protect itself from the toxic oxygen and ultimately to utilize that oxygen for more efficient metabolism than could ever have been possible in the environment of early Earth. It is this evolutionary path that led to the development of the animal kingdom and eventually to us.

We will discuss the interaction of life and the atmosphere further in Chapter 12, in a comparative study with the evolution of the atmospheres of Venus and Mars.

STRUCTURE OF THE ATMOSPHERE

The density of air at sea level on Earth is about 0.001 g/cm^3, and the pressure is, by definition, 1 bar. This pressure drops off rapidly with elevation, reaching one-half its sea-level value at 5.5 km (Fig. 9.14). Water vapor is even more strongly concentrated near the surface, and it declines to half its sea-level value at about 2 km; this is why mountain air is dry and skiers and mountain hikers suffer so much from chapped skin. It is also the reason mountaintops are selected for observatory sites.

The lower 10–15 km of the atmosphere is called the **troposphere.** This region, which contains 90% of the mass, is the location of nearly all of the phenomena we call weather. The main characteristic of the troposphere is the convection of air within it. Heating by sunlight causes warmer air near the surface to expand and rise under its own buoyancy, to be replaced by downdrafts of cooler air from above. Convection in the troposphere maintains a steady decrease in temperature with altitude of 6°C per kilometer. The top of the troposphere, called the tropopause, is at a temperature of about −60°C or 213 K.

One of the most important cycles linking the ocean and the lower atmosphere is the hydrologic or water cycle. Water in the atmosphere forms clouds and precipitates as rain and snow on both the land and the sea. Some of the precipitation is stored temporarily in the ground or in lakes, but most finds its way to the greatest reservoir, the ocean. Evaporation back into the atmosphere completes the cycle. In the process of evaporation and condensation, moreover, the water can store and release large amounts of solar energy. This energy as well as direct heating by sunlight drives tropospheric winds and generates weather.

9.6 UPPER ATMOSPHERE AND MAGNETOSPHERE

Above the troposphere the atmosphere is too thin to sustain life. However, chemical reactions that occur at high altitudes are critical to the maintenance of the environment near the surface. Still higher, the tenuous outer fringes of the atmosphere are influenced by the Sun and subatomic particles from the solar wind, linking our planet directly with the plasma that fills interplanetary space.

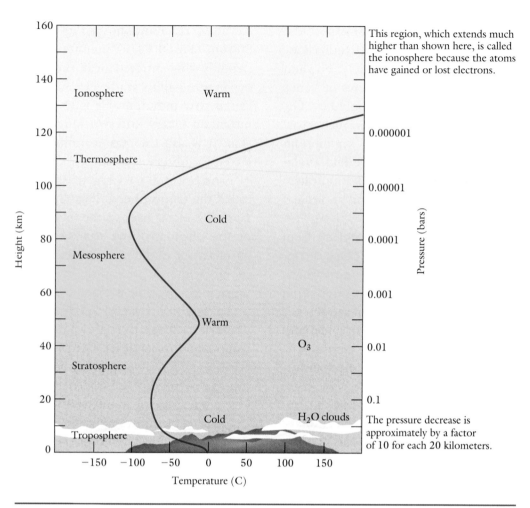

This region, which extends much higher than shown here, is called the ionosphere because the atoms have gained or lost electrons.

The pressure decrease is approximately by a factor of 10 for each 20 kilometers.

Figure 9.14 The vertical structure of the Earth's atmosphere showing the approximate altitudes of the principal regions and the change in pressure and temperature with altitude.

THE STRATOSPHERE

Above the troposphere is the **stratosphere,** in which very little vertical mixing takes place and the temperature is nearly uniform at about −60°C. In the stratosphere, water is almost entirely frozen out and the atmosphere is very dry. In the upper part of the stratosphere, the absorption of solar ultraviolet light creates a layer of ozone, or O_3 (see Section 3.5). The ozone layer, sometimes called the ozonosphere, extends from about 30 to 50 km. Without it, solar ultraviolet light would penetrate to the surface, doing irreparable harm to nonmarine living organisms, both plant and animal.

Because ozone is so important to the health and even the survival of life on the Earth, we are especially concerned about the damage to the ozonosphere caused by the industrial chemicals called chlorofluorocarbons (**CFCs**). These inert, nontoxic gases have been manufactured in huge quantities for use in air conditioning and refrigeration systems and for cleaning electronic components. Unfortunately, CFCs have the unintended property of interacting with sunlight at high altitudes to destroy O_3. Beginning in the early 1980s, an increasing quantity of the stratospheric ozone over the Antarctic continent has disappeared every spring to generate the Antarctic "ozone hole." Each year the hole has expanded, and by the 1990s significant ozone depletion was observed in temperate latitudes as well, in both hemispheres. This ozone depletion is especially scary because the lifetime of CFCs in the atmosphere is more than a century. The manufacture of CFCs has been severely curtailed by international agreement, but it will take a long time for the upper atmosphere to return to its pre-CFC condition.

THE IONOSPHERE

A convenient if somewhat arbitrary altitude to mark the beginning of the tenuous upper atmosphere is 100 km. This altitude represents the highest level at which meteors are seen and the lowest level at which an Earth satellite can complete an orbit without being brought down by atmospheric friction. It is also about the level at which solar ultraviolet light breaks apart molecules into atoms more rapidly than they can recombine. Finally, and perhaps most important, 100 km represents approximately the lower boundary of the ionosphere of the Earth.

The **ionosphere** is the region of the upper atmosphere in which many of the atoms are ionized—that is, broken apart into positively charged ions and negatively charged electrons. The atmosphere in this region is therefore a plasma, albeit a weak one, since only a small fraction of the atoms are ionized at any given time. The ionosphere was discovered early in the twentieth century by its ability to reflect radio waves transmitted from the ground.

The degree of ionization of the atmosphere is measured by the density of electrons. Below 80 km, ions and electrons recombine as quickly as they form, and there is no ionization. The electron density rapidly increases with altitude, forming a first maximum at 100 km of a little more than 10^5 electrons/cm^3. A stronger ionization peak occurs at about 140 km, where the electron density reaches somewhat more than 10^6 electrons/cm^3. Above 140 km, the electron density falls off gradually with altitude. At each altitude, the electron density represents a balance between ionizing solar ultraviolet and x-ray radiation, recombination of the ions and electrons, and the total amount of gas available. Above 500 km most of the gas is ionized, and the declining electron density with altitude simply reflects the thinning of the atmospheric gas.

Within the ionosphere, between altitudes of about 150 and 400 km, occurs one of the most beautiful of natural phenomena: the **aurora** or polar lights. Named for the Greek goddess of the dawn, the auroras are regularly seen at polar latitudes but only rarely grace the night sky in temperate zones. The displays can take on a variety of forms, with the most common being pale rays or curtainlike sheets of green or red that silently sway and dance across the night sky. Viewed from space, the auroras are seen to be concentrated in rings centered on the north and south magnetic poles (Fig. 9.15). Auroras are the result of electric currents

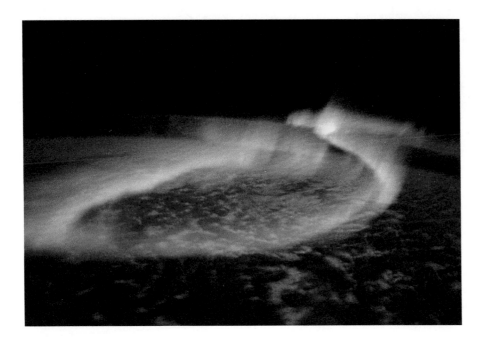

Figure 9.15 Auroral arcs in the upper atmosphere, photographed from space.

flowing through the ionosphere, stimulating the gas to glow, much as an electric current produces light from a fluorescent lamp. The green color comes from the fluorescing oxygen, while red is contributed by hydrogen atoms. The currents themselves originate at higher altitudes, in the magnetosphere, to which we now turn our attention.

THE MAGNETOSPHERE

The Earth's **magnetosphere** was discovered in 1958 by instruments on board the first U.S. Earth satellite, Explorer 1 (Fig. 9.16). The scientist who built the high-energy charged-particle detectors on Explorer 1 was University of Iowa professor James Van Allen, and his name has been given to the primary features of the inner magnetosphere, the Van Allen belts. These are regions of space that contain large numbers of protons and electrons, trapped in the magnetic field of the Earth.

Unlike the ions and electrons of the ionosphere, the magnetospheric charged particles are highly energetic, spiraling back and forth within the magnetic field at speeds of thousands of kilometers per second. When they strike solid mate-rial, such as the skin of a spacecraft, they produce additional subatomic particles and gamma and x radiation, leading to the misnomer "radiation belts." But it is not radiation that is trapped in the Van Allen belts, it is energetic electrons, protons, and other ions.

The configuration of the magnetosphere is the result of interactions between the solar wind and the magnetic field of the Earth. From the direction of the Sun, the charged particles of the solar wind stream toward the Earth at a speed of about 450 km/s. As they near our planet, however, these particles are deflected by the Earth's magnetic field. This deflection distance, which represents the point at which the magnetic field strength is just sufficient to balance the pressure of the solar wind, is at about ten Earth radii, or 60,000 km. The cavity within which the planetary magnetic field dominates over the solar wind is the magnetosphere. It is shaped like a wind sock or a stubby comet tail, pointing away from the Sun. Downstream, it extends just about to the distance of the Moon (Fig. 9.17).

Figure 9.16 From right to left: Werner von Braun, James Van Allen, and William Pickering holding aloft a model of the Explorer 1 satellite that discovered the Earth's radiation belts in February 1958. Von Braun organized the development of the launch vehicle, Van Allen designed and built the instruments, and Pickering was the director of the Jet Propulsion Laboratory, which built and operated the satellite.

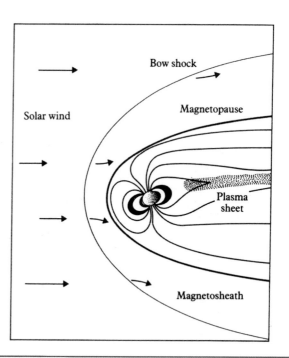

Figure 9.17 Diagram of Earth's magnetosphere. The planet's magnetic field forms a shield that staves off the solar wind. The tilt of the field lines close to the planet results from the inclination of Earth's rotational axis. The belts of electrons and protons (discovered by James Van Allen) surround the Earth within the magnetosphere.

Charged particles from three different sources contribute to the magnetosphere of the Earth. First are the solar wind particles. Second are escaping atoms from the atmosphere, which contribute most to the population of the inner magnetosphere. Third are particles generated from impacts on the atmosphere of the very high energy galactic cosmic rays, which contribute primarily to the middle belt a few thousand kilometers up. To balance the sources of charged particles, there must be corresponding sinks: the escape of particles from the downstream side of the magnetosphere back into the solar wind, and the impact of charged particles on the atmosphere of the Earth. Both sinks are triggered when the boundary between the solar wind and the magnetosphere becomes unstable as the result of fluctuations in solar wind pressure. These fluctuations, generated by outbursts on the Sun, produce the bright auroral displays and ionospheric disruptions associated with periods of intense solar activity.

9.7 CLIMATE AND WEATHER

Weather and climate are both aspects of the closely coupled systems of air, land, and water on our planet. *Weather* refers to the state of this system at a given place and time and its short-term variations, whereas *climate* is concerned with long-term trends. While there is no sharp line of demarcation between the two, climate is generally thought of as referring to changes that take place over a period of 10 years or longer. More rapid changes are considered part of the phenomena of weather.

HEATING BY SUNLIGHT

The source of energy to drive motions in both the atmosphere and the water is the Sun. Of the energy incident on the planet, about 30% is reflected back to space from the atmosphere or surface. The principal reflectors are water and ice clouds, atmospheric dust, surface snow and ice, and unvegetated deserts. Thus, the basic energy balance of the planet is affected, through changes in reflectivity, by the amounts of dust in the atmosphere, the degree of forestation of the land, and the size of the polar deposits of snow and ice.

The 60% of incident sunlight absorbed by the planet corresponds to an average power of 240 watts for each square meter of surface, or a total input of more than 100 billion megawatts (1.2×10^{17} W). (Imagine the consequences of more efficient ways of harnessing this solar energy!) If the climate is to be stable, an exactly equal amount of energy must be radiated from the Earth back into space, primarily at infrared wavelengths. Part of this infrared radiation escapes directly from the surface, but most is absorbed and reradiated within the troposphere.

The average surface temperature of the Earth is about 10°C, about 25°C higher than would be the case without an atmosphere. This increase in surface temperature is the direct result of the blanketing effect of the atmosphere, primarily caused by the carbon dioxide greenhouse effect.

THE GREENHOUSE EFFECT

The **greenhouse effect** is named after the gardener's greenhouse, which provides a simple means of passively heating a room. The glass in the greenhouse roof allows visible sunlight to enter and be absorbed by the plants and soil within. These objects then heat up and radiate at infrared wavelengths, just like the Earth itself. But glass is largely opaque to infrared radiation. It acts like a color filter, letting short wavelengths through but limiting the passage of longer-wave thermal radiation. Since most of the heat can't get out, the interior of the greenhouse warms up, until enough infrared radiation escapes through the glass to balance the energy coming in. A similar effect occurs in a car left out in the sunlight on a hot day. Once again, sunlight passes easily through the glass in the windows, heating the upholstery and metal of the car's interior. Infrared radiation is trapped inside, and soon the temperature inside the car is much higher than that of the surrounding air.

The same thing happens in a planetary atmosphere, with the atmospheric gas playing the role of the glass in retaining infrared energy radiated by the surface. The magnitude of the greenhouse effect is determined largely by the infrared opacity of the gas. Both water vapor and carbon dioxide are effective at blocking thermal emission, and they are thus sometimes called greenhouse gases. Other trace constituents of the Earth's atmosphere, including CFCs and some hydrocarbons, are also effective greenhouse gases.

On the Earth, carbon dioxide dominates the greenhouse effect. Even at only 0.036% of the atmosphere, the CO_2 is sufficient to raise the surface temperature by about 25°C. As we will see in the next chapter, Venus has a massive atmosphere composed primarily of CO_2, and its greenhouse effect is much greater than that on our own planet.

CIRCULATION OF THE ATMOSPHERE

The Earth is heated more near the equator than in polar regions, and therefore the average surface temperature decreases with increasing latitude. However, this temperature decrease is less dramatic than it would be if there were no oceans or atmosphere to redistribute the solar energy. Air that is warmed near the equator tends to rise and move toward the poles, carrying much of this energy with it

(Fig. 9.18). Transport of energy from warm regions to cooler ones generates the most important forces that drive the circulation of the atmosphere.

Because the rotation axis of the Earth is tilted by about 23°, we have seasons on our planet. (It is surprising how many people think that the seasons are caused by changes in the distance from the Sun to the Earth, apparently forgetting that Northern Hemisphere winter occurs at the same time as Southern Hemisphere summer.) Each year the familiar cycle is repeated, as the latitudes of greatest solar heating move north and south, bringing with them the succession of spring, summer, autumn, and winter. If the axis were not so tilted, the atmosphere might be able to sustain a single pattern of atmospheric circulation. The constantly shifting deposition of energy encourages the circulation pattern to break up into the cyclones and storm fronts and other phenomena we call weather.

Near the equator the weather is simplest. Between 35° N and 35° S latitude, the large-scale winds blow toward the equator near the surface and away at higher altitudes, carrying warm moist air

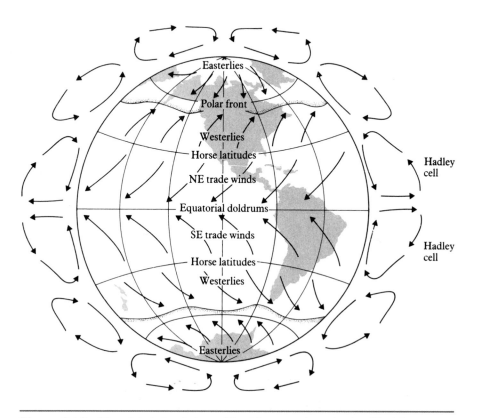

Figure 9.18 The global circulation of Earth's atmosphere in simplified form.

upward near the equator and recirculating cooler, dryer air back to complete the cycle. The rising air generates a region of high precipitation within 10° of the equator to produce the tropical rainforests of the Earth. In contrast, most of the deserts of the Earth are at latitudes between 15° and 35°, where the dry air sinks and flows back toward the equator.

Poleward of 35°, in the temperate and polar regions of the Earth, the basic character of the circulation changes. Here the predominant winds are east-west rather than north-south. Instead of a smooth flow of warmer air toward higher latitudes and cooler air back toward the tropics, a series of large wavelike patterns develops, with the primarily east-west winds veering north and south. As the waves develop, masses of cool air are carried toward the tropics, and masses of warm air penetrate toward the poles. At any one time, about a dozen of these waves, each a couple of thousand kilometers across, are likely to exist in the temperate regions of each hemisphere.

Planetary Rotation and the Coriolis Effect

The rotation of a planet, as well as thermally driven atmospheric motions, is important in determining its weather. If it were not for the relatively rapid spin of the Earth, we would not have the rotating weather systems that are variously called cyclones, hurricanes, and typhoons. All of these phenomena are generated when the motion of air is deflected by the **Coriolis effect.**

To understand the Coriolis effect, we must imagine ourselves riding along on a north- or south-moving wind, perhaps as passengers suspended from a hot air balloon. Suppose we start our trip at 30° N latitude and the wind is initially blowing northward. The air, of course, shares the rotation of the surface, which at 30° N corresponds to an eastward speed of 1200 km/hr. As we move north, we continue going east at this speed, but we soon notice that the landscape below us is moving east more slowly because it has less distance to go to spin once around in 24 hours. Thus, our extra eastward momentum causes us to turn toward the right. In a similar way, a balloonist moving south from the same starting point is deflected to her right because her initial eastward speed is slower than that of the ground nearer the equator. If both are deflected toward the right, the result is a counterclockwise circular motion (Fig. 9.19).

The same line of reasoning will convince you that in the Southern Hemisphere the direction of turning is reversed, and that air diverging from a

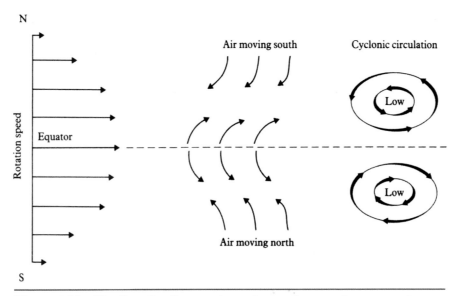

Figure 9.19 The Coriolis effect on air moving north or south on a rotating planet. You can experience this same force by trying to walk toward or away from the center of a spinning carousel.

point is deflected toward the left. A similar mental exercise shows that air blowing inward toward a center is deflected in the opposite way from air blowing outward. You may even have experienced the Coriolis effect directly, if you have ever tried to walk quickly from the outside edge to the center of a carousel.

As a result of the Coriolis effect, air converging toward a center sets up a left-handed spin in the Northern Hemisphere and a right-handed spin in the Southern Hemisphere. This sense of motion, corresponding to a low-pressure region, is called **cyclonic.** A cyclone is a circular storm system moving around a low-pressure region. Diverging air spins the opposite way and is called **anticyclonic.** Because high-pressure regions do not generate clouds and rain, however, we rarely refer to an "anticyclone" when discussing weather.

A low-pressure cyclone has the potential to become a major storm. When the inward-moving air is moist, it may cool as it nears the center of the system and is carried to higher altitudes. The resulting condensation of water vapor not only produces rain but also releases energy to increase the speed of the wind. (Since it requires the addition of energy to water to produce steam, it follows that the opposite phase change, when water vapor condenses back to liquid, must release energy.) A situation can then develop in which the growing storm feeds upon itself, drawing in more and more moist air and releasing more and more energy. The result is a tropical storm, more commonly known as a hurricane or typhoon. Only on Earth, where water is abundant, can self-sustaining storms of this type be formed. When such storms move over land, they immediately begin losing their strength because their source of water (and hence energy) is cut off.

CHANGES IN CLIMATE: THE ICE AGES

Weather on Earth is an extremely complex phenomenon, difficult to understand and almost impossible to predict with any high degree of accuracy. Climate presents fewer problems, in that changes are small. However, change in the climate of the Earth has much more far-reaching consequences than any storm.

We know that the climate on Earth has changed, even during the relatively brief time human records have been kept. At the time civilization was developing in Mesopotamia (Iraq), rainfall in that region was higher than it is now, and vast forests supported lions and other animals now confined to sub-Saharan Africa. Then, this part of the world was called the "Fertile Crescent." About a thousand years ago, when Norse seafarers founded colonies in Greenland and present-day Canada, the climate in these regions was warmer than it is today, and when a global cooling took place around 1400, these colonies did not survive. Even more recently, there is evidence of recurring droughts in the plains of North America at approximately 22-year intervals.

More dramatic climatic changes on our planet are associated with the great ice ages of the past million years. At intervals of about 100,000 years, the average temperature of the Earth has dropped by about 3°C, sufficient to produce vast ice sheets up to 3 km thick over much of the Northern Hemisphere landmasses. During an ice age, the sea level drops and atmospheric circulation patterns alter significantly. The last such glacial period ended only about 10,000 years ago, and the Earth today is in an unusually warm period. It is so warm, in fact, that the sea ice that still covers the Arctic Ocean and the thick ice cap on Greenland are in danger of melting. The measured thickness of the arctic winter ice has decreased by a remarkable 40% since the 1960s.

The primary cause of the great ice ages is now believed to be changes in the orbit of the Earth and the tilt of its rotation axis. Astronomers have calculated the changes expected from the gravitational influence of other planets over the past million years, and the pattern of ice ages follows the orbital changes closely. Even though the resulting variations in solar heating are small, they seem to be sufficient to shift the Earth from an interglacial equilibrium, such as we have at present, to an ice age condition, with about equal intervals of time spent in each climatic state.

Even more mysterious are the events that seem to have occurred about a billion years ago when the entire ocean surface froze, leading to the phenomenon called "snowball Earth." There

may have been several snowball events, each lasting millions of years. Fortunately, only the top of the oceans froze, so that life continued in the liquid water beneath. There is some indication that life itself may have played a role in restoring a warmer climate on Earth by suppressing the production of oxygen and allowing carbon dioxide to build up, increasing the greenhouse effect.

GLOBAL WARMING

One of the consequences of the Industrial Revolution of the nineteenth and twentieth centuries has been the release of immense quantities of carbon dioxide into the atmosphere from the burning of fossil carbon (coal, oil, and natural gas). Increasing population has also led to massive deforestation of the land, with associated burning of biological carbon. As a result, the CO_2 content of the atmosphere and oceans has risen and continues to do so. It is inevitable that this change in atmospheric composition will have some effect on global climate.

Figure 9.20 illustrates the CO_2 content of the atmosphere as measured from a special atmospheric observatory high on the slopes of Mauna Loa in the middle of the Pacific, far from any local sources of pollution. Calculations show that such increases in CO_2 will lead to an enhanced greenhouse effect and an average increase of the Earth's temperature, relative to preindustrial values, of about 2°C. These calculations appear to be confirmed by direct measurements of temperatures, which reached all-time highs in most land areas of the Earth during the 1990s. However, such effects are difficult to distinguish from shorter-term variations in temperature, such as the global cooling that followed the volcanic eruptions of El Chicón (Mexico) in 1982 and Pinatubo (Philippines) in 1991.

In order to assess the consequences of continued oxidation of carbon, especially of the fossil fuels that have lain in the ground for longer than a hundred million years, we need to predict the changes in the greenhouse effect that will follow the release of CO_2 into the atmosphere. Most models suggest that global temperatures can be expected to rise another several degrees by the middle of this century, but there are substantial uncertainties in these numbers. One of the most important contributions to society that can be made by planetary scientists is to improve these greenhouse models. The future of our industrial society depends in part on our ability to predict the impact of our activities on global climate.

Note that in the debates concerning global warming some commentators seem to question the reality of the greenhouse effect. This is foolish. The greenhouse effect is very real, and without it the Earth would be frozen. The questions concern

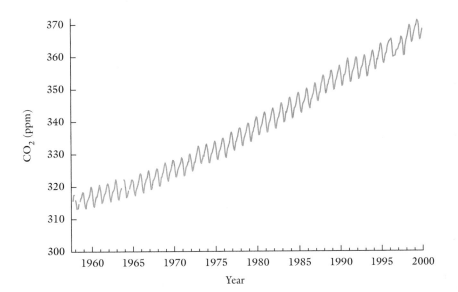

Figure 9.20 Variations over time in the CO_2 content of the Earth's atmosphere, as measured from an atmospheric observatory on the upper slopes of Mauna Loa. The CO_2 content has been rising by about 1% per year, mostly as a consequence of consumption of fossil fuels in the advanced nations and destruction of tropical forests in the developing nations. The small periodic wiggles are a seasonal effect.

not the reality of the greenhouse effect, but rather the degree of greenhouse warming to be expected from the production of additional CO_2. That there will be some global warming is inevitable, but the question of just how much is crucial.

There is a great deal that scientists do not yet understand about both climate and weather. Billions of dollars are spent each year in collecting data and attempting to make accurate predictions. The economic impact of even a single major storm can reach billions of dollars, and the consequences of a shift in climate are almost incalculable. One way of helping to understand our own planet, of course, is to compare it with other planets. We will frequently emphasize such comparisons as we look at the atmospheres of the other worlds.

Summary

The Earth is the planet we know best, and it therefore is an appropriate object with which to compare the other worlds studied in this book. We did not begin with the Earth only because of its complexity, which led us to start with the smaller, simpler planets Mercury and the Moon. Now, with our own planet, we must look at a much wider range of phenomena related to its active geology, its oceans and atmosphere, and the unique influences of life on its evolution.

The various layers of the Earth, above and below its surface, have given their names to similar divisions on other planets. Thus, we must understand how the interior is divided into crust, lithosphere, mantle, and the liquid and solid cores. Similarly, the atmosphere is divided into the troposphere, stratosphere, and ionosphere, with the magnetosphere extending still farther out to the edge of interplanetary space.

The complex geology of the Earth can be understood in terms of plate tectonics, the process by which heat released in the interior causes convection currents in the mantle, in turn exerting forces on the lithospheric plates floating on the top. These plate motions result in the formation of new oceanic crust at rifts, the destruction of the crust at subduction zones, and the generation of earthquakes and volcanoes along faults where one plate scrapes against another. In this process, the ocean crust is recycled on a timescale of about 100 million years, while the floating granite continents are moved about and occasionally compressed and raised up to form great folded mountain ranges.

Above the crust are the oceans and the atmosphere. The short-term variations called weather and the longer-term conditions that constitute climate both result from varying solar heating coupled with the large-scale motions of ocean and atmosphere. We looked at atmospheric circulation and saw the way Coriolis forces give rise to cyclones and anticyclones in the temperate regions of the Earth. Also considered were the processes that can affect the climate, such as changes in the inclination of the Earth's axis of rotation and the shape of the orbit, or the injection of dust from volcanoes.

Atmospheric chemistry is another complex topic. Our present atmosphere is the result of outgassing from the interior, primarily through volcanic eruptions, combined with changes due to interaction of the gas with the crust, the oceans, and especially with life. It is life that has removed almost all of the carbon dioxide (now mostly in the form of carbonate deposits) and has generated free oxygen through photosynthesis.

The Earth is a remarkable planet, like the others in many ways but also uniquely influenced by its liquid water oceans, its oxygen-rich and carbon dioxide-poor atmosphere, and its abundant and varied lifeforms. As we will see in the next chapter, Venus is the other planet most like the Earth, yet it has evolved to a strikingly different condition.

Key Terms

anticyclonic	greenhouse effect
aurora	ionosphere
bar	lithospere
caldera	magnetosphere
composite volcano	ozone
Coriolis effect	Phanerozoic period
cyclonic	plate tectonics
fault	Precambrian period
granite	rift

sedimentation   subduction
seismometer   tectonic
shield volcano   troposphere
stratosphere

REVIEW QUESTIONS

1. In simple terms, how does the Earth rate as a planet? Compare its size, density, composition, and other basic properties with those of its neighbors. How would you expect the study of other planets to help us understand our own?

2. Distinguish between evolutionary processes and cyclic processes. What examples of each have we encountered so far in this book? Why are cyclic processes more common on the Earth than on the Moon?

3. Make sure you understand how plate tectonics works. How does this process give rise to mountains, volcanoes, deep-sea trenches, and earthquakes? What does plate tectonics tell us about the fate of deep-sea sediments, including the carbonates that trap most of the carbon dioxide that would otherwise be in our atmosphere?

4. The changes in geological thinking that resulted from the acceptance of plate tectonics and continental drift constituted one of the major scientific revolutions of the twentieth century. Why do you think these ideas were so slow in being widely adopted? You may wish to compare this scientific revolution with others of the past century: Darwin's discovery of the role of natural selection in biological evolution; Pasteur's proof of the role of germs in causing disease; the insights into human psychology provided by Freud; and the revolution in physics represented by relativity, quantum mechanics, and the discovery of the nature of the atom.

5. Compare the ocean crust and the continental crust in terms of their composition, origin, and evolution. Are these two divisions of the terrestrial crust at all analogous to the division of the lunar crust into highlands and maria? Explain.

6. Discuss the chemical balance that exists among the land, the ocean, and the atmosphere. Imagine what would happen to the other two if any one of these were dramatically changed.

QUANTITATIVE EXERCISE

1. Typical motions of plates in the Earth's crust amount to about 5 m per century. At this rate, how long will it take Los Angeles to move up next to Seattle? How much older is the Hawaiian island of Kauai than the currently active island of Hawaii, about 300 miles away?

ADDITIONAL READING

Lamb, H. H. 1995. *Climate, History and the Modern World*, 2nd ed. New York: Routledge. Popular-level overview.

Leggett, J. 1999. *The Carbon War: Global Warming and the End of the Oil Era*. New York: Penguin Books. Discussion of the controversies surrounding global warming, including political and economic issues as well as scientific.

Lunine, J. 1999. *Earth: Evolution of a Habitable World*. Cambridge: Cambridge University Press. Comprehensive discussion of the history of the Earth, with emphasis on the role played by life on our planet. (Written as a senior-level textbook.)

Miller, R. 1983. *Continents in Collision*. Alexandria, VA: Time-Life Books. Beautifully illustrated popular account of the fundamentals of plate tectonics.

Sullivan, W. 1985. *Landprints*. New York: Times Books. Popular book by a leading science journalist on the role of plate tectonics in shaping the geology and landscape of the Earth.

10 VENUS: EARTH'S EXOTIC TWIN

The Birth of Venus by Sandro Botticelli (1455–1510). Botticelli's life overlapped that of Copernicus. One could imagine them discussing the origin of Venus over a glass of wine, from rather different points of view.

Earth and Venus are more nearly twins than are any other pair of planets. They have essentially the same diameters (12,756 and 12,104 km, respectively), nearly the same densities (5.5 and 5.3 g/cm³), and presumably very similar bulk compositions. As we might expect from their large size relative to Mars or Mercury, these two planets also have the most active geology in the inner solar system. Both planets also have major atmospheres.

The twins also differ in important ways. Earth has a relatively large Moon; Venus has none. Earth rotates directly in 24 hours; Venus rotates retrograde in 243 days. Earth experiences active plate tectonics; Venus does not. Earth has extensive oceans of liquid water; Venus is dry. And perhaps most dramatically, Earth has a climate that can support abundant life, whereas Venus has developed an oppressive atmosphere, sulfurous clouds, and blistering surface temperatures. Twins these two planets may be, but certainly not identical twins. Understanding the reasons for their divergent atmospheric and surface evolution is one of the outstanding problems of planetary science, with important implications for the future of the Earth.

10.1 UNVEILING THE GODDESS OF BEAUTY

Venus is a fascinating planet in its own right, in addition to the insights it may provide about the Earth. With its surface hidden under a perpetual blanket of cloud, Venus remained largely mysterious while other planets were yielding their secrets to telescopic observers and the reconnaissance of spacecraft. To understand the geology of Venus and its history, it is necessary to penetrate the clouds with radar, providing the resolution and coverage necessary to reveal the planet's geology.

BASIC PROPERTIES

Venus is the nearest planet to the Earth. When we see it glowing brightly in the twilight skies as the morning or evening star, we can understand why this planet was named for the goddess of love and beauty. Venus owes its brilliance both to its proximity to the Earth and Sun and to its layer of clouds, which serves as an excellent reflector of sunlight. The Moon, with a cloudless surface of gray rock, reflects only 11% of the sunlight that strikes it, whereas Venus has a reflectivity of about 75%.

Both Venus and Earth have substantial atmospheres and brilliant clouds. However, Venus is completely covered by its clouds, unlike the ragged canopy of Earth. The clouds prevent astronomers from seeing the surface of Venus, and they frustrated early attempts to measure even so basic a property as the rotation rate (Fig. 10.1). The temperature of those clouds, first measured in the 1930s, is 230 K, similar to the temperature that would be measured in our own stratosphere, and astronomers of that time generally assumed that the clouds of Venus were composed of water, like those of Earth.

Less than 40 years ago, the only gas that had been detected from spectra of the atmosphere of Venus was carbon dioxide (Fig. 10.2). Spectral observations, however, did not preclude the existence of other gases, since observational difficulties could not rule out undetected amounts of nitrogen, oxygen, and water vapor. Therefore, it seemed possible, near midcentury, that the atmosphere of Venus was primarily nitrogen, like

Figure 10.1 A picture of Venus taken in visible light with the Palomar 5-m (200-in.) telescope when the planet was relatively close to us. It shows no detail, indicating an atmosphere filled with featureless clouds.

our own, and that carbon dioxide might be a relatively minor component. All of these observations and inferences suggested to those who wished to believe it that Venus might have oceans of water beneath its brilliant clouds and a climate conducive to the existence of life.

SURFACE TEMPERATURE

The optimistic picture of Earth-like surface conditions changed dramatically in the late 1950s when radio telescopes were first used to measure the thermal radiation from Venus. Unlike visible light, radio waves easily penetrate clouds. (You know this from your experience in listening to radio broadcasts or watching television on cloudy and rainy days.) Furthermore, any object that has a temperature above absolute zero (0 K or $-273°C$) radiates some energy at all wavelengths, including radio wavelengths.

It is a basic physical law, known as Wien's law, that the hotter an object is, the more the maximum energy output shifts toward shorter wavelengths. The outer layers of the Sun are at a temperature (5800 K) that puts this maximum in the range of visible light. A planet with a temperature like

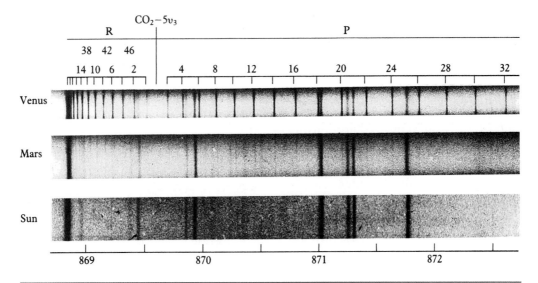

Figure 10.2 A comparison of portions of the spectra of Venus (recorded photographically at the McDonald Observatory in Texas), Mars, and the Sun. Carbon dioxide absorptions appear strongly in the Venus spectrum, weakly in that of Mars, and not at all in the spectrum of the Sun. We conclude that even Mars has more CO_2 in its atmosphere than our planet does, while Venus must have an enormous amount of this gas.

Earth's is radiating most of its energy in the infrared, at wavelengths near 10 μm (0.001 cm). But some of the thermal energy from both Sun and Earth is radiated at much longer wavelengths in the centimeter to meter range, the region of the spectrum where radio telescopes operate.

Since Venus is closer to the Sun than we are, astronomers expected its temperature to be higher than the Earth's. On the assumption that both planets rely on solar radiation alone to warm them, it is possible to calculate what temperature Venus should have. Its greater proximity to the Sun is countered to some extent by its high albedo; its bright clouds reflect most of the incident sunlight back into space. Such calculations showed that the equilibrium temperature of Venus should be roughly 15°C warmer than that of our planet, or about 280 K.

In 1958, however, radio astronomers found that the amount of thermal energy emitted by Venus at radio wavelengths implied a surface temperature higher than 600 K. Venus was radiating more than twice as much energy at radio wavelengths as had been expected. This result was so surprising that at first it was not accepted, and scientists sought alternative explanations for the high intensity of the radio radiation. One proposal was that emission from a dense ionosphere on Venus gave the appearance of a high surface temperature.

EARLY RADAR OBSERVATIONS

One of the primary scientific objectives of the first interplanetary spacecraft was to test whether the radio emission from Venus was thermal radiation, arising from the surface, or whether it might instead be ionospheric in origin. In 1962, Mariner 2 successfully made the 14-week flight to Venus and radioed back the critical result: The radiation really did come from the surface. At about the same time, radar astronomers succeeded in bouncing radar waves off the surface of Venus and found not only that the ionosphere was transparent but also that the planet was rotating backward incredibly slowly. By defining the location of the planet's surface relative to the top of the cloud layer, the radar data also indicated that the atmosphere was massive, with a surface pressure at least 50 times higher than the 1-bar pressure on the Earth.

Subsequent radar observations have determined the exact rotation period of 243.08 days in a retrograde direction. In other words, Venus rotates on its axis in the opposite direction from the course of its motion around the Sun. On Venus, if you could see it through the clouds, the Sun would appear to rise in the west and set in the east.

The length of a day on Venus requires further definition. Though not quite so peculiar as Mercury, Venus also has a day that is longer than its year. The rotation period, measured with respect to the stars, is 243.08 days—about 19 days longer than the period of revolution around the Sun, 224.7 days. This "day" is the length of time required for Venus to make one rotation about its axis. But as we saw for Mercury, the other way to determine the length of a day is to measure the time between two successive "noons" (or sunsets, sunrises, and so forth). The time between successive

noons—the solar day—is much shorter, amounting to 116.67 Earth days, with the Sun moving across the sky in the wrong direction, from west to east (Fig. 10.3).

THE VENERA MISSIONS

The next major step in the exploration of Venus was taken by the USSR in 1967 with the deployment of a probe called Venera 4 into the atmosphere (*Venera*, the Russian word for "Venus," gave its name to a long series of Russian missions to this planet). This probe successfully entered the atmosphere of Venus and transmitted measurements to Earth as it descended by parachute. Although the Venera 4 probe demonstrated that the atmosphere was thick and that CO_2 was the major gas, it failed to reach the surface. The last signals were received from the craft at an altitude of 23 km, with the atmospheric pressure already up to 20 bars and the temperature higher than 500 K. Many Soviet scientists thought that Venera 4 had reached the surface, but the real problem was spacecraft failure. The requirements for the early Venera probes had been set by a Russian scientist who did not believe the radio observations and had confidently assumed that the spacecraft would never have to withstand a pressure higher than 20 bars. The probe had been designed to fail at 20 bars of pressure.

Incidentally, if you are wondering why the first U.S. mission was numbered Mariner 2 and the first successful Soviet probe was Venera 4, it is because the first numbered craft in each series failed. In the early days of space exploration, more missions failed than succeeded.

In 1970, Venera 7, with a redesigned probe, successfully landed on the surface of Venus. As was expected by then, the pressure was 90 bars and the temperature a sizzling 740 K.

The discoveries of the late 1950s and 1960s changed forever our concept of our sister planet. No longer could the surface of Venus be imagined, even in science fiction, as a lush jungle populated by exotic creatures. Continuing optical and radio observations from Earth, in addition to direct measurements carried out by Soviet spacecraft, left no doubt that the entire surface of

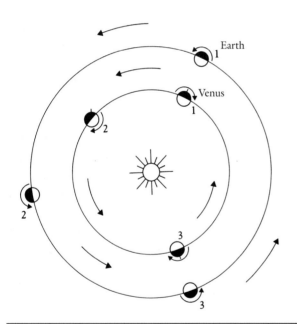

Figure 10.3   Venus rotating as it revolves around the Sun. At position (1), Venus is at inferior conjunction, with a feature on its surface pointing toward the Earth; 486 days later, the two planets are at the positions labeled (2). Venus has undergone two complete retrograde rotations and slightly more than two revolutions around the Sun. Earth has undergone 486 rotations and 1⅓ revolutions around the Sun. Ninety-eight days later, the two planets have moved to position (3). Venus is again at inferior conjunction 584 days after position (1), and the surface feature has almost rotated into position to face the Earth.

Venus was at temperatures hot enough to melt lead, tin, and zinc. We now know that Venus is the hottest planet in the solar system, despite the fact that Mercury is closer to the Sun. Indeed, the picture of Venus that has emerged is reminiscent of traditional concepts of hell.

10.2 THE ATMOSPHERE AND THE GREENHOUSE EFFECT

Why is Venus so hot? The explanation comes from the greenhouse effect, which we discussed in Section 9.7. The challenge posed by Venus is that the greenhouse effect is so large that traditional theories were unable at first to explain it.

THE GREENHOUSE EFFECT

When a planet's atmosphere contains gases that are opaque to infrared radiation, they inhibit the outward flow of heat and increase surface temperatures. Recall that the troposphere is defined as the part of an atmosphere where convective circulation takes place. Tropospheric convection establishes a temperature gradient, with higher temperatures near a planet's surface and cooler temperatures aloft. The equilibrium temperature for Venus, which is about 230 K, applies to the cloud layers high in the atmosphere, while below the clouds temperatures are much higher.

Although the clouds contribute to the greenhouse effect, it is primarily the infrared opacity of the gases that causes the lower atmosphere and surface to heat up. The infrared is the part of the spectrum where the bulk of the energy is radiated. The tiny amount of energy escaping directly from the hot surface of Venus in the form of radio emission is too small to affect the energy balance.

Not all gases are opaque to infrared radiation. For example, nitrogen and oxygen are virtually transparent at these wavelengths, so they cannot contribute significantly to an atmospheric greenhouse effect. Water vapor and carbon dioxide are very good infrared absorbers, and even the relatively small amounts of these two greenhouse gases in our own atmosphere are sufficient to raise Earth's average temperature some 25°C above the value our planet's surface would have if there were no atmosphere, or if only nitrogen and oxygen were present.

The greenhouse effect on Venus is obviously much more significant than the one on Earth. This is clear not only from the high surface temperature, but also from the fact that the temperature is uniform to within a few degrees all over the surface. The entire planet acts as if it were encased in a heavy blanket.

We have already noted the presence of substantial amounts of carbon dioxide on Venus together with a relative lack of water vapor. It is clear that the CO_2 must be the major source of infrared opacity. The much greater greenhouse effect on Venus compared with that on Earth requires both that CO_2 be the major constituent of the atmosphere and that the atmosphere itself be massive.

Carl Sagan, the Cornell University astronomer who later became one of the best known scientists of our time, began his research career working on theoretical calculations of the Venus greenhouse effect. At that time, about 1960, the large mass of the atmosphere was not suspected, and early calculations failed to explain the temperatures that were measured by the radio astronomers. During most of the decade of the 1960s, Sagan and his student and later colleague James Pollack produced theoretical models of increasing sophistication, using new measurements of the atmosphere and more and more powerful computers for their work. Only after a decade of this effort were they able to match the magnitude of the observed effect.

MASS OF THE ATMOSPHERE

As Sagan, Pollack, and others struggled to interpret the evidence that our sister planet had an extremely high surface temperature, they were also learning how massive the atmosphere of Venus is. If you were standing on the surface of Venus at a pressure of 90 bars (in an air-conditioned, asbestos suit!), you would feel the kind of pressure a deep-sea diver deals with at a depth of 900 m. Venus has nearly the same surface gravity as the Earth, with 1 bar corresponding to the pressure

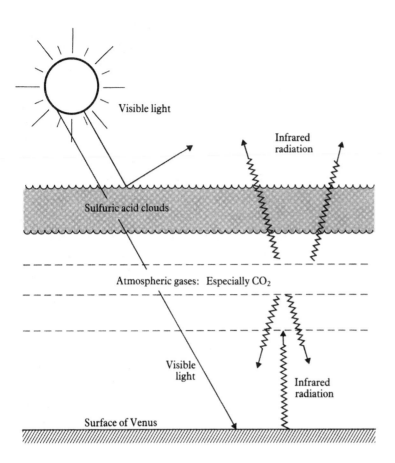

Figure 10.4 The atmospheric greenhouse effect on Venus. Although about 75% of the sunlight incident on Venus is reflected by the planet's brilliant clouds, some light still reaches the surface where it is absorbed. Infrared radiation from the warm surface tries to make its way out of the atmosphere but is strongly absorbed by carbon dioxide, causing additional surface warming.

exerted by a layer of water about 10 m deep. Heavy armor is required for a human to withstand such conditions. The high surface pressure adds to the infrared opacity of carbon dioxide, making it a much more efficient absorber of thermal radiation than the CO_2 in our atmosphere.

In order to maintain a greenhouse effect, at least some of the solar heat needs to be deposited at the surface. When scientists first realized the extent of the atmosphere and the thickness of the clouds, some questioned whether any light would reach all the way down to the ground. However, experiments on spacecraft that have reached the surface show that some sunlight does get through the cloud layer, providing an illumination about equivalent to a heavy overcast on Earth. This is sufficient to warm the ground and provide the necessary heating (Fig. 10.4).

COMPOSITION OF THE ATMOSPHERE

The discovery that the atmosphere of Venus is 90 times more massive than that of Earth and consists of 97% CO_2 surprised many planetary scientists.

Displaying a kind of Earth chauvinism, they had expected the atmosphere of Venus to be composed mainly of nitrogen. They also expected water vapor to be present, since water is so abundant on Earth. Even oxygen, a gas that might betray the presence of Earth-like life on Venus, was also anticipated.

Early searches for spectroscopic evidence of N_2, H_2O, and O_2 in the atmosphere of Venus were entirely negative. Nitrogen does not have any absorption bands in the part of the spectrum accessible to ground-based observers, and possible evidence of water and oxygen was blocked by the strong absorption by these gases in the atmosphere of the Earth. In the 1950s and 1960s, efforts were made to put telescopes on high-altitude airplanes and balloons to get above most of the absorptions in our own atmosphere. The results were still negative. Not until the late 1960s were very weak H_2O absorption lines found, followed by carbon monoxide (CO), hydrochloric acid (HCl), and hydrofluoric acid (HF). A tiny amount of O_2 was discovered spectroscopically a few years later.

Evidently Venus is much, much drier than Earth. At its high temperature, any H_2O should

Table 10.1   Composition of the atmosphere of Venus

Gas	Formula	Abundance
Main Constituents		
Carbon dioxide	CO_2	96.5%
Nitrogen	N_2	3.5
Trace Constituents		
Sulfur dioxide	SO_2	130 ppm[†]
Argon (40)	Ar-40	33
Argon (36)	Ar-36	30
Oxygen	O_2	30
Water vapor	H_2O	30*
Carbon monoxide	CO	0
Carbonyl sulfide	OCS	10*
Neon	Ne	9
Hydrochloric acid	HCl	0.6
Hydrofluoric acid	HF	0.005

*Abundances of these gases vary with altitude and latitude. They are not yet well defined.

[†] Parts per million

be in the atmosphere, yet only traces were found, instead of the equivalent of oceans. The absence of H_2O on our sister planet is one of the major mysteries of the inner planets, and we will return to its solution in Chapter 12.

Keep in mind that all of these observations were made from the Earth and refer only to the portion of the atmosphere of Venus that is above and just within the planet's ubiquitous clouds. These results were confirmed and extended by the space missions of the 1970s, when probes descended into the atmosphere and landed on the surface. Table 10.1 is a list of all the gases now known to be present in the atmosphere.

CLOUDS OF VENUS

Ever since astronomers first realized that Venus was covered by a cloud layer, they have wondered about the composition of those clouds. At first the clouds were assumed to be made of H_2O (either liquid drops or ice), like those of Earth. Then the discovery of the very low concentrations of water vapor in the planet's atmosphere led some scientists to seek

alternative explanations. Hydrocarbon droplets were briefly in favor, but no proof surfaced to support this conjecture. The liquid or solid droplets of a cloud (like the surfaces of asteroids discussed in Chapter 5) do not display the sharp, diagnostic spectral lines that would permit a definitive identification. Unfortunately for the astronomer, solids and liquids always present a more difficult problem for analysis than do the simpler gases.

The puzzle was finally solved by a series of observations in the 1970s. Improved data obtained from NASA's airborne telescope showed features in the infrared part of the spectrum corresponding to an unexpected material: concentrated sulfuric acid (H_2SO_4) (Fig. 10.5). The same conclusion had already been reached independently by observers studying the variations in brightness and polarization of sunlight reflected from the clouds in different directions. Results from both lines of evidence were announced at the same scientific meeting. At last, after decades of speculation and years of hard work, we knew the composition of the clouds of Venus, the brightest object in the sky after the Sun and Moon!

Sulfuric acid is produced by chemical reactions involving sulfur dioxide (SO_2) and water (H_2O). The most important cloud-forming process on Venus is **photochemistry**—chemical reactions driven by the energy of ultraviolet sunlight. Photochemical reactions are important in the upper atmospheres of planets, including the Earth, where the production of ozone from oxygen is an example of a photochemical process. On Venus, the H_2SO_4 and the unknown compound responsible for the ultraviolet-absorbing clouds (discussed later) are probably both produced and destroyed photochemically.

In addition to photochemical reactions, at least some of the basic cloud material may be supplied from below by active volcanism on the planet's surface. Observations over the past 25 years suggest that large fluctuations occur in the concentration of SO_2 in the atmosphere of Venus above the clouds. Sulfur dioxide is one of the products expected from volcanic eruptions; if there are occasional large eruptions of SO_2 on Venus, then we might understand the variations in the H_2SO_4 clouds.

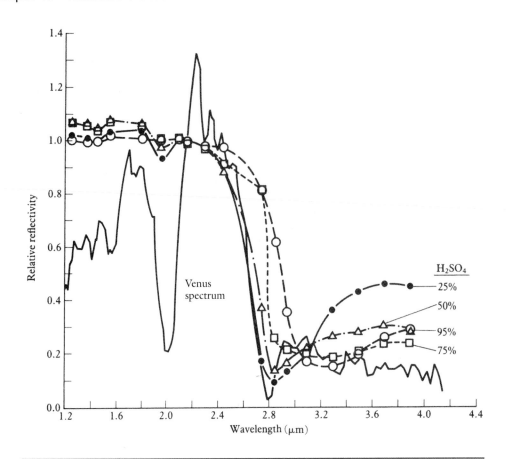

Figure 10.5 A comparison of the spectral reflectivity of solutions of concentrated sulfuric acid and the clouds of Venus. The spectra are similar, indicating that the clouds of Venus are made of sulfuric acid.

Space probes that have passed through the clouds gave us the picture shown in Figure 10.6 of a series of discrete cloud layers. Clouds are seen extending from 30 to 60 km above the surface. But what are these various cloud layers made of? Are they all sulfuric acid, like the topmost layers? Only the Venera probes attempted compositional measurements of the clouds, and their results have been contradictory. Sulfur or possibly chlorine compounds of some sort are indicated, but their exact identities are unknown.

RUNAWAY GREENHOUSE EFFECT

In view of their many similarities, we might well ask why Venus has such a different atmosphere from that of the Earth. How did it acquire its huge quantities of CO_2, sulfuric acid clouds, and remarkable surface temperature? Or, alternatively, how did the Earth avoid this fate?

In Chapter 9, we discussed how the development and proliferation of life on Earth extracted CO_2 from our atmosphere and enriched it in O_2. Life, which in turn depends on the presence of liquid water, may be the key to the difference between the two planets. Perhaps Venus was just a little too close to the Sun and therefore a little too hot to maintain liquid water and provide a suitable environment for life. We will discuss this possibility further in Chapter 12, after we have had a chance to look at Mars and see how it compares with both the Earth and Venus.

A second possibility is that Venus began with a more Earth-like climate and subsequently evolved the hellish conditions we see today. Suppose that Venus started with a surface temperature consistent with liquid water and a CO_2 atmosphere; is there a way it could have made the transition to its present state? Many scientists think it could, if its greenhouse effect got out of control. Such an

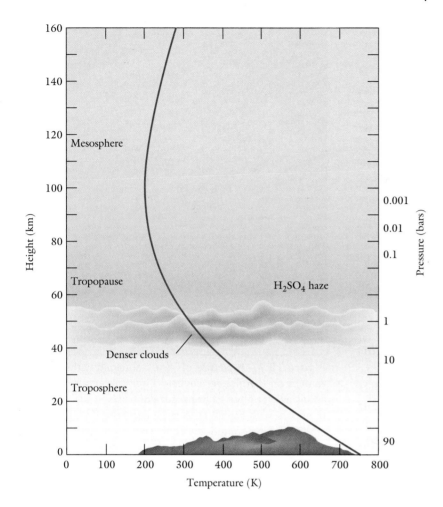

Figure 10.6   The vertical structure of the atmosphere of Venus. The clouds consist of several layers with different concentrations of particles. Below the clouds, the atmosphere is clear.

atmospheric instability is called the **runaway greenhouse effect.**

Imagine what would happen if we could move the Earth into the orbit of Venus. Being closer to the Sun, our planet would absorb more solar energy and its temperature would rise. At the distance of Venus (0.72 AU), sunlight would be delivering about twice as much energy to every square meter of the Earth's surface. Higher ocean temperatures would lead to increased evaporation and more water vapor in the atmosphere. Water vapor is an effective greenhouse gas, so the result would be a stronger greenhouse effect and still higher surface temperatures, and hence more evaporation from the oceans. We have established a positive feedback loop, in which an initial disturbance—increasing the Earth's surface temperature—produces consequences that lead to an enhancement of that disturbance. The cycle would continue until the oceans literally boiled away and all the water was converted to vapor. This is the runaway greenhouse effect.

At this point the atmosphere is so hot that water vapor can easily rise to great heights where it is exposed to solar ultraviolet light. This is a crucial step. On Earth, water is protected from escape by a natural cold trap in the atmosphere. The air at the top of the troposphere is so cold that water cannot diffuse upward to levels where it could be attacked by ultraviolet light. A runaway greenhouse effect can raise the temperature throughout the lower atmosphere, giving water free access to high altitudes.

Just as in the case of evaporating water molecules from an icy comet nucleus, the H_2O in the upper atmosphere is broken apart into H and O atoms. Because of the large mass and high escape velocity of Venus, only the light hydrogen atoms escape into space. The oxygen remains behind to combine with rocks on the planet's surface. Thus, the runaway greenhouse effect leads to the elimination of most water from a planet.

Note that the runaway greenhouse effect is not simply a large greenhouse effect. It is a process whereby a planet's atmosphere can evolve from one composition and temperature to another different composition and much higher temperature.

Is this really what happened to Venus? Is there evidence that Venus ever had oceans of water that were subsequently lost? Might conditions on our sister planet once (perhaps briefly) have been suitable for the development of life? There are no firm answers to these fascinating questions. We will return to these issues in Chapter 12.

10.3 WEATHER ON VENUS

The massive atmosphere of Venus produces uniform surface temperatures and inhibits winds near the surface. These surface conditions, together with the very slow rotation period of Venus, result in a relatively simple tropospheric circulation pattern. In the stratosphere, however, wind speeds on Venus are remarkably high.

THE UPPER ATMOSPHERE

Long ago, telescopic observations of the clouds of Venus showed that they are featureless in visible light, but dusky markings could be seen in pictures of the planet that were taken through filters that transmitted only ultraviolet light (Fig. 10.7). Studying such pictures in the 1960s, astronomers found that they could follow the motions of some of these dusky features long enough to see them move completely around the planet and back to their starting position. A complete circuit of Venus required only 4 days. The clouds moved in a retrograde direction, the same direction that was later found to hold for the planet's rotation. Thus, the atmosphere and the surface move in the same direction, although at very different speeds. This atmospheric motion corresponds to high-altitude winds blowing at a speed of 100 m/s (360 km/hr) from east to west.

The 4-day stratospheric jet streams on Venus are the result of a complex interaction between the rotation of the planet and heating of the high

Figure 10.7   A picture of Venus obtained with the camera on the Pioneer Venus orbiter spacecraft, using an ultraviolet filter to bring out contrasts in the clouds (compare with Fig. 10.1). A large recurring Y-shaped cloud marking, sometimes dimly seen from Earth, is clearly visible here.

clouds by absorption of ultraviolet sunlight. These winds are strongest at the planet's equator and taper off toward either pole, unlike the pattern of tropospheric east and west winds found at temperate latitudes on Earth. More detailed imaging from spacecraft, also using ultraviolet filters, has permitted scientists to map these high-altitude jet streams on Venus. However, we still do not know the chemistry of the thin, ultraviolet-absorbing clouds that make the dark patterns in these pictures.

STRUCTURE OF THE ATMOSPHERE

We are fortunate in the case of Venus to have had two competing programs of exploration carried out independently by the United States and the Soviet Union. As a result, numerous probes carrying a variety of instruments were launched toward Venus in the years from 1962 to 1985, providing a great deal of information on the planet's atmosphere.

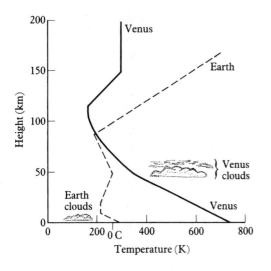

Figure 10.8 The atmosphere near the surface of Venus is much warmer than that of the Earth, but it is colder at higher altitudes.

The atmospheric temperature profiles of Venus and Earth cross; Venus is colder than Earth at high altitudes and much warmer near the ground (Fig. 10.8). There is a region in the atmosphere of Venus where the pressure is near the sea-level pressure on Earth and the temperature a balmy 30°C. This would be a shirtsleeve environment for astronauts in the gondola of a balloon floating in the cloudy skies of Venus, were it not for the fact that those clouds are mainly sulfuric acid and the atmosphere unbreathable carbon dioxide.

In 1985, a joint Russian-French project deployed instrumented balloons at an altitude of about 54 km. Instruments on the balloons measured pressure, temperature, and wind speed as part of a cooperative Soviet-French-U.S. experiment. The results indicated that winds at this altitude are more blustery than had been anticipated, with updrafts over the continent called Aphrodite. The balloons radioed back data for 48 hours, sufficient time for them to travel nearly halfway around the planet.

Near the surface of Venus, the wind speeds are low, with measured values ranging from 0 to 2 m/s (approximately 0–6 km/hr). The pressure and density are so great at these levels that the atmosphere behaves in many ways more like an ocean than the air we are familiar with on Earth. Like the deep oceans of our planet, the surface of Venus has a nearly uniform temperature, from pole to pole and noon to midnight.

ATMOSPHERIC CIRCULATION

Above the surface of Venus, there is a pattern of air rising near the equator and traveling north and south to descend near the poles (Fig. 10.9). This simple type of atmospheric circulation is called a **Hadley cell,** after the British scientist who first proposed it as a model for the circulation of the Earth's atmosphere. Though not describing our own planet very well, Hadley cell circulation turns out to be a very good model for the lower atmosphere of Venus.

At higher elevations on Venus, near the main cloud layers about 50 km above the surface, the strong retrograde (westward) rotation described earlier becomes apparent. At this altitude, of course, the atmosphere is as thin as that of our own planet, so high wind speeds are possible. At the poles, these winds form a vortex of descending air, rather like the pattern of water swirling down a drain.

Calculations carried out with powerful computers suggest that the key to the difference between the circulations of Venus and of Earth is the slower rotation of Venus. If the rotation of Earth were this slow, the high- and low-pressure systems that correspond to centers of fair and foul weather

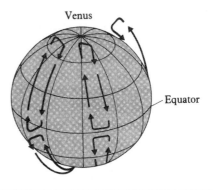

Figure 10.9 In a simple Hadley cell circulation, warm air rises at the equator of a planet and travels toward the pole, where it sinks and returns to the equator along the surface.

would fade away, the midlatitude stratospheric jets would disappear, the small Hadley cells now confined to the equatorial zone would expand all the way to the poles, and a globe-encircling wind would begin to blow at high altitudes.

The decrease in rotation does two things to produce these changes. First, it lengthens the duration of daylight, increasing the effects of heating in the daytime and cooling at night. Second, the Coriolis effect (discussed in Section 9.7) is less strong on a slowly rotating planet, and winds are less likely to be deflected into swirling, cyclonic motion. The combination of longer-term heating and a low Coriolis effect, aided by the larger total mass of the atmosphere, evidently produces a circulation pattern like the one on Venus.

The analysis given here is an example of the ways in which observations of other planets can help us understand the Earth. To evaluate the true importance of various forces at work in natural systems, a scientist would like to be able to perform an experiment, change the forces, and then study the effect of these changes. Since we obviously can't slow down the Earth, the next best thing is to use a computer to model the effect. We are still left with some uncertainty, however, since a planetary atmosphere is very complex. Maybe we left out something important in our model. We can test the model by looking at another planet where basic conditions governing circulation are really different. Slowly-turning Venus is one extreme; the giant planets with rotations more rapid than Earth's provide another. The best assurance we have that our models for the atmosphere of the Earth are accurate is their ability to deal correctly with these other worlds as well, where we find differences in rotation rates, distance from the Sun, and other basic parameters.

10.4 THE HIDDEN LANDSCAPE

Since we cannot photograph the surface of Venus through its clouds, we must use other techniques to map the topography. We have noted that the atmosphere is transparent to radio waves; thus, radar can be used to beam microwaves through the clouds from the outside. These signals are reflected by the planet's solid surface and pass back through the clouds to our receivers. Radar mapping can be done with transmitters and receivers on the Earth, but it is an especially powerful technique when used from a spacecraft in orbit about Venus. Being so much closer to its target, the radar on the spacecraft can afford to be much less powerful than its ground-based counterpart and still achieve higher resolution and more extensive surface coverage. In this and in the following two sections, we examine the surface of Venus as revealed by radar studies.

LARGE-SCALE TOPOGRAPHY

The U.S. Pioneer Venus spacecraft was the first to map the surface of Venus, using a simple kind of radar that measured the altitude of the spacecraft. As the spacecraft orbited from north to south, the planet turned underneath from west to east, building up complete surface coverage over a period of about 2 years (1978–80).

The Pioneer Venus map of surface elevations has a horizontal resolution of 50 km, only slightly better than that of the Moon seen without a telescope. Like the lunar appearance at this resolution, this radar map cannot show features as small as individual craters or mountains. Unlike the view of the Moon as seen without a telescope, however, the radar map measures topography, not color or reflectivity, yielding a global picture of large-scale features such as continents or mountain ranges.

Figure 10.10 shows the Pioneer Venus map and compares it with the topography of the Earth at the same resolution. In looking at the two maps, we are immediately struck by the absence of large continents or ocean basins on Venus. On Earth, most of the surface is either oceanic (typically several kilometers below sea level) or continental (from sea level up to 9 km in altitude). As we saw in Chapter 9, these two terrain types reflect deep-seated differences in the composition and thickness of the crust, and they are formed by different processes. Venus, in contrast, has no deep basins and a smaller area of highlands. Only 10% of Venus consists of highlands, compared with 45% continental surface on the Earth. As soon as the Pioneer Venus map was obtained, it

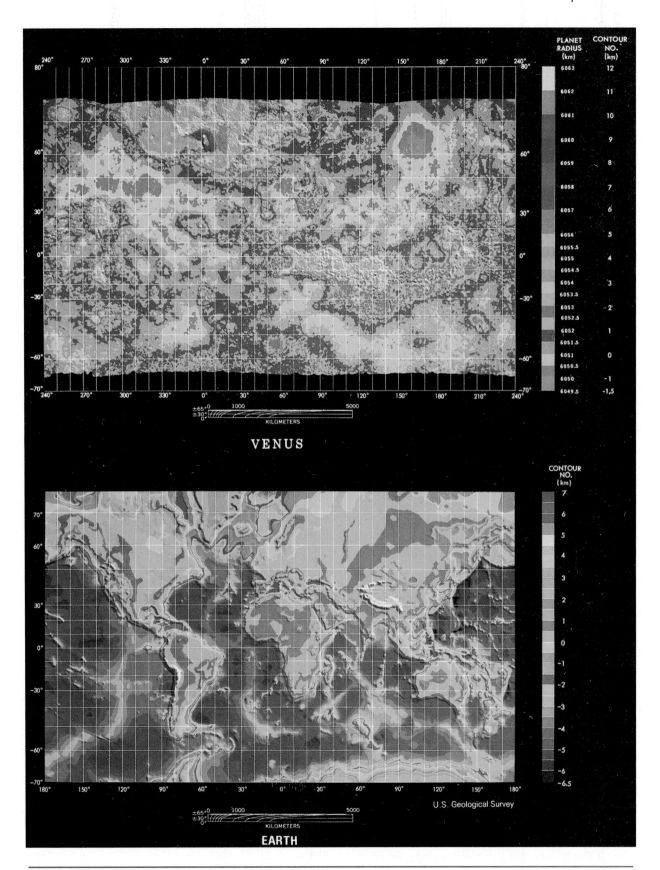

Figure 10.10 Comparison of the large-scale topography of Venus and Earth. The topographic map of Venus was obtained from radar altimetry carried out with the Pioneer Venus orbiter. The Earth's crust is divided about equally into continents and deep basins, whereas most of the surface of Venus consists of rather flat plains.

became evident that Venus is not experiencing the same kind of plate tectonics as the Earth, or if it is, the continental masses are much smaller and the rifts and subduction zones less well defined.

NOMENCLATURE

To discuss the processes that have been shaping the landscape of Venus, we must first gain some additional familiarity with the map in Figure 10.10. With few exceptions, the topographic features of Venus have been named after real or mythical women, appropriate to the one planet in our solar system with a female name.

The largest upland or continental region is called Aphrodite, the Greek name for the goddess the Romans called Venus. Ishtar, named after the Babylonian goddess of love and beauty, is the prominent upland in the north, at about the same latitude as Greenland on our own planet. Ishtar is bigger than Greenland but only a small fraction of the size of North America. It contains the highest elevations on Venus, named the Maxwell Mountains after the nineteenth-century (male) Scottish scientist who first formulated the laws of electromagnetic radiation.

As improved radar images have been obtained from spacecraft, opportunities have multiplied to name more features. The large craters of Venus include, for example, Ariadne, Callas, Cleopatra, Dickinson, Joliot-Curie, Mead, Meitner, and Stuart. The circular tectonic features called coronae bear such names as Artemis, Gaia, Nefertiti, Nightingale, Sacajawea, and Sappho.

10.5 CRATERS AND TECTONICS

Many fundamental geological questions were left unanswered by Pioneer Venus. Without measuring individual impact craters, we could not determine the age of the surface or assess the level of geological activity. Without detecting individual volcanoes, we could not determine whether volcanic eruptions still take place on Venus. And without being able to see rift or subduction zones, we could not decide whether Venus shared with Earth any form of plate tectonics. To address these questions, we require higher-resolution radar images.

MAGELLAN

The Magellan radar mapper began orbiting Venus in 1990 and completed its mapping mission in late 1992. In addition to radar altimetry like that of Pioneer Venus, which simply measures the altitude of the spacecraft above the surface, Magellan carried an advanced **imaging radar,** also sometimes called a synthetic aperture radar, or SAR.

In radar imaging, the microwave signal is beamed to the surface at an oblique angle, and reflections are obtained from a relatively broad area. As the spacecraft moves, the angle at which it views the surface continually shifts, and consequently the details of the returned signal vary as well. Processing these data by powerful computers permits a radar image to be constructed of the planet's surface along the spacecraft track. This radar image looks very much like an ordinary photograph taken with oblique illumination—just the kind of information we need to reveal

Figure 10.11 Geologist Ellen Stofan, the Magellan deputy project scientist, helped organize and interpret a vast amount of topographic and geographic data that has transformed our view of the geologic evolution of Venus.

topographic detail, as we described for the Moon (see Section 7.1). However, in this case brightness differences in the image correspond to variations in radar reflectivity, not color or albedo.

Magellan was not the first radar orbiter. The first radar imaging satellites at Venus were the Soviet Venera 15 and 16 spacecraft, which arrived in 1983 and mapped most of the northern hemisphere with a resolution of 2 km. Magellan, however, yielded 20 times the Venera resolution, mapping with a resolution of 100 m. Each day in orbit yielded a long, skinny image about 20 km wide and several thousand kilometers long. In its 2-year mapping program, Magellan returned more data to Earth than all previous missions to all planets combined (Fig. 10.11).

Radar is an ideal tool for the geologist. Radar waves reflect from the rock and are little affected by thin layers of soil or surface debris. In addition, of course, Venus has no vegetation or bodies of water covering the surface. Thus, underlying geological patterns are more readily apparent than they would be on the Earth, making Venus a geologist's dream planet (Fig. 10.12).

IMPACT CRATERS

Magellan discovered almost 1000 impact craters on Venus, ranging in size from Mead, with a diameter of 280 km (larger than Chicxulub in Mexico), to a few as small as 2 km (about twice the size of Meteor Crater in Arizona). The absence of larger craters (or of impact basins like Imbrium or Orientale on the Moon) tells us immediately that the surface of Venus does not date back to the time of heavy bombardment but must be at least as young as the lunar maria. The absence of smaller craters is a consequence of the thick atmosphere of Venus.

On Earth, the atmosphere filters out most incoming projectiles with diameters smaller than

Figure 10.12 Global radar image of Venus showing the bright, highly deformed bands near the equator that indicate prolonged periods of crustal compression. In this presentation, color is used to indicate elevation, with red showing the highest continental areas. The image is centered on the equator at 180° longitude. (Compare with Fig. 10.10.)

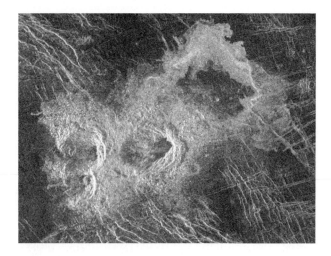

Figure 10.13   Triple crater Stein, a complex feature formed when the impacting projectile broke apart in the thick atmosphere of Venus. The projectile had an initial diameter of 1–2 km. Width of image is 30 km.

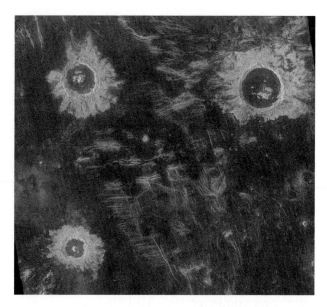

Figure 10.14   Three impact craters in the Lavinia region of Venus, the largest with a diameter of 50 km. The rough crater rims and ejecta appear bright because they are better reflectors of radar energy than are the surrounding smooth plains. Width of image is 400 km.

about 50 m, which fragment and burn up before reaching the ground. With its much larger atmosphere, Venus is protected against most meteoroids smaller than about 500 m across. An impact by a 500-m asteroid typically makes a crater 5–10 km across. Because of this atmospheric filtering, Venus has relatively few craters less than 10 km in diameter and none smaller than 2 km. Figure 10.13 shows the sort of complex, multiple crater that is formed when the projectile breaks apart just before striking the surface.

Three moderately large craters (30–50 km in diameter) are shown in Figure 10.14. The crater rims and ejecta are very bright, while the crater floors are dark. In a radar image, bright areas are rough and dark areas are smooth because rough materials are better reflectors of radio energy. It makes sense that the ejecta, which consist of rocky blocks and fragments, are rough, but why is the floor smooth? This may be the result of impact melting of rock near the point of impact; since the surface temperature of Venus is so high, less additional impact energy is required to melt the rock than would be the case on the Earth or Moon.

Long narrow flows of material are seen near some craters, extending beyond the ejecta blanket. These flows appear to be composed either of lava melted by the impact or of ejecta that flowed like a liquid along the surface. We will discuss evidence of similar fluidized ejecta around some martian craters in the next chapter. On Mars such features can be attributed to the abundant water in the crust; however, Venus is dry and so the presence of these flows is more difficult to understand.

One of the most striking and unexpected properties of the craters of Venus is their crisp, new appearance. Like the craters on the lunar maria, they look as if they formed just yesterday, yet we know that most of them must be millions of years old. In spite of the thick atmosphere, erosion rates on Venus are very low, with craters neither worn down nor filled in by processes of erosion and sedimentation.

CRATER DENSITY AND SURFACE AGE

Venus has a total of approximately 1000 craters spread over a surface area nearly equal to that of the Earth. Our planet, in contrast, has only about 150 craters, and some of these are so old and degraded that we would not recognize their equivalents on Venus. Therefore, we see that crater-retention times on Venus are longer than on the Earth, and the level of geological activity is correspondingly less.

On the Earth and Moon, the density of craters is strongly correlated with the type of surface terrain, with the most recently active areas having the fewest craters. Most lunar craters are found in the highlands, and most terrestrial craters are on the older continental cores. Is there a similar difference in crater density on Venus? Or, equivalently, are there regions of the venusian surface that differ significantly one from another in their ages?

It is difficult to answer these questions because the craters are sparsely spread over the surface of Venus. The continental areas of Ishtar and Aphrodite appear to have fewer craters, not more, than the lowland rolling plains. Within the plains, crater densities seem to be about the same everywhere. The grouping of three craters close together in Figure 10.14 seems to represent an anomalously high density, but this apparent clustering is probably just a statistical fluke.

The average crater density on Venus, in units of 10-km craters per 1 million km², is 15% of the value on the lunar maria, implying an age of about 500 million years. To within a factor of 2, all crater-retention ages on the plains of Venus are the same. Only Aphrodite and (perhaps) Ishtar are significantly younger. For comparison, remember that the age of the ocean basins on Earth is about 100 million years.

We know from the pristine appearance of the craters that erosion is not obliterating craters on a 500-million-year timescale. Some more fundamental geological process must be at work to renew the surface, either volcanic or tectonic or both. But whatever it is, this process does not erase craters gradually. For the most part, craters are either "fresh" or absent completely.

TECTONIC FEATURES

Tectonic geologic features are the result of either tension or compression in the crust of a planet. The cracks and ridges that result from tectonic forces show up readily in radar images. The many narrow bright lines in Figure 10.14, for example, are all part of a tectonic grid of cracks.

A remarkably regular tectonic pattern may be seen in Figure 10.15, which shows a region of the Lakshmi plains about 40 km across. The fine straight lines, spaced about 2 km apart, are cracks that result from stretching of the crust in one direction, while the less regular cross-pattern of ridges is caused by compression of the crust in a direction at right angles to the stretching forces.

Although there is evidence of tectonic modification of the surface everywhere, the crustal forces have been especially strong in certain areas. Figure 10.16 shows a detail from the equatorial region of Venus, within the radar-bright area of tectonically deformed terrain called Alpha. Many of the brightest regions of Venus are complex ridged terrain consisting of low mountain ridges spaced typically 10–20 km apart. These are areas of compression, where squeezing of the surface has caused it to wrinkle. Similar forces produce the mountains of our own planet, but it is rare on Earth to find belts of ridges or valleys on a scale approaching the ridged terrain of Venus.

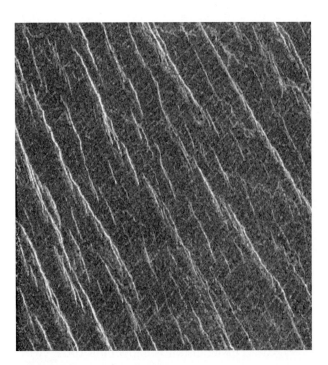

Figure 10.15 Regularly spaced tectonic features in the Lakshmi plains. The surface has been fractured to produce a grid of parallel cracks and ridges with spacing of 1–2 km.

Figure 10.16   Detail of the Alpha region of Venus, part of the tectonic band that circles most of the planet's equator. The surface shows a dense network of ridges and mountains, resulting from compressional forces in the crust.

ORIGIN OF CONTINENTS AND MOUNTAINS

The highest and most dramatic folded mountains of Venus are in the northern Ishtar continent, which consists of a central elevated plain called Lakshmi surrounded by ranges of mountains. The elevation of Lakshmi is 6 km, and the highest point in the adjacent Maxwell Mountains rises to an elevation of 11 km relative to the surrounding lowland plains. It is interesting to compare this area to the Tibetan Plateau (elevation 4–5 km above sea level) and the adjacent Himalayan Mountains (up to 8 km for the highest peaks) on Earth.

Recall that Tibet and the Himalayas are the result of the collision of the Indian subcontinent with Asia. This collision is happening today, with compressional forces maintained by the northward motion of the Indian plate. Without continuing application of force, the elevation of Tibet

and the Himalayas would gradually decline. There is reason to think the same processes are at work on Venus. The mountains surrounding the Lakshmi Plateau are the steepest of Venus, with average slopes as great as 30°. Calculations show that such mountains would quickly collapse if they were not maintained by compression of the crust. The Ishtar continent thus seems to be a region where crustal forces converge, but is it the result of plate tectonics like that on the Earth?

One factor that distinguishes terrestrial plate tectonics is the relative ease with which the lithospheric plates can slide over the mantle beneath. As a result, each plate moves as a unit, and most of the forces are exerted along the edges in fault and subduction zones. On Venus there may be some small movement of plates, but in general it appears that the tectonic forces are distributed throughout the crust. There are no well-defined tectonic plates. In some places, like the folded mountains of Ishtar, the geological consequences are virtually the same on the two planets, but more commonly the tectonic activity on Venus is spread over the entire surface.

10.6 VOLCANOES AND PLANET-SCALE CONVULSIONS

Volcanic activity is common on the surface of Venus. We don't know whether any volcanoes are active today, but certainly such activity is recent on a geologic timescale. Many of the volcanic landforms on Venus look very much like their terrestrial counterparts, but there are some surprises as well (Fig. 10.17).

VOLCANIC PLAINS

The most common landscape on Venus is volcanic in origin. About 80% of the surface consists of lava plains roughly similar to the lunar maria, presumably resulting from high-volume eruptions of fluid lava that spread across large areas. It is these plains areas that are most reliably dated by crater counts as being roughly 500 million years old.

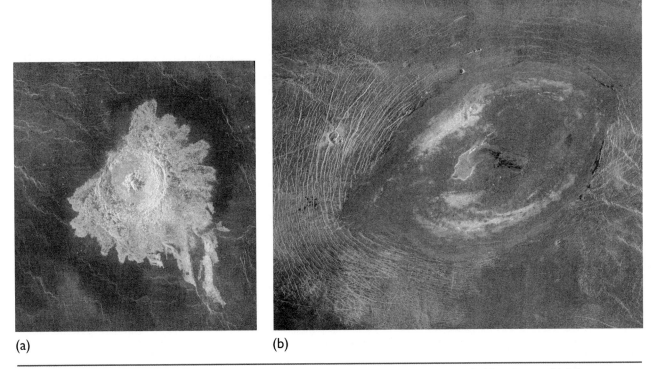

(a)   (b)

Figure 10.17   Crater comparison. **(a)** Crater Aurelia, a fresh impact crater surrounded by ejecta, which appears light in this radar image because it has a rough surface. **(b)** Volcanic caldera Sacajawea. At low resolution, this was mistaken for an impact crater but the Magellan radar image shows the surrounding tectonic features that identify it as volcanic in origin.

It is easy to estimate how much lava is produced on Venus and to compare this with the formation of new oceanic crust on the Earth. The total area of the plains is 400 million km², and we can estimate that the lava flows must be at least 2 km thick to have obliterated all preexisting craters. The total volume of plains lava is then about 800 million km³. From the age of 500 million years, we estimate that about 1.6 km³ of lava is deposited on the plains each year on the average. For comparison, the volume of new basaltic crust emplaced on Earth each year is about 10 km³. Thus, Venus is currently less active volcanically than our own planet. In fact, the calculated eruption rate for the plains of Venus is fairly similar to the volcanic rate for the continents of Earth, neglecting oceanic eruptions.

Can we determine any more about the origin or timing of the plains volcanism? If this volcanic activity were a continuous process (as implied by the

calculation of an annual eruption rate of 1–2 km³ of lava), we would expect to see many craters modified by lava flows. As many as 20%, for instance, should have one wall or part of an ejecta blanket missing as a result of some nearby eruption that took place since the crater was formed. However, only about 5% of the craters on the plains show evidence of subsequent destruction or modification by lava flows, and the other 95% are pristine.

If this logic is correct, then we cannot be observing a steady-state situation in the plains, with old craters removed by eruptions at the same rate new craters are formed from impacts. It appears that some time several hundred million years ago, the activity level was very much higher than it is today. Indeed, we can imagine most of the volcanic plains of Venus having been produced within a very short period about 500 million years ago in some sort of planetwide cataclysm, with a much lower level of activity continuing today.

VOLCANOES AND LAVA FLOWS

Many thousands of individual volcanic mountains are seen in the Magellan images. The largest (as is also the case on the Earth and Mars) are shield volcanoes, characterized by shallow slopes and a summit crater or caldera. On Venus, shield volcanoes can be several hundred kilometers across and up to about 5 km high, generally similar to the size of Mauna Loa on Earth (Fig. 10.18). Individual lava flows mark their slopes, and as judged from the radar images, these volcanoes are indistinguishable from their terrestrial counterparts.

Venus also has volcanoes of a very different form produced by the eruption of thick, viscous lava. Many of these, called pancake domes, are remarkable circular features as large as 45 km in diameter and 2–3 km high (Fig. 10.19). Terrestrial volcanic domes are never so symmetrical. Apparently all of the lava to make one of the venusian pancake domes is erupted at once from a single vent, rather like a large belch of material, and it then spreads out evenly to form a circular feature.

At the opposite extreme, Venus experiences eruptions of very low viscosity, fluid lava that can flow across the surface for great distances before congealing. The result is the formation of lava rivers of remarkable length (Fig. 10.20). About 40 such channels are longer than 100 km, and one, called Hildr, is 7000 km long—as long as the longest rivers on Earth, such as the Nile or the Mississippi. In contrast, the longest lava channels on Earth extend only a few tens of kilometers.

CORONAE

All of the tectonic and volcanic features we have discussed so far have their counterparts on Earth. Now we turn, however, to a unique class of features called coronae, which are found only on Venus. A **corona** is a circular or oval feature hundreds to thousands of kilometers across characterized by concentric and radial tectonic patterns and often by associated volcanic eruptions (Fig. 10.21). Approximately 400 coronae have been identified on Venus, with Artemis (diameter 2000 km) being the largest.

Each corona is characterized by a low central dome surrounded by a shallow trough and a great many concentric tectonic cracks. Associated volcanic activity may take the form of one or more shield volcanoes near the central dome. Apparently each corona is the result of a mantle hot spot—a plume of rising magma similar to the

Figure 10.18 Computer-generated radar image of Venus showing the large shield volcanoes Sapas Mons (center) and Maat Mons (on the horizon). The vertical relief in this image is exaggerated ten times, which accounts for the distorted appearance of the impact crater at right. The actual topographic relief is low over most of Venus. (Color is false.)

Figure 10.20 Small segment of the Hildr lava channel, which is about 7000 km long and about 2 km wide. Such lava rivers were formed by the eruption of highly fluid lava.

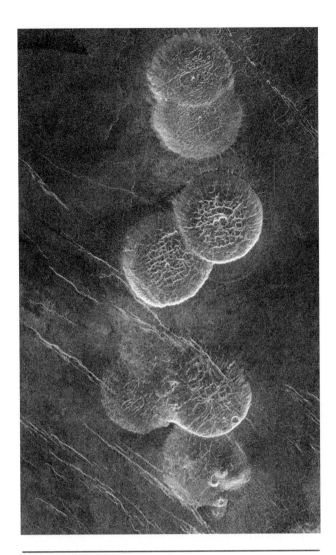

Figure 10.19 Pancake domes. Circular volcanoes about 45 km in diameter and 2–3 km high.

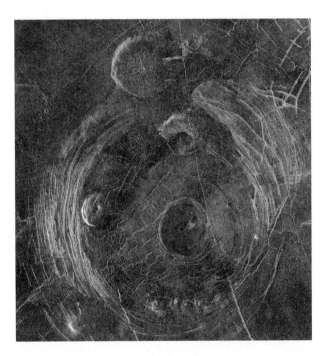

Figure 10.21 Aine Corona, a circular feature about 200 km in diameter. Such features, which are unique to Venus, appear to have been produced over rising plumes of magma in the mantle of the planet. (With a little imagination, you can recognize the face of "Miss Piggy" in this feature).

mantle plume that has created the Hawaiian Islands on the Earth. Many of the coronae probably represent failed hot spots—plumes that turned off before the magma broke through to the surface. Alternatively, coronae may represent a stage in plume development that will lead to later surface eruptions and the construction of large shield volcanoes.

GEOLOGICAL HISTORY

The Magellan data have revolutionized our knowledge of Venus. They clearly show that the planet is geologically active, with ongoing tectonic and volcanic modification of the surface. The energy for such activity is derived from convection currents in the mantle, as on the Earth. However, these currents have not led to planetwide plate tectonics, either because the plumes of rising magma are too small or because the lithosphere of Venus is unable to slide over the mantle as it does on the Earth. The resulting pattern of coronae and regional deformation of the surface has sometimes been called blob tectonics, to distinguish it from terrestrial plate tectonics.

Major questions are also associated with the possibly episodic formation of the lowland lava plains. We are used to thinking of geological processes as slow and continuous. On Earth, for instance, we know that the lithospheric plates move a few meters per century, and we confidently expect that they have been doing so for a long time in the past and will continue to move at this rate in the future. It is startling, therefore, to imagine that the entire surface of Venus may have been catastrophically resurfaced just 500 million years ago. Planets are not supposed to behave like that (or so we have been taught to think). This apparent planetwide resurfacing—and the question whether it was a unique event or part of a recurring pattern—remains the greatest post-Magellan mystery about our twin planet.

Whatever the answers to these particular questions, we now know that Venus is lacking in ancient surface material. It is unlikely that any rock on Venus is as old as the oldest rocks that make up the terrestrial continents. Venus probably has the youngest surface of any of the terrestrial planets, in the sense that its crust was most recently formed, with little if any memory of the way the planet was even a billion years in the past.

10.7 ON THE SEARING SURFACE

Among the most dramatic achievements of the space program of the former Soviet Union were the Venera spacecraft that successfully landed and operated on the hostile surface of Venus. These hardy robots allowed us to glimpse a place where no human will ever set foot.

THE VENERA LANDERS

To gain an appreciation for what conditions are like on the surface of Venus, you might begin by looking at the temperature control of a modern kitchen oven. You will find that the highest temperature the oven can achieve is about 500°F. The ground temperature all over Venus is within 15 degrees of 860°F. And remember this temperature is coupled with an atmospheric pressure 90 times the value on Earth.

Despite these extremely inhospitable conditions, direct measurements were successfully carried out at seven locations on the surface of Venus by Soviet landers. These landers were suitably armored against the high pressure and equipped with some internal cooling. The best of them survived nearly 2 hours, while a variety of investigations were carried out, including pictures transmitted back to Earth, four of them in color. These are the only close-up pictures we have of the surface of another planet except for the Viking and Pathfinder pictures of Mars and, of course, extensive surface photography of the Moon.

The Venera cameras provided panoramic views of the surface of Venus extending from the soil directly in front of the spacecraft out to the horizon. The cameras were positioned about 1 m above the ground to provide a perspective comparable to that of an observer sitting on the surface of Venus. The cameras did not gaze straight out at the landscape as you would; instead, they were directed at mirrors set at a 45° angle, and the mirrors were rocked back and forth to produce the desired panoramas (Fig. 10.22).

The immediate foreground of each picture shows the view toward the feet of the hypothetical observer, including a part of the spacecraft itself, with a scale reference provided by the triangular teeth spaced 5 cm apart. The white ladder-shaped boom extending out just left of center is about as long as your arm, while the distant white viewport covers (ejected after landing) are about the diameter of a human head.

IMAGES OF THE SURFACE

The first pictures of the surface of Venus, transmitted by Veneras 9 and 10 in 1975, showed rough, undistinguished landscapes dominated by loose rocks. These spacecraft both landed in regions of fractured lava plains on the lower slopes

ВЕНЕРА-13 ОБРАБОТКА ИППИ АН СССР И ЦДКС

(a)

ВЕНЕРА-14 ОБРАБОТКА ИППИ АН СССР И ЦДКС

(b)

Figure 10.22 (a) A view of the surface of Venus obtained by Venera 13. The triangular teeth in the foreground are about 5 cm apart. Note the fine-grained, soil-like material in the foreground. The bright crescent-shaped object in the center of the frame is the camera cover. The device to the left of the ejected camera cover measures the surface hardness. **(b)** Venera 14 was surrounded by flat, platelike rocks resembling pieces of a dried lakebed. One of them was overturned by the shock of landing revealing a bright surface (left foreground). In this case, the deployed camera cap unfortunately ended up under the surface-hardness measuring device, showing that Murphy's law also applies on Venus.

of the Beta volcanic complex; Venera 10 apparently photographed old basaltic lava flows, while Venera 9 landed near an area of more recent tectonic modification. The surface illumination, as we noted previously, was similar to that on a heavily overcast day on Earth.

The more detailed pictures transmitted by Veneras 13 and 14, which landed in March 1982, represent the best data available on the appearance of the surface of Venus (Fig. 10.22). The Venera 13 picture shows many small rocks and some fine-grained soil, demonstrating that erosional processes on Venus break up large blocks into smaller material. Alternatively, the fine-grained material might be wind-borne lava fragments (so-called volcanic ash) from a large volcanic eruption. We know from the Magellan images that erosion is a very slow process on

Venus. The pictures also show layering on rock surfaces, especially around Venera 14, perhaps indicative of lava flows.

SURFACE COMPOSITION

Five of the successful Venera landers carried devices to detect gamma rays emitted from radioactive isotopes of uranium, thorium, and potassium in the surface rocks. By analogy with the levels of radioactivity of terrestrial rocks, their levels suggest basalt at all five sites, although the exact chemistry varies from one location to the next.

On Veneras 13, 14, and 18, a more sophisticated technique measured the surface composition. A drill was deployed to gather samples and bring them inside the spacecraft for analysis. Using an x-ray source to stimulate emission from the sample, this instrument was able to detect additional elements. The Venera 13 sample indicated a composition typical of oceanic basalts on Earth, whereas the result from Venera 14 resembled a much rarer kind of basalt with a high percentage of potassium. Venera 18, which landed near the Aphrodite continent, also revealed a basaltic composition, this time unusually rich in sulfur.

All five of the measured locations are in the common volcanic plains, where we expect basalt to be the primary rock type. Venera 13 very likely measured one of the young lava flows visible in the Magellan image of the site, but the other four landers appear to be on older and less distinctive surfaces.

In all of these investigations of Venus we lack much of the critical information that helps us to understand the geology of the Earth and Moon. We have no seismic probes of the interior and no measurements of subsurface temperature or heat flow. Since there are no samples from Venus in our meteorite collections, we lack the chemical information and age measurements that have played such a large role in terrestrial and lunar studies. Venus has no detectable magnetic field, so even this indirect method of investigating the interior is not available. Thus, in spite of the fabulous success of the Magellan mission in providing detailed surface maps, Venus is likely to remain an enigma for a long time to come.

10.8 QUANTITATIVE SUPPLEMENT: THE GREENHOUSE EFFECT

Developing accurate models for the greenhouse effect is a challenging task that requires a detailed understanding of how all of the gases in an atmosphere interact with visible and infrared radiation. However, we can look at an idealized greenhouse model to get some sense of how the process works.

Consider a two-layer model as sketched in Figure 10.23. The planet is illuminated by sunlight falling from straight above. The atmosphere is represented here by a single isothermal layer

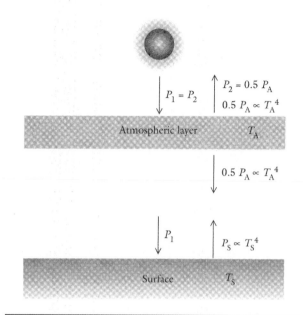

Figure 10.23   A planet with an atmosphere will have a surface that is heated both by direct radiation from the sun (P_1) and by radiation from the atmosphere ($0.5 P_A$).

that is completely transparent to sunlight and completely opaque to the infrared radiation emitted by the surface. The incoming solar power is P_1, and the power radiated back into space (all of which comes from the atmosphere because radiation from the surface is blocked) is P_2. In equilibrium, the power emitted by the planet must equal the incoming solar power, so $P_1 = P_2$.

The surface temperature is T_S, and by the Stefan-Boltzmann law it must be radiating a power P_S given by σT_S^4, all directed upward. The atmospheric temperature is T_A, and in this model it radiates a power P_A, half upward to space and half downward to the surface, both powers given by $0.5 \, \sigma T_A^4$, where $0.5 \, \sigma T_A^4$ is also equal to P_2, the total power radiated by the planet.

We can now write an algebraic expression for the energy balance at the surface. The power reaching the surface is P_1 (directly from the Sun) and $0.5 \, \sigma T_A^4 = P_2 = P_1$ (the back-radiation from the atmosphere). Thus, the total power reaching the surface (and radiated from the surface) is twice the power from the Sun alone. If T_E is the equilibrium surface temperature without an atmosphere, you can see that

$$\sigma T_S^4 = 2\, \sigma T_E^4$$

$$T_S = \sqrt[4]{2}\; T_E = 1.2\; T_E$$

Thus, if the equilibrium temperature (without a greenhouse effect) is 400 K, the surface temperature with the greenhouse effect is 480 K.

The next step in sophistication is to consider a multilayer atmosphere, with the energy balance at each layer calculated just as we have done here for the surface. Each layer of the atmosphere receives radiation from both above and below, and each layer radiates both up and down at its appropriate temperature. If you add up many such layers, you will find that the surface temperature can increase by more than the factor of 1.2 that we derived above. The exact value depends on the details of the infrared opacities of the atmospheric gases: how they absorb at different wavelengths and how these absorptions change with pressure and temperature. In the case of Venus, however, the observed surface temperature is more nearly twice the expected value rather than being enhanced by a factor of only the fourth root of 2. This is why the high temperature of Venus was such a shock to the early radio observers, and why it took many years of sophisticated modeling to understand what is really happening in this complex atmosphere.

SUMMARY

So similar to Earth in size and bulk composition, Venus is dramatically different in many other characteristics. It rotates slowly backward on an axis nearly perpendicular to its orbit; why, we do not know. Its surface is baking at a temperature of 740 K, at a pressure of 90 bars. The immense atmosphere that produces this high surface pressure is made mainly of carbon dioxide, which maintains the high surface temperature through the greenhouse effect. Some visible light from the Sun penetrates the thick sulfuric acid clouds and reaches the planet's surface.

The circulation of this atmosphere is considerably different from ours, owing to the planet's slower, retrograde rotation and closer proximity to the Sun. Hadley circulation, a simple equator-to-pole exchange, seems to dominate the flow. The thickness of the atmosphere maintains a nearly constant temperature over the entire surface of the planet.

Radar mapping has shown that the geology of Venus is substantially different from that of the Earth. Most of the planet (80%) consists of low, rolling volcanic plains produced by large eruptions of fluid basaltic lava. The crater densities on these plains are only 15% of the values for the lunar maria, indicating an age of about 500 million years. To our surprise, we find very few degraded or eroded craters, suggesting that the formation and destruction of craters are not in balance. It appears that active volcanic eruptions, or possibly even a general overturning of the planet's crust, formed most of the plains surface about 500 million years ago, with lower levels of volcanism (roughly 1 km^3 of new lava per year) since that cataclysm. The cause of this global-scale resurfacing remains one of Venus's greatest mysteries.

The crust of Venus has been extensively modified by tectonic deformation to produce a wide variety of cracks and ridges, as well as major belts of folded mountains. The highest parts of the surface are found in the Ishtar continent, where the Lakshmi Plateau and the Maxwell Mountains resemble the Tibetan Plateau and the Himalayan Mountains of the Earth, uplifted by compression of the crust. However, Venus is not experiencing terrestrial-type plate tectonics, perhaps because its lithosphere is too tightly attached to the mantle below. A variety of ongoing volcanic processes produce shield volcanoes, pancake domes, and lava rivers up to 7000 km long. Coronae are unique geological features that mark the top of subsurface plumes of magma rising through the mantle.

Both surface imaging and chemical analysis of the soil have been carried out by Venera landers. All of these spacecraft landed on the widespread lowland lava plains, and their analysis indicates the presence of basalt.

Many of the striking differences between Venus and Earth can be explained by our sister planet's closeness to the Sun. At the distance of Venus, water cannot remain on a planetary surface. A runaway greenhouse effect will ultimately lead to the breakdown of water molecules in the planet's upper atmosphere by solar ultraviolet light. At least we speculate that this may have happened on Venus. We will return to a comparative study of the evolution of the Earth, Venus, and Mars in Chapter 12.

KEY TERMS

corona

Hadley cell

imaging radar

photochemistry

runaway greenhouse effect

REVIEW QUESTIONS

1. Compare the interior structures of the terrestrial bodies studied so far: Earth, Venus, Mercury, and the Moon. Consider also what we know about their thermal evolution. How do structure and evolution depend on the size and composition of a planet?

2. Compare the atmospheric structures of the Earth and Venus. Does the Venus atmosphere have the same regions as those identified on Earth: troposphere, stratosphere, ozone layer, ionosphere, magnetosphere? Why are there differences between the two planets?

3. Compare the atmospheric circulation patterns of Venus and the Earth. What are the roles of solar heating and planetary rotation? How does the presence of water on Earth give rise to kinds of weather that would be impossible on Venus? What is the effect on atmospheric circulation of the much larger mass of the Venus atmosphere?

4. Contrast the compositions of the atmosphere and clouds of Venus with those of the Earth. Do you understand why the clouds on both planets do not have the same composition as one of the primary gases in the atmosphere, but instead represent a trace constituent of the gaseous atmosphere?

5. Explain the runaway greenhouse effect. How does this differ from the ordinary greenhouse effect? If Venus once had oceans of water and later lost them through the runaway greenhouse effect, what evidence today might reveal this past history? Consider both chemical and geological clues.

6. Describe the large-scale surface topography of Venus and compare it with those of the Earth and Moon. Does Venus have features analogous to the lunar highlands and maria? Or to the terrestrial continents and ocean basins? How does Venus differ from each?

7. Explain how craters are used to determine the ages of the surface features of Venus. How do these ages compare with those of the lunar highlands, the lunar maria, the terrestrial continents, and the terrestrial oceans? Does Venus have different surface units with very different ages, as do the Earth and Moon?

8. What features on Venus have a tectonic origin? What are the indications for and against plate tectonics on Venus? Can you think of reasons the Earth and Venus should differ with respect to their plate tectonics?

9. How does the volcanic history of Venus compare with that of the Earth and Moon? Are any of the volcanic features on Venus distinctive or unique? Do scientists understand the reasons for these differences?

10. What would it be like to stand on the surface of Venus (with appropriate protection from the heat and pressure)? What would the ground and sky look like? How might you tell if you were standing on volcanic plains, in a region of tectonic deformation, or in the Maxwell Mountains? What would the Earth look like from the surface of Venus?

QUANTITATIVE EXERCISES

1. Venus requires 440 days to move from its greatest elongation (apparent distance) west of the Sun to its greatest elongation east, but only 144 days to move from greatest eastern elongation to greatest western elongation. Explain why.

2. Magellan scientists estimate that volcanism on Venus generates about 1 km^3 of new lava per year. At this rate of eruption, how long would it take to cover the planet uniformly to a depth of 1 km? Generally, if we wish to obliterate a large crater,

we need to bury it to a depth of several kilometers. What time is required to cover Venus to a depth of 3 km?

3. The surface of Venus is nearly twice as hot as the surface of the Earth. At what wavelength is the maximum of its infrared radiation from the surface?

4. The terrestrial greenhouse effect produces an increase in the surface temperature of about 25 K, or a factor of about 1.1. Suppose that a huge impact generated a thick layer of dust at high altitude. In terms of the simple two-layer model discussed in Section 10.8, this would be equivalent to having the incident sunlight absorbed in the upper layer rather than at the surface. According to this model, what would happen to the greenhouse effect and the surface temperature?

ADDITIONAL READING

Chapman, C. R. 1982. *Planets of Rock and Ice: From Mercury to the Moons of Saturn*, Chapter 7. New York: Scribner's. Good general discussion of the atmosphere of Venus and the difficult road to understanding its true nature.

Cooper, H.S.F., Jr. 1993. *The Evening Star: Venus Observed*. New York: Farrar, Straus and Giroux. A popular account of the Magellan mission by the veteran space reporter of *The New Yorker*.

Grinspoon, D. 1997. *Venus Revealed: A New Look Below the Clouds of Our Mysterious Twin Planet*. Reading, MA: Addison-Wesley. Excellent overview of Venus, including the fascinating geology revealed by the Magellan radar images.

11

MARS: THE PLANET MOST LIKE EARTH

The heavily eroded martian canyonlands as imaged by Viking. The width of this image is about 50 km.

Of all the planets, Mars has the most romantic appeal. The Moon is closer, Venus brighter, Jupiter larger, and Saturn more beautiful, but Mars remains the one other world besides Earth with which humans feel an emotional connection. In the past, this planet was thought to harbor advanced forms of life, including intelligent creatures with civilizations more sophisticated than ours. Even today, Mars seems to be the most likely place beyond Earth to support indigenous life (probably microbial life), and we recognize its tremendous potential as an exciting target for robotic and ultimately human exploration.

Mars may have great potential, but today it is a forbidding world. With only one-tenth the mass of Earth or Venus, it has lower levels of geologic activity, with consequently less outgassing of volatiles from the interior. Being farther from the Sun, Mars is also noticeably cooler. It is a planet caught in a terminal ice age, with most of its water frozen in subsurface permafrost. The surface atmospheric pressure is so low that liquid water cannot exist, even if

259

the temperatures should rise above the freezing point. Solar ultraviolet light bathes the surface in a lethal glow, unimpeded by the thin atmosphere.

What makes Mars so interesting in spite of these difficulties is its evidence of very different past climates. Volcanoes once erupted vast amounts of lava and presumably released water and other gases from the interior. Riverlike channels give testimony of a time when water flowed over the surface and perhaps rain fell from the martian skies. Layered deposits in the polar regions indicate cyclical climatic variations that appear to continue today. Mars is the one place besides Earth where we see clear evidence of climatic cycles, in addition to long-term evolution of the surface and atmosphere. It is the only other planet where there is evidence that liquid water once ran and pooled on the surface. And because much of this surface is as ancient as the uplands on the Moon, there is an opportunity on Mars to study the early history of a volatile-rich planet, an opportunity that geologic activity has effectively eliminated on Earth.

Figure 11.1 shows the Mars of today—a world that has been extensively mapped by a flotilla of spacecraft. Indeed, we have more accurate and detailed maps of Mars than we do of most of the ocean floor that makes up the majority of Earth's surface. A century ago, however, Mars was a small and distant world, and the only tools of exploration were the human eye looking through a telescope.

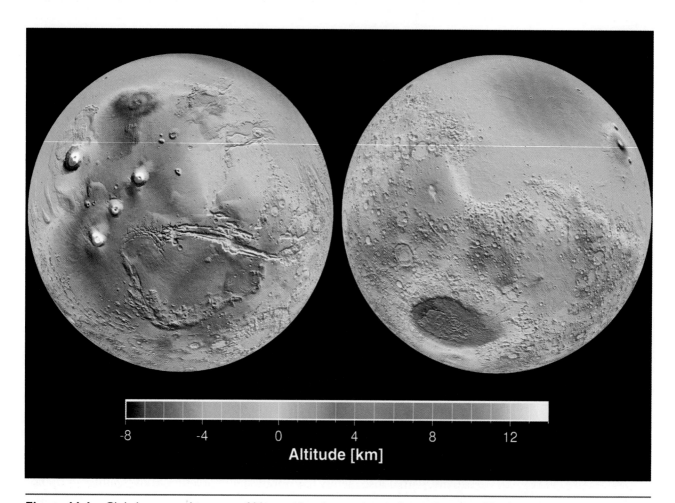

Figure 11.1   Global topographic maps of Mars, with color used to indicate elevation. The hemisphere on the left includes Olympus Mons, the highest mountain; the hemisphere on the right includes the Hellas Basin, which has the lowest elevation on Mars.

11.1 A CENTURY OF CHANGING PERCEPTIONS

Mars is among the brightest objects in the sky, and its red color sets it apart from the other planets. As we saw in Chapter 1, observations of the movements of Mars played a critical role in the development of Kepler's laws of planetary motion. Interest in Mars as a world, however, dates from the late nineteenth century.

MARS THROUGH THE TELESCOPE

As seen from Earth, Mars follows a 26-month cycle from one opposition to the next. When the two planets are far apart, Mars looks like an inconspicuous red star. Even in the best telescopes, little detail is visible on its surface. When the planets are close together, however, the distance from Earth to Mars can be as little as 55 million km. Then Mars is almost as bright as Jupiter, and telescopes can reveal features as small as 100 km across (Fig. 11.2).

As we noted in Section 7.1, a resolution of 100 km is not sufficient to reveal topographic features on a planet. In the case of Mars, the situation is worse because we always see the planet at nearly full phase. Recall how little topography is visible at full moon, even through a telescope (see Figs. 7.1 and 7.2). Thus, the features that are seen telescopically on Mars are markings that represent different colors and reflectivities of surface materials. Primarily, we map dark regions and light regions, corresponding to reflectivities near 15% and 30%, respectively.

In addition to these relatively permanent light and dark surface markings, Mars has bright polar caps that grow and shrink with the seasons, just as one would expect for deposits of ice or snow. Transient bright yellow or white clouds also appear, sometimes associated with specific regions on the surface. The presence of seasonal polar caps and transient clouds allowed the earliest visual observers to deduce an atmosphere on Mars. Not until the 1940s, however, when G. P. Kuiper first identified CO_2 spectrally, did astronomers begin to learn the composition of this atmosphere.

Figure 11.2   One of the best Earth-based photos ever obtained of Mars, taken with the Hubble Space Telescope in June 2001 when Earth and Mars were exceptionally close (68 million km). The resolution is about 20 km, much better than can be obtained with ground-based telescopes, but still insufficient to reveal much about the geology of the surface.

THE CANAL CONTROVERSY

The history of martian studies notes well 1877, the year the Italian observer Giovanni Schiaparelli recorded the linear markings he called *canali*, or channels. These faint dark lines, glimpsed near the limit of detectability, seemed to stretch for thousands of kilometers across the surface of the red planet. In English-speaking countries, the word *canali* was translated as "canals," a term implying construction by intelligent beings.

By the early years of the twentieth century, the conviction was widespread that the canals of Mars proved the existence of intelligent life on that planet. The most vocal advocate of this position in the United States was Percival Lowell, who for two decades dominated the American public image of astronomy. Most astronomers, in

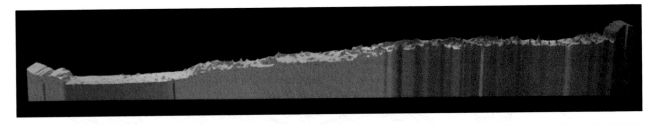

Figure 11.8   Topographic cross section of Mars along the zero meridian, from the north pole (left) to the south pole (right). The southern hemisphere has an average elevation about 5 km greater than the northern. Note the smooth low area in the north that may once have held a shallow polar ocean, and the "bumps" at each pole that are the permanent polar caps.

hemispheres, with the southern hemisphere higher than the northern by about 6 km. The lowest areas in the northern hemisphere (colored blue) are also remarkably flat—exactly what we would expect if this were the bed of an ancient ocean. A number of properties of Mars also differ in the two hemispheres: In particular, the higher southern regions are more heavily cratered (hence older) and are generally darker than the northern lowlands.

The difference between the northern and southern hemispheres of Mars is one of the most fundamental and mysterious aspects of the planet. If the heavily cratered uplands of the south represent the older crust of the planet, then something peculiar has happened in the north, lowering the elevation by several kilometers and reworking the surface by volcanic or other processes. Whatever happened, it was more than a simple stripping away of the surface material, since there is no place to hide such a large quantity of excavated soil.

The second major feature of the large-scale topography of Mars is the Tharsis bulge, a volcanically active region the size of North America that rises about 10 km above its surroundings (Fig. 11.9). Tharsis straddles the boundary between the southern uplands and the lower northern plains, and it is the least cratered and therefore the youngest part of the planet's surface.

GEOLOGIC FEATURES

Near the middle of the Tharsis bulge are three great shield volcanoes, each rising 18 km. A still larger volcano, Olympus Mons, rises 25 km above the northwestern slope of Tharsis. Olympus Mons

is nearly 700 km across at its base, about the size of France (Fig. 11.10).

Associated with the Tharsis bulge are the Valles Marineris (Mariner Valleys), the central parts of an interconnected system of east-west-running canyons approximately 4000 km long. The individual canyons are about 3 km deep, but in the center of the Valles Marineris, the depression is nearly 7 km and the width of the canyon more than 500 km. The Grand Canyon in Arizona could easily fit into one of its branches.

Geologists have concluded that the volcanoes, Valles Marineris, and the Tharsis bulge are

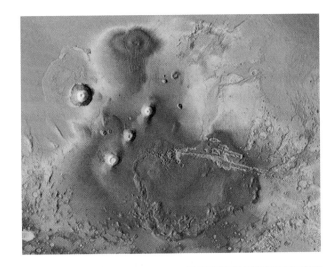

Figure 11.9   Topographic map of the Tharsis bulge and its surroundings. Note the four huge volcanoes in the upper left and the long Valles Marineris system of canyons extending horizontally to the right. The drainage from the canyons swings northward in a series of large channels that may once have fed water into the northern ocean.

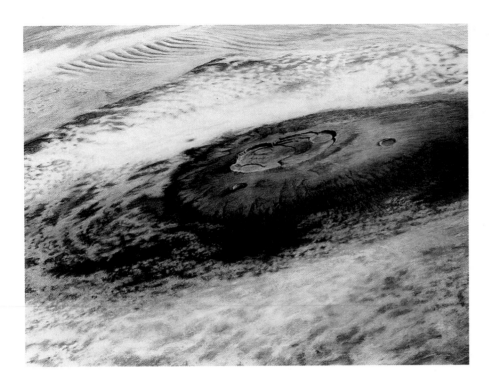

Figure 11.10 Oblique Viking view of the Olympus Mons shield volcano, partly surrounded by high-altitude clouds.

all expressions of a major, long-lived center of volcanic activity that has dominated this one side of Mars. We will explore these ideas further in later sections of this chapter.

IMPACT CRATERS AND BASINS

Like the Moon and Mercury, Mars has thousands of impact craters. On Mars, these craters are concentrated in the older southern highlands. Generally speaking, these craters look much like their counterparts on airless planets, with the same raised and terraced rims, flat depressed floors, and frequent central peaks. Presumably the same impact processes formed them. From the crater densities, which are substantially higher than those of Venus and Earth, we conclude that the martian highlands date back to the first billion years of solar system history.

Although martian impact craters themselves look like their lunar counterparts, their patterns of ejecta are unique. On the Moon, a typical large crater is surrounded by a rough, hilly ejecta blanket close to the rim, with radial streaks and chains of secondary craters farther out (see Figs. 7.15 and 7.16). In contrast, many martian

craters have smooth ejecta blankets with well-defined edges. There are no rays or streaks of material extending to great distances. These have been called **fluidized ejecta** craters or, more informally, splosh craters.

The most common type of ejecta is a multi-lobed pattern, called a flower form. Several ejecta layers are present, with each lobe bounded by a steep, smooth edge (Fig. 11.11a). These flower craters are found primarily in the equatorial regions, within 40° of the equator. In the second type (Fig. 11.11b), the ejecta look like a relatively smooth pancake of material with a slightly scalloped rim. We call these pancake craters, and they occur primarily at higher latitudes. Evidently the martian ejecta blankets were formed by fluid debris that flowed along the ground rather than following explosive aerial trajectories. This hypothesis of fluid flow also explains a number of cases where the ejecta flowed over and around obstacles. The most likely explanation for the fluidized ejecta of martian craters is the large quantities of water in the surface at the time the impacts occurred. If martian conditions then resembled those we see today, the water was in the form of ice until melted by the energy of the impact.

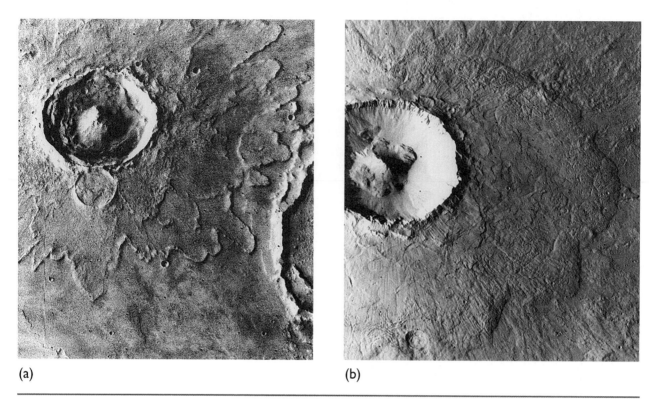

(a)   (b)

Figure 11.11   The unusual ejecta blankets of martian impact craters probably result from the presence of frozen water (permafrost) in the soil. **(a)** Yuty, an 18-km crater at 22°N, showing a central peak and several thin sheets of ejecta, with multiple lobes. **(b)** Arandas, a 28-km crater at 43°N, with a thick pancake of ejecta terminating in a flow front.

The largest martian impact basin is Hellas, about 1800 km in diameter and some 6 km deep. Unlike its lunar counterparts, Hellas has a simple form with a single rim of uplifted mountains. Its interior, which is frequently hidden by hazes of dust or water fog, is covered by wind-blown dust. Parts of the rim mountains are missing. The best-preserved large basin on Mars is Argyre, with a diameter of 700 km (Fig. 11.12). The rim consists of a broad rugged area more than 200 km wide made up of mountains thrown up in the basin-forming impact. In Figure 11.12, also note the 210-km-diameter crater Galle, lying on the rim of Argyre.

Mars has fewer large basins than the Moon, in spite of its much greater surface area. What does this mean? One possibility is that many of the basins that formed early in martian history have been degraded beyond recognition by subsequent geologic activity, including extensive erosion. Another way of stating this idea is that the martian crust stabilized later than that of the Moon, near

Figure 11.12   A part of the Argyre impact basin, about 700 km in diameter. This Viking photo also shows haze layers in the atmosphere of Mars. The 210-km crater Galle is visible superimposed on the basin rim.

the end of the terminal heavy bombardment. This would be consistent with the larger size of Mars, which led to slower cooling of the interior.

CRATER DENSITIES AND THE AGE OF THE MARTIAN SURFACE

If the craters on Mars and the Moon were caused by impacts with the same population of asteroids and comets, their cratering rates must have been roughly the same. To refine this result, actual impact rates must be calculated and allowance must be made for the differing impact speeds on individual planets. Shown in Table 11.2 are the results of one such cratering model, calculated by William K. Hartmann of the Planetary Science Institute in Tucson. Here cratering rates are expressed relative to those for the Moon.

Hartmann's model provides one way to date different martian terrains from their observed crater densities, if we assume the lunar cratering history, including a terminal heavy bombardment. We will use these age estimates throughout this chapter when discussing various features on Mars. However, note that cratering models developed by other scientists yield ages that differ by as much as a factor of two for the younger volcanic terrains. All investigators, however, agree that the oldest terrains date from about 4 billion years ago. Table 11.3 summarizes the results for some representative surface features.

A wide variety of crater densities exist on Mars. Some areas, such as the summits and slopes of the Tharsis volcanoes, have very few craters. The crater densities in these areas are substantially less than on the youngest lunar maria. In contrast, the densities on the older cratered uplands are higher than on the lunar maria, although still less than on the lunar

Table 11.3 Crater ages for martian surface features

Feature	Crater Density Relative to Lunar Maria	Crater-Retention Age (billion years)
Olympus Mons volcano	0.1	0.2
Arsia Mons volcano	0.1	0.2
Tharsis plains	0.5	1.6
Elysium plains	0.7	2.6
Chryse Planitia	1.1	3.2
Alba volcano	1.8	3.5
Hellas basin	1.8	3.5
Cratered uplands	10	4.0

highlands. This result dates the uplands to ages somewhere close to 4 billion years.

One fundamental difference between the Moon and Mars must be considered when interpreting crater densities: erosion. On the Moon, the only important agents that degrade old craters are further impacts. Mars, however, has an atmosphere. Even today, the erosive effects of martian windstorms are significant, and in the past it seems likely that the atmosphere was thicker. The effects of erosion are particularly evident in the most heavily cratered uplands (Fig. 11.13), where the large craters are shallow and subdued. In these areas, there are many fewer small craters than would be expected from the numbers of larger ones present. Figure 11.14 also displays this effect, plotting the density of martian craters of different sizes measured on both the uplands and the lower northern plains. The shortage of craters with diameters less than about 5 km is attributed to enhanced erosion early in the history of Mars. The fact that smaller craters are not in short supply relative to large ones in the younger terrains suggests that this era of high erosion rates terminated billions of years ago.

Venus also has few small craters, but in Chapter 10 we attributed this to the shielding effect of the atmosphere. Why is the explanation different on Mars? To produce a crater cutoff at 5 km

Table 11.2 Relative cratering rates for the terrestrial planets*

Mercury	Venus	Earth	Moon	Mars
2.0	1.0	1.5	1.0	2.0

*The higher value for Mars is due primarily to its proximity to the asteroid belt, while that of Mercury is a result of the Sun's gravity.

Figure 11.13   The cratered uplands that characterize most of the southern hemisphere of Mars. The craters are generally subdued by erosion, with a notable absence of small craters relative to similar views of the Moon and Mercury. This frame is about 500 km across.

requires an atmospheric pressure of about 1 bar (like Earth), which is unlikely but certainly not impossible for Mars at the time the uplands were being cratered. It is indeed possible that the atmosphere played such a role on Mars. However, the presence on Mars of shallow degraded small craters strongly suggests that erosion is the primary cause of the missing craters. This situation is in dramatic contrast to that of the craters of Venus, which show no erosion at all at the resolution of the Magellan radar images.

In the northern plains, the density of craters is much lower than in the southern uplands, as shown in the crater counts plotted in Figure 11.14. The best-studied northern plains area is Chryse Planitia, the place where Viking 1 landed (Fig. 11.15). Chryse is a flat basin of apparently volcanic origin. For craters with diameters between 1 and 10 km, the crater density on Chryse is virtually identical to that of the average lunar maria, while for larger craters the density on Chryse is somewhat greater. Similar crater densities apply to the other northern lowlands.

11.3 VIEW FROM THE SURFACE

Although no human has yet set foot on Mars, we have sent our surrogates in the form of the Viking and Pathfinder landers and the Sojourner rover. Through the eyes of these spacecraft we have looked more carefully at the surface of Mars than at that of any other planet beyond the Moon.

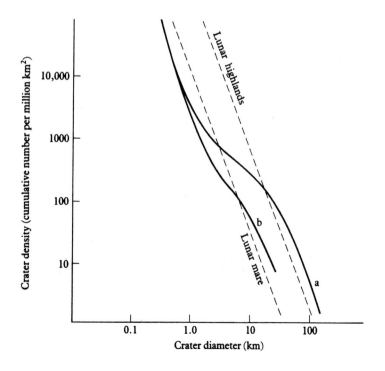

Figure 11.14   The distribution of crater sizes in the martian uplands and in the Chryse Basin, compared with crater densities on the Moon. Note the shortage of small craters in the older uplands of Mars, presumably the result of erosion. Curve a is for average martian uplands; curve b shows the data for Chryse.

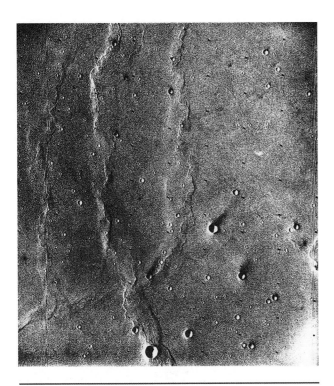

Figure 11.15 An orbital view of the site of the Viking 1 landing in Chryse, a low basin characterized by wrinkle ridges similar to those of the lunar maria.

THE VIKING AND PATHFINDER LANDERS

The Viking landers were each self-contained laboratories weighing about a ton and similar in size to a subcompact car. Powered by a radioactive electric generator, each lander could communicate with Earth either directly or via one of the orbiter spacecraft. The scientific experiments on board included those aimed toward a general examination of the immediate environment (cameras, chemical analysis devices, a meteorological weather station, a seismometer, and a 3-m-long mechanical arm to probe and manipulate the soil) as well as a complex biology package to look for evidence of microscopic life on Mars. In addition, a special group of instruments investigated the atmosphere of Mars during spacecraft entry and descent.

Twin cameras on each Viking lander could record pictures in either color or monochrome, and operating together they provided stereoscopic views of the surface. The weather station measured atmospheric pressure, temperature, and wind speed and direction. Chemical analysis of soil was carried out with an x-ray spectrometer. An even more powerful general analysis device called the GCMS (gas chromatograph–mass spectrometer, discussed in more detail in Chapter 12) measured the composition of the atmosphere and of volatile components in the soil, such as organic compounds.

Each Viking lander was operated by controllers on Earth, who could command it to poke and dig in the soil, push rocks around, pick up and deliver samples to its various instruments, and take pictures of its surroundings. The Viking landers lacked mobility, however. Fixed at the spots where they landed, they could not show us the view over the next hill or even behind a nearby rock.

The Sojourner rover provided the first demonstration of mobility, although it was designed to move only slowly around the immediate landing site. Usually one or two days were required to move a few meters from one rock to the next, under the control of operators on Earth. About the size of a microwave oven, Sojourner weighed 12 kg and was powered by solar cells (see Fig. 11.6). It contained a camera and a device for determining the elemental composition of rocks. The Pathfinder lander also carried its own camera and a set of meteorological instruments.

SELECTION OF THE LANDING SITES

The one imperative in selecting landing sites on Mars was safety. Just as had been the case for the first Apollo landings on the Moon, every effort was made to find dull, flat locations for the Viking and Pathfinder landers. This meant avoiding all the remarkable geologic features found by Mariner 9: No sites involving volcanoes, canyons, channels, large impact craters, or anything else of geologic interest were considered. In addition, it was necessary to select sites at relatively low elevations to make sure there was enough atmosphere to reduce the speed of the descent. The only places that met these criteria were the lowland plains of the northern hemisphere.

The site selected for the Viking 1 landing was in Chryse Planitia, a depression east of Tharsis. In Mariner 9 images, this region appeared flat and featureless, but the higher-resolution pictures from

(a)

(b)

(c)

Figure 11.16 The martian surface as seen by three landers, all of them targeted to the northern lowland plains because the spacecraft parachute systems were not effective at high-altitude locations. **(a)** Viking 1 landed on a desolate plain with wind-sculpted hills. **(b)** Viking 2 touched down in a plain heavily littered with rocks. **(c)** Pathfinder was targeted to a place where a variety of rocks was expected.

the Viking 1 orbiter revealed a much more complex topography, including features that appeared to be the result of water erosion. The Viking project manager rejected the preplanned site and began a search for a smoother area. It required nearly a month to identify a less rugged looking alternative, still in Chryse but several hundred kilometers northwest of the original site (22°N, 48°W).

Viking 2 was aimed toward a more northerly latitude of 44°, where seasonal effects might be stronger. Again, the original site proved to be unexpectedly rough, and a search was made for a smoother area at the same latitude. A spot in Utopia, at longitude 226°W, was found.

Twenty years later, the Pathfinder spacecraft was sent to Ares Vallis at 19°N, 34°W, a region near Chryse Planitia but closer to the outwash channels than the Viking 1 site. Unlike the Viking landers with their rocket-assisted gentle descent, Pathfinder came down hard, surrounded by huge air bags that allowed it to bounce over the landscape before coming to rest. At that point, the air bags deflated and the Sojourner rover was deployed. All three landing sites are shown in Figure 11.16, as imaged by the lander cameras.

Plans for the next set of missions call for more precision in targeting the landers and greater range for the rovers. A prime goal is to land in a region where orbital imagery has revealed the deposition of waterborne sediment, such as the interior of a crater that was once a lake.

THE PLAINS OF GOLD

The Viking 1 site in Chryse is illustrated in Figure 11.15. This orbiter view shows a marelike plain with a crater density suggestive of an age of about 3 billion years. Like the lunar maria, this area is probably volcanic, but with a surface subsequently modified by later floods of water (see Section 11.5) as well as by wind erosion.

As seen from the lander, Chryse is a desolate but strangely beautiful landscape not very different from some terrestrial deserts (Fig. 11.16a). The topography is gently rolling, and the surface is thickly strewn with rocks in sizes ranging from golf balls up to boulders. The soil among the rocks is fine-grained and has a relatively hard, crusty surface, like what are called desert duracrusts on Earth.

The ubiquitous rocks at the Viking 1 site appear to be volcanic in origin, and they probably are ejecta from impact craters, just like their lunar counterparts. The scene has been modified by erosion, particularly by windblown dust. Erosion takes place slowly, however, and only a keen observer can see any differences between photos taken years apart at the Viking 1 site.

UTOPIA

Viking 2 landed in the Utopia plain on ejecta from the 90-km-diameter crater Mie, located about 200 km to the east. As seen from the lander, this site is even rockier than Chryse (Fig. 11.16b). Indeed, we are fortunate that the lander was not wrecked on a protruding boulder as it settled to the surface. About all there is to see are the rocks. These are interesting enough to the geologist, however, with their distinctive angular shapes and wide variety of colors and textures. Most are heavily pitted, perhaps an indication of their volcanic origin. The finer material that must once have been present in the ejecta from crater Mie has been stripped away by wind, leaving the fragmented rocks.

While every effort was made to land both Vikings in flat regions with smooth surfaces, it is apparent from the pictures that this goal was not achieved. Both sites are rocky, with evidence of wind erosion. It may be that the two Viking sites are more typical than had been expected, and that a great deal of the surface of Mars consists of rock-strewn plains like these.

ARES VALLEY

The Pathfinder landing site indeed looked a lot like the locations visited by the Viking landers (Fig. 11.16c). Once again we find a rocky landscape, this time relieved by some distant hills. Sojourner visited some of the nearby boulders to determine their composition, and these particular rocks quickly acquired names such as Yogi, Barnacle Bill, and Scooby Doo so scientists could easily tell which object they were investigating.

The primary rationale for the Pathfinder site was to examine a variety of rock that (geologists hoped) had been carried from the highlands to this site by ancient floodwaters. Looking at the angular rocks in the surface photos and measuring how they were aligned, geologists concluded that their expectation had been met. However, the half-dozen chemical analyses carried out by Sojourner showed no differences in chemical composition: All the measured rocks were igneous (see below), although they were not the basalts that had been expected based on the two Viking landing sites.

THE SURFACE MATERIAL

Mars is the "red planet," but what accounts for its color? The Viking landers confirmed what was suspected for decades: The color of Mars is due primarily to iron oxides in the surface soil. The soil itself is largely fine-grained dust, which can be transported for great distances during major dust storms. Chemical analyses carried out by the Viking landers showed that the martian soil is similar in composition to some iron-rich clays found on Earth, but these spacecraft were not equipped to analyze the rocks themselves. This experiment was carried out by the Sojourner rover, which discovered that the rocks it investigated did not exhibit the basaltic composition that was expected. Their high silicon content is more similar to the granitic composition of the continental crust on Earth. This was a surprise to geologists because Mars does not exhibit the plate tectonic activity that has produced such rocks on Earth. Are these rocks the fragments of ancient martian crust? Or are they simply a local anomaly? More studies of martian rocks at different sites will be necessary before this puzzle can be unraveled.

Most of the fine dusty material is light in color, like the sand in the deserts of Earth. During dust storms, this light material is deposited over the surface, obscuring underlying rock, much of which is darker. Thus, after a major dust storm, the overall reflectivity of the martian surface is high. Subsequently, surface winds redistribute the lighter material, and part of the darker underlying terrain reappears. As we will see in Section 11.7, even small dust devils can strip off the thin layer of light dust to leave distinctive dark streaks on the ground.

This process of dust transport by wind, distributing fine material over the martian surface by seasonal circulation patterns of the atmosphere, accounts for the seasonal changes in the surface markings recorded by generations of telescopic observers. Among the first scientists to advocate this nonbiological explanation for the surface changes were Carl Sagan and James Pollack, whom we encountered in the last chapter when discussing the Venus greenhouse effect. It is somewhat ironic that Sagan, who is so closely associated with the search for life on other planets, should have played this role in the history of martian studies.

High-resolution photographs show literally millions of sand dunes composed of windblown sand and dust. Some of these pictures are exceptionally beautiful, such as the pattern of dark dunes shown on a lighter wind-rippled surface in Figure 11.17.

MARS SAMPLES

Although we have not sent astronauts or robotic spacecraft to collect rocks from Mars, we do have about 20 martian samples to study. These are the martian meteorites, which we briefly discussed in Section 4.5. Most of these meteorites are basalts with solidification ages near 1.3 billion years, presumably derived from one or more impacts in martian volcanic areas. One, however, is much older, dating back more than 4 billion years. We are confident that the meteorites come from Mars in part because they preserve tiny bubbles of gas that match perfectly the analysis of the martian atmosphere carried out by Viking. There is no explanation for the apparent difference in composition from the rocks measured by Sojourner in Ares Valley.

Like the lunar basalts, the martian meteorites are derived from the upper mantle of the planet by partial melting. Detailed chemical analysis of these rocks, together with Viking and Pathfinder data and remotely sensed information on the composition of the soil, permits some educated guesses concerning the chemistry of the crust of Mars. Ratios of isotopes of elements in these meteorites can

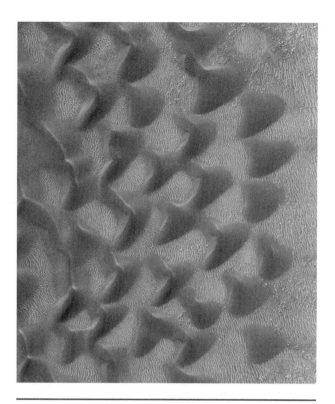

Figure 11.17   Dark sand dunes in Proctor Crater. Each dune in this high-resolution image is about I km across.

also be used to compare the bulk properties of Mars with those of Earth and the Moon.

The oxygen isotopes in the silicate minerals found in these Mars samples have different relative abundances from the value found identically on both Earth and the Moon. Mars clearly did not participate in the giant impact that formed our satellite. Furthermore, the oxygen isotopes have different abundances from the isotopes found in chemically bound water in these same rocks. This is completely different from the situation on Earth, where oxygen in water and in rocks is identical. Evidently there is much less mixing on Mars, again consistent with the absence of a vigorous system of plate tectonics such as we have on Earth. This lack of mixing is also manifested in the dryness of martian magmas, as revealed by these same basaltic meteorites. On Earth, surface water is repeatedly recycled through the upper mantle by subduction of the crust, and terrestrial magmas are consequently much wetter than those on Mars.

The martain meteorites have also been investigated for evidence of biological activity on the red planet and for the record of atmospheric evolution they must contain. We will return to these topics in Chapter 12.

MEMORIALS ON MARS

Each Viking lander had a nominal lifetime of only 90 days, just sufficient to complete the life-detection experiments described in Chapter 12. However, scientists and engineers alike hoped for much longer operations on the martian surface, and they were not disappointed. The Viking 2 lander operated in the Utopia plain until April 11, 1980, spanning two martian years. In spite of its far northern latitude, the spacecraft survived two cold martian winters to provide unique meteorological information.

The first lander performed even better, and at the end of the regular Viking mission it was still in excellent condition. NASA then placed the lander into an autonomous operating mode, so that it returned data to Earth about once a month. This lander was designated the Thomas Mutch Memorial Station, named in honor of the Viking imaging team leader who later became the associate administrator of NASA in charge of planetary exploration (Fig. 11.18a). It was hoped that the Mutch Memorial Station could continue operations until the 1990s, but it fell victim to a human programming error that misdirected its antenna away from Earth, breaking the vital communications link with controllers at JPL. The last data were received on November 5, 1982, six Earth years after landing on Mars.

In 1984, NASA formally transferred ownership of the Viking 1 lander to the Air and Space Museum of the Smithsonian Institution. It is the only museum exhibit on another planet. Someday perhaps it will be brought back to a place of honor in Washington, next to the Wright brothers' airplane and the Apollo 11 spacecraft. Or perhaps it will remain on Mars, for the benefit of the first human explorers and the generations that succeed them.

The Pathfinder mission was much shorter. The mission was originally designed to test new technologies, not as a science mission. Nevertheless, many scientific results were obtained during the 85 days before communication was lost for unknown reasons. This was nearly three times the expected lifetime of 30 days for the lander and 12 times the expected lifetime of seven days for the Sojourner rover. The lander was officially designated as the Carl Sagan Memorial Station, commemorating the scientist who did more than any other to popularize astronomy during his lifetime, while making many important contributions to planetary science (Fig. 11.18b).

(a)

(b)

Figure 11.18 **(a)** Thomas Mutch, leader of the Viking lander imaging team. The Viking 1 lander was named in his honor. **(b)** Carl Sagan (right) shown here with planetary geologist Bruce Murray, Director of JPL at the time of Viking. Sagan provided much of the scientific inspiration for Mars exploration, and the Pathfinder lander was later named in his honor.

11.4 VOLCANOES AND TECTONIC FEATURES

Like the other terrestrial planets, Mars has experienced extensive periods of volcanism. Most of the low-lying northern hemisphere appears to have been covered by mare-type flood basalts in the distant past. The most spectacular evidence of volcanism, however, is found in the shield volcanoes associated with the Tharsis bulge. They are the largest individual volcanic structures in the solar system (see Figs. 11.9 and 11.10).

SHIELD VOLCANOES

The largest volcanoes of Mars are all shield volcanoes, formed by the eruption of relatively fluid basaltic lavas, which build up massive domes with gentle slopes (see Fig. 9.6 for a terrestrial example). Most shield volcanoes on both Earth and Mars are also distinguished by summit craters or calderas, produced by collapse of the surface when the underground pressure of the magma is withdrawn.

The four highest volcanoes are in Tharsis (see Fig. 11.9). The story of their discovery is an interesting one. As we noted in Section 11.1, Mars was shrouded by a planetwide dust storm when the Mariner 9 orbiter arrived in 1971. At that time, there was no reason to expect large volcanoes on Mars. As the dust began to clear from the upper atmosphere of Mars, the first surface features to appear were the tops of these four highest mountains, each with its large summit crater. Thus, scientists faced an essentially featureless planet with four faintly visible large craters.

On the basis of just this limited information, Harold Masursky of the U.S. Geological Survey reached the correct conclusion that Mars had volcanoes and that these volcanoes were much larger than their terrestrial counterparts. He thought that if this were not the case, the coincidence of large craters being found at just the four spots that first emerged out of the dust was too unlikely. Note that this is just the mirror image of the reasoning used by Gilbert to argue against volcanic craters on the Moon: Gilbert concluded that since the craters of the Moon were not on the tops of mountains, they were not volcanic, whereas Masursky reasoned that if the craters were on the tops of mountains, they were volcanic. Subsequently, as the atmosphere cleared to reveal the great shield volcanoes, Masursky's insight was confirmed.

Each of the three central Tharsis shield volcanoes is about 400 km in diameter, and they all rise to the same height. The southernmost volcano, Arsia Mons, has the largest caldera, 120 km in diameter; the other two calderas are less than half as large. Olympus Mons, which rises from a lower area on the side of the Tharsis bulge, is the largest volcano on Mars or, indeed, on any planet. Its diameter is 700 km, and the volume of this immense mountain is nearly 100 times greater than that of the largest volcano on Earth, Mauna Loa. Its caldera, about 80 km across, shows multiple collapse craters with no sign of erosion (Fig. 11.19). The lava flows that make up its flanks slope downward at an angle of only about 4°. Individual lava channels about 100 m in width can be traced, together with many flow features that

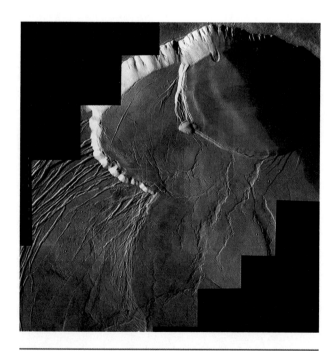

Figure 11.19 The caldera of Olympus Mons. With a diameter of about 80 km, this caldera consists of several interconnected, flat-floored depressions, similar to the calderas of terrestrial shield volcanoes. Note the lava flows extending radially out from the caldera.

are virtually indistinguishable from their counterparts in the Hawaiian volcanoes.

The four volcanoes we have been discussing are the largest on Mars, but even modest martian volcanoes are huge by terrestrial standards. Figure 11.20 shows two shield volcanoes in the Tharsis region, the largest about 100 km in diameter. The tectonic cracks also visible in this image will be discussed below.

Hundreds of small shield volcanoes can be identified on Mars as well. We also see many small steep-sided cones apparently formed by eruptions that were more explosive. However, there are no viscous eruptions such as the pancake domes of Venus, nor any examples of larger composite volcanoes like those of Earth.

TECTONICS AND THE THARSIS BULGE

How does the Tharsis bulge compare with the continental masses of Earth and Venus? In part, the Tharsis bulge results from the accumulation of volcanic lavas over billions of years, but in addition it appears to be the product of tectonic forces that have bent the crust and uplifted an area as large as North America or Aphrodite.

The extensive fracturing of the surrounding crust shown in Figure 11.20 provides dramatic evidence of the tectonic forces that shaped Tharsis. Most of these fractures, which can be hundreds of kilometers long and a kilometer or more in width, point away from the center of the Tharsis bulge. Detailed studies of the apparent ages of these tectonic features based on crater densities indicate that the uplift began about 3 billion years ago and that the surface fracturing tapered off about 1 billion years later. If these studies are correct, the uplift largely preceded the formation of the large Tharsis volcanoes.

The continents of Venus and Earth are the product of compressional forces, resulting in folded and uplifted mountains (like Maxwell or the Himalayas) and plateaus (like Lakshmi or Tibet). The Tharsis bulge on Mars more nearly

Figure 11.20 Two of the intermediate-sized Tharsis volcanoes are shown in this Viking orbiter mosaic of photos, together with a region of extensive tectonic fracturing associated with the formation of the Tharsis bulge. The width of this frame is about 600 km.

Figure 11.21   Valles Marineris. View of two of the canyons showing tectonic cracks and numerous landslides.

Figure 11.22   Noctis Labyrinthus, a region of intersecting closed depressions possibly caused by a combination of tectonic fracturing and collapse associated with the melting of permafrost. The image is 120 km across.

resembles the top of a mantle plume, perhaps analogous to an upscaled corona structure on Venus. This close association of tectonic uplift with massive volcanism is unique to Mars. Thus, Mars alone of the terrestrial planets has volcanoes, rather than tectonic mountains, as its tallest features.

The most spectacular tectonic features on Mars are the great equatorial canyons, especially the Valles Marineris (Fig. 11.21). These canyons, each as much as 100 km wide and up to 7 km deep, stretch a quarter of the way around the planet. The upper parts of the canyon system, near the summit of Tharsis, are called Noctis Labyrinthus (Labyrinth of Night) and consist of short intersecting segments that produce a labyrinth (Fig. 11.22). Farther down, the Valles Marineris descends the eastern flank of the bulge

and extends into the old, cratered uplands east of Tharsis (see the chapter-opening photo).

The term *canyon* is really a misnomer because the martian canyons were not carved by rivers. They do contain some features characteristic of running water, but they do not terminate in lowland basins. They are also distinct from the channels, which are discussed in the next section. The canyons are tectonic in origin, basically huge cracks in the planet's surface that subsequently have been widened and shaped by erosion. Figure 11.23 shows an area with several landslides tens of kilometers across. Note on the far wall that the upper part of the slide consists of blocky material, while the lower part of the slide shows evidence of a more fluid flow across the canyon floor, probably at speeds greater than 100 km/hr. Just as in the case of the fluidized impact crater ejecta, subsurface ice may have helped to lubricate these flows.

Landslides and possibly undercutting water springs have played a major role in widening the original fractures, but where has the eroded material gone? Some of it was subsequently deposited in side channels where water evidently pooled and

Figure 11.23 Landslides in Ophir Chasma in the martian canyonlands. Note the flow patterns in the large fan-shaped slides and the indication of a harder cap rock at the top of the canyon wall. The width of the frame is about 100 km.

dropped whatever sediment it carried. The resulting layers or strata are clearly visible in images obtained by the Global Surveyor spacecraft (Fig. 11.24). Perhaps some of the missing material was water ice, which is believed to make up a sizable fraction of the crust. The finest material would have been carried away by winds whistling down the canyons.

Figure 11.25 is one of the most spectacular images obtained from Global Surveyor, showing layered rock outcrops within Candor Chasma. These are small wind-eroded mesas that are hardly visible at all in the Viking images. The layers (nearly 100 visible) could be either sedimentary rock or lava flows deposited before the Tharsis bulge began its uplift, billions of years ago. Some geologists interpret this photo as evidence of deposit of sediment on the floor of an ancient martian lake or sea, in a dynamic underwater environment.

THE NATURE OF MARTIAN VOLCANISM

Geologists such as Michael J. Carr of the U.S. Geological Survey, who was leader of the Viking orbiter imaging team, have carried out detailed

Figure 11.24 Layering in the canyon walls. This Global Surveyor image of the wall of Ius Chasma distinguishes features as small as a few meters across. The cliff, which is about 1000 m high, shows evidence of about 80 layers or strata, each between 5 and 50 m thick. The lighter material in the image is wind-blown sand or dust.

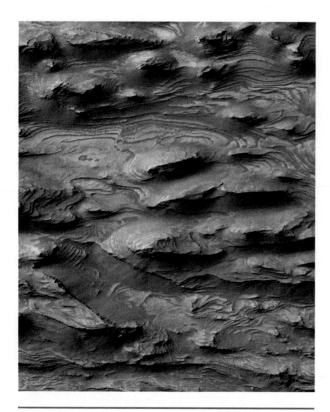

Figure 11.25 Is this sedimentary rock? As many as 100 layers of rock are visible in this high-resolution image of wind-eroded mesas within Candor Chasma. The width of this image is only 1500 m. Many geologists interpret this photo as evidence for layers of sediment deposited in martian lakes or seas billions of years ago, before the tectonic events that formed the Tharsis bulge and the martian canyon system.

studies of the martian volcanoes. Even without surface measurements of chemical composition, Carr and his colleagues have concluded from the shapes of the volcanoes and the behavior of their flows that they were produced by basaltic volcanism similar to that which formed the Hawaiian volcanoes. Individual flows extend for great distances, indicating that the lavas had fairly low viscosity and that the rate of flow during eruptions was large. However, there are no equivalents to the remarkable lava rivers of Venus.

Estimates of the ages of different flows suggest that the eruptions spread over vast time spans, with individual volcanoes active for hundreds of millions or even billions of years. Most of the martian meteorites, with their solidification ages of 1.3 billion years, presumably are

samples of some of these flows, probably from the Tharsis area. The scars from the impacts that ejected these samples from Mars must be visible, but of course we do not know which craters produced these remarkable rocks.

The concentration of volcanoes in the Tharsis region is consistent with the idea that the Tharsis bulge is the result of a mantle hot spot. In this case, the same forces that raised the bulge provided the heat to generate the volcanoes. It is also easy to see why the martian volcanoes are so much larger than those on Earth. If there was no plate tectonic motion, the crust remained stationary over the hot spot. Thus, each eruption was superimposed on the previous one, and the volcanoes just kept growing over billions of years.

Are any of the martian volcanoes still active? There is no evidence of eruptions taking place right now, but we know that active terrestrial volcanoes can remain dormant for decades. Carr and others have determined from crater counting that the youngest lava flows on Olympus Mons are less than 100 million years old. One hundred million years is a very short time on the scale of planetary history, and there is no reason to think a mantle hot spot that has been active for much of the past billion years or more will not continue to induce volcanic eruptions episodically in the future. Therefore, scientists conclude that Olympus Mons probably remains intermittently active, although the intervals between major eruptions could be thousands or even millions of years.

11.5 MARTIAN CHANNELS

No geologic discoveries on Mars have equaled the sinuous channels for either excitement or mystery. Impact craters, volcanoes, tectonic fractures, canyons—all of these might have been expected, and each has counterparts on other planets. The channels, however, speak to us of running water on Mars, a phenomenon previously thought to be unique to Earth. From a study of the martian channels, we hope to illuminate the larger questions of the history of water on Mars and the possibilities that the planet could once have supported water-based forms of life.

The two main types of ancient channels on Mars are called runoff channels and outflow channels. Both were formed long ago by running water, in contrast to the tectonic canyons discussed in Section 11.4. A third kind of channel, visible only in the highest-resolution photos, looks like recent gullies carved by intermittent flows of water that may be continuing today.

RUNOFF CHANNELS

The martian features most like terrestrial dry river valleys are the **runoff channels,** found exclusively in the heavily cratered uplands of the southern hemisphere. Over much of this old terrain, channels are common, usually consisting of simple valleys tens to hundreds of meters wide and a few tens of kilometers long, often on steep slopes such as crater walls.

More interesting than these simple valleys, because they are more clearly indicative of running water, are the *networks* of interconnected runoff channels. Dozens of major networks have been identified, typically several hundred kilometers in length. They have the characteristic form of terrestrial river systems, with small tributary channels merging into larger channels to provide drainage for a wide area (Fig. 11.26). Compared with terrestrial drainage systems, however, even the best-developed martian runoff channels show relatively few tributaries.

In some cases, the smallest channels have been obliterated by subsequent erosion, and there are also instances where the water originated from underground springs rather than from direct surface runoff. Similar looking channels are formed on Earth by this process of sapping from springs. The process may begin with water seeping out at the base of a cliff from a subsurface reservoir or aquifer. As the water escapes, it carries some soil with it, and the cliff face above the flow begins to crumble. Thus, the flow eats its way back into the surrounding terrain, forming a channel with a characteristically blunt appearance. The process continues until the groundwater reservoir is depleted or the spring is stopped by freezing. The gullies described further below may be an example of this sort of erosion.

Figure 11.26 Drainage network of runoff channels in the old highland terrain near 48°S latitude. These finely divided and branching erosional features were almost surely formed at a time when there was liquid water on the surface of Mars. The width of the image is 250 km.

ORIGIN OF THE RUNOFF CHANNELS

The shapes of the channel networks and the way they drain from higher into lower areas point inescapably to erosion by liquid water. Many geologists have sought alternative explanations involving lava, wind, and ice, but none has survived detailed criticism. Unless we are badly misinterpreting the data, Mars once experienced real rivers fed primarily by underground springs.

When did the rivers flow? An important clue is the restriction of these channels to the cratered uplands. There are no such channels in the lower northern plains, which formed at about the time of the lunar maria, and none in any of the volcanic areas such as Tharsis. Apparently, conditions have not been suitable on Mars for the formation of runoff channels during the past several billion years.

Detailed crater counts in the areas drained by the runoff channels confirm this timescale. The climatic conditions that permitted rivers to flow ceased before the formation of the northern plains or the emergence of the Tharsis bulge. According to the standard chronology, this puts the age of the rivers back about 4 billion years. Thus, the channels we now see were active near the time of the terminal heavy bombardment, which is also the period suggested for enhanced levels of surface erosion. This association makes sense. Running water requires a thicker, warmer atmosphere than Mars has today, which would also have contributed to more erosion than is taking place under present conditions.

OUTFLOW CHANNELS

Even larger than the runoff channels are the martian **outflow channels.** These tremendous valley systems are confined to the equatorial parts of Mars, primarily leading from the southern highlands into the northern plains. Like the runoff channels, they appear to have been carved by running water. However, in the case of the outflow channels, the flow was probably intermittent and catastrophic, and the source of the water remains somewhat mysterious.

The largest and best-studied outflow channels lie north and east of the martian canyonlands and drain into the Chryse Basin. These include Kasei Valley, Maja Valley, Simud Valley, Tiu Valley, and Ares Valley (where Pathfinder landed). Each major outflow channel is at least 10 km wide and hundreds of kilometers long, and each drops several kilometers in elevation along its length (Fig. 11.27).

Over most of their lengths, the outflow channels have cut multiple parallel channels that diverge and interconnect, each showing characteristic patterns of water erosion such as terraced walls, streamlined islands, and sandbars. In places, the flow seems to have broken out of these well-defined channels and spread over areas tens of kilometers wide.

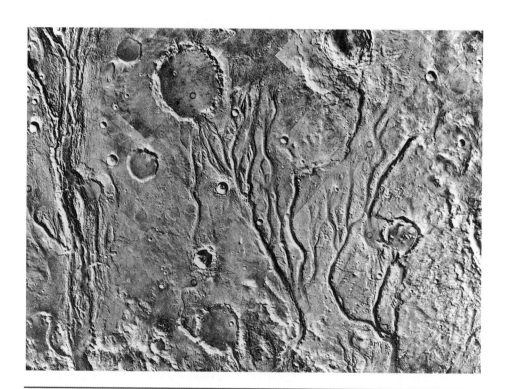

Figure 11.27 Several of the large outflow channels that drain from the southern highlands toward the northern basins. These appear to be the product of catastrophic water floods that took place about 3 billion years ago. The width of the image is 400 km.

Teardrop-shaped islands and other sculpted landforms characterize the mouths of the outflow channels, where they descend from the uplands into the Chryse Basin (Fig. 11.28). The islands are remnants of the plateau, shaped by the flood of water rushing out into the lowland plains. Some of the rocks visible in Pathfinder and Sojourner images show the rounded shapes commonly associated with water erosion on Earth.

CATASTROPHIC FLOODS

The scale of the martian outflow channels and the evidence of massive flows of water dwarf terrestrial rivers. Estimates of the rates of flow in the larger channels are a hundred times greater than the flow from Earth's greatest river, the Amazon. To seek comparable rates of flow on Earth, we must look to catastrophic floods caused by the sudden failure of natural dams holding back large lakes.

The best-studied example of such a flood on Earth is to be found in Washington and Montana, where 18,000 years ago an ice dam broke, releasing the water in a prehistoric lake called Lake Missoula. Within a few days the entire lake emptied westward across the Columbia Plateau, generating flows up to 120 m deep over a front tens of kilometers wide. The awesome force of this water erased previous surface features in western Washington State, cutting new channels up to 200 m deep through the bedrock. The resulting terrain, called the channeled scabland, is very different in appearance from that generated by slow, steady processes of water erosion. With its multiple parallel channels, teardrop islands, and huge bars, the channeled scabland is the closest analogue we have to the martian outflow channels.

Presumably the martian features were formed in a similar way—by brief catastrophic floods arising in the uplands and dissipating in the broad northern basins. But where could such vast quantities of water have come from, and when did these remarkable events take place?

CHAOS: THE SOURCES OF THE OUTFLOW CHANNELS

To find the origin of the martian floods, we must look to the sources of the outflow channels. These are strange landscapes called **chaotic terrain.** They resemble regions where the surface of the old cratered uplands has collapsed into a jumble of irregular, rugged hills and valleys (Fig. 11.29). Both the chaotic terrain and the great floods could have a common origin in the sudden release of huge reservoirs of underground water, mud, or ice.

When did these events take place? Because the outflow channels extend into the lightly cratered northern plains, we can see that they represent a later era than that of the runoff channels. Crater counts in Chryse indicate that the deposits from

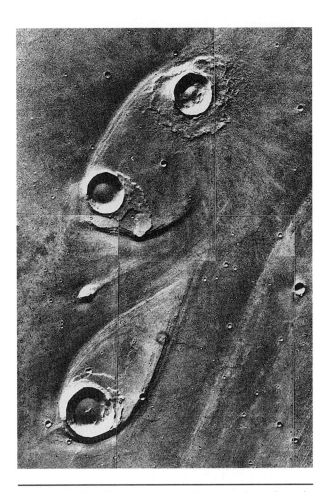

Figure 11.28   Flow features, such as teardrop-shaped islands, mark the region where several outflow channels spread out into the lowland basin of Chryse, near the site originally targeted for the Viking 1 landing on Mars. Each island is approximately 40 km long.

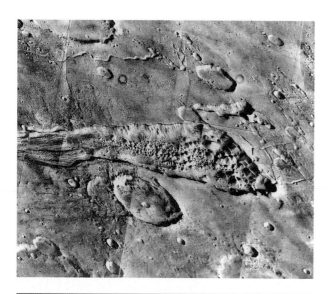

Figure 11.29   The chaotic terrain at the head of an outflow channel system. Evidently a local collapse of the surface was associated with the generation of a massive flood of water. The image is about 400 km across.

Figure 11.30   Is this evidence for sustained water flow? This Global Surveyor view of a part of Nanedi Vallis shows what appears to be a 500-m-wide meandering river channel within the 2.5-km-wide outflow channel. The image is about 10 km across.

channel outflows occurred there at just about the same time the volcanic plains were formed, which is probably about 3.5 billion years ago. This is also the time when internal forces were producing the Tharsis uplift.

While most geologists agree that the release of underground water generated the floods and produced the chaotic terrain, there is no consensus concerning the details of the mechanism. One possibility is the melting of subsurface ice by deep-seated volcanic activity. Another suggestion is the chemical release of water loosely bound to the clays of the martian soil. A third invokes the shifting of large quantities of underground water in response to the Tharsis uplift. Possibly all of these processes were involved.

There is intriguing evidence that some of these flows continued at a reduced rate after the major floods that carved the channels. Figure 11.30 is a high-resolution view of one branch of Nanedi Vallis. The valley is about 2.5 km wide. The upper part of this image shows a narrow twisting channel that resembles a terrestrial streambed rather than the result of a brief catastrophic flood. The existence of such features is an argument for the flow of "rivers" inside the flood-carved channels, perhaps rivers beneath a covering of ice.

Both the magnitude of the outflow channels and the huge depressions of chaotic terrain left behind indicate that water or ice must have made up a sizable fraction of the martian crust. Large quantities of ice are still present below the surface, but the conditions necessary to produce catastrophic floods ceased billions of years ago. We shall examine the fate of that water in Section 11.7.

GULLIES: WATER ON MARS TODAY?

The most exciting images of Mars obtained by Global Surveyor reveal freshly cut gullies on the steep inner slopes of some crater walls and runoff channels (Fig. 11.31). The comparative youth of

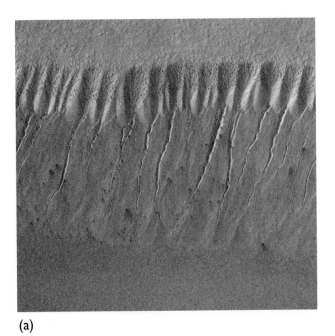

(a)

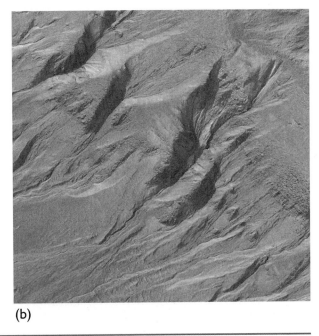

(b)

Figure 11.31   Gullies on Mars. With the greatly improved resolution of the Global Surveyor cameras, we can begin to see evidence for recent water erosion. These two images (each about 2 km across) show gullies of young geologic age, apparently carved by flowing water. **(a)** Gullies on a cliff near the south polar cap. **(b)** Gullies on the wall of Newton Crater.

these gullies is demonstrated by the crispness of their edges and by the fact that the debris aprons produced at their bases lie on top of windblown deposits that must themselves be relatively recent.

These gullies have been found only at high latitudes on the shaded walls of craters and channels, some of the coldest places on Mars. It is paradoxical that this apparent evidence for recent flows of liquid water should exist only in regions where water seems most unlikely to be present. However, one scenario for the formation of these gullies explains how this could happen. A key feature of this explanation is the realization that for water to cut these gullies, it must be released explosively, because a slow dribble will evaporate instantly into the thin martian atmosphere.

It is at high latitudes on Mars that we expect the permafrost to be closest to the surface. If a hot magma plume approaches the surface in this region of the planet, it will cause extensive melting of the permafrost, perhaps generating steam under pressure. Water will be impeded from upward movement by the overlying rock layers and will try to escape laterally as the pressure increases.

Approaching the crater or channel wall, it will readily sublime into the atmosphere from regions warmed by the Sun, but it may freeze into an impermeable ice plug behind the frigid walls that lie in perpetual shadow. Eventually the pressure of the heated water and steam will be sufficient to blast the ice plug loose and the liberated water will swoosh out, driving debris down the face of the wall to create the gully.

If this interpretation of these gullies is correct, it suggests that in certain underground areas on Mars, liquid water is present today. This new perspective changes our ideas about the possibilities for life on Mars.

11.6   THE POLAR REGIONS

The geologic features of Mars described in the previous sections of this chapter are generally confined to equatorial and middle latitudes. Above about 70° north and south, the surface takes on a different character strongly influenced by the polar caps.

PERSPECTIVE

THE FACE ON MARS

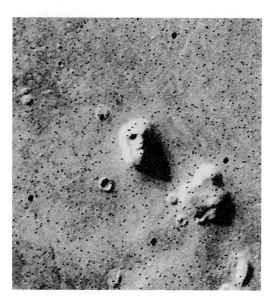

Figure 1

Scanning new images of the martian surface with a magnifying glass to look for a suitable landing site for Viking 1 in 1976, author Owen was startled to find a face on the planet looking back at him. This "face" consists of a natural rock formation, enhanced with a data drop that created a black "nostril" at just the right place (Fig. 1). It was an amusing sight and was delivered as such to the media.

The result was sadly predictable. This charming accidental creation was immediately interpreted by many eager Americans as a huge sculpture made by those same intelligent Martians who have become so firmly planted in popular mythology. More reasonable people pointed out the harmless human tendency to find images in natural landscapes, from the Great Stone Face in New Hampshire to the Crouching Lion in Hawaii. Such efforts at rationality were greeted with cries of conspiracy. NASA was accused of suppressing the true explanation for its own devious ends.

During the two-decade hiatus in successful Mars missions that followed Viking, a cult developed centered on the idea that this really was a human face carved by intelligent Martians, perhaps part of a scheme to communicate with Earth. With the passage of time, the story grew more convoluted, complete with ruined pyramids and cities. A few proponents assert that these martian features can reveal, to a chosen few, the mathematical foundation of an extraterrestrial technology. The story later expanded to include a crystal dome

➡

SEASONAL POLAR CAPS

The polar caps seen through a telescope—the white surface deposits that grow and shrink with the changing seasons—are called the seasonal caps (see Fig. 11.2). They are analogous to the seasonal snow cover on Earth. We do not usually think of the snow that covers the ground in winter as a part of Earth's polar caps. If we looked at our planet from space, however, the seasonally changing snow cover would be obvious, and the area involved much larger than that covered by the permanent ice caps. It is the same for Mars.

The substantial eccentricity of Mars's orbit carries the planet farther from the Sun during southern winter, resulting in long, cold winters and a large southern seasonal cap, extending down to about 55°S. The northern cap, half a martian year later, extends only to about 65°, although some surface frost was seen as far south as the Viking 2 landing site, at 48°N. During spring, many miles high photographed on the Moon by Apollo astronauts, and cult members pour over transcripts from old NASA missions looking for oblique references to the presence of alien artifacts. The makers of the face were also linked to crop-circles in Britain, which showed that the alien presence is with us today, at least in the United Kingdom.

According to the conspiracy cult, the face was the most important discovery of the entire space program. When NASA's Mars Observer spacecraft failed while approaching Mars in 1990, conspiracy believers announced that this was really a secret mission to contact the aliens, and its failure was faked. A few even organized a protest outside the gates of JPL.

The popular commotion about this innocent feature became so intense that scientists

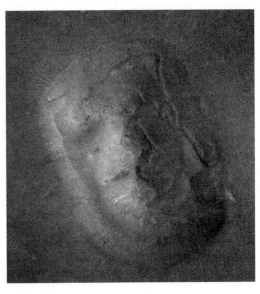

Figure 2

on the Mars Global Surveyor mission deliberately observed the feature under several different conditions of illumination during 1998. Even better photos were obtained in 2001. With the much higher resolution of the Global Surveyor camera, the natural character of the face on Mars became obvious to everyone (Fig. 2). Well, almost everyone. The conspiracy cult had invested so much into their beliefs that they could not accept the new results at face value. Some said that NASA had faked the image. More imaginative was the suggestion that the earlier Mars Observer's secret mission had been to drop a nuclear bomb on the face to destroy it before other Mars missions could reveal its true alien heritage. Why anyone imagines that an exploration agency like NASA would want to suppress information about life on Mars is another mystery, which we won't pursue here.

each cap retreats at a rate of about 20 km/day, reaching its minimum size in early summer.

Both seasonal caps are composed of CO_2 frost or dry ice. Since CO_2 is the main component of the atmosphere, it is always available to condense on the surface whenever the temperature falls below 150 K, the freezing point of CO_2. From changes in atmospheric pressure as the caps form and evaporate, the thickness of the seasonal deposit has been calculated to vary from a meter or so at high latitudes down to a few centimeters near its edge. This is less, but not much less, than the corresponding thickness of the seasonal snow cap on Earth.

PERMANENT POLAR CAPS

As the seasonal polar cap retreats during southern spring, it reveals a brighter, underlying permanent cap that persists through the summer months (Fig. 11.32). The diameter of this residual south polar

Figure 11.32   Residual south polar cap, 250 km in diameter and composed of both frozen CO_2 and H_2O.

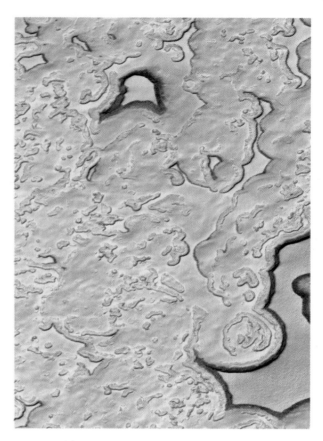

Figure 11.33   Detail of south polar residual cap.

cap is about 350 km, and it is located a short distance away from the true south pole, with its center at 30°W, 86°S. Measurements of its surface temperature made from orbit show that the residual cap remains at 150 K, the frost point of CO_2, throughout the summer, indicating that it is composed in part of CO_2. Water ice is undoubtedly present as well, perhaps in large quantities, but the temperature of the cap is controlled by the CO_2 ice, and this cold reservoir is sufficient to maintain itself in spite of the continuous summer sunshine.

The residual southern cap has a remarkable appearance, showing a spiral swirl pattern of dark lanes in the bright ice. These lanes are sunward-facing slopes and valley bottoms a few kilometers wide that heat sufficiently to lose their ice cover. The cap shows in the topographic profile of Figure 11.8, which suggests that the CO_2 deposit is roughly 2 km thick. Figure 11.33 shows a close-up of the southern residual cap in mid-summer, displaying a surface of evaporating CO_2. The dark bands are dust that has been left behind as the ice evaporates, accumulating at the base of slopes.

The permanent north polar cap is much larger than that in the south, never shrinking to less than a diameter of about 1000 km (Fig. 11.34). It shows a similar swirl pattern of exposed sunward-facing

slopes. Orbital temperature measurements show that in summer its temperature climbs above 200 K, far too warm for CO_2 ice to be stable. Simultaneously, the concentration of atmospheric water vapor rises sharply above the cap. Thus, we see that the northern residual cap cannot contain frozen CO_2 but instead is composed of ordinary water ice, rapidly evaporating in the summer warmth. With its diameter of 1000 km and a thickness of several kilometers, the residual northern cap may be one of the main storehouses for H_2O on Mars.

What is the reason for this major difference between the two martian polar caps? It is not simply a question of the summer temperatures, since the southern summers are hotter, yet it is the southern cap that stays cold enough to trap CO_2. The difference is probably associated with the global dust storms, which always take place during southern summer, when the northern cap is

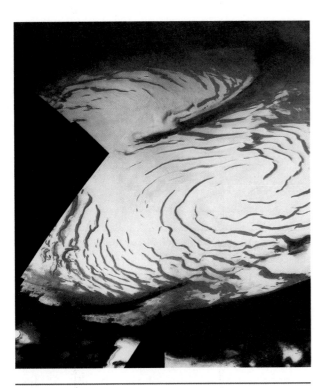

Figure 11.34 Residual north polar cap, about 1000 km across and composed of frozen H_2O.

forming; therefore, the northern cap becomes dusty. Being dustier, it is also darker, and therefore it can absorb more sunlight. It may be just this difference in the albedo of the ice that allows the northern cap to heat up and lose its frozen CO_2.

Both of the martian polar caps contain large quantities of H_2O ice, but they may represent just the proverbial tip of the iceberg for martian water. Calculations suggest that ice may be as stable as permafrost within the martian crust for all latitudes above about 40°. Nearer the equator, any ice originally present in the soil will eventually either melt or evaporate; the outflow channels may have originated as a by-product of this loss of subsurface ice. At higher latitudes, the permafrost layer is progressively thicker and nearer the surface until it emerges into view near the poles in the form of the residual polar caps. Thus, unlike the terrestrial ice caps, the martian polar caps are directly connected with a much larger underground reservoir of frozen water.

The long-suspected presence of subsurface ice on Mars was verified in 2002 by the Mars Odyssey spacecraft. This was accomplished by the gamma-ray spectrometer carried on the orbiter, which was able to detect gamma rays emanating from hydrogen and neutrons modulated by hydrogen trapped in martian soil to a depth of approximately 1 m. The most abundant carrier of hydrogen on Mars is H_2O, so the gamma rays are a good detector of water ice. In its first month of operation this instrument identified regions near the martian poles where 35% (by mass) of the subsurface soil (to 1 m depth) consists of water ice (Fig. 11.35).

POLAR TERRAIN

So far we have concentrated on the polar caps of Mars, but the underlying polar terrain is equally interesting. At latitudes above 80° in both hemispheres, deep-layered deposits of sediment cover the ground. Here the term *sediment* has its geologic

Figure 11.35 Evidence for subsurface ice on Mars. The regions where ice is abundant (up to 35% in the top 1 m) are shown as dark and light blue in this view of the south polar region. The data are from the first month of operation of Mars Odyssey's gamma-ray spectrometer.

Figure 11.36 Many layers on the edge of the north polar cap. This high-resolution view shows multiple layers, each several meters thick, revealed at the edge of the permanent ice cap.

meaning of finely divided material; sediment does not have to be produced or deposited by water. These deposits lie on top of the old cratered uplands in the south and overlie the volcanic plains in the north. In southern summer, the southern layered terrain is exposed by the retreating polar cap, while in the north, the residual cap continues to cover most of these deposits throughout the year.

The polar layered deposits blanket all but the largest underlying topography, indicating a depth of several kilometers. They are sparsely cratered and therefore relatively young. On the defrosted slopes within the residual polar caps, the layering is especially well shown (Fig. 11.36). The individual layers are between 10 and 50 m thick, distinguished by darker bands alternating with light and sometimes by terraces between the bands.

In the north, there is another special polar terrain, extensive sand dunes. Extending to the boundary of the layered deposits near the edge of the north residual cap, these dunes form a band about 500 km wide around the pole, between latitudes 70°N and 85°N. The other hemisphere has no similar band of dunes, but smaller dune fields are common inside craters at high southern latitudes.

FORMATION OF THE POLAR DEPOSITS

The polar terrains we have been discussing differ from most other parts of the martian surface in that they are depositional; they represent windblown dust transported from elsewhere. These areas may therefore be the final resting place of much of the material scoured out of the canyons and other eroded areas of the planet.

The layering of the polar deposits is particularly intriguing. Calculations of the amount of dust raised on Mars during a major dust storm suggest that the maximum annual deposit from this source is less than 1 mm/yr, which of course we cannot see; the layers must represent much longer timescales. To build up a thickness of 10 m would require about 10,000 years at this rate. Thus, the polar layering must represent climatic changes over tens of thousands of years. What might these be?

The most likely causes of such climatic changes are variations in the orbit and spin axis of Mars caused by the combined gravitational attractions of the other planets. (Earth undergoes similar but smaller cyclical variations, which are probably a major cause of the ice ages that our planet has experienced at more or less regular intervals.) One of these variations causes the role of the two martian poles to alternate with a period of 51,000 years, so that about 25,000 years in the future it will be the north pole of Mars, not the south, that will experience the hottest summers. Another effect is a periodic shift in the tilt of the martian axis of rotation. Calculations show that this tilt varies with a period of about 100,000 years. The eccentricity of the orbit also changes, this time with a period closer to a million years.

As a result of all these orbital and axial changes, conditions at the martian poles undergo

12

PLANETS, ATMOSPHERES, AND LIFE

Botticelli's painting *Primavera* ("Spring") shows a profusion of the life we take for granted in our everyday experience on Earth. Does this remarkable property of matter manifest itself on any other planet in our solar system? Mars seems the most likely candidate.

After four decades of space exploration, life on Earth is still the only life we know. This makes it very difficult for us to generalize about what kind of life we might expect to find elsewhere and how we might recognize it. The best approach is to begin discussing the search for alien life by considering terrestrial life and trying to understand its fundamental properties. We can then see whether the other inner planets ever offered environments in which these fundamental properties might have manifested themselves. The resulting perspective will give us a much better chance of realistically assessing the possibilities for finding life elsewhere in the universe (a topic explored in additional detail in Chapter 18, where we will discuss other planetary systems).

In our own solar system, Mars is the best candidate for an Earth-like planet that might support life. The Viking mission described in Chapter 11 included a set of experiments specifically designed to search for evidence of life. These Viking experiments yielded largely negative results, but interest in possible life on Mars still provides an incentive for the new wave of Mars missions under way at the beginning of the twenty-first century. Why is this so? And why do the atmospheres and surfaces of Mars, Earth, and Venus exhibit so much variety, despite the common composition of the planets themselves? We have already provided some answers, and we will now use the study of habitable planetary environments to emphasize the critical differences among the three inner planets that have atmospheres.

12.1 LIFE ON EARTH

Surely the existence of life is the most extraordinary characteristic of our planet. If there is life elsewhere in the solar system, it is well hidden, whereas on Earth life advertises its presence on a global scale. There must be some underlying attributes that make Earth uniquely suitable (in our planetary system) for the origin and continued existence of this remarkable property of matter. Having an appropriate and relatively stable surface temperature is certainly one essential attribute, as it permits the presence of the liquid water that is essential for life as we know it. However, to allow life to exist is not to ensure that it originates. Current ideas about life's origin stress the importance of the composition of the early atmosphere as well as the presence of liquid water. The chemistry that occurred in that atmosphere got life started, and life has been closely coupled to the atmosphere ever since.

THE ORIGIN OF LIFE

While no one has duplicated in a laboratory all the steps that originally led from inanimate matter to matter that is alive, it is essential that these events took place very early in Earth's history. The necessary organic building blocks for life could have been produced in abundance only in an environment (either on Earth or elsewhere) free of atmospheric oxygen. If this highly reactive gas had been present in the primitive atmosphere, the compounds necessary for the beginning of life would never have formed or, if they arrived from beyond the Earth, they would have been converted rapidly into oxides. Life has since evolved the capacity to protect itself from oxygen and even to use it to extract energy . . . but this is getting ahead of our story.

The primordial solar nebula, described in Section 4.1, must have been rich in simple compounds of carbon, hydrogen, oxygen, and nitrogen, the very elements that are the predominant constituents of life. We know this from studies of the molecular clouds that are the birthplaces of stars. We have learned from comets and primitive meteorites that substantial amounts of more complex organic compounds were also available in the early solar system (see Sections 4.4 and 6.2). Some of these materials must have been delivered by accreting planetesimals to Mars and Venus as well as to Earth at the time these planets were forming.

Furthermore, laboratory experiments have demonstrated that a mixture of gases containing compounds of carbon, nitrogen, oxygen, and hydrogen (but no free oxygen gas) will form amino acids and other prebiological organic compounds when it is subjected to an electric discharge, ultraviolet light, or other sources of energy. This happens because of the unique ability of carbon to form multiple bonds leading to many complex molecules—the subject that is called organic chemistry. Such experiments have been carried out to simulate conditions on the primitive Earth, starting with the fundamental work of Stanley Miller and Harold Urey at the University of Chicago in 1953 (Fig. 12.1).

The Miller-Urey experiment used a mixture of methane (CH_4), ammonia (NH_3), and water (H_2O) to imitate the early Earth. Evidence has accumulated since the original Miller-Urey experiment that the initial atmospheres of the inner planets were more likely composed of carbon monoxide (CO), carbon dioxide (CO_2), molecular nitrogen (N_2), and water, rather than methane and ammonia. However, an atmosphere based on carbon monoxide, nitrogen, and H_2O also yields many of the same prebiological organic compounds when energy is supplied to drive the chemical reactions.

Life has two fundamental capabilities: *metabolism* (the ability to utilize energy in its environ-

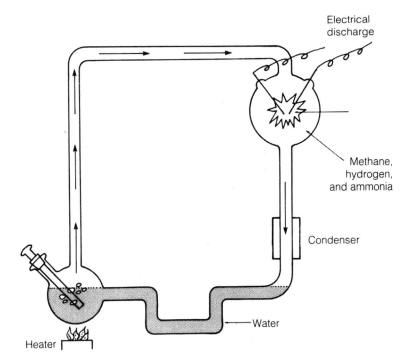

Electrical
discharge

Methane,
hydrogen,
and ammonia

Condenser

Water

Heater

Figure 12.1   Schematic diagram of the experimental equipment used by Stanley Miller and Harold Urey shows the 5-liter flask that contained water vapor, methane, hydrogen, and ammonia to simulate their concept of Earth's primitive atmosphere and oceans. The electrical discharge in the flask at the upper right reproduced the effects of lightning.

ment) and *reproduction* (a property that involves the coding and transmission of a great deal of information through the complex nucleic acids DNA and RNA). The greatest challenge to understanding how life began is reconciling these two requirements, each of which is quite complicated by itself. Which came first? Did the prebiotic molecules begin by extracting chemical energy and later develop a way to transmit this ability to offspring? Or was the essential first step the ability to store and transmit information through the genetic code or its simpler precursors? We don't know. Even though a great deal of research has revealed parts of the picture, we would be foolish indeed to claim that scientists today understand the actual origin of life on Earth. This remains one of the great scientific challenges of our time. Yet it did happen!

In our current model for the origin of life, there is nothing that singles out Earth. Both Venus and Mars should have begun with atmospheres similar to our planet's, and both also collected volatiles and organics from impacting comets and asteroids. While Venus rapidly became inhospitable, we know Mars had liquid water on its surface during the time the chemical reactions we just described were taking place on Earth. Hence, at this stage of our knowledge, we cannot exclude the possibility that life began on Mars at the same epoch that it originated on Earth, approximately 4 billion years ago.

THE FOSSIL RECORD

It would be immensely helpful if we could examine the rock record to see what was happening on Earth at the time life was first developing. As we saw in Section 9.3, however, this record has been irretrievably lost. The oldest rocks currently known solidified 600 million years after Earth formed, although individual mineral crystals called zircons have come to us from the first few hundred million years of history. Indirect evidence of life may be present in the chemistry of these oldest rocks and mineral grains, but this is an issue of scientific debate today. The earliest unambiguous record of life preserved as fossils dates from about 3.5 billion years ago. By that time life had already progressed to the level of complex colonies of microorganisms called stromatolites (Fig. 12.2). The bacteria that

Figure 12.2   Cross section of a fossil stromatolite, about 20 cm wide. The layered, domed structure consists of sediment trapped by this colony of micro-organisms as it slowly grew upward toward the sunlight. Some fossil stromatolites are as old as 3 billion years.

Figure 12.3   The trilobites that flourished in the seas at the start of the Phanerozoic era had two eyes and a rather complex body structure. This photograph shows a trilobite fossil about 2 cm long.

made the stromatolites are quite sophisticated life forms. It must have taken hundreds of millions of years for life to evolve to this level of complexity.

The further evolution to multicelled creatures required another 2.7 billion years. The first important step was the evolution of photosynthesis, the series of chemical reactions by which life can extract energy from sunlight. Photosynthesis was probably already in the tool kit of life by the time the stromatolites were formed. Later came the appearance of cells with nuclei and specialized subsystems for locomotion, which first show up in the fossil record about 1.5 billion years ago. Next came the development of multicellularity, and by the beginning of the Phanerozoic era, 590 million years ago, sexual reproduction had evolved, along with hard skeletons and other specialized tissues (Fig. 12.3). In another 100 million years, the backbone evolved, and shortly thereafter the colonization of the land began. From that point to the evolution of *Homo sapiens* was but a short interval on the cosmic calendar.

Until the middle of the 20th century, many scientists were skeptical that any fossils would be found from the eons of time before animals had hard shells and skeletons. Gradually, paleontologists have learned to read the records of earlier eras, recognizing the faint rock imprints of soft-bodied creatures and learning to find fossils of microbes (Fig. 12.4). Fossils, however, are just the

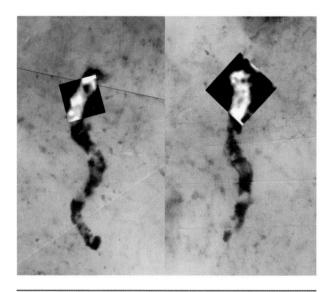

Figure 12.4   Fossil microbes from 3.5-billion-year old sediments of Western Australia. A portion of each of the two filaments is covered by an overlaid compositional image, where white shows the presence of organic matter—an indication that some carbon has survived the process of fossilization.

shadows of the creatures themselves, a mineralized sculpture in which the important organic materials (including DNA and RNA) have been replaced by silicates. Most of what we know about evolutionary history has been derived (until recently) from the careful analysis of such fossils. Now scientists are bringing new techniques from genetics and genomics to provide alternative ways to study the evolutionary record of life.

THE GENOMIC RECORD

The genetic foundation of life is wonderfully complex. The sequence of molecules that makes up our genetic signature is unique for each individual. This is why "DNA fingerprinting" has become so important, providing unambiguous evidence in forensics and enabling us to infer paternity—for example, to identify definitively the descendents of Thomas Jefferson. In spite of these individual differences, however, members of the same species clearly carry a common set of genetic information or genes that makes us what we are, whether a human or a great oak or a tiny paramecium living in the gut of a worm.

The transmission of the genetic code to our descendents is highly conservative; that is, it conserves information. You are genetically identical at the 99.9% level with Cleopatra or Napoleon or Newton. But over time, genetic differences accumulate and the genome evolves. By comparing two different genomes, we can measure how closely related they are and can estimate the time that has passed since they diverged genetically—the time since their "last common ancestor." Thus, for example, we can see that chimpanzees are closely related to humans because the genomes of the two species are 99% identical. We can also estimate the time (several million years) since the last common ancestor of people and chimps, a primate species that no longer exists but has left these two lines of descendents.

Until recently it has been impossible, or at least prohibitively expensive, to consider comparing entire genomes. But in the 1970s, microbiologist Carl Woese at the University of Illinois found a much simpler way to compare two

species by looking at just one ribosomal RNA sequence that is held in common by all known life. Because this gene sequence is part of every living thing and has evolved only very slowly with time, it provides both a quantitative measure of how closely related any two life forms are and an estimate of the amount of evolutionary time that separates the two. The display of these evolutionary time intervals is sometimes called the "universal tree of life."

Figure 12.5 is a simplified version of the tree of life. It looks very different from the way we usually think about Earth's biosphere. Almost all of the creatures in the plot are microbes, emphasizing that it is microbes, not the multicelled creatures we see in natural history museums, that dominate the history of life and exemplify its diversity. Look at the two "kingdoms" of plants and animals; they are right next to each other. You are much more closely related to the flowers than to the vast majority of life on Earth. The tree of life also suggests how much evolution took place before any of the "higher" forms of life appeared. Our planet's biosphere has been dominated by microbes, and these microbes had undergone billions of years of evolution before the first multicelled life appeared.

One of Woese's discoveries was the existence of a previously unsuspected domain of terrestrial life. Until the 1980s, only two domains were known: prokaryotes (cells without nuclei, also called bacteria) and eukaryotes (cells with nuclei, ranging from microbes to multicellular forms of life like us). Woese found that genetically the prokaryotes should be separated into two distinct domains, as different from each other as the bacteria are from mammals. These domains are now called bacteria and archaea. The tree of life shows that the archaea are more closely related to eukaria than to bacteria; that is, the earliest divergence from the last common ancestor was the split between the earliest bacteria and the line that eventually produced both eukaria and archaea.

The tree of life shows only currently extant species because we cannot carry out genomic analyses of fossils. In spite of this limitation, it does hint at the properties of the last common ancestor.

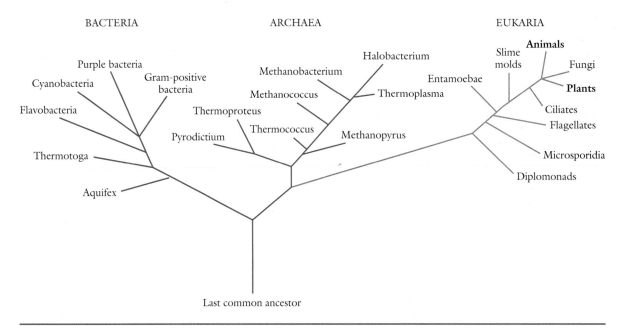

Figure 12.5 The universal tree of life, based on the similarities of ribosomal RNA sequences of different species. The distance between any two species, measured along the lines connecting them, is a measure of the divergence in their RNA sequences. The smaller this distance, the more closely related are the two life forms. The pattern is divided into the three domains of life first identified by Carl Woese: bacteria (red), archaea (blue), and eukaria (green).

The current life forms that most resemble these earliest ancestors are anaerobic (oxygen-avoiding), thermophilic (heat-loving) bacteria and archaea. Today these creatures survive in isolated environments, such as hot springs and deep ocean hydrothermal vents, where the chemical environment is free of oxygen. These are "living fossils," whose evolutionary trajectory has not taken them too far from their (and our) ancestors 4 billion years ago.

12.2 CO-EVOLUTION OF THE PLANET AND LIFE

Our human presence on this planet is actually a gift from the green plants, since it is oxygen-producing photosynthesis that created the environment required for the development of warm-blooded animals. Rocks much older than 2.2 billion years are suboxidized, meaning that they contain compounds that have not combined with all the oxygen they could (CO is suboxidized compared with CO_2). Such rocks must have formed in the absence of free oxygen, consistent with our model for the origin of life in an oxygen-free environment. The fossil record shows that stromatolites became common and widespread at about this time, indicating an increase in oxygen-producing photosynthesis by colonies of microorganisms. Oxygen is a waste product for green plants and must have been a poisonous gas for many life forms that existed in these early times. However, organisms that could use it flourished by taking advantage of the easily accessible source of chemical energy released by oxidation.

OXIDATION ON THE INNER PLANETS

We have seen that CO_2 is the dominant gas in the atmospheres of both Mars and Venus. On our planet, it is now a minor constituent, primarily because of the activities of living organisms. Marine shell-forming creatures—most of them microscopic—manufacture their shells from calcium and carbon dioxide dissolved in the ocean. As these shells are incorporated into sediment, great deposits of carbonate limestone are created at the expense of the atmospheric carbon dioxide. The quantity of CO_2 represented by these carbonate deposits today is truly staggering. If all this gas were put back into the atmosphere at once, the pressure of CO_2 alone would be 70 bars, or 70 times the total atmospheric pressure today. This is close to the 90 bars of CO_2 that is presently in the atmosphere of Venus, where it causes an extremely large greenhouse effect.

Looking at the climate of Venus today, we can see that if all Earth's CO_2 was ever in the atmosphere at one time, most of it must have been deposited as carbonate rocks even before life arose. Otherwise, Earth would have been too hot for the chemistry leading to the origin of life to occur. The key to the early removal of CO_2 from our planet's atmosphere was the presence of liquid water. Water dissolves CO_2, forming a weak acid that reacts with silicate rocks to make carbonates. As expressed in a highly simplified form by Harold Urey, who first described it, this reaction may be written as

$$CO_2 + CaSiO_3 \leftrightarrow CaCO_3 + SiO_2$$

It is therefore clear that with or without life, the early atmosphere of Earth would have evolved with time. Ultraviolet light from the Sun would continually break molecules apart in the upper atmosphere, a process called **photodissociation.** Heavy atoms like carbon and oxygen could recombine to form new compounds, a process called photochemistry, since it is driven by energy from light. Hydrogen atoms, having very low mass, would steadily escape from the atmosphere into interplanetary space. Thus CH_4 (if present at all) would be converted to CO_2, NH_3 would inevitably form N_2, and CO would be oxidized to CO_2. The oxygen that is essential for this process would be supplied by the photodissociation of H_2O, supplied to the atmosphere by evaporation from the warm wet surface of the planet. This transition to a fully oxidized environment must occur on any small inner planet in any solar system, regardless of the presence of life.

EVOLUTION ON THE EARLY EARTH

Oxidation of the environment is yet another threat to the survival of primitive forms of life on a young planet. Even the first, most simple organisms require food to supply the energy and raw materials for their continued existence. At first this food is free because it is produced spontaneously from the same chemical reactions that led to the origin of life itself. As we have seen, however, the character of any warm, rocky planet's atmosphere will steadily evolve in ways that presage the end of the free lunch.

To make the energy-rich organic compounds required for life, the gases in the atmosphere had to be suboxidized, on average. Free hydrogen generated from the reactions making the more complex molecules would help to maintain this suboxidized state, but Earth's gravity field is too weak to keep the hydrogen in the atmosphere. As hydrogen escaped, the free food would soon have stopped forming.

This is a critical bottleneck. One can imagine planets on which life begins, only to die out as the environment becomes oxidized. On other worlds, life may have progressed sufficiently before extinction by oxidation to evolve into organisms like the anaerobic bacteria we find on Earth today. Such microbes could survive in protected, suboxidized environments like mudflats or at great depths below the planet's surface.

On Earth, something more exciting happened: Life evolved photosynthesis. With this step, life connected with the Sun, a source of energy that will last for billions of years, continuously providing the oxygen required for more advanced organisms.

We conclude that the beginning of life on a planet provides no assurance that life will survive there. Furthermore, even if life survives, there is no guarantee that it will evolve into advanced forms capable of intelligent thought. It may remain at the level of microorganisms, exquisitely adapted to their local environments and capable of surviving through progressive generations for billions of years, but totally uninterested in poetry or music. It is also important to recognize that if a cosmic accident should destroy all living organisms on a planet after the atmosphere has become oxidized, there will be no new origin of life. Once the early, hydrogen-rich conditions are gone, they are gone forever.

THE IMPACT FRUSTRATION OF LIFE

Even before the atmosphere began to alter its chemistry, other aspects of the Earth's environment limited the development of life. As we described in Section 7.3, the Moon (and therefore the Earth) was still being heavily pelted by cosmic debris until about 3.9 billion years ago, when the bombardment dropped to its present much lower level. The end of this terminal bombardment is very close to the age of the earliest chemical evidence of life on Earth, which currently seems to be about 3.8 billion years ago.

We have no record on Earth of this period of terminal bombardment. Indeed, it may be the bombardment itself that helped to erase the geologic evidence from the first half-billion years of terrestrial history. We must therefore infer the Earth's early history from that of the Moon, which does preserve the scars of major collisions between 4.2 and 3.9 billion years ago. It is plausible, but not proven, that the impact rates were even higher prior to 4.2 billion years ago. To derive the terrestrial from the lunar impact rates, we must correct for the larger cross section of the Earth and for the effects of our greater gravity, both of which tell us that the Earth was much more heavily battered than the Moon.

Models of this earliest bombardment strongly suggest that the Earth was struck several times during its first few hundred million years by objects as large as several hundred kilometers across (Fig. 12.6). These isolated large projectiles were not part of the initial accretion of the Earth or of the cataclysmic events that formed the Moon. We are discussing here impacts that took place after the crust had cooled and the oceans and atmosphere

Figure 12.6   Artist's impression of the collision of a large asteroid with the Earth. An impact of this size would boil away the ocean and sterilize the planet. Earth may have collided with such objects during the first 600 million years of its existence, but fortunately such a calamity is no longer possible today.

had formed. Such large impacts would have boiled away the oceans to form a temporary steam atmosphere. The planet would have remained in this unhappy state for only a few decades before it cooled, but this is long enough to destroy any life forms that might exist.

It seems that the earliest life began in just such an environment of intermittent catastrophes. Several times over a period of a few hundred million years, the planet was violently sterilized. The last such sterilizing impact might have occurred as recently as 4.0 billion years ago. This scenario, in which early life had to struggle against a series of environmental catastrophes, has been aptly called "the impact frustration of life."

As the cosmic bombardment finally declined, we reached a point in history when there were no more planet-sterilizing impacts. For the first time, life that began would have a chance of long-term survival. Presumably our ancestors were the particular life forms that developed on Earth after the last sterilizing event.

Is there any actual evidence for this interesting but speculative scenario? There may be. As described above, the earliest common ancestor of life on our planet seems to have been a thermophilic (heat-loving) microbe. This does not necessarily mean that the first life was thermophilic. This situation would also arise if, following the last planet-sterilizing impact and the development of life, the planet suffered an impact that was not sterilizing but was large enough to boil away the top hundred meters of so of the ocean. Any microbes living near the surface would be destroyed, but those that had taken up residence at deep submarine vents or even under the land would survive. These are precisely the thermophilic microbes. According to this scenario, the surviving microbes from hot deep regions repopulated the planet and became the common ancestors of all life on Earth today.

IMPACTS AND EVOLUTION

We are no longer at risk from planet-sterilizing impacts, but the cosmic bombardment of our planet continues at a much slower pace. Occasional environmental catastrophes have always been a part of our planet's history, but only during the 1980s did scientists discover the critical role played by impacts in the evolution of terrestrial life.

The best documented such catastrophe took place 65 million years ago when a comet or asteroid 10–15 km in diameter having a mass of more than a trillion (10^{12}) tons hit in the Yucatan state of Mexico. This collision released an energy of about 10^8 megatons (100 gigatons) of TNT and blasted out a crater (named Chicxulub) almost 200 km in diameter, the largest crater so far positively identified on Earth.

The impact took place precisely at the chronological boundary between the Cretaceous and Tertiary periods, and the event is known as the **K/T event.** (If this seems a strange choice of letters, note that "Cretaceous" is spelled with a K in German.) The K/T boundary has long been recognized by geologists as one of the major breaks in the history of life on Earth. It corresponds to what is called a **mass extinction,** in which a great many species suddenly became extinct, to be succeeded by new species in subsequent strata. Extinctions are important, as 99% of the species that have existed on our planet are extinct. Mass extinctions are especially important because much of the evolution of new species happens during life's rebound from a mass extinction. In the K/T event, more than half of the major marine species that left their fossils were destroyed. At the same time the dinosaurs and many other land plants and animals also became extinct, presumably as a consequence of the same global environmental catastrophe.

The key to understanding the nature of the K/T event was the discovery in 1979 of a thin sedimentary layer of remarkable composition exactly at the boundary where the fossil record indicated the mass extinction (Fig. 12.7). In this boundary layer, which has subsequently been recognized all over the globe, the quantities of several rare elements, notably the metal iridium, are dramatically enhanced.

Iridium is rare on the surface of Earth because it dissolves readily in iron, and most of our planet's allotment of iridium is locked up in the Earth's metallic core. The relative concentrations of isotopes of the element osmium in the K/T

Figure 12.7 The K/T boundary layer. The arrow points to the thin clay layer that is enhanced in extraterrestrial material and contains shocked minerals from the impact. This site is in Colorado, more than 2000 km from the point of impact.

boundary layer are also peculiar, being typical of meteoritic materials rather than terrestrial. Both of these anomalies can be understood by the sudden addition of about a trillion tons of typical stony meteorite material to Earth's atmosphere, which would include 200,000 tons of iridium. The impact of a 10-km stony asteroid (or a 15-km comet) would do just that.

The energy of the explosion as the asteroid struck Earth is almost beyond imagining: the equivalent of 5 billion bombs of the size that obliterated Hiroshima and Nagasaki. The direct blast and associated tsunami (tidal wave) would have killed almost any living thing within a thousand kilometers, and the spray of molten rock fragments that fell back into the atmosphere ignited forest and range fires over much of the planet. This firestorm would have destroyed most of life on the land, but it did not have an effect on marine life. Devastating for the marine biota was the dust cloud raised by the explosion, the same dust that is responsible for the iridium-enriched layers that mark the K/T boundary. With its mass of a hundred trillion tons, this cloud blotted out the sunlight from the surface of Earth for months. Land temperatures dropped below freezing while marine life died from lack of sunlight. Probably 99% of all living things were killed over the whole Earth by a combination of conflagration followed by darkness and cold.

Horrible though this event may sound, it had profoundly beneficial effects for us. The new species that rapidly evolved to fill the ecological vacuum of this mass extinction gave rise to most of the life on Earth today. The tiny mammals that survived were our ancestors. Had it not been for the K/T impact, who knows what course evolution would have taken on Earth?

Half a dozen mass extinctions are recognized in the Phanerozoic era, the time that covers the evolution of multicelled creatures over approximately the past 600 million years. Several of these mass extinctions may be associated with impacts. We can be sure that the K/T impact was not unique. The crater densities on the lunar maria indicate that several impacts of objects 10 km or more in diameter have taken place on Earth during the Phanerozoic era. Surely these other Earth impacts have also placed their mark upon the planet and its biosphere (Fig. 12.8).

If impact-induced extinctions have played an important role in the evolution of life, then we must look at natural selection from a new perspective. It may be that the most important traits for long-term survival of a species have not been size or speed or even intelligence, but rather the ability to survive random global catastrophes of fire and ice. Natural selection works the way Darwin suggested, but the criteria for survival have changed.

FUTURE IMPACTS

If it happened to the dinosaurs, could we also succumb to an impact catastrophe? This seems entirely possible, although the near-term odds of such an event are extremely small. We know from the study of comets and asteroids, and from the graphic example of the collision of Comet Shoemaker-Levy 9 with Jupiter in 1994, that such collisions still take place in the solar system (see Fig. 5.11 for an estimate of the frequency of impacts of different sizes). The average interval between collisions of the magnitude of the K/T event is about 100 million years, but that is only an average, and as far as we know such events are random—meaning that one could happen any time with a probability of about 1/100,000,000 per year. This is a very low probability, but the consequences of such an impact are so great that we must consider it seriously.

From the perspective of contemporary hazards, we are more at risk from smaller but much more frequent impacts. As we discussed in the Perspective in Chapter 5, the greatest danger is from objects 1–2 km in diameter, which are too small to trigger a mass extinction but are large enough to cause worldwide crop failures from dust blasted into the atmosphere. For you as an individual, the risk of death as a result of such an impact is probably at least as great as the risk of death from such more common natural disasters as earthquakes and tornadoes. That is why governments are beginning to think about developing defense systems to deflect or destroy threatening asteroids before they can strike Earth.

12.3 COMPARING THE PLANETS

Somehow, during the first few hundred million years of our planet's history, the equilibrium among impacts, outgassing, chemical interactions with the rocks, and the escape of hydrogen left Earth with an atmosphere that contained enough CO_2 to keep the surface warm and enough air pressure to allow liquid water to exist at that warm temperature. This was the environment in which life began. Did the same series of events occur on Venus or Mars?

WHY YOU ARE NOT READING THIS ON VENUS

As we saw in Chapter 10, both Venus and Earth probably began with approximately the same inventory of volatile compounds. Venus, being

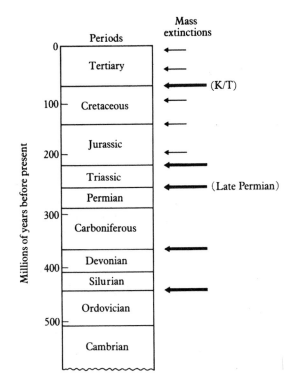

Figure 12.8 Geological timescale for the Phanerozoic era (the past half-billion years of Earth history). Arrows indicate mass extinctions. The names of the various time periods were assigned by nineteenth century geologists in part on the basis of the fossils they contain; thus it is not surprising that the boundaries of these time periods generally correspond to extinction events. The heavy arrows indicate the "big five" mass extinctions, in each of which more than half of all species were killed off. The most dramatic extinction was the Late Permian event 250 million years ago, in which more than 90% of species went extinct. An impact was clearly the primary cause of the K/T extinction 65 million years ago, and there is evidence (less compelling) of impacts associated with several other extinctions, including the Late Permian.

closer to the Sun, however, was subject to the runaway greenhouse effect described in Section 10.2. We may conclude that liquid water simply can't remain on the surface of a planet at this distance from its star.

Support for this conclusion comes from a study of the isotopes of hydrogen on Venus. Recall that in the runaway greenhouse effect, all the water on Venus gets into the atmosphere and heats it up so much that water vapor rises to great heights. At these altitudes, solar ultraviolet radiation photodissociates the water molecules, and a huge amount of hydrogen must then escape from the planet. Hydrogen has two stable isotopes. One, given the name *deuterium*, is twice as massive as the other. At a given temperature, less massive atoms or molecules in a gas move faster than heavier ones.

Since the common light isotope of hydrogen (H) will escape more readily from Venus, we expect to find evidence of this loss when we look at the relative abundances of the two isotopes of hydrogen in the H_2O that is still present in the Venus atmosphere today. And indeed we do. Measurements of the ratio of deuterium to hydrogen (D/H) by mass spectrometers on the Pioneer Venus probe and by ground-based spectroscopic observations of water have demonstrated that deuterium has been enriched to over 100 times its value in terrestrial oceans. A runaway greenhouse situation certainly occurred on Venus, leaving that planet in its present, highly desiccated state.

Without liquid water, life will not develop, and with no life and no liquid water, there is no way to remove carbon dioxide from the atmosphere. If Earth were as dry and lifeless as Venus, our planet would have a massive atmosphere dominated by CO_2 with a surface pressure of 70 bars and a surface temperature of several hundred degrees Celsius.

It is possible that during the first few hundred million years of its history, there were periods when Venus was cool enough for liquid water to exist on its surface. Such low-temperature epochs could have arisen because 4.5 billion years ago the young Sun was about 30% less luminous than it is today. As the luminosity of the Sun increased,

however, the climate inevitably changed, turning Venus into the hellish oven that we see today. To find a planet that may once have resembled Earth more closely, we must turn our attention to Mars.

Even though the martian environment is obviously very hostile today, is it possible that life could have begun during that earlier epoch when the channels were cut and liquid water was present on the martian surface? If not, we at least have the opportunity to study the first billion years of planetary history on Mars, an opportunity we do not have on Earth. We may be able to find a record of the composition and early evolution of the martian atmosphere preserved in the ancient rocks. And we might even find survivors, life that has adapted to the harsh martian environment.

THE SEARCH FOR LIFE ON EARTH

Let's begin our discussion of the search for life on Mars by asking ourselves what kind of evidence would convince a martian astronomer that there was life on Earth. Three obvious signs of life are detectable from a remote vantage point. First, a Martian might look for something in the physical appearance of Earth that demonstrated the presence of life, such as seasonal changes in surface markings that could be caused only by the growth and death of vegetation. The second clue is subtler. We know that life on Earth has had (and is continuing to have) a profound effect on the composition of the atmosphere. Hence, a spectral search for gases that are not in chemical equilibrium with each other or with the surface of the planet (like O_2 on Earth) would be a clever way to look for evidence of life. Finally, a Martian could use sensitive receivers to search for radio and TV broadcasts from a terrestrial civilization.

Until very recently, looking and listening would have failed to indicate the presence of life on Earth to a Martian who was using the same technology that we have. Clouds and the distorting effects of Earth's atmosphere would obscure our planet's surface sufficiently that detection of large-scale human constructions or seasonal changes in vegetation would be very difficult. Although a martian astronomer would be able to view the night side of Earth (as we can observe the dark hemisphere of Venus), the nighttime glow of our cities would not have reached the level of easy detection before the widespread adoption of electric lighting about a century ago. Similarly, no radio transmissions would have been detectable until humans began to generate large amounts of radio power in the late 1920s. The second test, however—studying the composition of the atmosphere—would have yielded positive results, even if no humans had ever appeared on Earth.

CLUES FROM ATMOSPHERIC COMPOSITION

Spectral observations would reveal a great number of oxygen molecules in Earth's atmosphere, along with a small amount of methane (CH_4). This situation would provoke a Martian's curiosity because the continued existence of methane on Earth implies the presence of a source that replaces the methane as rapidly as natural processes destroy it. Terrestrial methane is produced primarily by bacteria that live in swampy marshes and in the stomachs of grass-eating animals.

In an equally striking way, the large amount of oxygen in Earth's atmosphere would itself constitute a puzzle for our intelligent Martian, since oxygen is a highly reactive gas that rapidly combines with rocks in the crust and with magma that is continually being brought up from the interior. This process removes oxygen from our planet's atmosphere, so the continued existence of oxygen also requires a source—in this case, the presence of green plants that release oxygen through photosynthesis. Thus, a study of the composition of our planet's atmosphere could suggest to Martians that life exists on Earth, long before they thought about landing a spacecraft to sample the immediate environment.

Likewise, if astronomers on Earth had discovered large amounts of oxygen on Mars, they would have considered the probability of life to be very high. But after repeated efforts, they found that oxygen forms only 0.13% of the thin martian atmosphere, which has a total density less than 1% of our own. This tiny amount of oxygen is produced by photodissociation of some of the water vapor in the atmosphere. Methane and other hydrocarbon gases remain undetected on Mars, so the planet shows no signs of the chemical disequilibrium that mark Earth as biologically active.

REASSESSING MARS

When radio telescopes were developed in the 1950s, terrestrial astronomers used them to study Mars. These sensitive instruments detected the thermal radiation coming from the planet, allowing a determination of the average surface temperature. The results indicated that Mars at its best resembled a dry valley in Antarctica. These low temperatures agreed with earlier infrared measurements; unlike Venus, Mars held no surprises at radio wavelengths. And that was true for artificial broadcasts, too; no artificial transmissions have ever been detected from Mars.

In short, Mars does not exhibit two of the three signs of life we discussed: no unexpected atmospheric gases and no radio broadcasts. Seasonal changes were indeed observed on the surface, but as we saw in Section 11.3, these could be explained by dust blown about by martian winds. Yet the interest in Mars as a possible home for life remained high, partly in response to the detection of the dry river channels described in Section 11.5. Mars is the only planet besides our own on which we find evidence that liquid water flowed over a rocky surface, and liquid water is essential for the existence of life. In response to this discovery, humans during the 1970s put nearly a billion dollars' worth of effort into a direct search for organisms on the martian surface. The United States built the two Viking landers that each carried three experiments to test for the presence of martian microbes, plus an assortment of other instruments that could assess the habitability of the martian environment (Fig. 12.9).

12.4 THE SEARCH FOR LIFE ON MARS

Although the greatest excitement from a search for life on any planet would be the discovery of large, advanced creatures capable of communica-

Figure 12.9   A frosty morning in Utopia. This Viking photo beckons us to Mars with evidence that even today there is H$_2$O in the atmosphere and on the surface. The only problem is that there is no liquid water on the surface of Mars, although it surely was present in the past.

tion, the history and present status of life on Earth demonstrate that microbes far outnumber large life forms and are much more versatile in their adaptations to various environments. The dirt in your backyard contains more organisms than the number of stars in the Galaxy. There is no reason to expect another planet with life to differ from Earth in the overall development of living systems, and this is why the Viking project emphasized the search for martian microbes. Viking adopted a very general approach: It did not search for individual microbes or try to anticipate how they might be made. Rather, it looked for evidence of biological *processes* taking place on Mars.

A major difficulty in any such search is trying to guess how alien organisms would interact with their environment. The guiding principles have to come from life as we know it, which is carbon based and water dependent. We know water was present on Mars and is still there in the form of ice. What Viking needed to determine was the status of

carbon on Mars. Is there some indigenous food for martian organisms to eat? If so, we expect to find energy-rich, carbon-containing molecules in the soil in addition to the CO$_2$ in the atmosphere. Do the putative martian organisms transform carbon from one form to another? If so, we can search for the by-products of their metabolism.

The Viking landers each carried three specific life-detection experiments plus one general analytical instrument. The entire system was carefully sterilized so that no terrestrial microbes would be present to confuse the results. The output from all four of these instruments would be needed to answer the question of indigenous martian life.

VIKING RESULTS: THE SEARCH FOR ORGANIC COMPOUNDS

One approach used by the Viking landers was to analyze the soil with a highly sophisticated instrument called a gas chromatograph–mass spectrometer, or GCMS. An example of the GCMS detection of organic matter is shown in Figure 12.10a, which illustrates the results of an analysis of Antarctic soil on Earth with a laboratory version of the Viking instrument. Although this soil contains barely enough living organisms to give weakly positive results in the three Viking life-detection experiments, the GCMS found lots of organic compounds. In other words, the fingerprint of life can be seen more easily than the living creatures themselves. For instance, the Antarctic soil sampled in Figure 12.10a contains 10,000 times more carbon in dead organic molecules than in living microorganisms.

We must remember that not all organic compounds require living creatures to produce them. Figure 12.10b shows a GCMS analysis of a carbonaceous meteorite that contains amino acids, the building blocks of proteins. Again, carbon-containing molecules are present in abundance, but these organic molecules were not made by life. Thus, the detection of organic molecules in general would not prove that life exists on Mars because the simpler kinds of organic compounds could have been brought to Mars by meteorites, or might even have arisen from photochemical reactions in the martian atmosphere.

Antarctic soil   Murchison meteorite

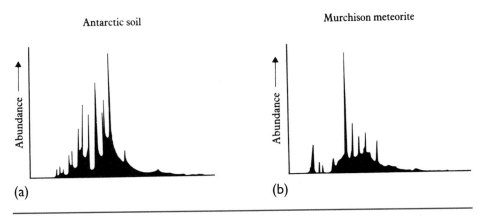

Figure 12.10   Viking GCMS test data. A test model of the GCMS was used to analyze **(a)** Antarctic soil and **(b)** a piece of the Murchison carbonaceous meteorite. The graphs indicate a rich variety of organic substances in each of these samples. Each peak in the graphs represents one or more organic substances. On this scale, the GCMS analysis of martian soil would be a straight horizontal line at zero, indistinguishable from the horizontal axis itself.

In practice, these potential ambiguities did not plague the GCMS experiment because it found *no organic compounds whatsoever* in the soil of Mars. On the scale of Figure 12.10, the martian soil results would be indistinguishable from the zero horizontal line of the graphs. These results apply to both landing sites and to two different samples at each site, including one from underneath a rock that might (so it was thought) have sheltered organisms from deadly ultraviolet light. The upper limits on all likely organic compounds in the soil, such as benzene and other hydrocarbons, fall at a few parts per billion. This is the kind of sensitivity that would allow detection of the proverbial needle in a haystack.

These negative results set powerful constraints on models of martian biology: How could life exist on Mars without leaving any trace of its presence? Could martian organisms be such efficient scavengers that no traces of their wastes, their food, or their corpses could be found, even with the high sensitivity that the Viking landers brought to Mars? The biologists involved in the Viking project considered this an extremely unlikely possibility. But before drawing any final conclusions, let's look at the results from the three specific life-detection experiments—the ones that looked for biological processes.

THE VIKING BIOLOGY EXPERIMENTS

After considering many alternatives, the Viking scientists selected three experiments to search for microorganisms on Mars. These experiments, called the gas exchange (GEX), the labeled release (LR), and the pyrolitic release (PR), all reflect scientific experience with life on Earth (Fig. 12.11). Thus, for example, all the organisms that we know derive their energy from two basic processes: oxidation (removal of hydrogen, combination with oxygen) and reduction (removal of oxygen, combination with hydrogen). The experiments were designed to measure these processes. The trick is to know what martian organisms would like to eat and drink, to supply these nutrients, and then to have a means to decide whether in fact the organisms consumed the nutrients or some nonbiological chemical reaction occurred.

The GEX and LR experiments both assumed that martian organisms would require water, which is essential for all life on Earth. The GEX supplied a mixture of several kinds of nutrients and then used a gas chromatograph to see what gases would be released by organisms that consumed this food. In effect, this was a search for evidence of metabolism. The LR experiment labeled its nutrient medium by using radioactive carbon in the organic compounds

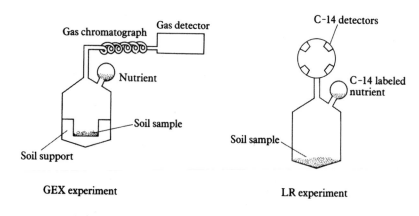

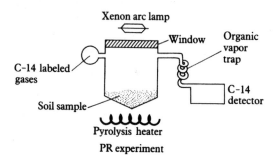

Figure 12.11 A schematic representation of the GEX, LR, and PR experiments shows that each of them analyzes the martian soil in a different way to test for the presence of life. The GEX experiment exposed a broth of nutrients to a few grams of soil and then looked for changes in the gas above the soil and nutrient mixture. The LR experiment tagged carbon-rich compounds with radioactive C-14 atoms in place of some of the usual C-12 atoms. These labeled compounds then dripped over the soil sample. Any biological processes should have caused some tagged compounds to appear in the gas above the sample. The PR experiment replaced the normal martian atmosphere with an equivalent set of gases labeled with radioactive carbon atoms. Any organisms that ingested some of these labeled molecules would produce a radioactive signal when the soil in which they lived had been roasted.

its nutrient mixture contained. CO_2 or CH_4 released by martian organisms could then be detected because of its radioactivity. This experiment was therefore looking for evidence of respiration.

The PR experiment was designed to imitate the martian environment as closely as possible. No nutrients were used; no water was supplied. The idea was simply to expose a sample of martian soil to a simulated atmosphere in which the CO_2 and CO were radioactive. The soil was illuminated by a lamp that matched martian sunlight as closely as possible. This experiment was thus a test for evidence of photosynthesis on Mars by seeing whether organisms in the soil would incorporate the radioactive CO_2 and CO from the artificial atmosphere above them.

All three of these experiments were tested extensively on Earth and were found capable of detecting microbial life in extreme conditions, such as the nearly sterile dry valleys of Antarctica. The stage was set for the search for life on Mars.

RESULTS FROM GEX AND LR EXPERIMENTS

On the eighth day after the first landing on Mars, the Viking 1 lander scoop dug a trench in the soil and distributed soil samples to the various experiments (Fig. 12.12). The GEX placed about a gram of soil into a tiny, porous container positioned above the nutrient medium. Two days later, the first analysis of the gas in the container showed an exciting result: A large quantity of oxygen had appeared in the chamber, 15 times the proportion in the martian atmosphere. The simple exposure of martian soil to the humidity in the test chamber produced by the nutrient-laden fluid had apparently been sufficient to liberate oxygen from the soil.

Was this an indication of life on Mars? After months of testing, the biologists concluded that they were observing not biological activity but merely the chemical interaction of martian soil

Figure 12.12 In addition to taking samples directly from the surrounding soil, the Viking boom pushed aside a rock (left) to gather a sample from the ground beneath it (right). The purpose was to obtain soil that had been protected from ultraviolet light.

with a higher pressure of water vapor than had been present on Mars for millions of years.

The day after the first GEX data, the LR experiment reported: again a positive result! After checking to be sure that the background radioactivity level was low, the LR added two drops of the radioactive nutrient material to the soil that had been brought into its chamber. A sudden rise in the radioactivity of the gases above the soil sample appeared, a more dramatic reaction than biologists had found in their comparison experiments with many life-bearing soils on Earth.

Unfortunately for those who hoped for proof of life, the Viking scientists soon realized that the radioactive gas, almost certainly carbon dioxide, could be produced from simple chemical reactions that involve oxygen compounds called peroxides. A second wetting of the soil showed no increase in the amount of radioactivity in the gas in the test chamber. Instead, the additional nutrient apparently absorbed some of the radioactive carbon dioxide that was originally released. Hence, the scientists concluded not that life exists

on Mars, but rather that the martian soil may contain chemicals such as peroxides that release carbon dioxide when exposed to simple organic compounds.

PR EXPERIMENT RESULTS

Since the first two experiments yielded information that seemed ambiguous, the Viking team eagerly awaited the results from the PR experiment, which did not use a water-based nutrient and thus avoided the primary agents that the biologists suspected of purely chemical reactions in the soil. Analysis of the initial experiment revealed that radioactive carbon had indeed become part of compounds in the soil (Fig. 12.13). Weak as this signal was, it seemed clearly positive: To the PR experiment, martian soil behaved much like Antarctic soil on Earth, nearly sterile but not entirely so. Yet even in this case, the Viking scientists were skeptical that they had found life on Mars.

The reason for this skepticism is that when the scientists arranged to heat the martian soil in separate experiments to 175°C and to 90°C for

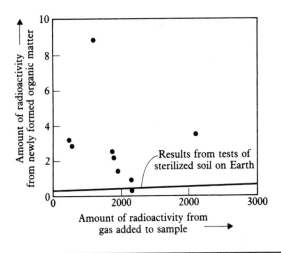

Figure 12.13 The PR experiment showed a rise in the amount of radioactivity in the gases above the soil being tested for living organisms after C-14–labeled CO_2 and CO were introduced. This suggested initially that living organisms had incorporated some of the radioactive carbon atoms from the simulated martian atmosphere.

several hours before the radioactive gases were injected, they still found positive results from the PR experiment. The higher temperature reduced the reaction by 90%, but a positive signature still emerged; the lower temperature had no effect. Since Mars's surface temperature is never higher than 30°C, the Viking scientists did not believe that any martian life could have adapted to survive 3 hours at 175°C, which no terrestrial organisms can do. Further testing has suggested that a small amount of ammonia in the first PR soil sample, contributed by leakage from the descent engines on the Viking lander, may have been responsible for the soil chemistry that produced the first, weakly positive answer.

Chemically Active Martian Soil

All three Viking biology experiments yielded some positive results, but they were not the results that had been hoped for. How can we explain this, and how can we resolve this ambiguity between biology and chemistry?

High-energy solar ultraviolet radiation easily reaches most of the surface of Mars and apparently enhances the chemical activity of the iron-rich soil. Because the temperature on Mars is so low, and because liquid water is entirely absent from the planet's surface, compounds formed in this way can remain in chemical disequilibrium for long periods of time. If a small amount of water is added to the soil, however, it produces chemical reactions that mimic some of the effects of biological activity.

The Viking missions have therefore demonstrated that martian soil probably contains no carbon-based living organisms, at least at the two landing sites. Perhaps the strongest arguments against a biological interpretation of the biology experiment results are provided by the "controls," in which heat-sterilized soil samples exhibited the same chemical reactions as the untreated samples. The absence of any organic compounds in the martian soil, to a level of parts per billion or less, is another compelling argument. These two findings effectively remove the ambiguity between chemistry and biology.

What Did the Vikings Miss?

What about life elsewhere on Mars? The similarity between the soil analyses at the two Viking landing sites, separated by more than 2000 km, argues against great variations from one place to another.

The soil particles on Mars are distributed over the entire planet during global dust storms. If every part of Mars eventually comes in contact with every other part, sampling the dusty material at one place should be equivalent to sampling at all locations. This argument gains strength from current models of the planet's surface, which suggest that the soil is turned over (gardened) by meteorite impacts, volcanic eruptions, and wind to depths of one to several meters, on timescales ranging from tens of thousands to tens of millions of years.

What about *oases* on Mars, regions where relatively high temperatures prevail, perhaps because of subsurface volcanism or some other cause of special microclimates? These exceptional environments are well known on Earth—in Yellowstone National Park in the United States, on the Kamchatka Peninsula in Russia, and near geothermal vents in the deep oceans. Such regions typically have very short lifetimes. As the sands shift to cover an oasis or dry spells exhaust it, as hot springs dry up in one place and appear nearby, seeds and microorganisms must arrive from somewhere else to repopulate them. Without this outside reservoir of life, any oasis must soon become lifeless. The problem with the oasis model for life on Mars is that there doesn't seem to be any hospitable "somewhere else" on this dry, cold planet.

The Viking experiments seemed so definitively negative that no subsequent organic chemistry or biology experiments were sent to Mars for 25 years. But the search for life on Mars is gaining new momentum from recent discoveries. Remember that the Vikings investigated only near-surface martian soil. They could not sample several potentially important repositories of organic material, such as the polar deposits, sediments buried in lakebeds, the interiors of rocks, and subsurface habitats irrigated by melting permafrost. Organics that were present early in martian history when water was prevalent on the surface might still be preserved in deeply

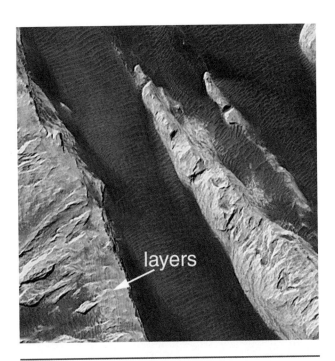

Figure 12.14 Possible sediments on Mars. This high-resolution image shows a detail of the so-called "white rock" within Pollack Crater—an eroded mound that is suspected to be sediment formed at a time when the crater contained an ice-covered lake of liquid water. The image width is 3 km.

buried sediments of ancient lakebeds, for example (Fig. 12.14). Such sediments would also be excellent sites to search for signs of ancient microbial fossils, like the ones that constitute the most ancient evidence of life on Earth (see Fig. 12.4).

LOOKING INSIDE ROCKS: MARTIAN MICROFOSSILS

None of the Viking experiments could have detected fossil life like the example in Figure 12.4. That search requires the microscopic examination of martian rocks.

In 1996, the public and scientists alike were excited by an announcement that evidence for fossil microbial life had been found in a martian meteorite. Since it had been the first meteorite found in the Antarctic in 1984, it is designated ALH 84001. It is the oldest of the martian meteorites, with a solidification age of more than 4 billion years—back when liquid water was present

on the martian surface. The meteorite itself is a volcanic rock, but it contains small quantities of sedimentary material deposited at some time in the past when liquid water came in contact with the rock. The evidence for life was found in minute carbonate deposits formed by this aqueous alteration more than 3 billion years ago.

The most spectacular original evidence was photos of alleged fossilized microbes from the carbonate deposits (Fig. 12.15). Even at the time of the announcement, however, there was widespread skepticism owing to the extremely small

(a)

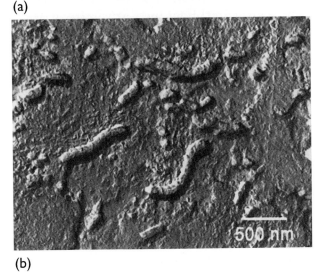

(b)

Figure 12.15 Mars meteorite ALH 84001. **(a)** Photo of the meteorite. **(b)** High-resolution scanning electron microscope image of a small part of the interior of one of the carbonate globules formed when water leaked into the rock. The tiny "worms" were originally claimed to be fossil microorganisms, but they were later recognized as more probably artifacts of the sample preparation process.

size of the alleged organisms, which were about the size of a terrestrial virus, an organism that cannot reproduce without access to the DNA in a bacterium. The current consensus among scientists is that the entities in Figure 12.15b are artifacts of the sample preparation process and not evidence for martian life.

Three other lines of evidence presented in 1996 also suggested a biological component in ALH 84001. Two of these were later interpreted as terrestrial contaminants that entered the rock during the 15,000 years it was trapped in the Antarctic ice. The final piece of evidence, however, cannot be easily discounted. This is the presence of chains of tiny crystals of magnetite that appear to be exactly like those formed on Earth by bacteria that use the magnetite as an internal compass for orientation in the water. The interpretation of this one remaining piece of evidence is still controversial as we write in 2002.

Whether or not the evidence for life holds up, ALH 84001 shows that indigenous organic material (carbonate) was carried into the rock by running water on Mars billions of years ago. This is an important verification of what had previously been speculation about the past climate of Mars. The current generation of rovers being sent to Mars will include drills to penetrate the eroded surfaces of rocks, but these rovers will unfortunately not be capable of detecting organics.

A serious search for life on Mars must be prepared to do much more than this. It will be necessary to land in places identified to have high potential for life or fossils, such as craters that contain ancient, undisturbed layers of waterborne sediments. The landers or rovers must be prepared to drill down several meters to obtain protected samples. Exploring those intriguing gullies will be even more difficult (Fig. 12.16). It will require vertical drilling through tens of meters of compacted rock layers or lateral drilling into canyon and crater walls at the sources of these outbreaks, or detailed investigations of the debris aprons at the bases of the gullies in the hope of finding material carried out from within the subterranean chambers where the water collected.

Figure 12.16 Mars gullies. These features, interpreted as evidence for recent liquid water, are cut into the wall of a small crater. The image is 3 km wide.

MARS SAMPLE RETURN

The experience with ALH 84001 has been instructive for the scientists involved. In spite of 6 years of hard work by dozens of highly skilled (and motivated) researchers, there is still no agreement on whether this small rock provides evidence of ancient microbial life on Mars. It is true that most scientists have decided that the original evidence is not compelling, that it does not meet the strict tests that would be required to convince the general scientific community that there is (or was) life on Mars. But the issue is not finally settled.

A few scientists still argue that the Viking experiments provided positive detection of life, although they are a small minority. Again, the evidence does not pass the strict requirements demanded by such an important question as the presence of life on

—— PERSPECTIVE ——

PLANETARY PROTECTION

Looking for life on another planet is challenging in many ways. One of the most difficult requirements is to make sure that we do not contaminate the other world with life carried from the Earth. At the very beginning of the spacecraft exploration of Mars, both the United States and the Soviet Union agreed, under international auspices, to avoid the contamination of Mars with terrestrial microbes. To meet these agreements, all spacecraft destined to land on Mars were carefully sterilized.

In the case of Viking, we know the sterilization was successful. Viking's failure to detect martian organisms implies as well that these experiments detected no hitchhiking terrestrial microbes. One of the motivations for sterilizing spacecraft is to ensure that we do not accidentally detect Earth life and mistake it for Mars life.

As we have learned more about the harsh conditions on the martian surface, the sterilization requirements have been somewhat relaxed. It is evident that no terrestrial microbes could grow on the martian surface, with its low temperature, absence of water, and intense ultraviolet radiation. Microbes might survive in a dormant, desiccated state (as they survived the worse lunar environment on the Lunar Surveyor spacecraft visited by *Apollo 12* astronauts; see Fig. 7.7), but they cannot grow and proliferate on Mars.

The problem of contaminating Mars will increase as we think about drilling down into the subsurface, where temperatures are higher and no ultraviolet light penetrates. Any spacecraft that is actively searching for martian life will have to meet stringent requirements. The situation is even more daunting if we consider human flights to Mars. Any humans will carry with them a multitude of terrestrial microbes of all kinds. It is hard to imagine how we can effectively keep the two biospheres isolated from each other if Mars has indigenous life. Perhaps the best situation is one in which the two life forms are so different that each is effectively invisible to the other—not recognized on a chemical level as living or as potential food.

The most immediate issue of public concern is not the contamination of Mars, but any dangers associated with returning Mars samples to Earth. NASA has committed itself to the complete biological isolation of returned samples until they are demonstrated to be safe. Even though the chances of samples containing viable martian organisms are extremely low, it is better to be safe than sorry.

Most likely, there is no danger, even if there is life on Mars and alien microbes inhabit some of the returned samples. Mars is sending samples to Earth all the time. We have found more than two dozen martian meteorites, and probably several more reach our planet every year. As a result of this natural transport of material, we have been exposed many times over to martian microbes, if they exist. Either they do not interact with the Earth's biota, or in effect our planet has already been inoculated against such alien bugs, or they do not exist.

Mars. But we aren't sure. And scientists have had a difficult time thinking of better ways to test for martian life. Today we could send highly sensitive instruments to Mars that could detect even extremely tiny quantities of DNA or RNA, but what if the martian microbes don't utilize DNA or RNA? Do we really know what we are looking for when we imagine searching for alien microbial life?

One thing that most of the scientific community agree on is that we can do a much better job of searching for life (and especially for fossil life) if we bring back martian samples for analysis in laboratories here on Earth. Even a modest university lab has instruments with greater sensitivity and versatility in analyzing small samples than we could hope to package in a Mars lander or rover. Once the samples are brought to the Earth, they can be examined by a much wider range of techniques. In addition, if the initial results are not satisfactory, it is probable that improved instruments will be available in the future. The best way to search for life on Mars is in our laboratories here on Earth.

DESIGNING MARTIAN MICROBES

Comparing Mars as we see it now with our own planet, it seems clear that no known terrestrial organisms, including the toughest and most sophisticated microbes, could grow in the present environment at the martian surface. The main problem is the absence of liquid water. Some terrestrial organisms could survive on Mars in a dormant state, but even the algae that create their own temperate microclimate inside rocks in the dry valleys of Antarctica require more water vapor than exists in the thin atmosphere of Mars.

The situation becomes very different when we look at subterranean life. At depths of a kilometer or more on Earth, scientists have found well-populated ecosystems of microbes (bacteria and archaea) that are thriving in the absence of oxygen or sunlight. All they apparently require is a small amount of liquid water and the minerals in the surrounding rocks. These recently discovered organisms are still under intensive investigation, and not all of their metabolic pathways are understood, but of all the organisms on Earth, these appear to be the ones most suitable for survival on Mars.

Finding a microbe many meters beneath the red sands of Mars would obviously be much less exciting than shaking hands with a small, green Martian. Nevertheless, it would be an astounding discovery! We stress again that at the present time, life on Earth is the only life we know in the entire visible universe. There are many serious scientists who feel that life is such an improbable phenomenon that life on Earth may well be all the life there is. Arguments rage endlessly on this subject and will never be resolved without additional information. We need to do the experiment; we need to go out and look for life. That is what draws us back repeatedly to Mars.

12.5 EARLY MARS: WHAT WENT WRONG?

Did life originate on Mars? If so, did it survive long enough to evolve into organisms like the ones we find on Earth today, deep beneath our planet's surface? We can approach an answer to the first question by returning to a consideration of the early history of Mars.

CONDITIONS ON ANCIENT MARS

We saw in Chapter 11 that there is ample evidence that liquid water existed on Mars during the first billion years of the planet's history. In Section 12.1, we found no reason why life couldn't have begun on Mars during this apparently favorable period, provided it lasted sufficiently long. What happened to change this promising beginning and how fast did this change occur? Why have the atmospheres of Mars and Earth evolved so differently, so that today Earth is teeming with life and Mars appears totally barren?

Mars is 1.52 times farther from the Sun than Earth, so the resulting decrease in solar illumination by a factor of 2.3 certainly decreases the solar heating. This means that the average surface temperature on Mars is significantly lower than the terrestrial value. One could even imagine a "runaway refrigerator effect": As the temperature drops, less water vapor is available in the atmosphere, so the greenhouse effect decreases, further lowering the average surface temperature. The resulting increase in the latitudinal extent of the polar snow deposits will increase the planet's reflectivity, further decreasing the surface temperature, and so on. Unlike the runaway greenhouse effect on Venus, however, there is an easy way to counteract this cycle: Simply increase the amount of CO_2 in the atmosphere. Adding the equivalent of only one or two Earth atmospheres of this gas (this is 1 or 2 bars, compared with the 90 bars of CO_2 on Venus) would create a greenhouse effect sufficient to warm the surface of Mars above the freezing point of water.

Is it reasonable to assume that Mars once had such an atmosphere? Based on the information currently at our disposal, the answer appears to be yes. Recall that our best model for the origin of Earth's atmosphere invokes a mixture of volatiles contributed by the rocks that built the planet and the delivery of volatiles by icy planetesimals. There is no reason to think that these icy planetesimals would have missed Mars on their way to Earth, especially when they seem to have endowed Venus with approximately the same volatile inventory as our own planet (Table 12.1). Thus the question is better phrased in another way: Why doesn't Mars have a thicker atmosphere today? What has happened to all those volatiles?

RECONSTRUCTING THE MARTIAN ATMOSPHERE: ISOTOPES AND NOBLE GASES

On Earth, we have recognized that the missing carbon dioxide is mainly bound up in carbonate rocks. We can therefore reconstruct the total mass of outgassed carbon dioxide on Earth by surveying the rocks on our planet. Unfortunately, we cannot do the same thing on Mars. None of the lander experiments has been designed to look for carbonates. However, Global Surveyor carries an infrared spectrometer capable of detecting carbonate rocks exposed on the surface. This investigation is being pursued further by Mars Odyssey, but to date (2002) no carbonates have been found, although there are some intriguing possibilities (see Fig. 12.14). Meanwhile, we will proceed indirectly in our efforts to understand whether indeed Mars ever had a truly dense atmosphere, and if so, when and for how long.

The approach we shall take is to look at abundances of the elements and their isotope ratios as we find them in the martian atmosphere today and to compare these values with similar data on the atmosphere of Earth. We will pay special attention to the **noble gases**—neon, argon, krypton, and xenon. These gases are called "noble" because they don't associate with other elements. In fact, they don't even associate with each other!

Table 12.1 Atmospheres of the inner planets

Gas	Earth		Venus	Mars	
	Now	*Reconstructed**	*Now*	*Now*	*Reconstructed†*
N_2	78.1%	1.9%	3.4%	2.7%	2%
O_2	20.9%	trace	trace	trace	trace
Ar-40	0.93%	190 ppm	40 ppm	1.6%	20 ppm
CO_2	360 ppm	98%	96.5%	95.4%	98%
Water‡	3 km	3 km	trace	trace	~0.9 km
Total pressure	1 bar	~70 bars	~90 bars	0.0065 bar	~7.5 bars

*For Earth, reconstruction removes the effects of water and life.
† For Mars, reconstruction removes the effects of impact erosion and escape.
‡ The abundance of water is given as the depth of a global layer that would contain all the water above the planet's surface.

They are chemically inert and form gases consisting of single atoms, rather than molecules like N_2, O_2, H_2, and so forth. Thus, once noble gases are introduced into a planet's atmosphere, they remain there; they don't combine with the rocks.

We met argon in our discussion of Earth's atmosphere, where we discovered that Ar-40 constitutes 1% of the air we breathe. This argon is produced by the decay of radioactive potassium in rocks. There are two other isotopes of argon that are far less abundant. They represent primordial argon that was once part of the original solar nebula. Together with the other noble gases, the abundance of this primordial argon in the atmosphere provides a measure of the quantity of volatiles delivered to the planet and subsequently outgassed to the atmosphere.

The relative abundances of argon, krypton, and xenon show the same pattern in the atmospheres of Mars and Earth. That suggests that other volatile elements, such as carbon and nitrogen, should also be present in similar relative abundances on the two planets. Knowing the ratio of total nitrogen and carbon to atmospheric argon on Earth and knowing the amounts of the noble gases on Mars, we can calculate how much carbon dioxide and nitrogen should be present there. This calculation yields a total surface pressure on Mars of 0.075 bar, about ten times the present value.

Other indicators of physical changes are the ratios of stable isotopes of common elements. For example, if a large amount of an element with more than one stable isotope has escaped from a planet's atmosphere, we expect fractionation of the isotopes to occur, just as in the case of the enrichment of deuterium on Venus. Hydrogen can certainly escape from Mars, and Earth-based spectroscopic observations have demonstrated that deuterium is enriched in atmospheric water vapor on Mars but only by a factor of 5.5 compared with ocean water on Earth. No runaway greenhouse effect occurred on Mars. This same enrichment was found in the bound water in some minerals examined in the martian meteorites. It is interesting that the lowest value of the ratio of deuterium to hydrogen (D/H) in these minerals is not the same as the terrestrial D/H.

Instead, it is equal to twice the value in our oceans, the same D/H that is found in comets. Evidently there are multiple reservoirs of water on Mars, and some of the water near the planet's surface may still show the effect of comet bombardment, undiluted by mixing with the larger reservoir of water from Mars's interior.

An equally interesting story is told by the isotopes of nitrogen. Nitrogen atoms cannot escape from Earth, but Mars has a weaker gravitational field, and there is a kind of photodissociation of N_2 that gives the atoms enough energy to escape from the red planet. Viking mass spectrometers determined that the light isotope of nitrogen is strongly depleted on Mars. Knowing the escape process and the present isotope ratio, we can calculate the original nitrogen abundance. The result is that Mars must have started out with about ten times the nitrogen we now find in the atmosphere. To estimate the total atmospheric pressure corresponding to this much nitrogen, we use the ratios of CO_2 to N_2 on Venus and Earth, finding an average of $CO_2/N_2 \approx 40$. The total surface pressure on Mars corresponding to the original nitrogen abundance then works out to be 0.065 bar.

Thus the two independent estimates of the total surface pressure—using the enrichment of the heavy isotope of nitrogen and the abundances of noble gases—overlap at a value of 0.07 bar, which we could use as the approximate total pressure of the early martian atmosphere. This is roughly a factor of ten more gas than we now find on Mars, but still 1000 times less than the volatile inventories on Earth and Venus. It seems that either Mars is fundamentally different from its two nearest neighbors in space, or we are still missing some crucial point about the early history of the red planet.

WATER ON MARS

The impression that Mars is different from Venus and Earth in some basic way is reinforced when we look again at the history of water on Mars. The geologic evidence examined in Chapter 11 makes it clear that liquid water once flowed over the martian landscape. A surface pressure of

0.07 bar is only barely adequate to permit this; it provides a very narrow temperature range within which water can be in the liquid state. Furthermore, the amount of water corresponding to this reconstructed atmosphere is too small. Again using Earth's volatile inventory for calibration, a 0.07-bar, CO_2-dominated atmosphere would be accompanied by an amount of water equivalent to a global layer only 9 m deep.

This may be compared with geologist Michael Carr's estimate of 500–1000 m for the amount of water required to cut the great flood channels on Mars. Another piece of evidence for substantial water on the early Mars comes from the Global Surveyor's detailed topographic mapping (see Figs. 11.1 and 11.8). The evidence for some sort of shallow polar sea that once partially filled the great northern basin seems convincing.

Apparently our attempt to reconstruct the martian atmosphere by determining the total inventory of volatiles produced by the planet since its formation has come up short. It is not simply that we are in sharp disagreement with what we find on Earth and Venus. Our reconstruction is internally inconsistent because we can't account for the amount of water necessary to cut the martian channels. Clearly we are still missing something that sets Mars apart from its two larger sisters.

IMPACT EROSION

We have stated repeatedly that early impacts by comets and icy planetimals were probably responsible for providing a significant fraction of the volatile inventories on the inner planets. But impacts can take away volatiles as well as supply them. We must worry about the loss of volatiles by impacts, especially by impacts from relatively massive asteroids. Obviously, the giant impact on Earth that formed the Moon would have had a catastrophic effect on our planet's early atmosphere. Subsequent bombardment by comets could have replaced the missing gases. Similar events must have occurred on Venus, not forming a satellite in that case, but leading to approximately the same volatile inventories on these two, similar-size planets.

Mars, being both smaller in mass (about 0.1 of Earth's mass) and closer to the asteroid belt, was much more vulnerable to atmosphere-stripping impacts. Even bombardment by smaller bodies would have had a devastating effect, and such impacts would have occurred more often. The consequent **impact erosion** of the atmosphere appears to be the missing process that accounts for the present low surface pressure on Mars.

The idea that Mars lost most of its atmosphere through successive impacts is supported by calculations of the intensity of this bombardment during the first billion years of solar system history. These calculations suggest that Mars must have had an atmosphere with at least 100 times the present surface pressure to end up with the atmosphere we find today.

There is an obvious test of this model: It predicts the near absence of carbonates on Mars. The missing CO_2 on Mars should not be bound up in massive deposits of limestone, as it is on Earth. If impact erosion was responsible, that missing gas has been blown right off the planet! No carbonate deposits have yet been found on Mars, but the search continues.

Some support for impact erosion already exists. The similarity between the total inventories of carbon dioxide and nitrogen on Earth and Venus suggests that Mars should have started with an inventory of these two gases that had the same proportions. Therefore, if CO_2 is removed from the atmosphere of Mars by the formation of carbonates, nitrogen should become increasingly abundant in the atmosphere, as it has on Earth (and on Titan, as we shall see in Chapter 15). This is obviously not the case: Both CO_2 and N_2 have been removed from the atmosphere of the red planet. Impact erosion would accomplish this.

12.6 GOLDILOCKS AND THE THREE PLANETS

As we consider the three largest inner planets, we seem to find ourselves in the classic situation of Goldilocks confronting the three bowls of porridge: One is too hot, one is too cold, and one is just right. Our original objective in this chapter was to try to understand what basic properties make Earth so different from its neighbors. Why only on our planet is the current climate "just right"?

THE TROUBLE WITH VENUS

Venus really is too hot. Being 41 million km closer to the Sun than Earth has destroyed the habitability of this planet. The resulting increase in the intensity of sunlight at the surface of Venus is about a factor of two compared with Earth. Under these conditions, liquid water cannot be stable and a runaway greenhouse effect is inevitable. The theoretical calculations that lead to this conclusion are brilliantly supported by the discovery of the huge enrichment of deuterium in the remaining water vapor in the atmosphere of Venus. An enormous amount of hydrogen must have left this planet to produce such an extreme fractionation of the hydrogen isotopes, and this is just what the runaway greenhouse scenario predicts.

When we correct for the losses, we find that the basic volatile inventory on Venus and Earth appears to be the same. The amounts of carbon dioxide and nitrogen now in the atmosphere of Venus are very similar to the total amounts of these gases Earth has produced over geologic time. Among the major volatiles, only the water is absent, and now we know why. All three of these planets (Venus, Earth, and Mars) should have started out with roughly the same proportions of carbon and nitrogen compounds, as well as water, relative to their individual masses. This realization helps us to reconstruct the early atmosphere of Mars.

EARLY MARS: WAS THERE AN ANCIENT EDEN?

The real problem with Mars is not that it is too cold—that is, too far from the Sun. The problem with Mars is being too *small* (Fig. 12.17). The 0.07-bar atmosphere we have reconstructed

Figure 12.17 Earth and Mars. This image shows, at the same scale, the two planets that are most likely to have been hosts for life on our solar system. The problem with Mars, relative to the Earth, is that it is too small to have retained most of its atmosphere.

appears to be nothing more than the volatile inventory left behind after the last major martian impact. That event has drawn a kind of screen over the early history of the atmosphere that we cannot penetrate with real confidence.

Once again we turn to the noble gases. If we assume that both Mars and Earth started out with atmospheres that were similarly dense, we should find the same amount of noble gases per gram of planetary rock on both planets today. We don't. We would need to add 165 times the amount of primordial argon that we find on Mars today to bring the ratio of gas to rock on Mars to the value we find on Earth. This is consistent with the factor of at least a 100 increase determined from the calculations of the effect of impact erosion. If we then add the carbon and nitrogen in their proportions on Earth, we would produce a CO_2-N_2 atmosphere on Mars with a surface pressure of about 7.5 bars, more than 1000 times the present value.

This high surface pressure may never have existed on Mars, just as Earth's surface probably never felt the weight of the 70 bars of CO_2 produced by the planet over geologic time. However, this inventory permits the existence of a 1- or 2-bar CO_2 atmosphere, sufficient for liquid water to exist on the martian surface. The inventory we have reconstructed also includes a ~0.9-km-deep global layer of H_2O, an ample amount to produce the channels and other erosion features we have discussed.

We are now able to reevaluate the problem of life on Mars. Perhaps no life exists on this intriguing red planet today, even though jets of water and steam may occasionally erupt into the atmosphere from subsurface reservoirs. But some future mission might still find traces of life's early beginnings there. It is arresting to realize that the oldest evidence of life on Earth formed at an epoch when the availability of liquid water on the martian surface was just coming to an end. In addition to producing the spectacular channels and the smaller branching drainage systems, the water on Mars also seems to have collected in some places long enough to produce lakes and perhaps even a northern polar sea, allowing deposition of layered sediments (Fig. 12.18). Probably these

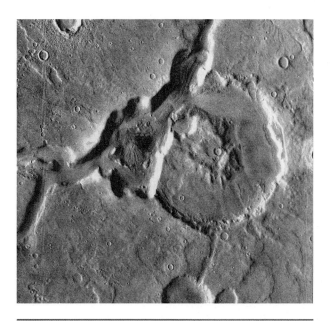

Figure 12.18 Flood channel intersecting martian craters. Evidently water from the channel has poured into the large crater through a breach in the wall. Many craters may have once held ponds of liquid water, perhaps with ice crusts.

bodies of water were covered with ice, at least in the later stages as Mars lost its dense atmosphere. By 3.5 billion years ago, life on Earth had evolved to a level of complexity that enabled the production of macroscopic fossils. The same thing might have happened on Mars, but it will be a difficult challenge to find out.

COULD WE BE MARTIANS?

Let's end this chapter with an intriguing speculation. Suppose that both Earth and Mars had the right chemical and environmental conditions for the formation of life prior to 4 billion years ago. On Earth, the limiting environmental consideration at that time was probably rare planet-sterilizing impacts. We can imagine that life might have formed and then been wiped out by such environmental catastrophes. Mars, in contrast, is less likely to have been hit by such large projectiles, simply because it presents a smaller target. It is likely that conditions for the stable existence of life developed on Mars before they did on Earth.

PERSPECTIVE

THE NEW SCIENCE OF ASTROBIOLOGY

Occasionally in science we reach a place where a new field or discipline is born. Usually this happens when breakthroughs or novel investigative tools cause a previously minor branch of science to assume a greater importance. Planetary science was such a new field, formed in the 1960s at the beginning of the space age. The prospect of sending scientific spacecraft to visit the Moon and planets was so exciting that many scientists from other fields wanted to participate, as well as students still in training. A new field of planetary science suddenly made sense—not as a part of astronomy or geology, but as a topic that could stand on its own.

In the middle of the 1990s another new field was recognized called *astrobiology*. The subject of astrobiology is life in the universe (or, as some prefer, the living universe). Many opportunities to study life in a cosmic context seemed imminent: advanced spacecraft to fly to Mars and even bring back martian samples for analysis in terrestrial labs; indication of a global ocean on Jupiter's moon Europa; discovery of planets circling other stars; new information about the conditions under which life began on Earth; and plans for a space station where astronauts could carry out biological experiments in microgravity. As it had with planetary science 30 years earlier, NASA took the initiative in funding the field of astrobiology.

Most of what we have been discussing in this chapter could be considered astrobiology. Part of Chapter 18 dealing with conditions in other planetary systems also fits within the definitions of astrobiology. We don't use the term very much in this text, but thinking about astrobiology has become a major theme in the exploration of our planetary system.

It is interesting to record, at the birth of this new field, what scientists think astrobiology is. These ideas are, of course, just a starting point. The field will change and adapt as some directions turn out to be blind alleys, while other paths open up exciting new discoveries. The following is a summary of the *NASA Astrobiology Roadmap*, prepared by about 400 scientists in 1998 under the leadership of author Morrison.

Now think about the exchange of material between the two planets. We know from the existence of about two dozen martian meteorites that even today Mars rocks can survive blast off from that planet and a long cold journey through space to reach the Earth. In the past, when impact rates were higher, even more Mars rocks would have reached our planet. The reverse—Earth rocks reaching Mars—is possible but much less likely. Because of Earth's larger gravity, it is much more difficult to launch a rock into space without also vaporizing it.

If environmental conditions stabilized first on Mars and if fragments of its crust were transported frequently to Earth, we can imagine that some of those rocks contained microbes. Following the last sterilizing impact on Earth, our planet might have been seeded by these traveling martian microbes. If an already functional life form arrived from another planet, it would likely have spread on Earth and consumed the available raw materials, precluding the emergence of native-born life. If this actually happened, then

Astrobiology addresses three fundamental scientific questions. The questions are old, but many opportunities to investigate these questions are new:

- Where did we come from (that is, how does life begin and evolve)?
- Are we alone (that is, does life exist elsewhere and can we find it)?
- Where are we going (that is, what is life's future on Earth and beyond)?

There are ten broad scientific goals in the *Astrobiology Roadmap*:

1. Understand how life arose on Earth.
2. Determine the general principles governing the organization of matter into living systems.
3. Explore how life evolves on the molecular, organism, and ecosystem levels.
4. Determine how the terrestrial biosphere has co-evolved with the Earth.
5. Establish limits for life in environments that provide analogues for conditions on other worlds.
6. Determine what makes a planet habitable and how common these worlds are in the universe.
7. Determine how to recognize the signature of life on other worlds.
8. Determine whether there is (or once was) life elsewhere in our solar system, particularly on Mars and Europa.
9. Determine how ecosystems respond to environmental change on timescales relevant to human life on Earth.
10. Understand the response of terrestrial life to conditions in space or on other planets.

None of these goals will be easy to reach. Answers will be derived slowly, as a result of hard work by many scientists from different backgrounds. But the results of this effort should be exciting, and perhaps revolutionary, in defining our place in the universe.

---

terrestrial life had its origin on Mars, and we would all truly be Martians.

12.7 QUANTITATIVE SUPPLEMENT: ATMOSPHERIC ESCAPE

A molecule or atom can escape from a planetary atmosphere when its energy is sufficient to overcome the gravitational attraction of the planet. Consider a molecule (or atom) of mass μ in the atmosphere of a planet with radius R, located sufficiently high that if it begins moving outward with sufficient speed, it will not encounter any other molecules. Its kinetic energy is given by the formula

$$E = \frac{1}{2}\mu v^2$$

where v is the speed of the molecule. The condition necessary for escape is that this energy be greater than the molecule's potential (or gravitational) energy, given by

$$E = \frac{GM\mu}{R}$$

where M is the mass of the planet and G is the universal constant of gravitation ($G = 6.7 \times 10^{11}$ newton m^2 kg^{-2}).

If we equate the kinetic and gravitational energies, we can solve for the speed at which the molecule is just able to leave the planet, called the escape velocity, v_e:

$$v_e = \sqrt{2\,GM/R}$$

Note that the mass μ of the molecule is not a part of this equation, since it appears in the expressions for both kinetic and gravitational energy. Thus, the escape velocity for a rocket, a baseball, or a molecule will be the same on the same planet. On Earth, this escape velocity is 11 km/s.

In a planetary atmosphere, the speeds achieved by atoms and molecules depend on their masses and temperatures. From thermodynamics, we know that in a mixture of gases each species will have the same kinetic energy. If the energy is the same, then the less massive molecules in a mixture must be moving faster, while the more massive ones must be moving more slowly. The average thermal velocity, v_t of a molecule of mass μ is given, according to an expression derived by the nineteenth-century Scottish physicist James Clerk Maxwell, by

$$v_t = \sqrt{3R^*T/\mu}$$

where T is the local temperature and R^* is the universal gas constant ($R^* = 8.31$ joule deg^{-1} mole^{-1}).

If a planet is to retain its atmosphere over the lifetime of the solar system, we require that the average thermal velocity of the molecules be considerably less than the escape velocity, or

$$v_t \ll v_e$$

Massive molecules at low temperatures have the best prospect of remaining bound to their planet, since their thermal velocities will be lowest. We can show this using the expressions already given for the thermal and escape velocities:

$$\sqrt{3\,R^*T/\mu} \ll \sqrt{2\,GM/R}$$

Since there is a distribution of molecular velocities about this average value, at any given temperature some of the faster molecules will still escape. Calculations that allow for the range of speeds present indicate that a gas will be retained over billions of years if it meets the condition

$$\sqrt{3\,R^*T/\mu} < 0.2\sqrt{2\,GM/R}$$

Summary

In spite of dedicated searches on other worlds, Earth remains the only inhabited planet we know. Current ideas for the origin of life indicate that our planet's early atmosphere must have contained no free oxygen. Instead, compounds such as N_2, CO, and H_2O would have permitted the chemical reactions necessary for the origin of life. Just how life began is not yet clear, but we find no reason at this stage to exclude the possibility that early Mars possessed the same conditions that led to the origin of life on Earth.

Life apparently began on Earth in an era when the planet was still being bombarded by debris, with the largest impacts capable of sterilizing the planet. Even as the impact rate has dropped, our biosphere must still deal with catastrophic events that can cause mass extinctions. Life, however, has survived to become a global phenomenon. As it evolved, some microbes developed the capability to perform photosynthesis, thereby connecting life on Earth with a source of energy that will last for billions of years. This connection with the Sun eventually led to a change in the composition of Earth's atmosphere, as green plants produced abundant free oxygen. The discovery of several percent of oxygen in another planet's atmosphere would be a sure indication of the existence of life, even without observing evidence of plant growth or detecting signals from a technically advanced civilization. Life on Earth would have revealed its presence through its effects on the atmosphere for the last 2 billion years. No comparable evidence for life exists on Mars.

In fact, the current environment at the surface of Mars is extremely hostile to life as we know it, with short wavelength ultraviolet radiation reaching the planet's surface, temperatures that sink more than 100°C each night, and, most important, no possibility of liquid water. Nevertheless, as we recognize that there are other, more favorable environments on the planet, the search for life on Mars that began with the Viking landings in 1976 is continuing.

The Viking spacecraft found no evidence for organic compounds in the martian soil at levels less than one per billion. Although each of three Viking biology experiments appeared to give a positive response, the same result was obtained from heat-sterilized martian soil, indicating that chemistry, rather than biology, was responsible. An exciting announcement of fossils of microbial life in a martian meteorite has proved to be inconclusive, so we still have no convincing evidence that life ever existed on Mars.

We can mentally reconstruct an early atmosphere on Mars that was ten times more massive than the present one by studying the abundances of the noble gases and the isotopes of nitrogen. This would not have been enough atmosphere to provide the water necessary to cut the giant outflow channels on the martian surface.

An additional loss factor of 100 can be derived from considerations of the early bombardment of the planet, causing impact erosion of the atmosphere. Mars, with its small size and its proximity to the asteroid belt, was especially vulnerable to that process. Venus, in contrast, was large enough to have a stable atmosphere, but it is sufficiently close to the Sun that liquid water cannot remain on its surface, and a runaway greenhouse effect ensues. The enormous enrichment of deuterium in the tiny amount of water left on Venus today is a result of this runaway greenhouse effect.

Comparing Venus and Mars with Earth provides a sobering reminder of just how special our planet is. Earth is the right size, and its nearly circular orbit is at the right distance from the Sun. It is now simply up to us, its human inhabitants, to avoid the destruction of its habitability.

KEY TERMS

impact erosion

K/T event

noble gases

Phanerozoic era

photochemistry

photodissociation

REVIEW QUESTIONS

1. What conditions were necessary for the origin of life on Earth? How did life begin?

2. Describe the interplay between life and the atmosphere. How does life affect Earth's atmosphere today?

3. How important were impacts for the evolution of life on Earth? How important are impacts for life on Earth today?

4. How would you go about finding life on another planet? Imagine yourself trying to find life on Earth. Which of your tests would work now? Which would work 1000, 1 million, or 1 billion years ago?

5. Describe the Viking GCMS experiment. What did it show? Why were these results important in the search for life on Mars?

6. Describe each of the three Viking biology experiments, how they worked, and what results they obtained. Did they prove or disprove the existence of life on Mars? Explain.

7. What are the noble gases? How can we use them to understand the thinness of the present martian atmosphere?

8. Why was impact erosion of the atmosphere more significant on Mars than on Earth? When did it occur? What was happening on the Moon at that time?

9. Why is the atmosphere of Venus so much thicker than that of Earth? Cite some evidence from the isotopes of hydrogen in the present atmosphere of Venus that supports your argument.

10. Explain what is meant by the phrase "Goldilocks and the three planets." How does our experience with life on Earth and the evolution of the atmospheres of Mars and Venus make you feel about the prospects for finding life elsewhere in the universe?

QUANTITATIVE EXERCISES

1. We have repeatedly stated that Earth has the equivalent of 70 bars of CO_2 stored in carbonate rocks. Suppose you took a column of CO_2 that produced a surface pressure of 70 bars on Earth and moved it to Mars. What surface pressure would you measure on Mars?

2. Imagine that a comet with a diameter of 20 km suddenly struck Mars, with a velocity sufficiently low that everything stayed on the planet: the pre-impact atmosphere plus all the volatiles contributed by the comet. The present water vapor content of the martian atmosphere is equivalent to a global layer of water 20 mm thick. Would the water from the comet make a noticeable contribution to this inventory?

3. If the average temperature in the atmosphere of Earth is 260 K, find the mass m of the lightest atom that can be retained over the age of the solar system. Repeat the calculation if the actual temperature of the upper atmosphere where escape takes place is 600 K.

4. Calculate the smallest molecular mass m that would remain on Neptune's satellite Triton for 4.5 billion years, if its atmosphere has the same temperature as the surface, 37 K. What gases could you suggest that would satisfy this constraint?

ADDITIONAL READING

Cooper, H. 1980. *The Search for Life on Mars.* New York: Holt, Rinehart and Winston. A description of the Viking mission by a science writer for *The New Yorker* magazine.

Davies, P. 1999. *The Fifth Miracle: The Search for the Origin and Meaning of Life.* New York: Simon & Schuster. Wide-ranging and up-to-date discourse on the nature and origin of life by physicist and leading science writer Paul Davies.

Goldsmith, D. 1997. *The Hunt for Life on Mars.* New York: Dutton. Popular discussion of life on Mars, with particular attention to the story of Mars rock ALH 84001 and its possible microbial fossils.

Goldsmith, D., and T. Owen. 2002. *The Search for Life in the Universe,* 3d ed. Sausalito, CA: University Science Books. Popular textbook on the origin of life on Earth and the search for life in our solar system and beyond.

Horowitz, N. 1986. *To Utopia and Back.* New York: Freeman. An exciting account of the Viking quest for signs of life on Mars and its significance, written by the principal investigator of the PR experiment.

Schopf, J. W. 1999. *Cradle of Life: The Discovery of Earth's Earliest Fossils.* Princeton, NJ: Princeton University Press. Entertaining account of deciphering the ancient records of life on Earth.

13

JUPITER AND SATURN: THE BIGGEST GIANTS

High-resolution Cassini image of Jupiter in true color.

When ancient astronomers named the planet Jupiter for the king of the gods in the Roman pantheon, they had no idea of the planet's true dimensions. The name is entirely appropriate, however, because Jupiter is more massive than all the other planets combined. It has a faint system of rings and more than 40 known satellites. Of these, Ganymede and Callisto are larger than the planet Mercury, Europa has a global ocean of liquid water, and Io is wracked by active volcanism. Jupiter itself has an internal source of heat; it is radiating about twice as much energy as it receives from the Sun. This giant also has the strongest magnetic field of any planet, which sustains a huge magnetosphere; at some radio frequencies, the magnetosphere can radiate more energy than the Sun itself.

Saturn is not far behind in these superlatives, even ahead in some. Most famous for its beautiful, intricate system of rings, this planet also has an internal source of energy, more intense than Jupiter's. Saturn has about as many satellites as Jupiter; one of these, Titan, is only slightly smaller than Ganymede.

Although these two huge planets are both made of mostly hydrogen and helium, their appearances are quite different. The visible surface of Jupiter is a deck of clouds that exhibits pastel shades of various colors in addition to the white condensation clouds one would expect. These clouds are organized into bands that run parallel to the planet's equator, producing a pattern of alternating dusky belts and lighter zones that is visible with even a small telescope. The planet Saturn is rather bland compared with Jupiter. The structure in Saturn's cloud deck is much less pronounced, and the color of these clouds is more uniform.

13.1 MISSIONS TO THE GIANT PLANETS

Knowledge about the Jupiter and Saturn systems increased dramatically in the period 1974–1981 as a result of two sets of spacecraft: Pioneers 10 and 11 launched in 1972, and Voyagers 1 and 2 launched in 1977 (Fig. 13.1). The Galileo orbiter and atmospheric probe that arrived at Jupiter in 1995 provided much additional information about this planet, and the similar Cassini orbiter mission is approaching Saturn as we write this book. Most of what we know about Jupiter and Saturn is derived from these highly successful space missions.

THE CHALLENGE AND THE PIONEERS

It is considerably more difficult to travel to Jupiter and Saturn than to Mars or Venus. The greater distances the spacecraft must traverse mean longer trip times and require greater reliability of all the spacecraft systems. In the inner solar system, electric power can be obtained by means of solar panels that convert sunlight to electricity. But even at Jupiter's distance, the sunlight lacks the intensity to generate the power required to operate a spacecraft. Hence missions to the outer solar system must include small electric generators powered by the heat released by the radioactive fuel they carry.

The large distances also create difficulties in communication. Just as sunlight decreases in

Figure 13.1 The launch of Voyager 1 on September 5, 1977, the first of two Voyager spacecraft to make the journey to Jupiter and Saturn. It has crossed the orbit of Pluto and is heading toward the Oort comet cloud.

intensity with the square of the distance, so do the radio waves that carry messages to and from the spacecraft. That means huge antennas and sensitive receivers must be used. During part of the Voyager 2 encounter with Neptune, for example, the signals from the spacecraft, traveling at the speed of light, took 4 hours to reach Earth. At such large distances, a spacecraft must be much more autonomous than if it were visiting Venus, where commands and responses can be sent in a matter of minutes.

Despite all these difficulties, the early missions to the outer solar system were very successful. The first missions were Pioneers 10 and 11, which encountered Jupiter in 1974 and 1975. Before these pathfinder missions, it was uncertain whether spacecraft could be sent safely to the outer planets. There was fear of destructive impacts with dust in the asteroid belt, which must be crossed to reach the giant planets. The Van Allen belts of the inner jovian magnetosphere also posed a hazard. Asteroid dust proved unimportant, but the data from Pioneer helped in the design of radiation-resistant electronics for subsequent spacecraft that could survive the rigors of the jovian magnetosphere.

In addition to their role as scouts for the Voyagers, Pioneers 10 and 11 carried the first scientific instruments to the outer solar system. Well instrumented to map out the charged particle belts, the Pioneers made numerous discoveries about Jupiter's magnetosphere as they passed through the system.

The trajectory of Pioneer 11 allowed a gravity-assisted deflection by Jupiter across the solar system for a close flyby of Saturn in 1979 (see Section 1.5). This passage led to the discovery of a new narrow ring as well as to the first description of the planet's magnetic field and belts of trapped charged particles. Again, the Pioneer served as pathfinder, passing through the plane of the rings at the same distance from Saturn required by Voyager 2 for its gravity-assisted trajectory to Uranus and Neptune. There were no adverse effects on the spacecraft, showing that the way was open for Voyager.

THE VOYAGER ENCOUNTERS

The Voyager spacecraft rivaled the two Viking landers in size and complexity (Fig. 13.2). Launched in 1977, Voyagers 1 and 2 reached Jupiter in March and July of 1979. Each encounter actually stretched over many weeks, as the spacecraft made close flybys of the four large satellites while approaching the planet. Gravity-assists at Jupiter then sent both spacecraft on to Saturn, where Voyager 1 arrived in November 1980, followed by Voyager 2 in August 1981.

The Voyager 1 encounter with Saturn was targeted toward its largest satellite, Titan. This meant that the spacecraft could not pass close enough to Saturn to get the necessary boost required to reach Uranus or Neptune. However, the trajectory of Voyager 2 was chosen so that the spacecraft could fly on to the Uranus system in January 1986, where a third gravity-assist directed it to Neptune, where it arrived in August 1989 (Fig. 13.3).

As the description of their trajectories indicates, these were flyby missions, and no instruments were sent into the planets' atmospheres. A complement of sophisticated equipment on the spacecraft made measurements of the magnetic fields and the properties of charged particles in

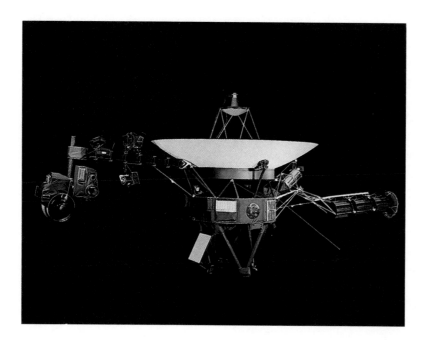

Figure 13.2 A photo of the Voyager spacecraft showing the location of the various instruments. The big antenna at the top is about 4 m in diameter. Most science instruments are mounted on the boom extending to the left. The nuclear thermal electric generators are on the opposite boom.

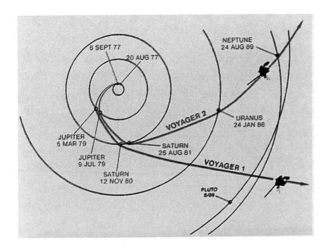

Figure 13.3 Because of a favorable lineup of the planets that occurs only once in 126 years, the Voyager spacecraft launched in 1977 could fly from Earth to Jupiter, Saturn, Uranus, and Neptune in less than 12 years. The decision to make a close flyby of Titan with Voyager 1 prevented this spacecraft from going on to Uranus and Neptune.

Figure 13.4 The 3-ton Galileo spacecraft being launched from *Space Shuttle Atlantis* in 1989. The white cylinder contains the rocket stage used to propel the spacecraft from Earth's orbit out to Jupiter.

the immediate environment, and used remote-sensing techniques to study the radio, light, and infrared radiation reflected and emitted by the planets, satellites, and rings.

GALILEO AND CASSINI HUYGENS

Successful as they were, the Voyagers, which were flybys, by no means represent the end of our efforts to understand the outer solar system. The next step in exploration requires the use of orbiters and probes. The Galileo spacecraft, launched in 1989, went into orbit around Jupiter after deploying a probe into the planet's atmosphere in December 1995 (Fig. 13.4). Sadly, the large antenna on the Galileo orbiter did not open completely after launch, thereby curtailing communication with Earth. Nevertheless, the spacecraft returned stunning images of the planet, its satellites and rings, as well as a vast amount of information about the magnetosphere, during an active lifetime of more than 6 years.

The probe had a lifetime of only 57 minutes, during which it successfully entered Jupiter's atmosphere, jettisoned its heat shield, and deployed a parachute that slowed its rate of descent to about 100 km/hr (Fig. 13.5). The instruments on board included a lightning detector, a mass spectrometer, and a nephelometer that could test for the presence of clouds as the probe passed through them. The probe survived to a depth in the atmosphere where the pressure was 22 bars and the temperature was 450 K. At that point, communication was lost, but the probe must have continued to descend until it reached depths with temperatures capable of vaporizing it some 9 hours after entry.

A Saturn orbiter called Cassini with a probe named Huygens was launched in October 1997 as a joint mission between NASA and ESA, the European Space Agency. It will arrive in the Saturn system in July 2004 and deploy the probe into the atmosphere of Saturn's satellite Titan in January 2005. This probe may survive for 30 minutes or longer on Titan's frigid surface, perhaps on the shore of a hydrocarbon lake, after a leisurely 2-hour descent through the atmosphere.

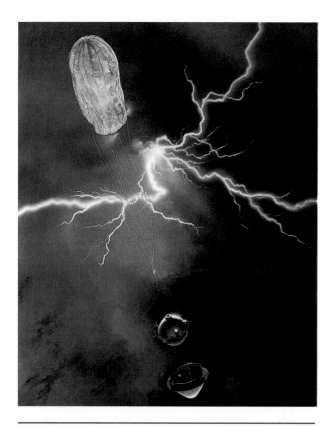

Figure 13.5 Artist's impression of the descent of the Gaileo probe into Jupiter's clouds. In this view, the glowing heat shield has just been jettisoned.

13.2 INTERNAL STRUCTURE OF TWO FLUID PLANETS

We can see the basic difference between the outer and inner planets by looking at their respective densities. The density of Jupiter is only 1.3 g/cm³, about the same as that of the Sun, and the density of Saturn is even lower: 0.7 g/cm³. For comparison, recall that the densities of the inner planets are more than three times greater, and that densities tend to increase with the planets' masses (except for anomalous Mercury) because gravitational compression compacts material in the interiors. A jovian-mass planet made of the same rock and metal as Earth or Venus would be even more dense from self-compression. The low densities of Jupiter and Saturn, therefore, are a sure indication that they are made of something far less dense than metal and rock.

CONSTRUCTING THEORETICAL MODELS

The explanation for the low densities of Jupiter and Saturn, as we noted in Section 3.1, involves the two most common elements in the universe, which are also the two lightest and the two most difficult to condense: hydrogen and helium. Early studies of these two planets quickly demonstrated that they could not be composed of pure hydrogen. If they were, their densities would be even lower than they are. Adding helium to the models in about the same proportion that it is found in the Sun and other stars makes up the difference. There must also be a dense core at the centers of these planets to account properly for the observed details of the planets' gravitational fields.

It is difficult to go further in our description of the interiors with great confidence. We have no seismic data to tell us about properties of the interiors, nor can we expect to get any, since neither of these planets has a solid surface. Instead, we must use the information we can obtain from remote investigations—mass, radius, rotation rate, heat balance, atmospheric composition, gravitational effects on satellite orbits—to construct models for the distribution and state of matter deep inside Jupiter and Saturn. This challenge is in many ways similar to that of modeling the interiors of the terrestrial planets (see Section 8.5), without the advantages of the seismology data we have for Earth and the Moon.

A basic problem in constructing such models of the giant planets is our limited knowledge of the behavior of the planets' principal constituents, hydrogen and helium, at appropriately high pressures and temperatures. The central temperature of Jupiter must be about 25,000 K to be consistent with the emitted thermal radiation, and the pressure there may be as great as 100 million bars, compared with just 4 million bars at the center of Earth.

Even before these extreme conditions are reached, we know that hydrogen will first liquefy and then assume a metallic state: The pressure squeezes the hydrogen atoms so much that the electrons are no longer bound to the nuclei. The liquid hydrogen then has the conductivity of a metal. This transition occurs at a depth of about

20,000 km, or about 75% of the distance out from Jupiter's center; the exact range of pressures and temperatures for this transformation is not well determined. Above this zone, hydrogen is in the molecular form of two atoms linked together (H_2), but both the molecular and metallic states are fluids. At still greater distances from the center, the hydrogen assumes a gaseous state, but most of the mass of the planet is fluid. Looking at Jupiter from the outside, we simply see the upper layers of the deep gaseous atmosphere.

THE CORES OF THE GIANT PLANETS

Continuing our journey to the centers of these planets, we would find that beneath the deep layer of liquid metallic hydrogen is a core of rock and ice. Here the terms *rock* and *ice* are used very generally to denote compounds of silicon, oxygen, metals, and the heavy volatile elements. We have no idea exactly what compounds these elements form. We know only that there seems to be a strong concentration of mass at the center of each planet, and that this mass must consist of elements heavier than hydrogen and helium.

Both Saturn and Jupiter appear to have cores of about the same size: 10–15 times the mass of Earth. There is still considerable uncertainty about the properties of these cores, however, with some models suggesting either higher or lower masses. Since Saturn is smaller than Jupiter, its core is larger compared with the planet as a whole. In other words, Saturn is depleted in hydrogen and helium, relative to Jupiter; if Saturn had its full quota of these two elements, it would be as large as Jupiter.

Calculations based on theories for the behavior of matter at the temperatures found in the giant planets indicate that Jupiter is just about as large (in diameter) as a planet can be. If we perform the thought experiment of adding more mass to Jupiter, we find that self-compression, due to the increased mass, would lead to a smaller diameter. Eventually, adding enough mass would cause the internal temperatures and pressures to increase to the point where nuclear fusion reactions would

begin. At this point, corresponding to 70–80 times the current mass of Jupiter, a star is formed. A smaller mass, with approximately 13 times the mass of Jupiter, could generate some internal heat from the fusion of deuterium into helium in its core, a possibility we will return to in Chapter 18. Once internal energy is generated by nuclear reactions, a new adjustment of internal structure takes place, allowing the object to grow in size. Meanwhile, it is noteworthy that a supergiant planet with 10 or even 50 times the mass of Jupiter would actually be smaller than Jupiter itself.

Figure 13.6 illustrates the conditions one would encounter on voyages to the interiors of these two planets. There are no well-defined solid surfaces or even gas-liquid boundaries. Such transitions are sharp only under relatively low pressures. Instead, we find immensely deep atmospheres in which the gases gradually become compressed to the point where they liquefy. In leaving Mars to venture into the outer solar system, we have left behind the planets on which it is possible to stand (except for Pluto—and don't forget the satellites, many of which seem more friendly to us than these giants).

TWO OVERHEATED GIANTS

Both of these giant planets radiate more energy than they receive from the Sun: Jupiter twice as much, Saturn nearly three times. If we think of each planet as a giant light bulb, Jupiter is radiating a total of 4×10^{17} watts, and Saturn is glowing with half this power, 2×10^{17} watts. These planets are therefore equivalent to 4 and 2 million billion 100-watt light bulbs, respectively, but the energy they radiate is not emerging as visible light. The outer, visible layers of both planets are far too cold for that. Instead, the energy emerges as infrared radiation.

It is possible to calculate what the temperature of Jupiter should be from a knowledge of the local intensity of sunlight and the albedo of the planet. The answer is about 107 K ($-166°C$). Observations of Jupiter at infrared wavelengths show that the amount of radiation it produces

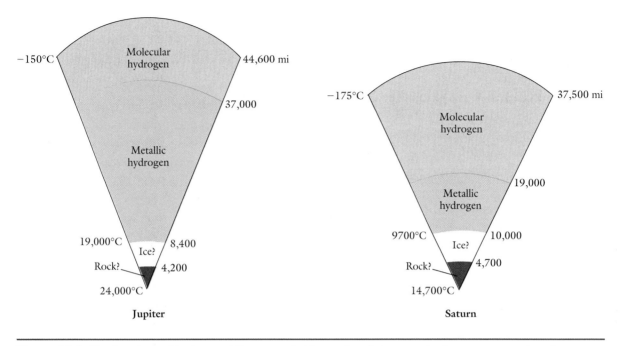

Figure 13.6 The interiors of Jupiter and Saturn probably consist of the same basic components, but they are distributed in different ways, as these simplified diagrams indicate. The interior of Saturn is not as hot as that of Jupiter, allowing helium to condense. Both planets have cores at their centers with masses approximately equivalent to 10–15 Earths.

corresponds to a temperature about 20° warmer than this calculation. Since radiated power increases as the fourth power of the temperature, this means that Jupiter radiates twice as much energy as it gets from the Sun, as we stated earlier.

What is generating this extra energy? As we have seen, the answer does not lie in nuclear reactions, the ultimate source of sunshine and starlight. So instead of invoking nuclear physics, let's go back to Isaac Newton. If an apple falls from a tree, it acquires energy of motion from Earth's gravitational field, which it gives up abruptly when it hits the ground. In the case of Jupiter, it is the contraction of the entire planet from an initial extended cloud of gas and dust that produces most of the observed energy. This cloud was a condensation in the primordial solar nebula. Newton's apple may now be thought of as a piece of Jupiter itself, moving inward under the influence of the planet's gravitational field. The resulting energy of motion is transformed into faster motions of the gas molecules—an increase in the temperature of the gas. (This is the

same process once invoked to power the Sun, before thermonuclear fusion was discovered; see Section 2.4.)

During the final stages of its formation process, when matter was streaming into the planet from considerable distances, Jupiter must have been very hot indeed, probably hot enough to produce a visible reddish glow. About 4.5 billion years ago, this hot young planet had a profound effect on its forming satellites, as we shall see in Chapter 15. Jupiter today is much quieter, slowly releasing primordial energy from that bygone era, possibly still contracting very slowly at a rate no greater than 1 mm/year.

HELIUM RAIN: A DIFFERENT ENERGY SOURCE

The simple contraction scheme just described does not account for all of the power generated by Jupiter, and it doesn't work for Saturn at all. Saturn is sufficiently small that it never reached the red-hot stage in its early youth; hence, there has been ample time in the ensuing 4.5 billion years for this planet to cool down much further than its giant neighbor. Why then is it still emitting so much energy?

The answer, ironically, is found in the *lower* internal temperature that Saturn exhibits today. At sufficiently high pressures and temperatures, liquid helium dissolves in liquid hydrogen, in the same way that a cook can dissolve large amounts of sugar in hot water. But just as the cook has trouble stirring sugar into cold water, at the lower temperatures in Saturn's interior, helium does not dissolve. Droplets of helium form in the liquid hydrogen, and, being denser, they move toward the center of the planet, like vinegar separating from the olive oil in a salad dressing. This very slow helium rainfall, deep in Saturn's interior, takes us back to Newton's apple: Once again, gravitational energy is converted into energy of motion, which is transformed into heat and ultimately radiated into space. The same thing is happening on Jupiter as well, but on a smaller scale. The interior of Jupiter is warmer, so more helium can stay in solution.

This theoretical model for processes occurring far down inside these giant planets can be tested by measuring the ratio of helium to hydrogen in their atmospheres. If helium is indeed raining out in the interior, it must be disappearing from the atmosphere. Therefore we expect to find the abundance of helium relative to that of hydrogen to be smaller in Saturn's atmosphere than in Jupiter's, and also Jupiter's He/H ratio should be smaller than the starting value in the solar nebula.

The fraction of the mass of the initial nebula that was helium was 0.28. On Jupiter, measurements by two different instruments on the Galileo probe found the helium mass fraction to be 0.24. But on Saturn the Voyager spacecraft found that the helium fraction was only 0.21. The amount of missing helium on both planets is consistent with a rate of helium precipitation that would produce the extra energy that Saturn and Jupiter radiate. Hence, there is good consistency between the model for this internal energy source and its two observable consequences: the amount of helium remaining in the atmosphere and the excess power radiated by the planet.

13.3 ATMOSPHERIC COMPOSITION AND STRUCTURE

The first gas to be identified in the atmospheres of Jupiter and Saturn was methane. More than 65 years ago, strong absorption bands in the spectra of both planets were found to coincide with absorptions in laboratory spectra of methane gas (Fig. 13.7). Ammonia was discovered on Jupiter in the same way at about the same time, but whether it was present on Saturn remained a controversial problem until the late 1960s, when conclusive evidence for it was obtained. Hydrogen and helium, the most abundant gases in the atmospheres, were much more difficult to detect, but they were also identified in the 1960s.

At the pressures and temperatures in the regions of these atmospheres that we see, ammonia can condense, just as water vapor condenses in the lower part of our own atmosphere. Depending on the circulation of the atmosphere, we expect to find both cloudy regions and clear regions when we look at these planets. The deeper we can see in the cloud-free regions, the more gaseous ammonia vapor we find, just as we have much more water vapor below the clouds on Earth than in the atmosphere above them.

ABUNDANCES OF ATMOSPHERIC GASES

Methane and ammonia dominate the spectra of Jupiter and Saturn (Fig. 13.7), but they are only minor atmospheric constituents, in smaller proportion on Jupiter than argon is in our own atmosphere. Although 500 times more abundant than methane, hydrogen has a much weaker absorption spectrum because it is a molecule of two identical atoms that interacts only weakly with light. Nitrogen and oxygen are similar identical-atom molecules, and they are also transparent to visible and infrared light—hence, their lack of importance for the terrestrial greenhouse effect as well as the fact that we can see long distances through our atmosphere when it is free of clouds. Helium is even more difficult to detect because its only absorptions at the temperature of Jupiter's atmosphere occur at very short wavelengths in the ultraviolet, which cannot be observed from Earth.

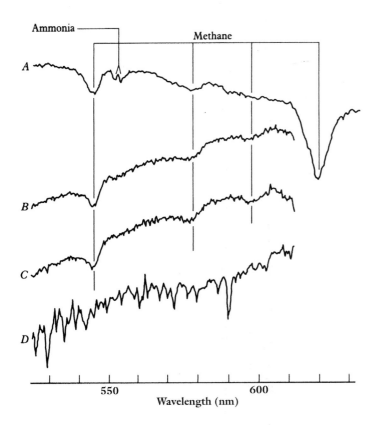

Figure 13.7 Spectra of Jupiter (A), Saturn (B), and Titan (C), showing methane and ammonia absorptions. These spectra have been divided by a spectrum of the Sun (D) to remove solar absorptions. If the planets had no atmospheres, their divided spectra would simply be gently curving lines. Note that there is more ammonia on Jupiter than on the other two objects.

Even from space, direct detection of helium is not easy because these short wavelengths are screened by hydrogen in Jupiter's atmosphere.

Despite these difficulties, we now have rather good estimates for the relative abundances of hydrogen and helium, as well as many other gases that were measured by the Galileo atmospheric probe or by spectrometers. The instruments on the Voyager spacecraft succeeded in determining the helium abundance by measuring the effect of this gas on the observed spectrum of hydrogen, but the definitive value for the helium abundance of Jupiter was provided by two instruments on the Galileo probe. The two instruments gave identical results: the mass fraction of 0.24 for He that we have already quoted. As Figure 13.8 illustrates, the probe mass

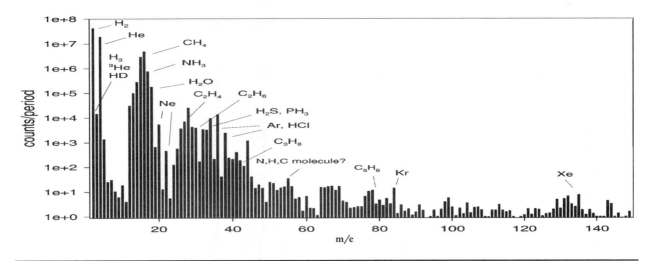

Figure 13.8 This mass spectrum was obtained by the mass spectrometer on the Galileo probe as it descended through the 18-bar region of Jupiter's atmosphere. The logarithmic ordinate gives the measured abundance of each species, while the abscissa reveals their atomic masses.

spectrometer was able to identify a large number of other gases that remote measurements could not detect: Ne, Ar, Kr, Xe, and H_2S. Table 13.1 presents a summary of our current knowledge of abundances in the atmospheres of Jupiter and Saturn.

EXPECTED AND UNEXPECTED GASES

Table 13.1 is divided into two parts to distinguish the major components of these atmospheres from trace gases. Our expectation is that all the abundant elements (except for the noble gases helium argon, neon, etc.) should combine with hydrogen to make molecules. These should be the simplest possible molecules, given the huge excess of hydrogen and the relatively low pressures and temperatures in the upper atmospheres of these planets. The resulting compounds are indeed present and are shown in the upper part of the table. The noble gases He, Ne, Ar, Kr, and Xe do not form compounds (nobility implies a lack of interest in associating with others!) and thus appear in the table as elements.

This table lists a large number of trace constituents as well, such as carbon monoxide (CO), acetylene (C_2H_2), and other hydrocarbons (carbon-hydrogen compounds). Evidently, sources of energy other than the local heat are acting to produce these molecules. In the upper atmosphere, there are photochemical reactions in which solar ultraviolet radiation breaks down methane, and its fragments form acetylene and other hydro-

Table 13.1   The compositions of the atmospheres of Jupiter and Saturn (by volume)

Gas	Formula	Jupiter	Saturn
Main Constituents (percent)			
Hydrogen	H_2	86.3	89.7
Helium	He	13.5	9.9
Water vapor	H_2O	0.36 (?)	0.8 (?)
Methane	CH_4	0.18	0.4
Ammonia	NH_3	0.06	0.1
Neon	Ne	0.006	?
Hydrogen sulfide	H_2S	0.0007	?
Argon	Ar	0.00014	?
Trace Constituents (parts per billion)			
Acetylene*	C_2H_2	80	200
Ethylene*	C_2H_4	7	trace
Methyl acetylene*	C_3H_4	2	0.6
Ethane*	C_2H_6	40,000	3000
Benzene*	C_6H_6	2	?
Phosphine	PH_3	700	3000
Carbon monoxide	CO	2	1
Carbon dioxide	CO_2	3	0.3
Germane	GeH_4	0.7	0.4
Arsine	AsH_3	0.22	3
Krypton	Kr	6	?
Xenon	Xe	0.6	?

*Photochemically produced hydrocarbons

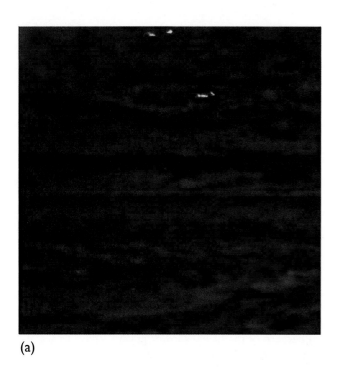

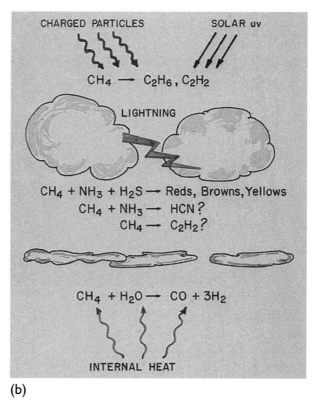

(a) (b)

Figure 13.9 **(a)** This Galileo image of the dark side of Jupiter shows the planet's cloud deck dimly illuminated by sunlight reflected from satellite Io. Within this background it is possible to see several bright flashes caused by lightening in deep—perhaps water—clouds. **(b)** In addition to lightning and the bombardment of charged particles that cause the aurora, ultraviolet light from the Sun and internal heat from the planet are available to drive chemical reactions.

carbons. These compounds are denoted with an asterisk (*) in Table 13.1. This is the same kind of sunlight-driven chemistry that produces ozone in Earth's atmosphere and creates radicals like NH_2 and C_2 in the heads of comets.

In the convective region of the atmosphere, where we see the clouds, lightning discharges (detected by the Voyager spacecraft and confirmed by Galileo; Fig. 13.9) can contribute to these chemical processes and may add some acetylene. Still deeper, at temperatures around 1200 K, carbon monoxide is made by a methane–water vapor reaction, which can be written as

$$CH_4 + H_2O \rightarrow CO + 3 H_2$$

Since the CO produced in this way is observable, vertical currents must be sufficiently strong to bring it up to the visible layers of the atmosphere where it can be detected.

Table 13.1 makes Jupiter and Saturn look rather similar, but there are some significant differences. The smaller proportion of helium on Saturn can be explained by the condensation of this gas deep in the planet's interior, as discussed in the previous section. The higher proportion of methane also seems to be real and may reflect a difference in the mass of Saturn's core relative to the mass of the entire planet. A relatively larger core would be expected to contribute more heavy elements. We see another manifestation of this enrichment in the abundances of ammonia and phosphine, but germane is actually about equally abundant on both planets. Evidently, there are differences in atmospheric chemistry that we still don't understand.

THE CHANGE IN TEMPERATURE WITH ALTITUDE

One of these puzzles concerns the abundance of water. The Galileo probe was able to establish only a lower limit on the abundance of H_2O on Jupiter,

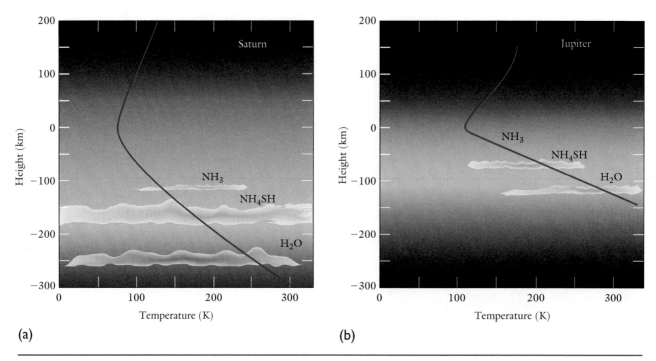

(a)   (b)

Figure 13.10   The atmospheres of both **(a)** Saturn and **(b)** Jupiter contain layers of clouds at different altitudes. Ammonia can condense over a greater altitude range on Saturn because of the planet's lower gravity. These diagrams show the approximate locations of major cloud systems on the two planets. The estimated position of the water clouds critically depends on the assumed abundance of H_2O in the atmosphere. The minimum temperature occurs at the tropopause.

even though we expect water vapor to be the most abundant gas (after H_2 and He) in the planet's lower atmosphere. All the abundances shown in Table 13.1 refer to the upper layers of the planets' atmospheres. Except for the photochemical products, we expect that the information obtained in this region would apply throughout the atmospheres for those gases that do not condense. The difference between the ammonia abundances above the clouds on Jupiter and Saturn is attributable to condensation. The visible layers of Saturn's atmosphere are colder than those of Jupiter, therefore they are depleted in ammonia gas. Observations made at radio wavelengths allow us to examine the atmospheres beneath the clouds, and here we find the amount of ammonia on Saturn to be about twice Jupiter's value. But there is no comparable information yet about water.

To understand this problem of condensation and mixing, we need to know how the temperature varies with altitude (or pressure) in these atmospheres. The best pre-Galileo information

about the change of temperature with altitude was derived from Voyager and Pioneer occultation measurements. As these spacecraft passed behind each planet, the radio signal they sent to Earth had to pass through thicker and thicker atmospheric layers (see Section 11.1).

The Galileo probe gave a much more accurate temperature profile over its descent trajectory on Jupiter, and this has been added to the occultation results in Figure 13.10. Comparing the two planets, we see that Saturn is distinctly cooler than Jupiter at similar pressure levels, owing to its smaller size and greater distance from the Sun. Approximate locations of various cloud layers in the atmospheres are marked. It is interesting that on both planets, temperatures above the freezing point of water occur at pressures just a few times greater than the sea-level pressure on Earth. An astronaut in the gondola of a balloon could float quite happily here in shirtsleeves and scuba gear, if one could just find a suitable gas to put in the balloon! These balmy conditions are a consequence of the planets' internal energy, although some warming would occur simply from an atmospheric greenhouse effect.

The cloud layers sketched in Figure 13.10 are located at the temperature levels expected for the condensation of the different gases known to be present in the atmosphere. However, the Galileo probe encountered quite different conditions during its descent (Fig. 13.11). Instead of discrete layers of thick clouds, it detected only thin haze. Furthermore, the mass spectrometer indicated that both ammonia and hydrogen sulfide were under-abundant, with too little to form the expected clouds. The scientists concluded that the probe had entered an unusually dry region of the atmo-sphere, perhaps caused by an enormous downdraft of cold dry air from Jupiter's stratosphere.

This same effect is well known on Earth. Pictures of our planet obtained from space inevitably show the Sahara Desert free from clouds (see Fig. 9.1). This is a consequence of descending dry air over subtropical latitudes (see Fig. 9.18), which also produces the Kalahari Desert and the Australian Outback in the Southern Hemisphere. As we shall see, such "desert regions" are more local on Jupiter rather than the natural consequences of global circulation.

The dry descending air may provide the answer to the water riddle: In this particular region on Jupiter, the water clouds must have been present at

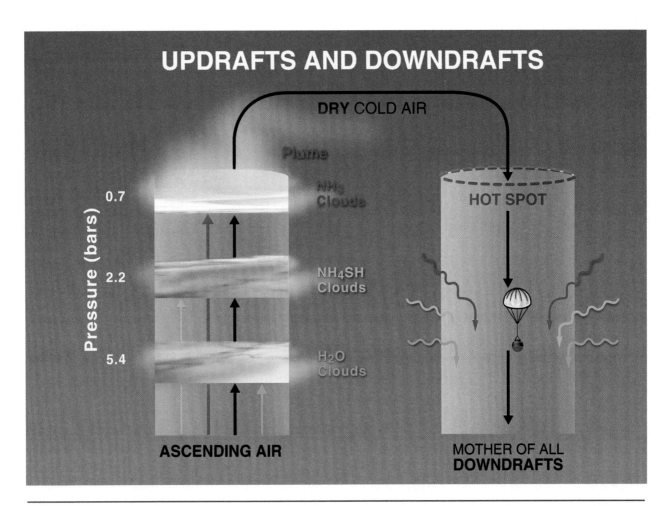

Figure 13.11 Artist's conception of a cloud-filled region of Jupiter's equatorial belt (left), where clouds of differing composition are formed progressively as condensation occurs in cooling, ascending air. At right, we see a region of downdraft corresponding to the hot spot where the probe entered the lower atmosphere. No clouds form in the descending dry air, but vapor from adjacent regions get progressively dragged into the downward-moving column.

a much lower level than Figures 13.10 or 13.11 predict. Only by passing through this condensation level could the probe expect to measure the high abundance of H_2O that would correspond to the predicted ratio of oxygen to hydrogen on Jupiter, but that was too deep for survival of the probe's instruments and radio transmitter.

A NEW TYPE OF PLANETESIMAL

The predicted abundances of the elements were based on the idea that carbon, oxygen, and sulfur should have the same proportions on Jupiter as they do in the comets that were thought to be the building blocks for Jupiter's core. As we discussed in Section 6.5, comets exhibit the same relative abundances of C, O, and S that are found in the Sun, but comets are deficient in nitrogen and argon. Thus, it was a great surprise when the Galileo probe revealed that both nitrogen and argon, as well as C, S, and every heavy element it measured besides neon and helium, were present on Jupiter in the same relative abundances as found in the Sun (Table 13.1). Neon is deficient on Jupiter because it dissolves in the helium droplets that precipitate in the planet's interior. When compared with the abundance of hydrogen, all the other heavy elements are enriched on Jupiter by the same factor of 3 ± 1 compared with the solar composition (Fig. 13.12).

We know of no solid objects in the planetary system today that exhibit this solar composition of the elements. Yet our models for giant planet formation tell us that the first step in this process is the accretion of icy planetesimals to form a core of heavy elements. If this model is indeed correct, we must postulate the existence of icy planetesimals that formed at temperatures below 38 K in order to trap gases like argon and nitrogen in solar proportions. This low temperature is surprising because it corresponds to the outer solar nebula, near Neptune's orbit, rather than to the temperatures expected at Jupiter's distance from the Sun. Given that the cores of all four giant planets are 10–15 times the mass of Earth, the icy planetesimals of solar composition that formed them must have been the most abundant form of solid matter in the early solar system.

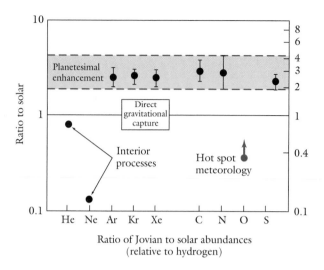

Figure 13.12 If Jupiter had formed as a giant condensation in the solar nebula and no internal differentiation had taken place, all the elements in its atmosphere would exhibit the same abundances relative to hydrogen that they have in the Sun (ratio to solar abundances = 1). However, helium and neon are precipitating out in the interior, and everything else appears to be enriched, although oxygen has not yet been measured.

13.4 WEATHER AND CLIMATE

Even a modest telescope shows much detail on Jupiter. The region of the planet's atmosphere that we see contains several different kinds of clouds. Individual pictures from spacecraft provide snapshot views of these clouds at particular instants in time (Fig. 13.13). Although such pictures may suggest that all the clouds we see are at the same level in the atmosphere of the planet, Figure 13.10 reminds us that this is not the case. Many of the color differences may be the result of seeing to different depths in different regions.

Even at telescopic resolution, which is about 2000 km for Jupiter, changes in the visible cloud systems can occur in a few hours. At the much higher resolution of spacecraft cameras, the planet presents a constantly shifting pattern of clouds, but an underlying pattern of currents flowing parallel to lines of latitude has maintained its stability for decades. On Saturn, cloud changes are usually much more difficult to see (Fig. 13.14), but there is also a stable underlying circulation pattern. It has been traditional to describe the appearance of these planets

Figure 13.13 A high-resolution image of Jupiter taken by Voyager 1 at a range of 4 million km. The dusky band running diagonally through the white clouds defines the north temperate current with wind speeds of about 120 m/s.

Figure 13.14 Saturn as viewed by Voyager 2. Even with enhancement of contrast, not much atmospheric structure is evident.

(especially Jupiter) in terms of alternating bright zones and dark belts, but the underlying currents have a greater persistence than this cloud pattern.

ROTATION PERIODS AND WIND SPEEDS

Three rotational periods have been established on Jupiter. The two periods labeled Systems I and II in Table 13.2 are average values and refer to the apparent average speed of rotation at the equator and at higher latitudes, respectively. These periods are defined by observed motions of features in the planet's cloud layers, which move with the local currents. Since there is no solid surface, we must look elsewhere for an absolute standard by which to judge these motions.

Studies of the jovian radio emissions show an unchanging periodicity that refers to the rotation of the planet's magnetic field. This deduction was

Table 13.2 Rotation periods for Jupiter and Saturn

	Jupiter	Saturn
Equatorial (System I)	$9^h50^m30^s$	10^h14^m
High latitudes (System II)	$9^h55^m41^s$	10^h40^m
Deep interior (System III)	$9^h55^m30^s$	$10^h39^m24^s$

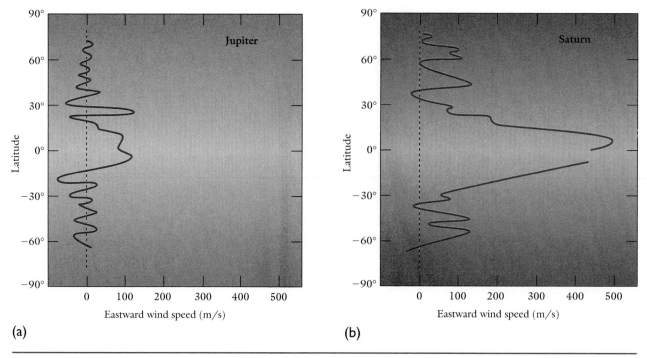

Figure 13.15   **(a)** A map of wind currents on Jupiter. There is a strong equatorial jet, with currents at higher latitudes alternating between easterlies and westerlies, roughly following the pattern of boundaries between dark and light belts and zones. **(b)** The equatorial jet on Saturn is both stronger and wider than that on Jupiter. This planet does not have the same pattern of alternating currents as its larger neighbor, except at latitudes above 40°.

verified by direct measurements of the magnetic field carried out by spacecraft. Because this field is generated deep in Jupiter's interior, the radio period, called System III, is identified with the true rotation of the planet itself. Judged against this standard, Jupiter exhibits an eastward-flowing equatorial jet stream (System I) with a relative velocity of about 360 km/hr, comparable to the velocities of jet streams on Earth. At higher latitudes in each hemisphere, an alternating pattern of easterly and westerly winds occurs (Fig. 13.15).

The visible clouds on Saturn are much more uniform than those on Jupiter. Only rarely do spots appear that can be seen in telescopes from Earth (Fig. 13.16). The Voyager cameras were able to define enough discrete features to map out Saturn's circulation, which is distinctly different from that of Jupiter (Fig. 13.15b). Once again an equatorial jet is apparent, but now it extends to 40° on either side of the equator, and its peak velocity is a remarkable 1600 km/hr. At still higher

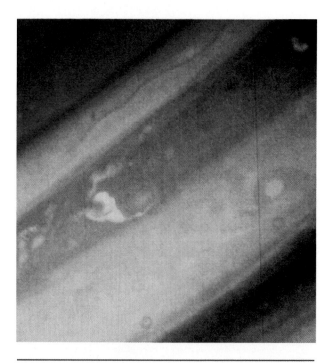

Figure 13.16   A close-up view of northern latitudes on Saturn. The contrasts have been greatly enhanced in this picture to reveal details that would otherwise be hard to see.

latitudes, there is again an alternating pattern of eastward and westward currents. Just as for Jupiter, the rotation of Saturn's magnetic field establishes the rotation period of the planet's core.

GENERAL CIRCULATION OF THE ATMOSPHERES

The reasons for the basic differences in atmospheric circulation between Jupiter and Saturn are not yet clear. One interesting possibility suggested by planetary meteorologist Andrew Ingersoll of Caltech is that the circulation patterns actually extend to very deep layers. The currents that we see may represent the outermost edges of concentric cylinders of gas, turning around the planet's rotational axis (Fig. 13.17). The sizes and shapes of these cylinders would then be determined by the internal structures of the planet, so the relative sizes of the cores of Jupiter and

Saturn may play a role in determining the nature of the atmospheric currents. This is another illustration of just how different these hydrogen-helium giants are from the rocky inner planets.

We can compare the meteorology on these planets with the global circulation of Earth's atmosphere (see Section 9.7). On Earth, huge spiral cloud systems often stretch over many degrees of latitude and are associated with motion around high-pressure and low-pressure regions. Clouds on Earth can move in latitude as well as longitude, as the migration of hurricanes from the Caribbean along the East Coast of the United States dramatically demonstrates every autumn. This kind of motion simply doesn't happen on Jupiter and Saturn.

Planetary meteorologists have simulated the general circulation of the giant planets by starting with a computer model of Earth and increasing the size of the planet and its rotation rate (in the

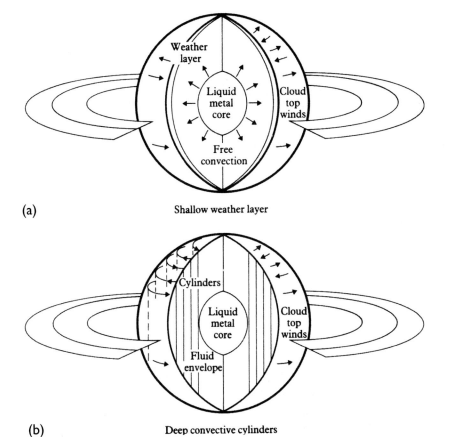

(a)   Shallow weather layer

(b)   Deep convective cylinders

Figure 13.17 Two models for the global circulation on Saturn apply to Jupiter as well: **(a)** suggests that the circulation is driven in a rather shallow layer in the visible part of the atmosphere; **(b)** invokes deep convection following a pattern of concentric cylinders.

model—not on the planet). Disruptive eddies are smaller relative to the size of the planet, leading to bands of east-west winds that are extremely stable. Thus, computer experiments play an important role in understanding the atmospheres of the outer planets, just as they do in tracing the thermal evolution of the inner planets and the operation of the greenhouse effect.

High rotation rate is certainly one factor that makes the circulation on the giant planets very different from that on Earth. Another is the lack of a solid surface. Weather on Earth is often closely tied to the local environment, which in turn is determined by the varied nature of our planet's surface, especially the land-water dichotomy. On Jupiter and Saturn, the absence of physical boundaries contributes to the stability of the atmospheric circulation patterns. The fluid interior short-circuits the atmosphere by eliminating the equator-to-pole temperature differences that generate the disruptive eddies on Earth.

THE GREAT RED SPOT

The permanence of underlying circulation patterns in the presence of constant change is most obvious on Jupiter, where one can see more features and variations in the cloud layers. The most famous of these features is the Great Red Spot (GRS), which has existed for at least 170 years and perhaps longer than 300 years (Fig. 13.18). Its dimensions are about 18,000 by 12,000 km, making it large enough to accommodate the Earth with plenty of room to spare. This huge size is probably responsible for the feature's longevity.

The true nature of this giant cloud system is still unknown, despite extensive observations. The reddish clouds within the spot exhibit a counterclockwise rotation with a period of about six days. (Remember that counterclockwise motion in the southern hemisphere of a planet is anticyclonic; see Section 9.7.) This cloud system thus appears to be an enormous anticyclone, a vortex or eddy whose center must be a region of locally high atmospheric pressure. Nevertheless, Voyager and Galileo pictures failed to reveal any evidence of upwelling at the spot's center. The clouds here seem remarkably

Figure 13.18 A close-up view of Jupiter's Great Red Spot. This giant storm system is large enough to swallow planet Earth. The white oval on the right had existed for more than 40 years when this photo was taken. Color contrasts are exaggerated in this and most other Voyager photos of Jupiter.

tranquil. Cyclones and anticyclones are also found on Earth, but on our planet they are very short-lived. Furthermore, the clouds associated with them are white water clouds like all the others that form on Earth.

The huge lateral dimensions of the GRS are associated with a considerable vertical extent as well, allowing it to project well above and below the adjacent cloud deck. The upward projection has been verified by direct observation, but the lower reach of this enigma has not been established. We also don't know what chemicals are responsible for its color.

OVALS, DARK BROWN CLOUDS, AND HOT SPOTS

On Jupiter, cloud features much smaller than the GRS are also persistent and localized. Three white ovals about as large as the planet Mars persisted for half a century at a latitude just south of the Great Red Spot (Fig. 13.19), before merging with each other in 2000. Dark brown clouds, which are evidently deeper layers glimpsed through holes in the nearly ubiquitous tawny cloud layer, are found

(a)

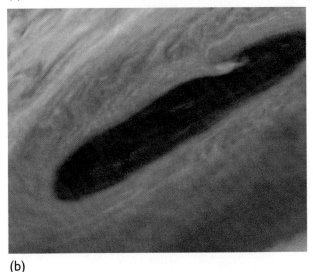

(b)

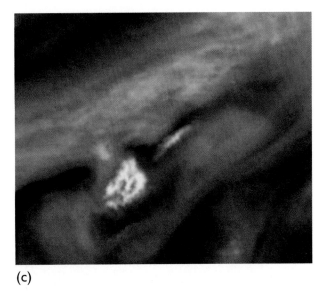

(c)

Figure 13.19 Weather on Jupiter. **(a)** Enhanced color image of the three white ovals on Jupiter in 1998, shortly before they merged—the most significant change in Jupiter's weather in the last half century. **(b)** Brown oval, probably an opening in the upper ammonia clouds. The width of the image is about 3000 km. Note the high, tawny cloud moving over the darker area. **(c)** Thunderstorm. False-color image of convective water clouds northwest of the GRS. The tower of white clouds is 1000 km across and stands about 25 km above the surrounding clouds.

almost exclusively at latitudes near 18°N. The blue-gray areas from which the strongest thermal emission is detected (hence their common name, "hot spots") occur only in the equatorial region of the planet.

How does Jupiter "know" that it should have one type of cloud feature in only one location? These equatorial hot spots must be relatively clear regions that allow us to look deep into the planet's atmosphere. We know that because of the intense thermal radiation we can detect there. Pictures of Jupiter made by recording just this escaping energy reveal these equatorial windows as the brightest areas on the planet (Fig. 13.20). The Galileo probe descended into one of these equatorial hot spots and consequently found no thick cloud layers as it drifted down through the atmosphere. There was no way to aim the probe, and energy considerations

required an equatorial entry. Hence, the chances of encountering a hot spot were quite good, and that is exactly what happened. (You may find it confusing that we describe the place where the probe entered as both a "hot spot" and a region of cold dry air. If you were riding on the probe, you would have found the surrounding air to be both cooler and dryer than the average. But the greater transparency of this dry air would let you see the warm glowing regions deep in the atmosphere, regions that are normally hidden beneath the clouds. Astronomers call such a region a hot spot because they can see deeper into the atmosphere, where it is indeed hot; Fig. 13.21.)

Despite the many successes of the Galileo mission, we still do not know why certain meteorologic features occur only at certain latitudes on Jupiter. One aspect of this puzzle seems fairly

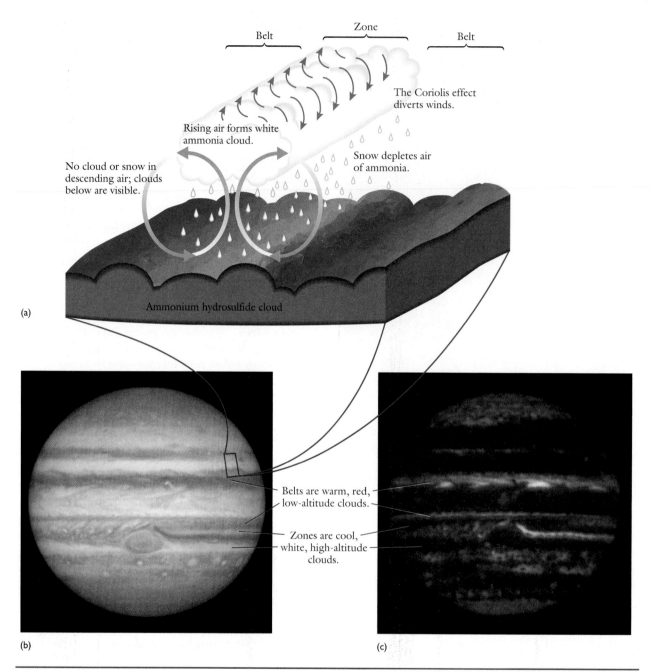

Figure 13.20 Images of the same hemisphere of Jupiter in visible light, **(b)**, and infrared light at a wavelength of 5 μm, **(c)**. The bright regions in the infrared picture are places where heat is escaping from the interior of the planet. The Great Red Spot appears dark in this image, indicating that it is opaque to this internal heat radiation. The diagram, **(a)**, shows a simplified picture of convection on Jupiter, with rising air more cloudy than descending currents.

clear, however. The larger the disturbance, the longer it lasts. On Earth, our most powerful storms—hurricanes and typhoons—die out (albeit with highly destructive effects) when they cross over land. The reason is that these storms get their energy from the condensation of water vapor that comes from the oceans. (Evaporation of water requires energy—that's why perspiration cools us—whereas condensation liberates energy, hence the destructive power of thunderstorms and hurricanes.) When hurricanes and typhoons move off the ocean onto a continent, they no longer have an unlimited source of fuel, and they gradually dissipate. On Jupiter, there are no continents, and

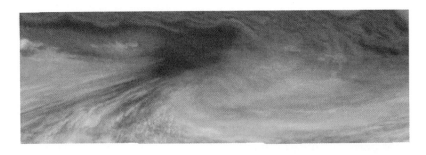

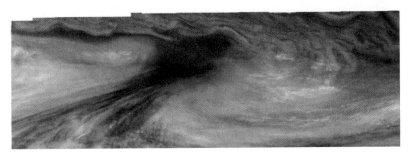

Figure 13.21 True and false color views of a Jupiter hot spot from the Galileo spacecraft. These images cover an area of about 30,000 by 10,000 km. Differences in coloration are due to the composition and abundances of trace chemicals in Jupiter's atmosphere. The bottom image shows variations in cloud height and thickness. Bluish clouds are high and thin, reddish clouds are low, and white clouds are high and thick. The light blue region to the left is covered by a very high haze layer. The hot spot itself is the darkest feature near the upper center of each image.

once these storms grow large enough to overcome the effects of encounters with smaller systems, they can evidently persist for a very long time.

Since both Jupiter and Saturn have major internal sources of energy, one might expect their weather to show no sign of seasons. This appears to be the case on Jupiter according to measurements by the Galileo probe. The probe entered the atmosphere near Jupiter's high-speed equatorial jet, so it was no surprise that the measured winds were high near the level corresponding to the tops of the visible clouds. What had not been anticipated were the continued high wind speeds as the probe descended to a depth of more than 20 bars. On Earth the absorption of sunlight drives circulation in the troposphere, but this is apparently not true on Jupiter. Instead, the new results raise the possibility that atmospheric winds are powered primarily by the internal heat of the planet. This is presumably true on Saturn as well.

13.5 CLOUDS, COLORS, AND CHEMISTRY

There are three basic kinds of clouds that can form in a planetary atmosphere. In our own skies, we are most familiar with white *condensation clouds*, formed when water vapor condenses into droplets or ice crystals. Dry areas of Earth also produce

dust clouds, swept up by surface winds into the atmosphere. As we saw in Chapter 11, this is the most prominent form of cloud on Mars. Finally there is the *photochemical haze* or smog produced by chemical reactions that are mainly powered by the absorption of solar ultraviolet light. Having no solid surfaces, the giant planets can produce only photochemical and condensation clouds.

PHOTOCHEMICAL SMOG

On Earth, we are accustomed to the formation of smog layers over our cities because it is the cities that furnish the gases on which sunlight can act to produce the smog particles. But on the outer planets, the hydrogen-rich atmospheres themselves contain gases that are easily converted to compounds that form hazes. Because of these organic gases, photochemistry plays a very important role in the outer solar system, in spite of the large distances of the planets from the Sun.

The stratospheres of both Jupiter and Saturn are strongly influenced by the absorption of sunlight. The increase in temperature just above the tropopause illustrated in Figure 13.10 is caused by the absorption of ultraviolet sunlight by gases and aerosol particles in this part of the atmosphere. Because temperature normally decreases with height in a planetary atmosphere, a region like this, in which the inverse occurs, is known as a **thermal inversion.** All four of the giant outer planets exhibit this effect; it is similar to the warming in Earth's upper atmosphere produced by the ozone layer (see Section 9.6). It is in this region of the atmospheres of Jupiter and Saturn that hydrocarbons form by photochemical reactions. This solar ultraviolet radiation cannot penetrate to deeper layers, just as on Earth. Despite the dramatic differences between our oxygen-rich environment and these hydrogen atmospheres, the physics of molecules dictates that most gases are excellent absorbers of ultraviolet radiation—hence, the similarity in atmospheric structure between these very disparate planets.

Both Jupiter and Saturn are very poor reflectors of blue and ultraviolet light, indicating that the atmospheres of these planets contain layers of smog that absorb this radiation. There is so much excess hydrogen on these bodies that the hazes remain thin: As the particles drift downward, they are heated and converted back to simple substances. In the low-temperature, hydrogen-poor atmosphere of Saturn's satellite Titan, however, there is a smog layer so thick that we cannot see Titan's surface. These thin hazes of photochemical smog are found in the planets' stratospheres, since the high-energy ultraviolet light from the Sun does not reach lower altitudes. The light is absorbed by methane and ammonia in the process of splitting these molecules apart. Below the tropopause, we encounter the thick condensation clouds that appear in the photographs of these planets.

Condensation Clouds

The highest, coldest condensation clouds are the white ammonia cirrus. The next clouds we encounter, glimpsed through breaks on the ammonia cirrus, are no longer white but tawny in color. Evidently a change in chemistry has occurred. There is no direct determination yet of the composition of these tawny clouds, but chemistry tells us that ammonia and hydrogen sulfide will combine to form ammonium hydrosulfide (NH_4SH). Since there is more nitrogen than sulfur in a cosmic (or jovian) mixture of the elements, this compound can use up all the H_2S and leave an excess of NH_3. But ammonium hydrosulfide itself is white, so some additional chemistry must be happening here to give the tawny tint that we observe. Because these clouds were not present in the dry hot spot where the Galileo probe made its measurements, we don't have any confirmation that these clouds are really composed of ammonium hydrosulfide.

At much lower levels in the atmosphere, water clouds should form. Once again, these will not be pure water. They will certainly contain ammonia and may well resemble a very dilute solution of household ammonia, and other soluble gases will also be involved. One of the indirect consequences of convective water clouds might be lightning discharges, and these were photographed on the night side of Jupiter by the Voyager cameras. The Galileo probe included a lightning detector that picked up discharges from distant thunderstorms. Later the Galileo orbiter photographed directly regions where lightning storms were taking place. These data provide indirect support for the presence of water clouds. But as we have seen, the entry point for the Galileo probe precluded it from reaching the level of the water clouds in this dry region, so the basic questions of the oxygen abundance, the corresponding water abundance, and the altitude of the water clouds elsewhere on Jupiter all remain unanswered.

Chemical Clouds

With our present knowledge, it is already obvious that Jupiter and Saturn have remarkable atmospheres. Imagine floating on a world with colored clouds! The views would be magnificent, even if the aromas leaking into your spaceship might leave something to be desired. If you succeeded in ignoring the disagreeable odors, however, you would find that the gases present in the atmospheres produce a complex and interesting chemistry.

The reason for the color of the tawny clouds may lie in the ability of sulfur to combine with itself to form compounds that are yellow or brown.

Other sulfur compounds may also be responsible for the dark brown clouds on Jupiter. Red phosphorus has been proposed as the substance responsible for the coloration of the GRS, but the theoretical calculations that led to this prediction have not been substantiated by observations of the planet itself. Yellow phosphorus is produced from phosphine (PH_3) in laboratory experiments, and it may contribute to colors seen on both Jupiter and Saturn. The reddish organic compounds produced by reactions of methane and ammonia provide additional possibilities.

The above suggestions are based on theoretical studies as well as on a variety of laboratory experiments in which mixtures of gases are subjected to electric discharges, ultraviolet radiation, or bombardment by electrons and protons, simulating the planetary environments (see Fig. 13.9). Such experiments often produce colored materials; the problem is that there are usually no discrete absorption features in the spectra of these materials (the way there are for gases) that can lead to their identification on the planets. We may have to wait for direct analyses by future probes to discover what the chemical agents are that produce the colors we observe.

IMPACT SCARS IN THE JOVIAN ATMOSPHERE

In July 1994, about 20 separate nuclei of Comet Shoemaker-Levy 9 smashed into Jupiter, as described in Section 6.3. Each of these nuclei penetrated below the visible ammonia clouds before it disintegrated and exploded with an energy of millions of megatons. From the post-impact observations that showed only limited amounts of water injected into the jovian stratosphere, we can conclude that the explosions took place above the dense water clouds, probably at a pressure level of about 5 bars.

Each of the cometary impacts generated a fireball that consisted of vaporized cometary material mixed with the gases of the jovian atmosphere. These materials were carried into the stratosphere and spread over thousands of kilometers (Fig. 13.22). These dark clouds included substantial

Figure 13.22 A sequence of images of Jupiter recorded by the Hubble Space Telescope in July 1994 reveals several regions of black debris left from the explosive impact of the fragment called G of Comet Shoemaker-Levy 9. The time span is 5 days, from bottom to top.

quantities of carbon and sulfur from both the comet itself and the jovian atmosphere. Twisted and dissipated by the strong jovian winds, the clouds gradually dispersed until they were no longer visible a few months after the impacts, although some evidence of enhanced haze at the latitude of impacts persisted longer than a year.

13.6 MAGNETOSPHERES AND RADIO BROADCASTS

In our discussion of Earth, we learned that our planet's magnetic field is strong enough to create a magnetosphere. This teardrop-shaped barrier to the flow of the solar wind contains the belts of trapped charged particles discovered by James Van Allen. We may reasonably expect to find similar environments around other planets that have strong magnetic fields. The giant planets satisfy this criterion; all four have well-developed magnetospheres, with Jupiter's being the biggest and most complex.

JUPITER'S RADIO BROADCASTS

Jupiter was the first planet found to be a source of radiation at radio wavelengths. This radiation was detected in 1955 at a frequency of about 20 megahertz, corresponding to a wavelength of 15 m. For comparison, a radio station on Earth that you tune in at 1000 on the AM dial is broadcasting at 1 megahertz, or a wavelength of 300 m. If you prefer FM, a station at 100 on the dial of your radio is broadcasting at 100 megahertz, or 3 m.

These signals from Jupiter are not news reports or rock music; they are simply radio noise produced by the interaction of charged subatomic particles—primarily electrons—with the planet's magnetic field and its ionosphere. These radio emissions, caused by electrons and ions moving at very high speeds, are called **nonthermal radiation** because they are not caused by the normal emission of energy associated with the heat of the source. The intensities of the jovian nonthermal noise bursts are occasionally great enough to make Jupiter the brightest object in the sky at these long wavelengths, except for the Sun during its most active periods.

Nonthermal radio bursts from Jupiter provided the first indication of a magnetic field on any planet other than Earth. Subsequent observations at wavelengths shorter than 1 m revealed that Jupiter is also a source of steady radio emission.

It has become customary to refer to these two types of emission in terms of their characteristic wavelengths: decameter radiation (the erratic bursts) and decimeter radiation (the continuous source). Except at the very shortest radio wavelengths, the decimeter radiation is generated nonthermally, primarily by electrons spiraling at very high speeds around magnetic lines of force. In addition to these trapped electrons, Jupiter is also surrounded by a doughnut-shaped distribution of protons (Fig. 13.23). In other words, this giant planet has belts of trapped charged particles analogous to the Van Allen belts surrounding Earth (see Section 9.6), but far larger and much more densely populated.

THE INFLUENCE OF IO

The noise storms at decameter wavelengths are strongly influenced by the position of the satellite Io in its orbit, as viewed from Earth. An example is an enhancement that occurs when Io and Jupiter form a 90° angle with Earth. Evidently, a cluster of magnetic field lines (called a **magnetic flux tube**) links Io to the planet. As Io moves in its orbit, the foot of this flux tube sweeps across Jupiter, rather like the shadow of a satellite that is causing a solar eclipse (Fig. 13.24). Electrons spiraling around the field lines and interacting with the jovian iono-

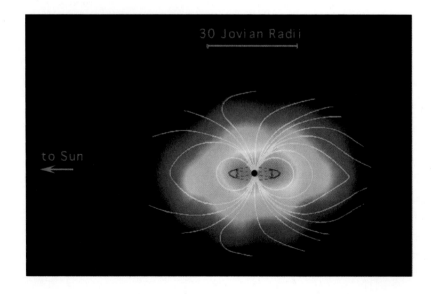

30 Jovian Radii

to Sun

Figure 13.23 Jupiter's enormous magnetic field has trapped electrically charged particles, mostly electrons and protons, from the solar wind. When the electrons spiral around the magnetic field lines, they produce radio waves that can be detected from Earth or from spacecraft. This is an image taken by the Cassini spacecraft, with magnetic lines of force superimposed.

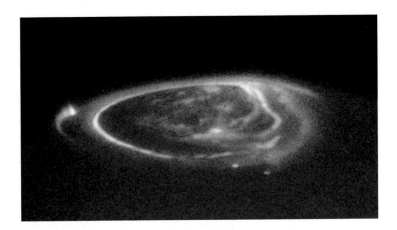

Figure 13.24 The north polar aurora on Jupiter, photographed by the Hubble Space Telescope. The bright point on the left is an "image" of Io, showing where the Io flux tube intersects the atmosphere of Jupiter.

sphere cause the observed bursts of radio noise. Another way to think about this is to consider the flux tube as a wire along which an electric current (the spiraling electrons) passes. This 5-million-ampere current is generated by the motion of Io through the planet's magnetic field, just as a wire (armature) passing through a magnetic field generates a current in one of our power stations on Earth.

The influence of the flux tube on jovian radio bursts is not the only way Io interacts with the magnetosphere. As we will describe later, Io and its volcanoes are the primary source for heavy ions in the inner magnetosphere of Jupiter. These heavy, energetic particles are concentrated in a doughnut-shaped volume, the Io plasma torus, which surrounds Jupiter near the orbit of this satellite. This Io torus is the strongest part of what we might think of as the jovian Van Allen belts. It has been photographed from Earth in the glow from the various atoms and ions it contains, as shown for neutral sodium in Figure 13.25.

MAGNETIC FIELDS

The deductions about the magnetic field of Jupiter from these Earth-based observations were refined and extended by the spacecraft that penetrated the jovian magnetosphere. The magnetic field of Jupiter is dipolar (like a bar magnet, the same type as Earth's) and is generated by a natural dynamo driven by convection within the electrically conducting layers of the planet's interior. Instead of molten iron, as on Earth, the jovian dynamo is produced by the liquid metallic hydrogen we described earlier. The intrinsic strength of this field is 19,000 times greater than that of Earth, leading to a field strength at the equator of 4.3 gauss, compared with 0.3 gauss at Earth's surface. (The surface field

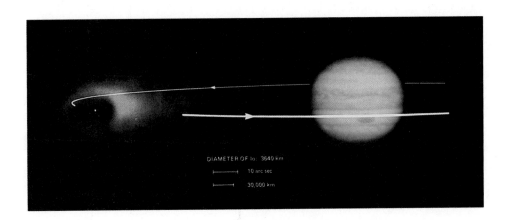

Figure 13.25 The orbit of Io is surrounded by a doughnut, or torus, of atoms and ions, as shown in a picture obtained from Earth in the light from the sodium atoms it contains. Images of Io and Jupiter have been added at their appropriate positions relative to the sodium cloud.

strength is reduced because Jupiter's volume is so much greater than Earth's.) The orientation of the jovian magnetic field is opposite to the present orientation of Earth's field, so a terrestrial compass taken to Jupiter would point south instead of north.

Saturn's magnetic field is somewhat weaker than Jupiter's, but it exhibits the same orientation. We have no idea whether these fields, like that of Earth, reverse direction from time to time. The field strength at the equator of Saturn is only 0.2 gauss. The most unusual characteristic of Saturn's field is that it is not inclined to the planet's axis of rotation, whereas the fields of Jupiter and Earth both have inclinations of about 10°. This lack of inclination is surprising, since most theories for the generation of planetary magnetic fields require such a tilt. Perhaps there is a conducting region around the field-generating core in Saturn's interior that modifies the external appearance of the field.

SATURN'S RADIO BROADCASTS

Another difference between Jupiter and Saturn is the lack of strong radio emission from Saturn. It was not until the first Voyager spacecraft approached the planet in 1980 that long-wavelength radiation was detected. This radio noise occurs at frequencies from 3 kilohertz to 1.2 megahertz, overlapping our familiar AM radio broadcasting band, with a peak intensity at wavelengths near 2 km. Radio signals in this frequency range are reflected by Earth's ionosphere, so our AM broadcasts do not get out and the Saturn radiation cannot get in—hence the need for a spacecraft to detect them.

The energy for this radio emission from Saturn is supplied by electrons from the impinging solar wind, so changes in solar wind pressure or speed produce large changes in the power radiated by Saturn. Since the interaction with the solar wind is related to the configuration of the planet's magnetic field, it has been possible to determine the rotation period of the field—and hence of Saturn's interior—exactly as in the case of the System III period for Jupiter. The resulting period of 10 hr 39.4 min is the standard against which the wind velocities are measured.

THE VAN ALLEN BELTS OF JUPITER AND SATURN

Both Jupiter and Saturn are surrounded by huge seas of plasma. The protons, ions, electrons, and neutral atoms in these plasmas come from three distinct sources: the solar wind, the atmospheres of the planets, and the surfaces of the satellites. Earth's magnetosphere is not populated from this latter source, which is one of the major differences between our Van Allen belts and the magnetospheres of Jupiter and Saturn. The satellites contribute the heavy ions of sulfur and oxygen that are stripped of one or more electrons. Being charged, they are under the control of the planetary magnetic fields.

Oxygen ions have been found around both Jupiter and Saturn. Sulfur and oxygen ions around Jupiter come from both the volcanoes and the surface of Io, which must also contribute the neutral sodium that is visible from Earth (Fig. 13.25). Some of these neutral atoms and ions escape directly from the volcanic eruptions on Io, which we will discuss in Chapter 15, and others are released from the surface by a process called **sputtering.** Sputtering results when an impacting ion from the plasma torus has sufficient energy to eject additional ions from a solid surface, almost like an atomic version of impact cratering. The process is important on Io because the more heavy ions that are injected into the plasma torus, the more "ammunition" there is to sputter other ions from the surface.

The heavy ions in the Io torus make the environment around this satellite lethal to human beings and very hostile to spacecraft. Pioneer 10 nearly died as it passed through this region, so considerable effort and expense were lavished on the Galileo spacecraft to protect its electronic components from this threat. In its extended mission, Galileo survived a total radiation dose more than three times greater than its design goals, but not without losses. By the end of the mission, the electronics had become so shaky and unreliable due to radiation damage that it was no longer possible to collect much useful scientific information.

In the Saturn system, sputtering from the surfaces of the icy satellites and the rings furnishes oxygen ions, but nitrogen ions are supplied by the atmosphere of Titan. The details of the interactions of the electrons, protons, and heavy ions with the ices in the main rings are still not understood. Some charging of small solid particles in both Saturn's and Jupiter's rings may take place, leading to observable effects.

Having seen where the ions and electrons in the magnetospheres originate, we now ask how they are lost. There are three major sinks for ions and electrons in the Jupiter and Saturn magnetospheres, compared with two in the simpler magnetosphere of Earth. Just as on Earth, some escape from the outer part of the magnetosphere, and some are lost in collisions with the upper atmosphere of the planet. The third sink is the surfaces of the satellites and rings, which act as both a source (through sputtering) and a sink (through absorption of particles that strike their surfaces). The absorptions at Saturn are stronger than those in Jupiter's magnetosphere owing to the large surface area of the rings and the closely spaced inner satellites. These combined surfaces are more effective at absorbing charged particles than are the inner satellites and thin rings of Jupiter, leading to a major difference between the distributions of plasma in these two giant magnetospheres.

DIMENSIONS OF MAGNETOSPHERES

As we saw for Earth, a magnetosphere is bounded on the upstream side (facing the Sun) by the equilibrium between the internal pressure exerted by the planet's magnetic field and the external pressure from the solar wind. The magnetic fields of Jupiter and Saturn are much stronger than Earth's, but the solar wind pressure is less at these greater distances from the Sun. The net result is that the two giant planets have magnetospheres so large that they exceed the size of the Sun (Fig. 13.26).

Since the intensity of the solar wind is not constant but changes with the activity of the Sun, the upstream boundaries of these magnetospheres also change. A strong solar wind pushes the boundary closer to the planet. At such times, Saturn's satellite Titan moves outside the magnetosphere when its orbit brings it around to a position between the planet and the Sun. At other times, Titan remains within the magnetosphere throughout the course of its orbital journey.

In the downstream direction, the magnetosphere stretches out into a so-called **magnetotail,** coaxed along in the anti-Sun direction by the solar wind (like the plasma tail of a comet—hence the name). Jupiter's magnetotail is so extended that it appears to reach all the way to Saturn's orbit, a distance almost equal to the distance of Jupiter from the Sun.

ENERGY TO POWER THE MAGNETOSPHERES

Given the sources and sinks for the charged subatomic particles in the plasmas, what supplies their energy? Ultimately, it is the rotation of the parent planets. Rotation provides the energy to generate the planet's magnetic fields. Charged particles spiraling along magnetic field lines must revolve around the planet at the same rate as the field lines. This is the rotation period of the planet itself, as we discussed earlier.

As a result of this coupling of the plasma to the magnetic field, we have the curious situation in which these spiraling electrons, protons, and ions are moving around the planet as if they were attached to it. They exhibit what we call rigid body rotation rather than following Kepler's laws of planetary motion as they would if their motion were dominated by gravity instead of by magnetic forces. This means that a particle at Io's distance from Jupiter must follow the radio rotation period of System III: 9 hr 56 min. This is much faster than the period of revolution of Io, which is 42 hr 28 min. Hence, the plasma flows past the satellite in the same direction as Io's motion, leading to the unusual result that a wake *precedes* the satellite in its orbit about the planet. The same thing happens with Titan in the magnetosphere of Saturn.

The charged particles become so energetic because of their rapid motion that when they strike the surface of a satellite, they can knock out an ion that then joins the co-rotating plasma. Forcing this plasma to rotate with the field and to generate the radio noise that is equivalent to 10^{14} watts costs the planet energy. It means that its rotation rate must be steadily decreasing, but at an immeasurably slow rate.

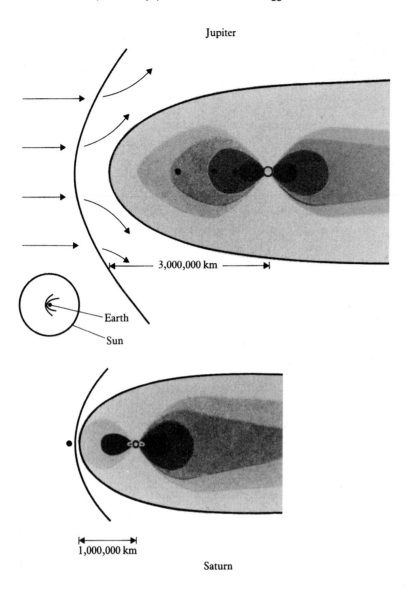

Figure 13.26 The small circle to the left in this diagram represents the Sun; inside it, to the same scale, is Earth and its magnetosphere. The magnetospheres of Jupiter and Saturn are shown to scale for comparison.

AURORAS

By analogy with Earth, one would expect that charged particles from the magnetosphere will sometimes follow the magnetic field lines right into the planet's atmosphere. On Earth, this charged particle "precipitation" produces auroras in the north and south polar regions. Indeed, the same phenomenon has been detected on Jupiter, both by direct photography and by ultraviolet and infrared spectroscopy (see Fig. 13.24). Not only is light produced by these interactions, but new compounds are also formed in the planet's polar atmosphere as molecules of methane break apart and the fragments recombine.

Ultraviolet auroras are also present on Saturn, but not the strong infrared emission from H_3^+

that is seen in Jupiter. Voyager's attempt to take pictures of Saturn's aurora was unsuccessful because the night sky on Saturn is never really dark. Sunlight scattered back into the dark hemisphere by the rings always leaves the planet in a twilight glow. This phenomenon is illustrated in Figure 13.27, which shows a view of Saturn we cannot have from Earth. Looking back at the planet from Voyager 1, we see the shadow of Saturn crossing the magnificent system of rings. An observer floating in the atmosphere on the night side of the planet would see the bright rings arching up from the horizon and disappearing in shadow overhead. Ring light would illuminate the surroundings.

The illustration also reminds us that in leaving Saturn, we have now visited all of the planets

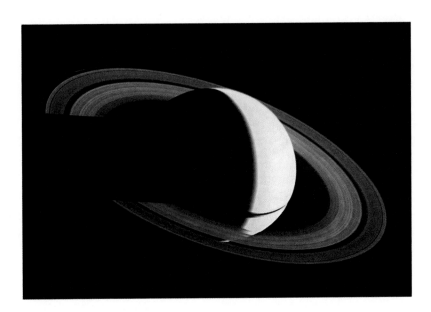

Figure 13.27 A dramatic picture looking back at Saturn, taken by Voyager 1 after its spectacular encounter with the ringed planet. In a view we can never see from Earth, Saturn appears as a crescent with its dark shadow crossing the magnificent ring system.

known to the ancients. As Voyager 2 ventured forward on its way to Uranus and Neptune, it was heading toward worlds unknown to Galileo, Shakespeare, Shah Jihan, Newton, Basho, and Bach.

13.7 QUANTITATIVE SUPPLEMENT:
BUILDING A GIANT PLANET

The main difference between Jupiter and Saturn on the one hand and the inner planets on the other is the ability of these giants to retain hydrogen and helium. Suppose we try to transform Earth into a giant planet by adding those missing gases until the composition is the same as that on the Sun. How big would this new giant be?

We begin by referring to Tables 2.2 and 3.2. These emphasize that the greatest deficiencies of our planet are its lack of hydrogen and helium. Let us adopt silicon (Si) as our index element and add enough hydrogen and helium to yield a solar ratio of these elements to silicon. The solar ratio of H to Si (by the number of atoms) is about 2×10^4. However, we must convert this to a mass ratio by multiplying the numbers of H and Si atoms by their respective atomic masses. This gives us a ratio of H to Si in the Sun of about 850. A similar calculation for helium gives a mass ratio of He to Si of about 250.

We could continue this process for each element, but the fact is that the others add very little, relative to hydrogen and helium. Even though their individual masses are greater than those of atoms of H and He, their abundances are lower by an even larger factor.

To build our planet, we begin with the mass of silicon in the Earth, which is 7×10^{23} kg. We then add the appropriate ratios of H (850) and He (250) and multiply each by the mass of Si: $(1 + 850 + 250)7 \times 10^{23} = 8 \times 10^{26}$ kg. This is equivalent to 120 times the mass of Earth. In other words, by restoring the missing light elements to Earth, we have created a planet with a mass greater than that of Saturn.

We should notice, however, that this giant planet we have conjured up has only one Earth mass of heavy elements, whereas Jupiter and Saturn have cores of 10–15 Earth masses. Evidently, the heavy elements are enriched on those giants through the planet-forming process.

SUMMARY

The two largest planets with their extensive retinues of satellites resemble miniature solar systems. Jupiter and Saturn are so massive that they have been able to retain thick atmospheres (and even more massive fluid envelopes) of hydrogen and helium throughout their existence. No solid surfaces exist, so there is no science of geology on these two giants.

Besides hydrogen and helium, these atmospheres contain other simple hydrogen-rich molecules. Some of these, such as methane and ammonia, are expected to form readily at the local temperatures and pressures. Others, like acetylene, require some additional source of energy such as solar ultraviolet radiation, lightning, or heat coming from the planetary interiors.

Both Jupiter and Saturn radiate more energy than they receive from the Sun. Jupiter is primarily radiating primordial heat, the result of its contraction from a giant protoplanetary phase. Smaller Saturn must have some other source of energy. Precipitation of helium deep in its interior provides the explanation, which is supported by the observed deficiency of helium in the planet's atmosphere. Both planets have layers of metallic hydrogen surrounding possible cores of rock and metal. These cores appear to be similar in size, about 10–15 times the mass of Earth.

The Galileo probe entered a desert hot spot in Jupiter's atmosphere and thus was unable to observe the expected cloud layers. However, it sent back a rich harvest of other data, some of which demonstrate that the elements heavier than helium (except for neon) are enriched by a factor of about 3 compared with solar abundances, relative to hydrogen. This enrichment was apparently caused by the formation of the core from icy planetesimals formed at temperatures below 35 K. If these low-temperature planetesimals built the cores of the other giant planets as well, they must have been the most abundant form of solid matter in the early solar system.

Despite the continuous motion of currents and clouds, some features in the atmospheres of Jupiter and Saturn have remained remarkably permanent. Outstanding among these is Jupiter's Great Red Spot, which has lasted more than 300 years. Its longevity is attributable to its size, which is large enough to swallow the planet Earth. The color of this giant high-pressure region, like the other colors found among the clouds in these atmospheres, is caused by traces of chemicals whose identities remain unknown. Hypotheses about the origin of the color of the GRS range from organic compounds to red phosphorus, and sulfur is regarded as a possible component of some of the brown and tawny-colored clouds.

Both Jupiter and Saturn are surrounded by giant magnetospheres. Their magnetic fields, which are thousands of times stronger than Earth's, trap energetic plasmas. The atoms and fragments of atoms that make up the plasmas originate from the solar wind, from the atmospheres of the planets, and from the satellites and rings. Jupiter is an intense source of nonthermal radio noise at both long and short wavelengths, but Saturn radiates only at very long wavelengths. This difference is attributable to the presence of Saturn's extensive system of rings, which serves as an effective absorber of charged atomic particles near the planet.

The plasmas surrounding these giants differ from the environment found in Earth's Van Allen belts in that they contain large numbers of heavy ions. These ions are contributed by the surfaces of the satellites through a process known as sputtering and, in the case of Io, by volcanic eruptions. Precipitation of charged particles from the radiation belts causes auroras, which so far have been seen only on Jupiter.

KEY TERMS

magnetic flux tube	sputtering
magnetotail	thermal inversion
nonthermal radiation	

REVIEW QUESTIONS

1. Which spacecraft have visited Jupiter and Saturn so far? When were they launched and when did they arrive? What are the prospects for future exploration of these planets?

2. Why are Jupiter and Saturn called "giant" planets? What are they made of? How do their interiors differ from the interior of Earth? How do we know?

3. Both Jupiter and Saturn radiate more energy than they receive from the Sun. What are the sources of

that energy? Are the sources the same on both planets?

4. What gases are present in the atmospheres of these planets? What do you think would happen if an astronaut lit a match while visiting Jupiter? Why?

5. What sources of energy are available to form compounds in Jupiter's atmosphere? How does the chemistry going on there differ from what took place on the primitive Earth? What are the visible signs of this jovian chemistry?

6. How does the global circulation of Saturn differ from that of Jupiter? Why do scientists think this difference exists?

7. Both Jupiter and Saturn are broadcasting radio waves even though there are no advanced, technical civilizations with radio transmitters on either planet. Explain.

8. Compare the sizes of the magnetospheres of Jupiter and Saturn with each other and with the size of the Sun. If you could see Jupiter's magnetosphere with your unaided eyes, how big do you think it would appear in our nighttime skies?

QUANTITATIVE EXERCISES

1. Use Table 2.2 to show that the addition of missing carbon and nitrogen to the artificial planet described in Section 13.7 has only a small effect on its calculated mass.

2. Referring to Section 13.7, show that hydrogen and helium could not escape from this new planet once it formed, even though it is closer to the Sun and hence much warmer than Jupiter or Saturn. What does this result tell you about the origin of the differences in composition between the terrestrial and giant planets?

3. How long would it take astronauts in a balloon floating in the atmosphere of Jupiter to circumnavigate the planet if the balloon moved at the same speed as the equatorial jet? Repeat this calculation for Saturn.

4. Calculate the wind speed at the outer boundary of the GRS, given its size and six-day rotation period. Compare this speed with the east-west (zonal) wind speeds.

ADDITIONAL READING

Beatty, J. K., C. C. Peterson, and A. Chaikin, editors. 1999. *The New Solar System*, 4th ed. Cambridge, MA: Sky Publishing Corp. An excellent compilation of beautifully illustrated articles by experts in the field, including fine reviews of the Voyager and Galileo discoveries.

Beebe, R. 1995. *Jupiter: The Giant Planet.* Washington, DC: Smithsonian Institution Press. Excellent pre-Galileo overview of the planet and its satellites, combining results from both astronomical and spacecraft investigations.

Burrows, W. E. 1990. *Exploring Space: Voyages in the Solar System and Beyond.* New York: Random House. History and politics, including extensive discussions of the tortured history of the Galileo mission to Jupiter.

Fischer, D. 2001. *Mission Jupiter: The Spectacular Journey of the Galileo Spacecraft.* New York: Copernicus Books, New York. Description of the troubled Galileo mission and of its many discoveries.

Morrison, D., and J. Samz. 1980. *Voyage to Jupiter.* NASA Special Publication 439. Washington, DC: U.S. Government Printing Office. Popular account of the Voyager flybys of Jupiter.

14

IN DEEP FREEZE:
PLANETS WE
CANNOT SEE

Family portrait of all four outer planets—Jupiter, Saturn, Uranus, and Neptune—as imaged by the Voyager spacecraft. Earth is shown at the bottom for size comparison.

Compared with Jupiter and Saturn, Uranus and Neptune are smaller, have higher densities, and exhibit a much higher proportion of methane in their atmospheres. Uranus and Neptune are not giants like Jupiter and Saturn; indeed, their masses are intermediate between the two giants discussed in Chapter 13 and the terrestrial planets like Earth and Venus. Since their masses are so similar, it makes sense to discuss Uranus and Neptune together. When we look closely, however, we find differences between these two mini-giants that are even more profound than those that distinguish the other planetary "twins," Earth and Venus.

The distance from Saturn to Uranus is about the same as the distance from Earth to Saturn. The next step, to reach Neptune, is only slightly larger. Crossing that gap, we find a planet that is more dynamic than Uranus, with a strong internal heat source and an active meteorology in its visible atmosphere. The reasons for this difference between mild-mannered Uranus and active Neptune are not yet clear, but the effect on the Voyager mission was to give the Neptune encounter a special edge of excitement that no one had anticipated based on the results from Uranus.

363

Pluto, the last planet in our system, is a different object altogether. Far from being a giant, Pluto is smaller than our Moon. It resembles one of the large icy satellites (discussed in Chapter 15) and the largest members of the Kuiper belt. It may be similar to the building blocks for the giant planet cores.

14.1 DISCOVERIES OF THE OUTER PLANETS

Beyond Saturn, only the unchanging stars were known until late in the eighteenth century. In extending our perspective to Uranus, Neptune, and Pluto, we enter new territory. These planets were discovered, in a literal sense: one by accident, one by calculation, and one as the result of a diligent search. In each case, the discovery generated at least as much public interest as when one of our spacecraft visits a planet for the first time.

Direct exploration of Uranus and Neptune by spacecraft did not occur until the late 1980s. Voyager 2, in its grand tour of the outer planets, reached Uranus in January 1986 and Neptune in August 1989. Unfortunately, Pluto could not be visited on this magnificent journey because the alignment of the planets did not allow either Voyager spacecraft to reach it (Fig. 14.1). Thus Pluto is now the last planet we perceive only in the ancient manner—as a moving point of light in the night sky rather than as a real, round world.

DISCOVERY OF URANUS

The night of March 13, 1781, was clear and dark over the town of Bath in southwest England. Pursuing his astronomical hobby that evening, a professional musician named William Herschel was continuing a project to chart all stars down to the eighth magnitude. He was using a 6-inch aperture telescope he had built himself, and on this particular evening he was examining the stars in a portion of the constellation Gemini. As he recorded in his notebook, one of the stars that came into view as he moved the telescope from one place to the next seemed unusual. Herschel described it as "a curious either nebulous star or perhaps a comet." Instead of a point of light, it appeared as a small disk.

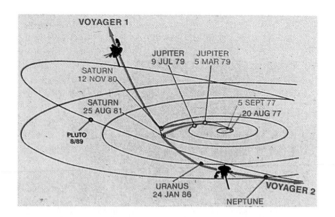

Figure 14.1 This perspective view of the Voyager trajectories clearly shows how far both spacecraft were from Pluto, even though Voyager 2 did reach Uranus and Neptune.

Subsequent observations revealed that this object moved (unlike a star) but that the motion was too slow for a comet. Soon its path had been mapped well enough to establish that it was a new planet orbiting the Sun at a distance of 19 AU (nearly 3 billion km). This is exactly where the next planet, if it existed, was expected. Earlier in the eighteenth century, astronomers had tried to work out a formula for the spacing of the planets, which is a regular progression of increasing separation as we go out from the Sun. The most popular numerical scheme was called the Titius-Bode rule, and this formula suggested that the next planet would be at a distance of about 20 AU. The discovery of Uranus strengthened the belief of contemporary scientists that this rule must have some deep, causal meaning. Today we recognize that the Titius-Bode rule has little if any physical significance, but in Herschel's day it was taken very seriously by astronomers.

A new world had been added to those known since antiquity, and the size of the solar system had suddenly doubled. In recognition of his accomplishment, Herschel received a knighthood and a royal pension, and went on to become one of the most productive astronomers of his era.

These days we have become rather blasé about new astronomical discoveries—volcanoes on Io, black holes, colliding galaxies—but 5 years after the United States declared its independence from Britain, the world was a very different place.

Adding a new planet to the solar system was an extraordinary event. Perhaps the most famous comment on the effect of this discovery is in John Keats's beautiful sonnet "On First Looking into Chapman's Homer," where the poet uses Herschel's discovery as an image for his own feelings on being introduced to Homer's great epic:

Then felt I like some watcher of the skies
When a new planet swims into his ken.

Herschel named this new planet "Georgium Sidus" after his king and patron, George III of England. We have been spared the humor associated with having a planet named George's star by the suggestion of Herschel's contemporary Johann Bode (the same person responsible for the Titius-Bode rule) that the ancient tradition for planetary names be continued by using the name Uranus. In Greco-Roman mythology, Uranus was the father of Saturn, who was in turn the father of Jupiter. The name thus seemed appropriate for the most distant known planet.

DISCOVERY OF NEPTUNE

Uranus is bright enough that even a small telescope can reveal it easily; in fact, it can just be glimpsed without a telescope in a dark, clear sky. After Herschel's discovery, astronomers looked through records of previous observations and found some 20 cases where Uranus had been recorded. It had been mistaken for a star because the earlier observers did not have telescopes as good as Herschel's and could not see the disk. With these additional observations (stretching back to 1690), it was possible to develop a very accurate path for the new planet.

Astronomers used these measurements to calculate the orbit of Uranus, but even when all of the gravitational effects from the known planets were accounted for, it was impossible to fit the observations with an elliptical orbit. By 1830 the discrepancy was 15 seconds of arc, which meant that the predicted position of Uranus was more than four times the planet's apparent diameter from its observed place in the sky. Even in Herschel's time, astronomers could easily measure a discrepancy this large. The validity of Newton's laws of planetary

motion was universally accepted, so it seemed likely to astronomers of the time that this discrepancy was the result of a **perturbation** caused by the gravitational pull of some unknown object. A perturbation is a small deviation from normality—in this case, a minor but cumulative change from a simple Keplerian orbit around the Sun.

Since scientists had now accepted one new planet, it was reasonable to suggest that perhaps another might be present, one orbiting the Sun at a still greater distance and perturbing the motion of Uranus by its gravitational attraction. How to find this unseen planet among the millions of stars? The first clue astronomers used was the Titius-Bode rule. As Uranus very nearly obeyed this rule, perhaps the new planet would lie at the distance from the Sun predicted by this example of numerology, about 39 AU. With this starting point, it was then a question of solving some formidable mathematical equations that described various orbits for the hypothetical planet of unknown mass and position.

Two gifted mathematicians succeeded in solving this problem independently: John Couch Adams in England and Urbain Jean Joseph Leverrier in France. Although Adams obtained the first solution in September 1845, he was unable to persuade any of his countrymen to initiate a telescopic search. After Leverrier published the preliminary results of his calculations in June 1846, the British astronomers realized that these agreed well with Adams's work. They then began to look for the object but, lacking adequate charts of this region of the sky, were unsuccessful.

Meanwhile, Leverrier finished his analysis in August 1846 and sent a letter to the German astronomer Johann Galle in Berlin suggesting that he try to find the new planet. The Berlin observatory had a newly completed set of star charts covering the region of Leverrier's prediction. Using these charts, Galle was able to find the planet on the evening of September 23, 1846, on his first attempt. It was just 1° (twice the apparent diameter of the Moon) from the position predicted by Leverrier and 2.5° from Adams's prediction. Galle could not discern the planet's tiny disk, but he identified it by its appearance as a wanderer among the fixed stars. Neptune is only about one-tenth as bright as Uranus, much too faint to be seen without a telescope.

The discovery of Neptune represented a stunning triumph for Kepler, Newton, and gravitational theory. Neptune was not a triumph for the Titius-Bode rule, however, for it was not at the distance from the Sun predicted by that numerical sequence.

Once again, a search was made for prediscovery observations, and once again several were identified, including one observation by the hapless English astronomer who first set out to find the planet two months earlier. But perhaps the most interesting early observation is one attributed to Galileo. In a sketch of observations of Jupiter made in December 1612 and January 1613, Galileo recorded a "star" that modern astronomers have been able to identify as Neptune by extrapolating the position of the planet backward in time along its orbit (Fig. 14.2). Even though Galileo's telescope was powerful enough to show Neptune, without charts he had no hope of identifying this tiny starlike object as a planet.

There was understandably some dispute over the proper credit for the discovery of Neptune. Matters were not helped by the intervention of the French astronomer François Arago, who suggested that the new planet be named Leverrier. Bowing to international disapproval, Arago later withdrew this idea and instead proposed Neptune, the Roman god of the sea. Happily, Adams and Leverrier managed to disassociate themselves from these disputes, and the two have been credited jointly with the discovery. Note that the discoverer is considered to be the person who made the mathematical calculations, not the one who actually first saw Neptune through the telescope. (We will see in Chapter 16 that three prominent features in the rings of Neptune are now called Adams, Galle, and Leverrier, commemorating all three of these astronomers together at last.)

Discovery of Pluto

Continued tracking of both Uranus and Neptune during the rest of the nineteenth century suggested that yet another planet was required to account satisfactorily for their motions. Although the evidence was only marginally convincing, a

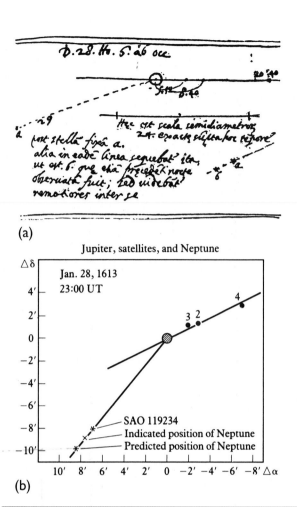

Figure 14.2 **(a)** An entry from the notebook in which Galileo recorded the relative positions of Jupiter, three of its moons, and two stars on January 28, 1613. **(b)** A reconstruction of Galileo's observations by Charles Kowal. One of the "stars" was actually Neptune.

number of astronomers undertook calculations similar to those of Adams and Leverrier. The most persistent of these individuals was Percival Lowell, already familiar to us from his work on Mars (see Section 11.1). Lowell not only calculated an orbit for this hypothetical planet, but also initiated some systematic searches for it at his observatory. These were unsuccessful, but they provided valuable experience and established the quest for a new planet as part of Lowell's legacy.

A special wide-field telescope was built at Lowell Observatory in 1929, 13 years after Lowell's death, and the search for a new planet was put in the hands of Clyde Tombaugh, a young astronomer from Kansas. On February 18, 1930, Tombaugh discovered the trans-Neptunian planet

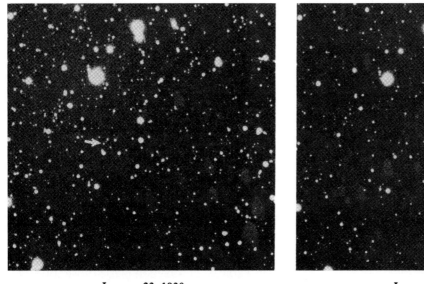

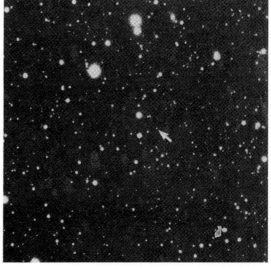

January 23, 1930

January 29, 1930

(a)

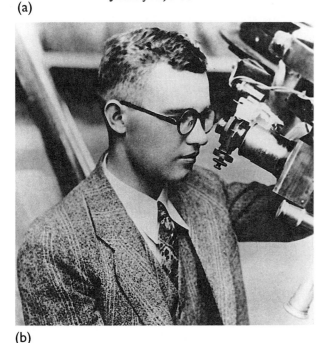

(b)

Figure 14.3 Discovery of Pluto. **(a)** The two photographs from the Lowell Observatory that led to the discovery of Pluto. In comparing the two, taken six days apart, Clyde Tombaugh noticed on February 18, 1930, that one of the "stars" (shown here with an arrow) had moved. **(b)** Clyde Tombaugh in 1930, the year he discovered Pluto.

Pluto was the Greek god of the underworld, a fitting title for a planet at such a great distance from the Sun. The symbol for this planet, ♇, has the additional attribute of commemorating the initials of Percival Lowell, the main mover behind its ultimate discovery. This is as close as one might properly come to naming a planet for a human rather than a god.

Pluto is so small and so far from the Sun that it is 10,000 times too faint to see with the unaided eye. It is only within the last 20 years that we have learned anything more about it than Tombaugh knew 60 years ago.

14.2 THREE DISTANT WORLDS

Before we discuss Uranus, Neptune, and Pluto in detail, it is useful to put them in perspective by comparing them with planets we already know. This comparison is summarized in the "family portrait" that opens this chapter and in Table 14.1, where we see that Uranus and Neptune are

on photographs he had taken with the new telescope the previous month (Fig. 14.3). The ninth planet was announced to the world on March 13, Percival Lowell's birthday.

Although this discovery might seem to be another triumph for gravitational theory, it was actually the product of meticulous astronomical observations. Pluto's mass is far too small to have caused the supposed perturbations in the motions of Uranus and Neptune. Recent studies of this problem suggest that the difficulty lies in the observations of these two planets. The reported perturbations do not, and never did, exist.

Table 14.1   Basic properties of Uranus, Neptune, and Pluto

Planet	Distance from Sun (AU)	Diameter (Earth = 1)	Mass (Earth = 1)	Density (g/cm³)
Jupiter	5.2	11.2	317	1.3
Uranus	19.2	4.11	14	1.3
Neptune	30.1	3.92	17	1.6
Pluto	39.4	0.19	0.002	2.1

very nearly twins. Although they can be classed as giant planets, with masses roughly 15 times that of the Earth, they are much smaller than either Jupiter or Saturn.

DENSITY AND COMPOSITION

In Section 13.2 we saw that Saturn was both less massive and less dense than Jupiter, which is consistent with the two bodies having the same bulk composition. With a smaller mass, Saturn suffers less internal compression; hence, the overall density is lower, only 0.7 g/cm³. If Uranus and Neptune had the same composition as Jupiter and Saturn, we would expect these bodies to have densities even lower than that of Saturn because they have only about 15% of its mass. Instead, Uranus has about the same density as Jupiter, while Neptune has a higher density. This leads us to the conclusion that each of these planets contains a higher proportion of heavier elements than Jupiter and Saturn.

Ice and rock are the logical candidates for the dense material in Uranus and Neptune, as inspection of a table of cosmic abundances will convince you (see Table 2.2). Oxygen is the most abundant reactive element after hydrogen, and these two combine readily to form water ice at low temperatures. What we call rocks are materials composed primarily of the common elements silicon and oxygen (see Section 3.3). Scientists who have constructed models for the interiors of these two objects find that they must contain massive cores composed of the common heavy elements that form ice and rock. Surrounding these cores are relatively thin envelopes of liquid and gaseous hydrogen and other gases. Both Uranus and Neptune are too small to achieve the pressures and temperatures required for the formation of metallic hydrogen (Fig. 14.4).

Some scientists are currently suggesting that both of these planets may have thick layers of water clouds in their lower atmospheres. There is no direct evidence for the existence of these clouds, since

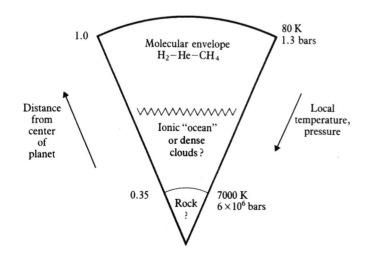

Figure 14.4   A schematic idea of what the interiors of Uranus and Neptune may be like. Pressures are not great enough for metallic hydrogen to occur (cf. Fig. 13.6).

water vapor has not yet been detected on either planet, but there are some chemical arguments in favor of this idea. In particular, the apparent deficiency of ammonia gas in both atmospheres could be explained if the ammonia were dissolved in these water clouds below the observable layers. As we shall see, however, there are alternative explanations for this apparent deficiency of ammonia.

Our knowledge of the interior structures of these two planets is still in a primitive state, so that even the existence of discrete cores is open to question. The only certainty is the marked concentration of heavy elements that gives both planets their relatively high densities.

DIFFERENCES BETWEEN URANUS AND NEPTUNE

Closer inspection of Table 14.1 indicates that the initial appearance of twinship may be incorrect, just as we found for Venus and Earth. The densities of Uranus and Neptune are distinctly different, indicating some internal structural or compositional differences. Probably even more significant is that Uranus lacks a significant internal heat source, but Neptune (like Jupiter and Saturn) derives much of its heat from internal sources. (One of us, Morrison, was part of the small team that discovered this difference in internal heat. The contrast was immediately striking when the two planets were measured at an infrared wavelength near 25 μm; the two planets had essentially the same temperature, even though Neptune was so much farther from the Sun.) Models for its interior suggest that, like Jupiter, Neptune could still be radiating primordial heat. Despite its small mass, the high proportion

of rock and ice to total mass leads to very slow cooling.

Perhaps the most puzzling enigma posed by Uranus is the bizarre orientation of its rotational axis, which is inclined by 98°, placing it practically in the plane of the planet's orbit (Fig. 14.5). The orbits of Uranus's satellites and rings are in the planet's equatorial plane, just as they are for Jupiter and Saturn, so the whole uranian system is tipped on its side. It is often suggested that a glancing impact by another planet-sized object, which then added to the mass of Uranus, could explain this strange orientation, but there is no detailed theory yet that satisfactorily describes such an encounter and its effects.

PLUTO: A SPECIAL CASE

The above discussion has focused on Uranus and Neptune because these two planets are similar in many respects to Jupiter and Saturn. When we turn to Pluto, we are dealing with a different kind of object altogether. This planet is smaller than our own Moon and has a density of 2.1 g/cm^3. This combination of small size and low density suggests that Pluto is composed partly of water ice and probably resembles Triton, the large icy satellite of Neptune. In no way can it be considered one of the jovian planets, nor is it a displaced inner planet, roaming the cold and dark domains of space beyond Neptune. At best, it might be representative of the kinds of objects that collided with one another to form the icy cores of the giant planets, a fate it somehow avoided. As such, it is both the smallest planet and the largest of the Kuiper belt objects (KBOs) discussed in Section 6.4.

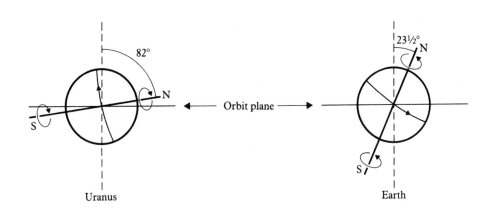

Figure 14.5 The axis of rotation of Uranus lies nearly in the plane of the planet's orbit. At the time of the Voyager 2 encounter, the axis pointed almost directly at the Sun.

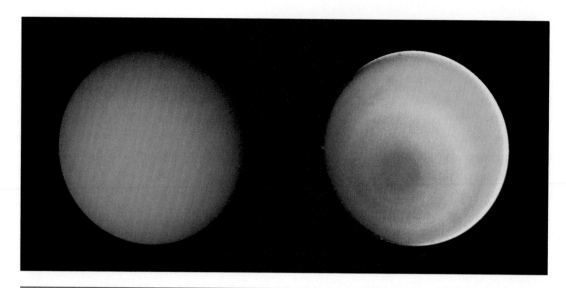

Figure 14.6 (left) Uranus as it would have appeared to someone riding along on the Voyager 2 spacecraft. (right) This image has been specially processed to reveal a banded polar haze. (The rotational pole is nearly facing us.) The colors in this image are totally artificial.

14.3 ATMOSPHERES OF URANUS AND NEPTUNE

Through a telescope, the tiny disks of Uranus and Neptune appear greenish because the upper atmospheres of these planets are generally free of clouds. Although beautiful in their way, the pictures of Uranus obtained by Voyager 2 are singularly free of detail (Fig. 14.6). Neptune reveals some scattered white clouds, but the underlying cloud deck in which storm systems appear is very low in the atmosphere (Fig. 14.7). Sunlight penetrates to great depths in the atmospheres of both planets before it is scattered back to space, and all along its path it is absorbed by methane gas (CH_4), giving both planets a greenish-blue hue.

COMPOSITION

The same methane absorption bands that are present in spectra of Jupiter and Saturn appear in the spectra of Uranus and Neptune, but they are very much stronger, and many new absorptions are also evident (Fig. 14.8). The hydrogen (H_2) absorption lines on Uranus and Neptune are also stronger than in spectra of Jupiter and Saturn, but the increase is not as great as for methane. In other words, the proportions of methane in the atmospheres of Uranus and Neptune are much

higher than in the atmospheres of Jupiter and Saturn (Table 14.2). This is consistent with the higher proportion of ices in the total masses of these smaller planets. Uranus and Neptune are clearly deficient in hydrogen and helium compared with Jupiter and Saturn, and this difference

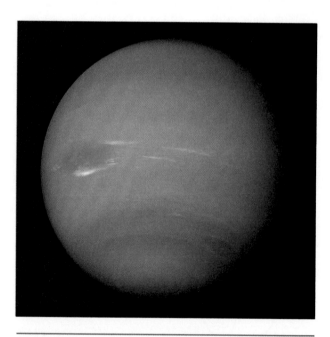

Figure 14.7 As Voyager 2 approached Neptune in the spring and summer of 1989, it took images of the planet that revealed a large, dark, oval storm system and several white clouds. The color is exaggerated.

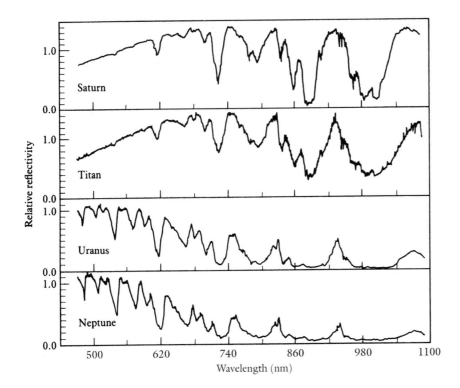

Figure 14.8 The spectra of Saturn, Titan, Uranus, and Neptune, each divided by the spectrum of the Sun. If these objects had no atmospheres, the spectra would be smooth, nearly flat curves. The deep absorptions are caused by methane, which is producing a much larger effect on Uranus and Neptune than on Titan and Saturn.

shows up in the composition of their atmospheres as well as in their average densities.

At a level in their atmospheres where the pressure is the same as the sea-level pressure on Earth, the temperatures on both Uranus and Neptune are about 73 K ($-200°C$), or about the temperature of liquid nitrogen. This is much colder than conditions on Jupiter or Saturn at the same pressure level (see Section 13.3). Ammonia will be

frozen solid at this temperature but may be present as a gas at lower, warmer levels; at still greater depths there should be water vapor as well. There is no hope of detecting water vapor with infrared spectra, but microwave measurements might be able to discern its presence.

The Voyager results indicate that there is a tenuous haze in the upper atmosphere of Uranus centered over the rotational pole. Evidently, this haze is formed as a result of the steady irradiation of the planet's upper atmosphere by solar ultraviolet light and thus is concentrated in the sunlit hemisphere. Even at this immense distance from the Sun, there is some atmospheric photochemistry forming a very thin smog layer that preferentially absorbs short-wavelength light.

The Voyager infrared spectrometer found that the ratio of the abundances of helium and hydrogen in the atmosphere of Uranus is even more similar to that in the Sun than is the case for Jupiter. This result is consistent with the absence of a metallic hydrogen layer in the interiors of these planets. There is no medium from which dissolved helium could precipitate, as happens on Jupiter and Saturn. The hydrogen-helium atmospheres are the natural result of the two-step

Table 14.2 Atmospheric compositions of Uranus and Neptune

Gas	Formula	Uranus	Neptune
Main Constituents (percent)			
Hydrogen	H_2	84	84 (?)
Helium	He	14	14 (?)
Methane	CH_4	2	2
Trace Constituents (parts per billion)			
Acetylene	C_2H_2	200	present
Ethane	C_2H_6	?	present
Hydrogen cyanide	HCN	<100	100
Carbon monoxide	CO	<40	1000

model for the formation of the giant planets that we mentioned in Chapter 13. First, a huge core of ice and rock, 10–15 times the mass of the Earth, accumulates, and second, an envelope of gas (mostly hydrogen and helium) from the solar nebula collapses around the core.

In fact, the two steps are not so sharply separated. As the core forms, it will produce a secondary atmosphere of its own as a result of the outgassing of accreting material. This atmosphere will be composed primarily of nitrogen, carbon monoxide, and methane. As the core grows, it will begin to attract hydrogen and helium from the solar nebula, and will finally develop an atmosphere that is dominated by these two gases. The core contributes the excess methane because the CO will also be converted to methane, but the bulk of the atmosphere is representative of the solar nebula. Apparently, Uranus and Neptune acquired less massive envelopes of solar nebula gas than Jupiter and Saturn, and therefore they have a higher proportion of heavy elements.

ATMOSPHERIC TEMPERATURES

At infrared wavelengths, where both Jupiter and Saturn exhibit strong emissions from methane and products of methane photochemistry, the spectrum of Uranus is essentially blank. In contrast, Neptune has emissions from both methane and ethane (Table 14.2). Why this difference? Evidently, there is a strong temperature inversion in the atmosphere of Neptune but only a weak one on Uranus. The structures of the atmospheres of all four giant planets are compared in Figure 14.9.

When temperature increases with altitude in such an inversion, absorption bands produced by gases in a normal atmosphere become emission bands, as seen from the outside. This is because the local gas temperature in the inversion region is higher than that of the lower levels of the atmosphere. Emission bands produced in inversion regions can be detected with spectrometers sensitive to the infrared, typically at wavelengths near 10 μm, where there is little reflected sunlight to interfere.

As we saw in the discussion of Jupiter and Saturn, it is in the upper atmospheres of these planets that we expect trace gases to be produced by photochemistry and bombardment by charged particles trapped in the planet's magnetospheres. Furthermore, many gases have some of their strongest emission lines in the infrared, thereby making it easy to detect them at these wavelengths, even in tiny quantities. This combination of circumstances has led to the discovery of several trace constituents in these inversion regions above the tropopause, where the temperature is increasing with altitude.

It is useful to notice just how cold the tropopauses on Uranus and Neptune are (Fig. 14.9). At 55 K ($-218°$C), our own atmosphere would condense to a mixture of ices! Hydrogen, helium, and neon can remain in the gaseous state, and nitrogen, methane, and carbon monoxide still have significant vapor pressures (like water vapor over ice on our planet). This is a small list of gases from which to make an atmosphere.

In the case of Neptune, we find evidence for all of these gases except neon, which is very difficult to detect. Hydrogen and methane were discovered by ground-based spectrometry, and helium was detected by Voyager 2 using the occultation technique. Carbon monoxide emission lines from the region of the thermal inversion were detected in 1991 by astronomers using a radio telescope on Mauna Kea. The same set of observations revealed emissions from HCN, indicating that there must be a source of nitrogen in Neptune's upper atmosphere. N_2 seems the logical choice because ammonia would be frozen out.

Presumably, the CO and N_2 on Neptune are convected up into the visible part of the atmosphere from the deep interior. Apparently, the reactions that efficiently convert CO to CH_4 and N_2 to NH_3 (ammonia) on Jupiter and Saturn are not working as well on Neptune. In fact, the amount of CO in Neptune's atmosphere is a thousand times greater than a chemical equilibrium model using these reactions would predict. Such a model uses the best current estimates for conditions in a planet's interior together with data for all relevant chemical reactions to predict what substances will

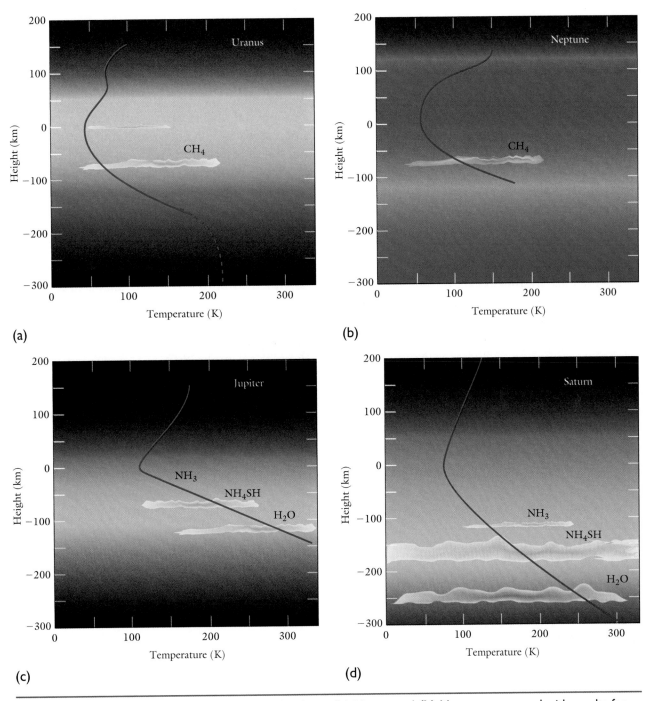

(a)

(b)

(c)

(d)

Figure 14.9 The variation of temperature with height on **(a)** Uranus and **(b)** Neptune, compared with results for **(c)** Jupiter and **(d)** Saturn from Chapter 13. Temperature increases more slowly with depth on Uranus than it would if the planet had a large internal heat source, as Neptune does.

form from the elements that are present. One such reaction is $N_2 + 3 H_2 = 2 NH_3$ in which nitrogen combines with hydrogen to produce ammonia. But this reaction requires a **catalyst** to proceed efficiently, and the problem on Neptune may be that an appropriate catalyst is not available. A catalyst is a substance that helps a reaction to proceed but does not take part in the reaction itself. In the case of nitrogen and hydrogen, iron compounds can serve as effective catalysts.

TEMPERATURES MEASURED AT RADIO WAVELENGTHS

Going to longer wavelengths, we find more puzzles in observations made with radio telescopes. No indications of nonthermal radio emissions were detected from either of these planets until Voyager reached them. Telescopes on Earth are not able to record noise bursts at decameter wavelengths, or steady signals from spiraling electrons in the decimeter region as they did for Jupiter. What the Earth-based radio observations do show, however, is very interesting indeed.

At short radio wavelengths, we can detect thermal radiation from deep within the atmospheres, so the corresponding temperatures should be higher than those detected elsewhere in the spectrum. In fact, one expects the measured temperatures to be higher as the wavelength of the observations increases because the atmospheric gases become increasingly transparent to radiation of longer wavelengths. This is how the high surface temperature of Venus was discovered (see Section 10.1). The expected increase in temperature with increasing wavelength is found on Jupiter, Saturn, and Neptune, but once again Uranus is the exception. Evidently, the temperature in Uranus's atmosphere does not increase steadily with depth, as it does in the atmospheres of the other giants.

Unlike some of the other peculiarities of Uranus we have discussed, we can explain this unusual temperature structure. Recall that Uranus has no detectable source of internal energy. In other words, it doesn't have an excessively hot interior. There is not enough energy to maintain convection in the lower atmosphere, so the increase in temperature with depth at these deep levels is very gradual. At higher altitudes, sunlight provides the energy to maintain convection. The situation resembles that in our oceans, where only the upper layers participate in the seasonal cycle driven by sunlight, while the murky depths remain at essentially the same temperature throughout the year. This combination of external sunlight and a weak internal heat source

leads to the unusual temperature profile shown in Figure 14.9.

The lack of a strong internal heat source probably also explains the absence of N_2 and CO in the upper atmosphere of Uranus. Without strong, vertical convection from the planet's interior, these gases will not rise to the part of the atmosphere that we can see.

The radio spectra of Uranus and Neptune reveal one other unusual characteristic. Radio brightness measurements for Jupiter and Saturn agree with calculations for the thermal emission expected from these planets, but the Uranus and Neptune data do not. The discrepancy is particularly marked in the 3–10-cm region of the spectrum, where the planets radiate more energy than the calculations would suggest. This means that the real atmospheres are more transparent at these wavelengths than the model atmospheres used in the calculations. This higher transparency indicates a smaller amount of ammonia (NH_3) because this is the gas that absorbs at these wavelengths. Why is ammonia depleted? Perhaps the answer lies in those hypothetical water clouds described in the previous section. Since ammonia dissolves readily in water, it might be kept out of the atmosphere. Alternatively, perhaps the production of NH_3 from N_2 is inhibited. Or there might be a deficiency of nitrogen relative to sulfur, leading to clouds of NH_4SH that consume the NH_3. We need more information before we can expect to find an explanation for this puzzle.

14.4 CLIMATE, CLOUDS, AND WEATHER

We have seen that the upper atmospheres of both Uranus and Neptune are remarkably clear. To study the local meteorology, we need to be able to observe clouds as tracers of global wind patterns. This proved to be especially difficult on Uranus, where we anticipated a climate radically different from that of every other planet owing to the extreme inclination of Uranus's axis of rotation (Fig. 14.10).

Figure 14.10 Voyager image of a crescent Uranus—beautiful but not very informative about conditions in the atmosphere.

THE GENERAL CIRCULATION OF THE ATMOSPHERES

Neptune experiences seasons similar to those we enjoy on Earth because its axis is inclined by 27°, but the seasons are nearly 165 times longer than ours because Neptune's period of revolution is 165 years. Now imagine the seasons on Uranus, whose axis is tipped over nearly in the plane of the planet's orbit. This planet requires 84 years to circle the Sun. This means that for 42 years, one pole receives some illumination from the Sun while the other pole is in total darkness. The pole-on configurations represent the winter and summer solstices, while the equinoxes occur when the equator faces the Sun.

One might think that this orientation of the rotation axis should affect the circulation of the atmosphere. When one of the poles is facing the Sun, it receives continuous illumination that is independent of the planet's rotation. One would expect the circulation to consist of warm air rising at the pole that points toward the Sun and then moving past the equator to the other pole, where it would sink. This is the simplest kind of Hadley cell circulation (see Section 10.3). However, Voyager found that the general circulation on Uranus is dominated by the planet's rotation, and the clouds therefore exhibit a banded pattern parallel to latitude lines, similar to that of Jupiter and Saturn.

This was not an easy conclusion to reach, since very few clouds were detected on Uranus. These discrete clouds are probably condensed methane, rather than the ammonia or ammonium hydrosulfide clouds we encountered at Jupiter. They appear to form at an altitude where the temperature is 80 K and the pressure is 1.3 bar. Figure 14.11 compares the atmospheric wind speeds for all four of the giant planets.

WEATHER ON NEPTUNE

Unlike Uranus, Neptune has well-defined, high-contrast white clouds in its upper atmosphere (Fig. 14.12). These clouds are so prominent that their existence was deduced from Earth-based observations long before Voyager 2 arrived. The spacecraft also found evidence for dark clouds in the lower cloud deck that provides a boundary to the visible atmosphere. The most prominent of these is the Great Dark Spot, a large elliptical vortex that is strongly reminiscent of Jupiter's Great Red Spot.

In the Great Dark Spot we are again dealing with a giant eddy, in this case about the size of Earth. As with the GRS, Neptune's Great Dark Spot occurs in the planet's southern hemisphere and represents an anticyclonic disturbance, with winds blowing counterclockwise around an area of high pressure. Smaller cyclonic/anticyclonic storms also occur at higher latitudes, again reminiscent of jovian weather patterns but on a smaller scale. While the white clouds continue to be seen in different shapes and locations, subsequent high-resolution Earth-based images reveal that the Great Dark Spot has disappeared.

The colors of the Neptune clouds give no hints that can be used to determine their composition.

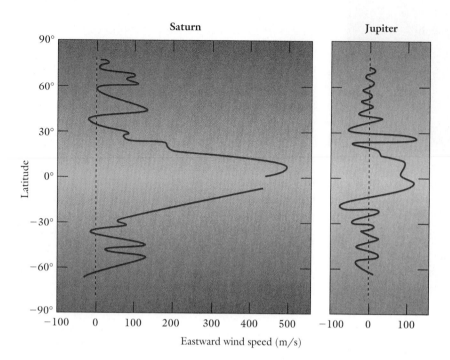

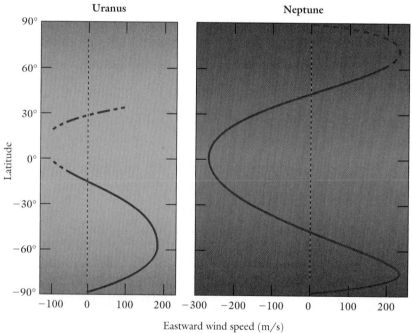

Figure 14.11 Comparative wind speeds for all four giant planets. When the speeds of planetary winds are plotted as a function of latitude, we find that the general circulations of the atmospheres of Uranus and Neptune are essentially identical. Neither planet exhibits the strong equatorial flow characteristic of both Jupiter and Saturn.

The white clouds certainly look like some form of cirrus—clouds of ice crystals—and the ice is probably methane, but we don't know for certain. The lower cloud deck is even less well defined, with suggestions for composition ranging from methane droplets to hydrogen sulfide ice crystals.

ROTATION PERIODS AND WIND SPEEDS

An accurate determination of the rotation period of Uranus was not possible before the Voyager encounter in 1986. The spacecraft pictures revealed clouds that could be tracked to indicate a period

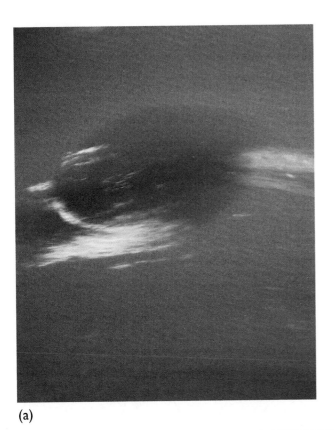

(a)

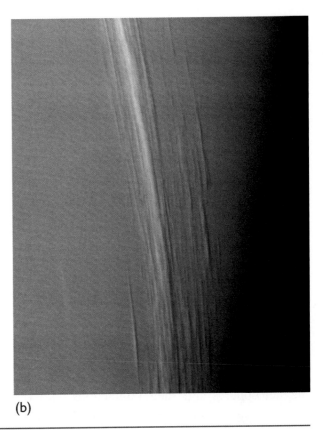

(b)

Figure 14.12 Neptune is the only other blue and white planet in our solar system besides Earth. But here the white clouds are made of frozen methane instead of water ice, and the large oval feature is not a continent but a giant vortex in this planet's thick atmosphere. **(a)** Close-up view of Neptune's Great Dark Spot showing the associated white clouds that occur at higher altitudes. **(b)** A band of high methane clouds, casting its shadow on the smooth cloud deck below. The widths of the cloud streaks range from 50 to 200 km, with a height of about 50 km, determined by measuring the shadows.

near 16 hours. As in the case of Jupiter or Saturn, however, what is actually measured by such observations is the sum of the underlying rotation and of the wind speeds at the location of the clouds. To disentangle these two effects, one must have an independent measurement of the rotation period of the deep interior, determined from observations of the magnetic field of Uranus to be 17.2 hours. Only then can the true wind speeds be determined. As shown in Figure 14.11, the speed varies with latitude on Uranus, but in a markedly different way than on Jupiter and Saturn.

On Neptune it has been easier to determine a rotation period because the clouds are much more prominent. Again, the rotation period of the magnetic field provides the reference: Neptune's period of rotation is 16.8 hours, very similar to that

of Uranus. This similarity extends to the pattern of wind currents on the planet as well. Uranus lies on its side, while Neptune's inclination is much more like Earth's, yet the change in wind speed with latitude is virtually identical on these two planets.

There is a curious paradox implicit in this result. The basic physics that governs moving air tells us that because the atmosphere of Uranus is getting most of its heat at the pole, wind speeds should increase toward the equator. However, the reverse is true: The fastest winds are at high latitudes. Furthermore, the Voyager measurements show that the temperature of the atmosphere of Uranus is the same at the north pole, which has just experienced 21 years of total darkness, as it is at the south pole, which has been heated for 21 years. Evidently, some dynamic processes—not yet understood—are redistributing the heat deposited from the Sun in ways that can account for both the unusual temperature distribution and the wind pattern. It is hard to escape the idea that something in the internal structures of these two planets is determining the wind speeds we see in the visible regions of their atmospheres.

14.5 MAGNETOSPHERES

As in the case of global circulation patterns in their atmospheres, the magnetospheres of Uranus and Neptune were assumed to be rather different prior to the Voyager encounters. Experience with Earth, Jupiter, and Saturn suggested that the axis of a planetary magnetic field should always be roughly aligned with the axis of the planet's rotation. Since the inclinations of the rotational axes of Uranus and Neptune are radically different, scientists expected the configurations of their magnetospheres and their interactions with the solar wind to differ as well. In fact, this turned out to be true, but in a different way from the original expectations.

A TILTED FIELD AT URANUS

The first surprise came at Uranus, where the magnetic field was found to be inclined 60° to the axis of rotation and to be significantly displaced from that axis as well. It is the close alignment of these two axes on Earth that leads inhabitants of our planet to say that a magnetic compass "points north." The magnetic pole is not too far from true north, as defined by the axis of rotation. But imagine the plight of a trekker on Uranus trying to establish directions on a cloudy day. There the magnetic axis is not only inclined by 60° but also offset from the rotational axis by one-third of the planetary radius (Fig. 14.13). Hence, the direction that a compass needle would point on Uranus (relative to the orientation of the rotational axis) would vary greatly with position on the planet. The strength of the field would also vary, from 0.1 to 1.1 gauss (compared with a nearly constant 0.3 gauss on Earth).

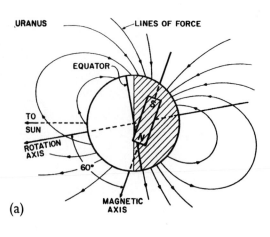

(a)

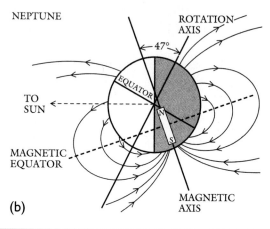

(b)

Figure 14.13 **(a)** The magnetic field of Uranus behaves as if there were a bar magnet inside the planet, tipped at an angle of 60° to the axis of rotation and offset as shown in the diagram. **(b)** To the surprise of most scientists, the magnetic field of Neptune exhibits a configuration very similar to that of Uranus.

Uranus has a magnetosphere similar in size to that of Saturn. However, the composition of the atomic particles in the magnetosphere is simpler than in Saturn's. It consists almost entirely of protons and electrons derived primarily from hydrogen escaping from the planet's atmosphere. The magnetosphere also shares the high inclination of the planetary magnetic field to which it is bound. A magnetotail stretches out tens of planetary radii behind the planet, but because of the inclination of the magnetic field, the magnetotail rotates like a corkscrew as the planet turns on its axis.

The entire sunlit side of Uranus is bathed in a glow of ultraviolet light emitted by escaping hydrogen atoms. This electroglow, as it is called, was also seen on Jupiter, Saturn, and Titan. As on the two giant planets, Uranus exhibits auroral activity produced by the collision of magnetospheric ions and electrons with the planet's upper atmosphere. The aurora consists of ultraviolet emission lines detected by the spacecraft and infrared emission from H_3^+, which was discovered by Earth-based observations in 1992. On Uranus, of course, the aurora is not located near the north and south poles. Instead, it occurs near the equator because this is where the magnetic poles are found. The Uranians (if there were any) would speak of the beauties of the "tropical lights" instead of the "polar lights" that we extol.

NEPTUNE IMITATES URANUS

One might naively assume that the peculiar orientation of the magnetic field of Uranus has something to do with the high inclination of the planet's axis of rotation, but Neptune shows that this is not the case. Here the inclination of the rotational axis is only 27°, close to that of Earth, but the configuration of the magnetic field relative to this axis is similar to the situation on Uranus. Neptune's magnetic dipole is offset from the axis of rotation by a little over half of the planet's radius (for Uranus it is one-third) and is tilted relative to that axis by 47° (Uranus's tilt is 60°) (Fig. 14.13b). The strength of the magnetic field is about half that of Uranus, also varying over the globe in response to the offset of the dipole.

How can we account for the strangely oriented magnetic fields of Uranus and Neptune, which are similar to each other but different from those of Jupiter, Saturn, and Earth? The best idea at present is that they are generated in mantles of pressure-ionized "ice" surrounding the dense "rocky" cores of these two planets. (We use quotation marks here because the exact nature of these materials is not well defined.) What this means is that compounds of H, C, O, and N could become ionized at high pressure so they become conducting. They would then pro-

vide the conducting fluid in which the magnetic field is generated. The separation of "icy" and "rocky" compounds would be consistent with our present ideas about the internal structures of these planets, but these ideas are still in a rather preliminary state. The point is that we need a conducting fluid to generate these fields, and models for the interiors of Uranus and Neptune suggest that this configuration is reasonable. Much more research is needed before we can claim a complete understanding of exactly how these fields are generated or why they have such peculiar orientations.

14.6 PLUTO AND ITS MOON

Is Pluto really a planet? Scientists have been asking this question ever since the discovery of this unusual object. You might think that since it moves about the Sun in a roughly circular orbit, it must be a planet by definition. Comets and asteroids also orbit the Sun, however, and Pluto's orbit is the most eccentric and highly inclined of any of the planets. Pluto is also very small, only two-thirds the size of our Moon. Its rotation period is unusually long; at 6.4 days, it is exceeded only by those of Venus and Mercury.

Pluto's orbit crosses that of Neptune, so there has been speculation that Pluto is actually an escaped satellite of this much larger planet: Perhaps a close encounter between Pluto and Triton when both were satellites of Neptune sent Pluto out into its own orbit about the Sun and gave Triton its puzzling retrograde orbit. Or perhaps a third body passed through the Neptune system, expelling Pluto and perturbing Triton. However, the orbital motions of Pluto and Neptune are in a resonance that prevents these planets from getting closer than 17 AU, making these intriguing ideas rather unlikely.

PLUTO'S MOON

The discovery in 1978 that Pluto has a satellite of its own adds further weight to the current consensus that this planet had an independent origin. The

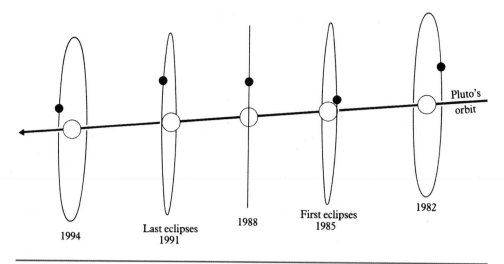

Figure 14.14 As Pluto moves along its orbit around the Sun, the aspect of Charon's orbit changes as viewed from Earth. From 1985 through 1991, a series of eclipses and occultations occurred.

satellite is named Charon (the *ch* is pronounced hard, like the *ch* in "chaos") for the mythical boatman who carried the dead into the realm of Pluto.

Beginning in January 1985, the orientation of the orbit of Charon caused the satellite to pass alternately in front of and behind Pluto (Fig. 14.14). Careful observations of these eclipses and occultations, which continued through 1991, provided data on the relative sizes and brightnesses of the two objects, as well as on the orbit of Charon around Pluto. Improved infrared cameras and spectrometers are also coming into play, so we are steadily learning more and more about these distant objects even without a space mission.

Charon orbits Pluto at a distance of just under 20,000 km, closer than any other natural satellite to its planet except Mars's Phobos. The period of Charon's revolution about Pluto is exactly equal to the period of the planet's rotation. Here is another case of tidal resonance in the solar system, and one that is different from all the others we know. It means that Charon is in a synchronous orbit that always keeps it above the same region on Pluto. Like two ballroom dancers, Pluto and Charon face each other as they waltz around the Sun. We deliberately give such geosynchronous orbits to some of our artificial satellites so they can function as televi-

sion relays, beaming down programs to fixed antennas on the ground. We don't expect a thriving television industry on Pluto, however. Instead, it appears that Charon was able to pull Pluto into a tidal lock because of its large relative mass, about one-tenth the mass of Pluto. This is the largest satellite/planet mass ratio in the solar system, with Moon/Earth being second at 1/80.

MASS OF PLUTO

Until the discovery of Charon, the mass of Pluto could be estimated only indirectly by the apparent effect of Pluto on the motions of Uranus and Neptune (see Section 14.1). These estimates were very uncertain, with values ranging all the way from 0.2 to 7 Earth masses. The corresponding uncertainties in the density of Pluto made it impossible to estimate the composition of this planet.

By studying the motions of Pluto and Charon, however, it is possible to use Newton's laws to determine Pluto's mass. The result is a mass of 1/400 that of Earth, much smaller than any of the previous estimates. When this mass is combined with the occultation-derived diameter (2390 km), a density of 2.1 g/cm³ is derived, essentially identical to the density of Triton. Evidently, both of these distant worlds are made of the same kinds of material.

PERSPECTIVE

IS PLUTO A REAL PLANET?

We saw in Chapter 6 that Pluto can be considered as the largest member of the Kuiper belt of icy planetesimals. The next largest KBOs are Varuna and Quaoar, with diameters near 1000 km, similar to Ceres, the largest of the asteroids. Pluto's satellite Charon does not confer planetary status because a surprising number of KBOs are turning out to be binary objects, as are many asteroids. There has been serious speculation that large KBOs were originally far more abundant than they are today, the present number having been reduced by collisions within the belt. Pluto is then seen as unique in having avoided these multiple collisions. It also avoided being incorporated in the core of Neptune, where many of Pluto's contemporaries must now reside. Triton, which we shall meet in the next chapter, represents another potential endpoint in the histories of these objects.

So is Pluto a planet or not? Well, what about Mars? Is it a planet or just a large asteroid? A simple definition suggested by astronomer Jack Lissauer is based on mass: The object should be massive enough that its gravitational field has formed it into a sphere (about 0.0020 times the mass of the Earth) but not so massive that nuclear reactions would begin in its interior (about 13 times the mass of Jupiter). Pluto just satisfies the lower limit of this definition and thus may be comfortably considered a planet, despite its generic association with the Kuiper belt.

Meanwhile, the issue of Pluto's planetary legitimacy continues to arouse passions. The American Museum of Natural History in New York has led those who would demote Pluto to asteroid status. In response, tens of thousands of school children have written letters or signed petitions asking that Pluto remain a planet. Hundreds of editorials have been written arguing the issue one way or the other. Some astronomers who study asteroids and KBOs want to add Pluto to the list of minor planets and give it a number just like other asteroids. Other scientists who would like more funds to study Pluto feel that its status as a planet will help keep the money flowing. Perhaps the strongest motivation comes from those planning for a spacecraft mission to Pluto. As long as it is a planet, they can argue for a mission to study "the only planet not yet visited by a spacecraft." They fear that as an asteroid or KBO, Pluto would have lower priority as a mission target.

This is all rather silly: Why demote Pluto and force millions of textbooks to be revised, for no obvious gain? Surely there are more important issues to debate, even within the rather esoteric field of planetary studies.

The orbit of Charon defines the inclination of Pluto's axis of rotation, since it is reasonable to assume that a system that is tidally locked in this fashion will have the satellite's orbit in the equatorial plane of the planet. This was another surprise: The inclination of Charon's orbit to the orbit of Pluto around the Sun is 112°. This means that Pluto, like Uranus, has its rotational axis very nearly in the plane of its orbit (Uranus is at 98°). Three planets then rotate backward: Venus, Uranus, and Pluto.

SURFACE AND ATMOSPHERE

Long before Charon was discovered, the rotation period of Pluto was determined very accurately through observations of periodic changes in the planet's apparent brightness. The global reflectivity of Pluto varies from 0.3 to 0.5 as the planet rotates, compared with the constant values of 0.1 for our Moon, a rocky object, and 0.8 for Triton, Pluto's twin.

The first crude spectral measurements of Pluto in 1975 by author Morrison and his colleagues revealed the presence of methane ice on the planet's surface. Observations made with improved instrumentation in 1992 by author Owen and colleagues show that ices of carbon monoxide and nitrogen are present as well (Fig. 14.15). Thus the composition of the surface of Pluto seems similar to that of Triton, with the difference that there is more dark material on Pluto, but no CO_2, and the methane to nitrogen ratio is higher than on Triton. Careful study of the eclipses and occultations of Pluto and Charon by each other has permitted the development of a crude map of the distribution of dark material on the planet's surface.

The similarity in the surface ices on Pluto and Triton means that the atmospheres of the two objects must also be similar. Although we don't have any spacecraft data for Pluto that would allow us to establish this case, the fact that nitrogen ice is present means that this gas must dominate the atmosphere: Nitrogen has the highest vapor pressure, so it will be the most abundant gas. This conclusion is consistent with the observations of a stellar occultation that took place in 1988. The dimming of the star's light before it disappeared behind the planet proved that a thin, distended atmosphere was present. If nitrogen is indeed the major constituent, the structure of the atmosphere will be similar to that of Triton, leading to a surface pressure in the 1–20-microbar range for a surface temperature of 35–40 K.

Observations of Pluto and Charon together and Pluto by itself (with Charon in eclipse or occultation) have allowed a determination of the surface composition of the satellite (Fig. 14.16). The surprising result is that Charon is coated with water ice. This common substance is difficult to detect in spectra of Pluto, although it is clearly present on Triton. There is no evidence (yet) for solid methane on Charon. Evidently the process(es) responsible for the formation of this satellite heated it sufficiently to drive off the more volatile compounds, just as the giant impact theory for the origin of our Moon explains its lack of volatiles (see Section 8.7). Alternatively, the methane on Pluto and Triton may have been made by impact heating, and the heat released in such events on smaller Charon was itself sufficient to drive off the methane. New observations of other small bodies in the outer solar system may provide the answer.

The eccentricity of Pluto's orbit must have a large effect on the atmosphere. At perihelion, Pluto's distance from the Sun is 29.7 AU, bringing it closer than Neptune, which occupies a nearly circular orbit at 30.1 AU. At the other extreme, aphelion, Pluto is at a distance of 49.7 AU. This means that the intensity of sunlight at the surface of the planet varies by a factor of about three around the orbit. Since we are dealing with an atmosphere that is nothing more than vapor in equilibrium with its ice, the resulting decrease in the temperature of the ice will essentially freeze out the atmosphere. We are very fortunate to be studying Pluto when it is so close to perihelion, which occurred in 1989.

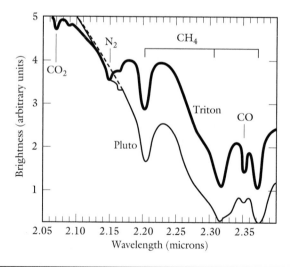

Figure 14.15 This spectrum of Pluto was recorded in 1992 with the United Kingdom Infrared Telescope on Mauna Kea. It shows the presence of ices of CH_4, CO, and N_2.

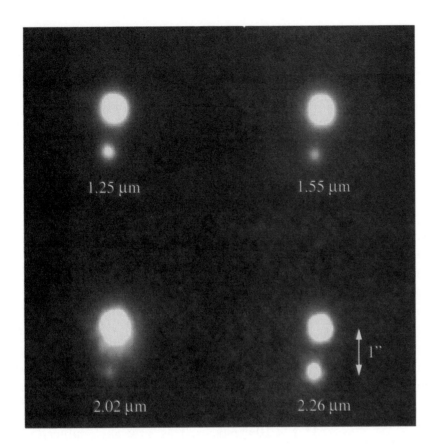

Figure 14.16 These images of Pluto and Charon were taken with the Canada-France-Hawaii Telescope on Mauna Kea and a special camera that employed adaptive optics to minimize the effects of Earth's atmosphere. Here we see the two objects clearly separated, as viewed through different filters that correspond to different absorption bands: 1.25 μm: no absorption; 1.55 μm: weak water ice absorption; Charon looks dimmer; Pluto bright. 2.02 μm: strong water ice absorption; Charon very dim. 2.26 μm: strong methane ice absorption. Pluto looks dimmer; Charon bright.

AT THE EDGE OF THE PLANETARY SYSTEM

In Pluto we have found an outer planet that we could stand on, instead of being limited to exploration of a thick atmosphere from a balloon. The sky overhead would be black, even at midday, because the atmosphere is very thin and the Sun very far away. Seen from this vast distance, our star would be some 1600 times fainter than it appears from Earth, too small even to show a disk without optical magnification. Charon would be visible from only one hemisphere of Pluto, as a consequence of the synchronous rotation of the planet.

Pluto is the last large body in the solar system. Beyond this frigid world, there are only comet nuclei. The closest and largest of these are relatively nearby in the Kuiper belt, still confined roughly to the plane of the planetary system. A thousand times farther out, the entire system is enveloped by the huge, spherical cloud of comets, first conceived of by Jan Oort.

It is as if we are on the beach of the outermost island of an archipelago, looking out to sea. There are a few rocks just offshore, their dark heads rising above the waves. Beyond them, the water stretches out unbroken to the horizon. From time to time, some pieces of driftwood wash up on our shore, but we know from examining them that these random wanderers began their journeys in the forests of the islands behind us. They do not signal the presence of new continents just over the horizon. The closest other land is very far away, and with the primitive means of transportation we have at our disposal, we will never reach it.

Meanwhile, we can try to find out more about Pluto. Improvements in ground-based and orbiting telescopes during the next 10 years are sure to provide some new insights. The real increase in our knowledge will come from a spacecraft mission, currently still in the planning stage. If this dream is realized, we will have completed a reconnaissance of our planetary system by the year 2020.

14.7 QUANTITATIVE SUPPLEMENT:
DISCOVERING A PLANET

When Galle first saw Neptune, it looked like a fuzzy, blue-green star because his telescope was not able to show the planet clearly as a disk. Later observations with more powerful telescopes revealed that Neptune has an apparent angular diameter of 2.3 arc seconds. (This may be compared with the apparent diameter of the Moon, which is 1865 arc seconds, or about 30 arc minutes.) Meanwhile, how could Galle decide whether this new object was really a planet and not simply an uncharted star? The answer, of course, is that he could see whether it moved, like a true wanderer among the background stars.

If this is a new, distant planet, it must be traveling slowly in a nearly circular orbit about the Sun. We can determine the period by watching Neptune for a few weeks. We then measure the length of its apparent path and calculate the angular speed of the planet in seconds of arc per day. If we divide this into 360°, the angular distance of a full traversal of the orbit, we obtain the period $P = 164.82$ years.

To determine the true size of this planet, we need to know how far away it is. We can use Kepler's third law to obtain this information. The relationship between distance D and period P can be written in the form $D = P^{2/3}$ if D is in astronomical units and P is in years. We know P, so we obtain $D = 30.1$ AU. If the orbit is circular, the minimum distance of Neptune from Earth is then $30.1 - 1 = 29.1$ AU.

Now that we know how far away Neptune is and how big it appears to be (2.3 seconds of arc), we can determine its true size using the small angle equation (Fig. 14.17):

$$\theta = \frac{2r}{(D-1) \times 1.5 \times 10^8 \text{ km}}$$

where $2r$ is the true diameter of Neptune and $(D - 1)$ is the distance of Neptune from Earth, which we convert from AUs to kilometers by multiplying by 1.5×10^8 km/AU. In this case, the angle must be converted to radians—that is, fractions of an arc equal to the distance to that arc. (The circumference of any circle of radius r is $2\pi r = 360°$, so 1 radian $= 360°/2\pi = 57.2958°$.) Converting the apparent angular diameter of Neptune to these units, we find 2.3 arc seconds $= 11.2 \times 10^{-6}$ radians. This is the angle subtended by Neptune's disk. Using the small-angle equation, we then find

$$2r = \theta(D - 1) \times 1.5 \times 10^8 \text{ km}$$
$$2r = 11.2 \times 10^{-6} \times 29.1 \times 1.5 \times 10^8 \text{ km}$$

or

$$2r = 49,000 \text{ km}$$

roughly four times the diameter of Earth.

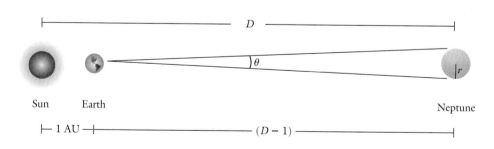

Figure 14.17 The distance of Neptune from the Sun is D, so Neptune's distance from Earth is $(D - 1)$, if D is expressed in astronomical units (AU). The diameter of Neptune is $2r$; seen from Earth, the planet subtends an angle $\theta = 2r/(D - 1)$ in the sky.

Summary

The three outermost planets were not known in antiquity; they had to be discovered. The stories of their discoveries illustrate an interesting variety of methods. Uranus was found accidentally, at a time when no one expected any planets beyond the five visible to the unaided eye. The search for Neptune involved theoretical predictions based on gravitational theory and the most sophisticated mathematical techniques of the nineteenth century; its discovery was a triumph for Newtonian celestial mechanics. The search for Pluto was similarly inspired by theoretical work, but ultimately the planet was found as the result of a very careful observational search. We now know Pluto is far too small to have perturbed the motions of other planets in a measurable way.

Uranus and Neptune form a pair of planets, substantially smaller than Jupiter or Saturn but much larger than the terrestrial planets. Their composition is also intermediate between that of the giants, with their nearly cosmic proportions of the elements, and that of the rocky, volatile-depleted terrestrial bodies. Uranus and Neptune are composed primarily of the common ices mixed with silicates and metals, with deep atmospheres dominated by hydrogen and helium, but with proportionately more methane than we find on Jupiter or Saturn.

Uranus is notable for its rotation, with its axis tipped on one side, and for its lack of a major internal heat source. The atmosphere of Uranus is largely free of clouds, but that of Neptune exhibits rapidly changing high-altitude clouds. It is surprising that the circulation and temperatures in the Uranus atmosphere do not seem to be influenced by the peculiar orientation of its rotation axis because Neptune exhibits a nearly identical circulation pattern.

Both of these planets have eccentric magnetic fields whose orientations are remarkably similar, inclined at 60° and 47°, respectively, to the axes of rotation. Charged particles trapped in their magnetospheres plunge into the atmospheres along magnetic field lines, creating auroras. The locations of these auroras are unusual from our terrestrial perspective because the magnetic field axes on these planets are so displaced from the planets' rotational axes.

Pluto is the last planet in the solar system. Its frigid surface is covered by ices of nitrogen, methane, and carbon monoxide. A tenuous atmosphere of vapors in equilibrium with these ices must mostly freeze out when Pluto is at aphelion in its eccentric orbit. Pluto's disproportionately large satellite Charon is in a remarkable synchronous orbit. Unlike Pluto, Charon's surface is dominated by water ice, with no evidence for solid methane. We will learn little more about this distant system until a spacecraft flies past it.

Key Terms

catalyst   perturbation

Review Questions

1. How was Uranus discovered, and by whom? What gave scientists the idea that there might be another planet more distant than Uranus, and how did they try to find it? Who succeeded in seeing it?

2. How would you compare the discovery of Pluto with the discoveries of Uranus and Neptune? Consider the application of perturbation calculations versus luck.

3. Compare Jupiter and Saturn with Uranus and Neptune. Do you think all four planets have the same composition? Why or why not? How does Pluto fit into these two groups?

4. Methane and hydrogen have been detected spectrally in the atmospheres of Uranus and Neptune, but ammonia has not. Explain.

5. Describe the seasons on Uranus and compare them with the seasons on Earth. What is the reason for the difference?

6. Why was it so difficult for the Voyager spacecraft to determine the atmospheric general circulation of Uranus?

7. Describe the difference in the internal heat sources of Uranus and Neptune. What effect does this

difference have on the structure (temperature versus pressure) of the atmospheres of these two bodies and on their respective weather patterns?

8. Describe the magnetic field orientations of Uranus and Neptune, and compare them with the fields of Earth and Jupiter. At what latitudes would you expect to find auroras on these two planets?

9. Why didn't scientists know the mass of Pluto before Charon was discovered? How did they then determine it?

10. Consider a thought experiment in which you move Pluto and Charon to an orbit at the position of Mars. Describe what would happen to Pluto's atmosphere. What would happen to Charon?

QUANTITATIVE EXERCISES

1. Calculate the acceleration of gravity on Neptune and on Earth. How do you explain the fact that the two values are so similar?

2. How many days would Herschel have needed to wait before Uranus moved a distance in the sky equal to the diameter of the full moon?

3. Calculate the speed of Neptune in its orbit in units of kilometers per second. Compare this with the orbital speed of Earth. Why are they so different?

4. Use the table of solar abundances (Table 2.2) to calculate the mean molecular weight of an atmosphere formed of H_2, He, CH_4, and NH_3, where the elements have solar abundances. Now compare this with a model for Neptune's atmosphere in which C/H is enriched 25 times relative to the solar value and all the nitrogen is present as N_2 instead of as NH_3. What change in He/H (compared with the solar value) would give the same mean molecular weight? Which possibility seems more likely to you, increased He/H or nitrogen as N_2? Explain how each possibility could occur.

5. The resolving power (in seconds of arc) of an optical telescope is given as R = 13.84 arc seconds/D, where D is the diameter of the telescope in centimeters. How large a telescope would you need to see a 20-km-wide crater on the surface of Pluto?

ADDITIONAL READING

Grosser, M. 1962. *The Discovery of Neptune*. Cambridge, MA: Harvard University Press. An engrossing account of the events leading up to the discovery of Neptune, including the discovery and early observations of Uranus.

Littman, M. 1990 (revised printing). *Planets Beyond: Discovering the Outer Solar System*. New York: John Wiley & Sons. A popular account of the outer planets that includes the discoveries made during the Voyager encounters with both Uranus and Neptune.

15

WORLDS OF FIRE AND ICE: THE LARGE SATELLITES

Family portrait of the four large moons of Jupiter. Io, Europa, Ganymede, and Callisto are shown to scale in this composite Galileo photo.

We have now described all the objects in the solar system that move in orbits around the Sun: the inner and outer planets, the asteroids, and the comets. We have seen that these objects can be roughly characterized as rocky, icy, or gas-rich, with giant planets incorporating heavier elements in solid or liquid form in addition to their deep hydrogen-rich atmospheres.

In this chapter, we focus on the six largest satellites of the outer planetary system: Callisto, Ganymede, Europa, Io, Titan, and Triton, each of which is a complex and interesting world in its own right. These satellites could easily be considered planets if they orbited the Sun; three are about the size of Mercury and one has an atmosphere denser than Earth's. Fortunately, we have been able to study them rather easily, since one spacecraft can visit several satellites in a single trip. The Pioneer, Voyager, Galileo, and Cassini missions have all taken advantage of the opportunity to study multiple worlds for the price of one. We will discuss the small satellites and the rings of the giant planets in Chapter 16.

15.1 IMPACT CRATERS AND SURFACE AGES

Before describing each of these satellites individually, let's look at some of the ideas that we will use in evaluating their histories. We have already discussed many of these topics in the context of the inner solar system, but now we must extend our vision to the colder and more distant realm of the giant planets.

The most important property of each satellite is its bulk composition. Much of a satellite's geologic history is predetermined by its interior composition and structure. This is especially true in the outer solar system, where a wide variety of ices are potentially available in addition to the silicates and metals that make up the inner planets. The dominant ice is, as we would expect from cosmic abundances, water ice (H_2O). Many of these satellites are roughly half ice and half rock, with correspondingly lower densities than we encountered for our Moon and the terrestrial planets. To work out the composition in a quantitative way, we must not only measure the mass and diameter accurately but also correct for self-compression, just as we did in estimating the uncompressed densities for the terrestrial planets (see Section 8.4). Ice is also detected in infrared spectra of the surfaces, as shown in Figure 15.1.

Table 15.1 lists the bulk properties of all six of the satellites described in this chapter.

IMPACTS AND CRATERING IN THE OUTER SOLAR SYSTEM

Craters are found throughout the planetary system. Let's ask ourselves (as did the Voyager scientists before the Jupiter and Saturn encounters) what cratering we should expect on these icy solid bodies in the outer solar system. Will numerous impacts have taken place, and if so, will the resulting craters have been preserved for us to see? Can we use crater densities to estimate the ages of satellite surfaces, as we did for the planets of the inner solar system?

Given the plethora of well-preserved craters on the Moon, Mercury, Mars, and Venus, it may seem odd that we worry about the preservation of

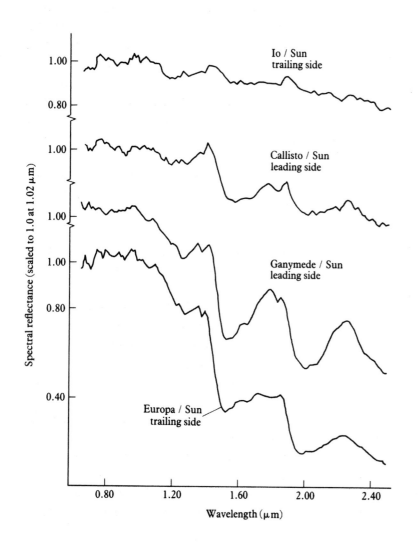

Figure 15.1 Long before spacecraft reached the Jupiter system, we knew that Callisto, Ganymede, and Europa were covered with ice. This deduction is made from infrared spectra such as these, obtained at Kitt Peak, Arizona.

Table 15.1   Large satellite summary

Satellite	Period (days)	Diameter (Moon = 1)	Mass (Moon = 1)	Density (g/cm^3)
Callisto	16.69	1.38	1.47	1.9
Ganymede	1.16	1.51	2.02	1.9
Europa	3.55	0.91	0.65	3.0
Io	1.77	1.04	1.22	3.6
Titan	15.95	1.48	2.00	1.9
Triton	5.88	0.84	0.30	2.1

craters on outer planet satellites. The reason is that we are now dealing with a very different material. The surfaces of almost all the outer satellites are composed of ice or of ice-rock mixtures, like the comets. We know that on Earth ice is a plastic material and that glaciers of ice can flow like a viscous fluid. If the same thing happens on these satellites, no craters would be preserved. However, surface temperatures are much lower on the satellites than on Earth, and at low temperatures ice becomes stiff and strong. How low is low enough?

At Saturn and beyond, ice should be as strong as rock, and craters should be preserved indefinitely unless there is local heating or erosion. For the jovian satellites, however, the ice may not be strong enough to preserve craters for billions of years. Thus, the craters of Saturn's satellite Rhea are nearly indistinguishable from those on the Moon, whereas the craters on Jupiter's Callisto have been considerably modified. In addition, there is a possibility of degradation and eventual destruction of craters by sublimation of ice and escape of water vapor into space, aided by the constant blast of subatomic particles from the magnetosphere.

What about the number of crater-producing impacts? If most of the impacts on Earth and the Moon were from asteroids, we might expect a much smaller flux near the outer planets. Alternatively, there should be more comets in the outer solar system, since Jupiter's massive gravity partially shields the inner planets from comets. Gene

Shoemaker, whose contributions to lunar cratering chronology were mentioned in Section 7.3, was the first scientist to extend these studies to the outer solar system. As a member of the Voyager science team, he was especially anxious to estimate cratering rates in the outer solar system. Since Shoemaker's death in 1997, others have continued these studies. The results summarized below are largely due to NASA scientist Kevin Zahnle.

In the inner solar system, there are two main sources of projectiles: asteroids and comets. For the Earth and the Moon, between 75% and 90% of craters are caused by asteroid impacts and the rest by comets. These conclusions refer to the present, of course; different sources may have been important in the past, and certainly a different and much larger population of impacting bodies was present at the time of the terminal bombardment, 3.9 billion years ago.

Very few asteroids are on orbits that permit collisions with Jupiter and its satellites, and virtually none can make it out to Saturn. Asteroids therefore contribute very little to crater formation on the moons of Jupiter and Saturn. At Jupiter the primary crater-forming impacts are from Jupiter-family comets, which are mostly comets from the Kuiper belt that have been diverted into orbits that bring them repeatedly near the giant planet and its satellites. Beyond Jupiter, the long-period comets are the primary source of impactors. Zahnle concludes that the total impact rates in the jovian system are about a factor of two lower than those in the Earth-Moon system, and at Saturn they are another factor of two lower.

There is another important difference between craters formed by comets and by asteroids. The size distributions of the two populations of impactors differ, with substantially fewer small comets. Perhaps there are fewer collisions among comets to break up the big ones and produce many small fragments. Or maybe small comets have short lifetimes and disintegrate or evaporate more rapidly than big ones. Either way, there is a significant deficiency in comets smaller than 1 km, and hence a deficiency in craters smaller than about 10 km.

VARIATIONS IN CRATERING RATES

Once the impact flux has been calculated, it should be possible to use crater densities on the satellites to calculate approximate crater-retention ages, as described in Section 7.3. Even if the uncertainties in the flux are substantial, crater densities should easily tell us whether a surface age is measured in billions, or hundreds of millions, or perhaps tens of millions of years. First, however, we must consider an important correction to the calculated impact rates.

We saw when studying cratering on the Earth and Moon that the size of the crater excavated by a projectile depends on the impact speed. Earth, being larger, attracts asteroids more strongly than does the Moon, and the resulting craters are larger for projectiles of the same size. In comparing crater densities on Earth and the Moon, we must allow for the gravity field of Earth.

In the outer solar system, both the satellite and its central planet are important, especially the latter. We can visualize each giant planet to be at the bottom of a gravity slope; impactors are attracted toward the planet, and the closer they come, the faster they move as they fall down this slope. The result is that a satellite located far down the slope receives more impacts, and they occur at higher speeds. This effect can be very large: Mimas, the innermost large satellite of Saturn, can expect 20 times the cratering rate of Iapetus, far from the planet. While the outer satellites of

Jupiter and Saturn should have lower cratering rates than Earth or the Moon, impact rates on the inner satellites of these planets should be higher.

CRATER-RETENTION AGES ON THE SATELLITES

Speaking very roughly, the effects of the local gravity field increase the cratering rate and compensate for fewer impactors in the outer solar system. Consequently, crater densities on many of the satellites are related to age in a manner not too different from that derived for the inner solar system. For example, we can conclude that a crater density on one of these satellites similar to that of the lunar maria indicates an age of several billions of years. Like the Moon, the satellites of the outer solar system also include surfaces with high crater densities. Such high densities cannot be explained from current cratering rates. Thus, we are drawn to the same conclusion described previously for the Moon: There was an early period when cratering rates were much higher than they are today.

This is an important conclusion, as has been emphasized by Laurence Soderblom of the U.S. Geological Survey. Soderblom was the deputy team leader and the senior geologist on the Voyager Imaging Team (Fig. 15.2). He played an important role in defining the standard cratering history for the inner solar system, with its late heavy bombardment ending about 3.8 billion years ago. Soderblom and his colleagues concluded that this unifying paradigm developed for the inner planets applies to the outer solar system as well. This heavy bombardment is therefore associated with the entire solar system, rather than limited to the terrestrial planets. With this perspective, we now turn to the individual satellites.

15.2 CALLISTO AND GANYMEDE, THE ICY MOONS

Callisto and Ganymede are twins in much the same way as Venus and Earth are. They have nearly identical sizes and densities, and they occupy similar regions of space. They were the first solid objects to be studied with a mass that is about half water ice,

Figure 15.2 Larry Soderblom (right) shares a humorous moment with Bradford Smith during the Voyager encounters with the Saturn system.

and it was with considerable anticipation that Voyager geologists awaited the initial high-resolution pictures of these objects in 1979. After a gap of nearly two decades, studies accelerated again in the late 1990s as the Galileo spacecraft was targeted for several close flybys of both of these moons.

CALLISTO: BASIC FACTS

Callisto has had the simplest geologic history of the galilean satellites, one dominated by impact cratering (Fig. 15.3). Its diameter is 4840 km, almost identical to that of the planet Mercury, yet its mass is less than one-third as great. From this comparison, we are immediately aware of a fundamental difference in composition. Detailed calculations for a body the size of Callisto with its density of 1.9 g/cm^3 suggest that water ice is about equally as abundant as rocky materials in its interior.

All the large satellites of the outer solar system should have differentiated early in their history.

They began their existence with a good deal of water ice mixed with rocky and metallic materials. Presumably the rocky component contained its fair share of radioactive materials as a heat source, while the water ice has a much lower melting temperature than rock. Inevitably, the deep interior temperature rose above the melting point of the ice, even if the satellite formed cold. The heavier materials should then sink to form a rocky or muddy core, while the water rose and refroze to form a crust and mantle of ice. It was therefore with great surprise and even some disbelief that scientists analyzing the details of Callisto's gravity field concluded that this big satellite was not fully differentiated. The motion of the Galileo spacecraft on its close flybys revealed that there was a modest increase in density toward the core of Callisto, but less than expected if all the ice had separated from the silicate and metal. Callisto seems to be a world that began to differentiate but then froze up part way through the process. This is one of the first surprises in dealing with a moon that had been supposed to be rather dull.

The surface temperature of Callisto varies from 150 K near noon to about 100 K at night.

Figure 15.3 Callisto has the most heavily cratered, and hence the oldest, surface of the galilean satellites.

Under such frigid conditions, water ice is stable over very long intervals of time, although several meters of ice were still expected to have evaporated from the surface during the past 4.0 billion years. A little closer to the Sun, in the main asteroid belt, an icy object could not survive.

Ice is observed directly on the surface of Callisto, where prominent infrared absorption bands reveal its presence (Fig. 15.1). This ice is by no means pure, however, and the reflectivity of Callisto is actually rather low, only 18%. If there has been no resurfacing with fresh ice for billions of years, the accumulation of meteoritic dust mixed with the original ice crust would be expected to be fairly dark.

GEOLOGY OF CALLISTO

Callisto is a good place to begin our detailed study of satellite geology because it is relatively simple to understand. Whatever happened (or did not happen) to the interior during its abortive attempt at differentiation, the mantle and crust are relatively primitive mixtures of ice and rock. There has been little of what we call geologic activity, and the surface should be saturated with craters. As expected, initial Voyager pictures revealed a heavily cratered world with a landscape that is almost exclusively the product of impacts (Fig. 15.4). Over most of the surface, these impact craters are nearly as densely packed as those in the lunar highlands. Expressed in the same units that we have used before, the crater density on Callisto is 250 10-km craters per 1 million km². Barring some improbable recent blizzard of impacts, it appears that the surface of Callisto is at least as old as the lunar maria. It also seems probable that many of these craters were formed at a time of higher cratering rates—the outer solar system equivalent of the late heavy bombardment on the Moon.

However, a closer look reveals some important differences between the cratered surfaces of Callisto and the lunar highlands (Fig. 15.5). The craters are not the same shape. Instead of having bowl-shaped profiles, Callisto's craters are subdued in topography, as though they had been flattened or the crustal material had undergone plastic defor-

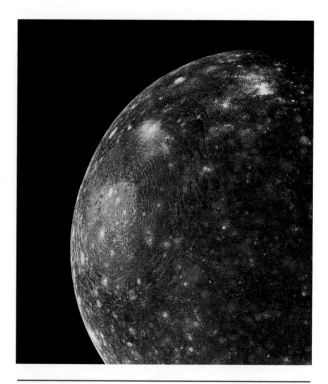

Figure 15.4 The ancient surface of Callisto as imaged by Voyager in 1979. The large bulls-eye structure is called Valhalla.

mation since the craters formed. Apparently, we see here one of the differences between a rocky and an icy object; at the temperature of Callisto's surface, ice, being less rigid, cannot preserve the sharp contours of an impact crater over hundreds of millions of years. Instead, it acts like a highly viscous fluid, which flows slowly, smoothing large-scale topographic features.

Callisto also lacks the big impact basins that we have encountered on the Moon, Mercury, and Mars. Instead, it has several large systems of concentric ridges, like a bull's-eye pattern; these are presumably the signature of the largest impacts. This difference is probably another result of the icy composition of this satellite. Large topographic features are harder to preserve in a plastic material than small ones, and probably any really big craters or basins would simply have subsided into the surface.

It was only when the Galileo spacecraft began to photograph the surface with resolutions of a few tens of meters that the most distinctive features of

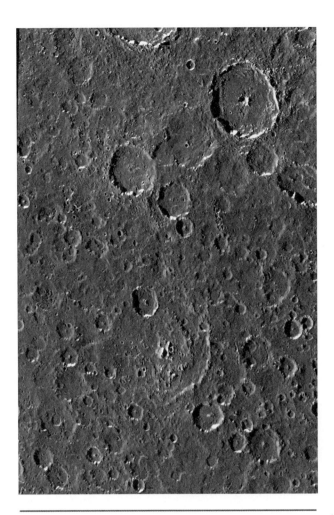

Figure 15.6 High-resolution view of the surface of Callisto taken by the Galileo cameras in 1996. The area shown is 13 km on a side, about the size of a city of a few million inhabitants. The surface has been extensively modified by sublimation and degradation of the dirty-ice surface, producing ice-capped peaks amid smoother plains of darker material left over when the ice evaporated.

Figure 15.5 A close-up look at an area near the equator of Callisto, approximately 200 km wide and 300 km long, the size of Connecticut. The craters have a subdued topography relative to those on the rocky terrestrial planets.

Callisto's geology became apparent (Fig. 15.6). Looked at closely, many of the craters reveal erosion that has left narrow sharp ridges and spiky looking peaks. The tops of these ridges and peaks are often brighter than the plains and valleys between them, giving Callisto the superficial appearance of a terrestrial landscape with snow on the mountaintops. But of course there is no atmosphere on Callisto, and certainly no snow. What has happened is that the upper layers of ice in this rock-ice surface have evaporated, leaving a dark surface like the dark lag-deposit on a comet. However, a fraction of the water vapor did not escape entirely into space. It recondensed on the cooler slopes, those that are at high latitudes or face away from the sunlight.

Different geologic processes work at different size scales. At large scales (tens of kilometers or more), plastic deformation of the crust has flattened the craters and reduced the topography on Callisto. But at small scales (hundreds of meters), sublimation has produced steep slopes and spiky mountains. A somewhat analogous situation occurs on Earth, where the massive mountain ranges such as the Himalayas are produced by tectonic forces, while the steep valleys and sharp peaks are the product of small-scale ice and water erosion.

GANYMEDE: A MOON WITH A HISTORY

Ganymede is the largest satellite in the solar system and one that has experienced a unique geologic history. Its density of 1.9 g/cm^3 is nearly the same as that of Callisto, indicating a similar bulk

Figure 15.7 A distant view of Ganymede obtained by Voyager 2 shows that the surface has both dark and light terrains reminiscent of the appearance of our own Moon to the naked eye. Many bright ray craters are visible.

composition. Here the results from tracking the Galileo spacecraft are unambiguous: Ganymede is a differentiated world, with a large dense core that is presumably composed mostly of silicates, and a mantle and crust that are primarily H$_2$O ice.

About half the surface of Ganymede resembles the dark, ancient, cratered terrain of Callisto, including such features as crater ghosts and concentric bull's-eye ridge patterns. These old terrains, however, are interspersed with lighter, less cratered areas that have been modified by internal activity. The crater densities on these modified areas are typically 100–200 10-km craters per 1 million km^2, significantly lower than those on the heavily cratered areas of either Callisto or Ganymede, although still indicating crater-retention ages of 1 to 2 billion years. Like the Moon, Ganymede has a combination of ancient terrains that probably date back to the early history of the solar system, plus other younger areas that have apparently remained geologically active (Fig. 15.7).

INDICATIONS OF INTERNAL ACTIVITY

Many of the younger areas of Ganymede have systems of parallel mountains and valleys (Fig. 15.8). In horizontal scale, these mountain systems on Ganymede resemble the low Appalachian Mountains of the eastern United States, with ridges separated by 10–15 km. These ridges are tectonic, but the mechanism of their formation is different from that of terrestrial mountains. Whereas the mountains of Earth and Venus were created by wrinkling of the crust subject to tectonic compression, the mountains of Ganymede appear to result from long cracks or faults separating strips of land that have been alternately lifted up or depressed (Fig. 15.9). Many of the depressed areas in turn look as if they had been flooded with liquid—presumably liquid water from the interior.

A compositional difference between the older and younger terrains on Ganymede can also be inferred. Both contain water ice mixed with dirty

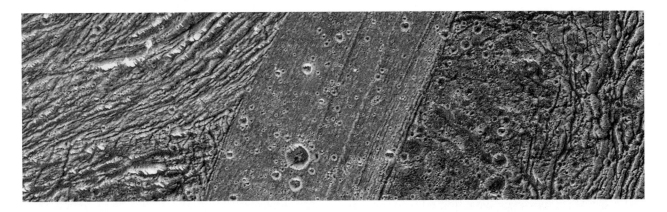

Figure 15.8 Several diverse terrains are shown in this Galileo image of Ganymede. The oldest area to the right consists of rolling hills with many craters. The area to the left has an intermediate age and shows a dense pattern of folded mountains. The smooth terrain in the center is the youngest, although even here the substantial numbers of craters suggest a surface at least 2 billion years old. Note that this is a single picture, not a montage of three separate images.

contaminants, but the geologically younger areas are lighter (higher albedo), suggesting less contamination of the ice. (Note that this is just the reverse of the case on our Moon, where the darker areas are younger.) The reflectivity of these

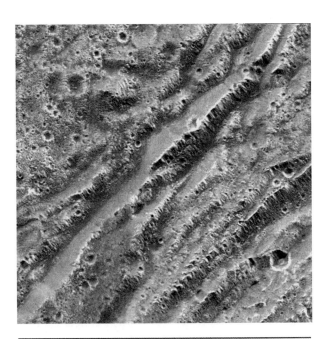

Figure 15.9 Ganymede up close. This ancient, heavily cratered terrain has been cut by a series of faults to form long parallel ridges. The smooth areas that fill the valley floors might have been flooded by water "lava." The area shown is 15 km on a side, similar to the area on Callisto shown in Figure 15.6.

younger terrains is about 40%, as compared with 25% for the heavily cratered areas. Ganymede is further distinguished by a few very bright craters with crater rays, which appear to have splashed nearly pure water across the surface (Fig. 15.7). No similar bright craters occur on Callisto, suggesting the presence of fairly clean ice below the dirty crust of Ganymede but not of Callisto. Remember that the crust of Callisto is a primitive mixture of ice and silicates, while the crust of Ganymede (which has fully differentiated) is nearly pure ice once we get below the surface layer of accumulated dirt.

Photographed at high resolution, some ridges and mountains of Ganymede resemble those of Callisto. Sublimation and erosion of the surface ice have sculpted fascinating patterns of exposed bright ice interspersed with drifts of dark material. Some of the features that initially appeared to be shadow patterns cast by mountains under oblique lighting are actually alternating drifts of bright and dark material. This separation into light and dark components is apparently stable. The dark material is warmer because it absorbs most of the sunlight, so no ice condenses there. The bright icy areas are colder and so attract more ice. Once the light and dark patterns arise, they are self-perpetuating.

The evidence from the surface of Ganymede suggests that this satellite experienced a series of internal upheavals sometime during the first 2 billion years of its existence—the age of the younger terrains as judged from crater densities. One suggestion is that these events were triggered by changes in the density and structure as the ice in the interior altered its crystalline form with the slow cooling of the core. The resulting expansions and contractions could have cracked the surface and resulted in the submersion of low-lying areas by a "lava" of liquid water. At one time the crust of ice may even have floated on a viscous mantle of slushy ice, resulting in an episode of icy plate tectonics analogous to that of Earth.

It is possible that Ganymede will experience more geologic activity in the future. One of the surprising discoveries from the closest Galileo fly-bys was the presence of a weak magnetic field produced by interactions between Ganymede and the jovian magnetosphere. This interaction tells us something about the interior of Ganymede. The most straightforward explanation is that there is an electrically conductive layer, most likely made of partially melted ice—perhaps a viscous slush—deep in the mantle. If this inference is correct, Ganymede may contain the seeds of future episodes of surface destruction and rebuilding.

CRATER CHAINS AND DISRUPTED COMETS

In 1992, two years before its spectacular collision with Jupiter, Comet Shoemaker-Levy 9 passed within just 35,000 km of the jovian cloud tops. At this proximity, jovian tides tore the comet's nucleus apart to form a string of about 20 fragments. (Astronomers later called them "a string of pearls.") Initially closely spaced, these fragments gradually drifted apart. By the time they collided with Jupiter in the summer of 1994, they formed a line millions of kilometers long, requiring nearly a week for all of them to hit the planet.

Presumably this tidal disruption of the comet was not unique. There are many Jupiter-family comets, and we have already noted that this group of comets is responsible for most of the craters on the galilean satellites. A small fraction of these will suffer the same fate as S-L 9. And a

Figure 15.10 Anatomy of a torn comet. This crater chain (called Enki Catena) on Ganymede was made by the impact of a disrupted Jupiter-family comet, such as Comet Shoemaker-Levy 9, that happened to crash into Ganymede just after its close encounter with Jupiter. The image is about 200 km on a side.

very small fraction of these disrupted comets will encounter one of the satellites immediately after breakup.

Figure 15.10 illustrates the result. This crater chain was formed when a just-disrupted comet hit the surface, while the individual fragments were still close together. Such crater chains had been found in Voyager photos but were not understood until after the S-L 9 crash. Additional examples were revealed by Galileo. Their presence nicely confirms that S-L 9 was not unique, and it is consistent with the conclusion that Jupiter-family comets are the main source of craters on the galilean satellites.

15.3 EUROPA, THE MOON WITH AN OCEAN

Even though their surfaces are made of ice, Ganymede and Callisto look reasonably familiar to us, with impact craters reminding us of our own Moon. Those craters tell us we are looking at very old surfaces. Nothing much has happened on these Mercury-size bodies for the last billion years. When we turn to the smaller galilean satellites, Europa and Io, we might expect them to be even more quiescent, since the "rule of thumb" is that the smaller the body, the more quickly it cools off and assumes a dormant state. It was therefore surprising to find that these smaller satellites are actually among the most geologically active worlds in our planetary system.

GEOLOGY OF EUROPA

Europa is the smallest galilean satellite (diameter of 3138 km), a little smaller than either Io or our Moon. Its density of about 3 g/cm^3 indicates a primarily rocky composition with only about 10% water ice, a much lower proportion than is present in Callisto or Ganymede. Yet the surface of Europa is the brightest of any of the galilean satellites (70% reflectivity), and its spectrum indicates a surface composition of relatively pure water ice (Fig. 15.1). In this respect Europa is like Earth, which is made of dense materials on the inside but has a coating of water and ice over most of its surface. What is special about the surface ice of Europa, relative to Callisto and Ganymede, is its comparative purity, suggesting a continuing process that resurfaces the planet with fresh material from below. Even more fundamental is the growing evidence that there is a large ocean of liquid water below the ice crust even today.

The visual appearance of Europa is different from that of any other planets or moons we have studied (Fig. 15.11). At large scales, there is little topographic relief; in fact, this satellite is the smoothest planetary object in the solar system. There were almost no impact craters visible at the Voyager resolution of a few kilometers, indicating that the record of early heavy bombardment by comets and asteroids has been erased. Even in the highest-resolution Galileo photos, craters are hard to find—in spite of the fact that the impact flux should be high this close to massive Jupiter. The

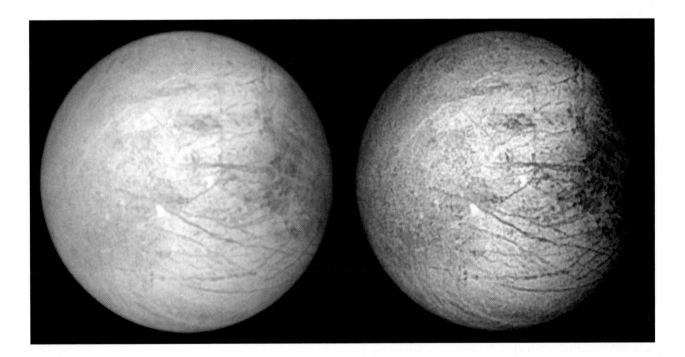

Figure 15.11 Europa is the smoothest satellite known. The absence of easily visible impact craters means that we are looking at a relatively young surface. Shown are two versions of a Galileo global view. The photo on the left shows natural colors, approximately as Europa would look to your eye. On the right the image has been enhanced to bring out more detail.

inferred age of the surface is only a few tens of millions of years (or even less in some crater-free areas), much younger than most of the surface of the Earth.

LINEAR MARKINGS

Although topographic relief is low, Europa displays thousands of linear features—long, narrow, light or dark lines stretching for hundreds or even thousands of kilometers across the landscape. Some of these lines are double or multiple, while others show a remarkable scalloped pattern. If this all sounds familiar, it is because Europa at low resolution actually resembles the fanciful Mars globes drawn by Percival Lowell at the beginning of the twentieth century. The straight canals of Mars have found their true home on a satellite of Jupiter!

The unique global scale of these linear features must be telling us something fundamental about the geology of Europa. We don't think they are canals built by intelligent creatures living in or on Europa. They are tectonic cracks, resulting from stresses in the crust. Evidently cracks, once started, can propagate in straight lines for huge distances. The implication is that the crust must be floating on a layer of low-viscosity material—as we might expect if the crust is water ice over a global ocean of liquid water or partially melted slush.

The features that follow a scalloped pattern reveal an even more remarkable story. Each arc has a shape that responds to the changing direction of tension in the crust caused by tidal forces acting on Europa, forces that vary with a period of one Europa day, or 3.5 Earth days. Each arc must have been formed in a single europan day, and the cracks evidently propagated through the crust at a speed of about 10 km/h, the speed of a marathon runner. If there are ten scallops, then the feature probably formed over a period of ten europan days, or about one terrestrial month—a very rapid event from a geological perspective.

Looked at closely, the typical linear feature is a double ridge with a depression running down the center (Fig. 15.12). The width is a few kilometers at most, and the heights of the ridges are up to few hundred meters. Their appearance sug-

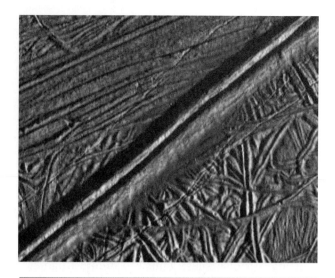

Figure 15.12   Close-up view of one of the linear ridges on Europa, showing an area of 14 by 17 km. This double ridge is the youngest feature shown in the photo, cutting across a variety of older terrain. It is about 3 km across and rises 300 m above the surrounding plains.

gests to geologists that a crack formed initially from tectonic tension, and subsequently either slushy material was injected from below or the closing of the crack under pressure formed the ridges. Similar pressure ridges form in ice packs on the Arctic Ocean, but at much smaller scales. On Europa, the crust is probably at least 10 km thick, in comparison with Arctic pack ice that is only about 10 m thick—a thousand times less. Where there are many sets of parallel ridges and valleys, we are probably seeing multiple events in which parallel cracks formed, as there are often multiple parallel faults between the tectonic plates in the Earth's crust.

CHAOS AND THE GLOBAL OCEAN

Most of the surface of Europa consists of smooth terrain crisscrossed by linear features (alternating low ridges and valleys) with an occasional impact crater. However, a few percent of the surface falls into the category of chaos, where the older surface seems to have been broken up and partially replaced by new material that lacks topographic features.

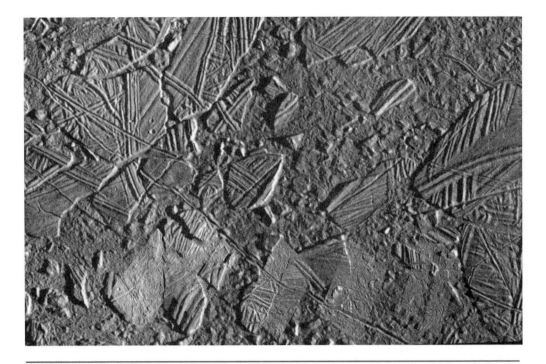

Figure 15.13 An example of "chaotic" terrain on Europa. The area shown is 70 km across and 30 km high. Many individual blocks of older grooved crust can be seen embedded in younger ice; some of them have been partially rotated or tilted as if they were icebergs floating in a liquid water sea.

Looking at the highest-resolution Galileo images, we can see that the pieces of older crust, which range from hundreds of meters up to a few kilometers in size, have rotated and drifted from their original positions, and some are also tilted (Fig 15.13). Their appearance is suggestive of large icebergs that have recently broken off from an ice crust, much as the great Antarctic ice sheets can spawn icebergs as large as 100 km across (Fig. 15.14). Of course, *recently* is a relative term. These europan icebergs are not free-floating today, and even if a part of the crust melted thousands or millions of years ago, today it is again solidly frozen.

Many topographic features can be interpreted in terms of a solid crust above a layer of liquid water or slush. The presence of the chaotic terrains further suggests that at some times parts of the crust melt, allowing liquid to reach the surface at least briefly. Geologists conclude that this crust is between 10 and 20 km thick, although it may be thinner in some places, as suggested by the chaos. These conclusions, however, relate to

Europa on geologic timescales. Is there any evidence for the presence of a liquid ocean today?

The strongest evidence for a contemporary ocean comes from measurements of the magnetic field of Europa made during two close flybys by the Galileo spacecraft. The pattern of magnetism indicates a coupling between the intense jovian magnetosphere and an electrically conducting layer in the upper 100 km of Europa. Liquid water with dissolved salts or other chemicals does the job nicely.

What is the source of the energy that maintains this liquid water at the great distance of Europa from the Sun? We will discuss this issue below when we look at Io, which has an even stronger internal powerhouse.

The inferred ocean of Europa is much deeper than the Earth's oceans, giving Europa the largest volume of liquid water in the solar system. We have frequently noted that liquid water is one of the essential ingredients of a habitable environment. The other requirement is an energy source.

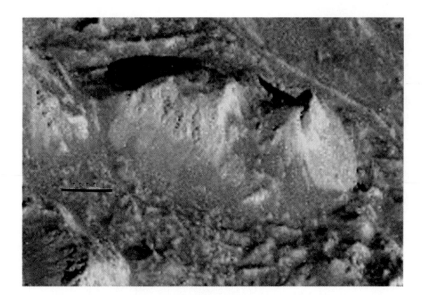

Figure 15.14 Europan icebergs up close. This view of Conamura Chaos (see Fig. 15.13) shows cliffs along the edges of high-standing ice plates, presumably parts of the older crust that were floating rather recently in liquid water or partially melted slush. The silhouette of the ocean liner Titanic indicates the scale of the image.

No sunlight penetrates into the europan ocean, so any life that might be present must extract its energy from disequilibrium chemical reactions that take place, ultimately deriving their energy from the heating of the interior. On Earth, we have recently discovered many microorganisms that can extract energy from chemical-laden hot water (for example, in hydrothermal vents on the ocean floor) without depending on photosynthesis. Indeed, some scientists are suggesting that such hot, dark environments may have had the right conditions for the formation of life.

Now that we have discovered a global ocean on Europa and found evidence of internal heat sources, this small moon of Jupiter has become a target of prime interest in the search for extraterrestrial life. However, it is a daunting challenge to imagine sending spacecraft to land on Europa and perhaps ultimately to drill (or melt) its way through the thick icy crust for direct exploration of its dark ocean depths.

15.4 Io, the Volcanic Moon

Io, the inner galilean satellite, is the most volcanically active object in the solar system. Heated by tidal forces, it is in a constant state of geologic change, with dozens of volcanoes erupting and vast lava flows spreading over its surface. Following the Voyager flybys in 1979, the name Io almost became a household word (except that no one can agree on how to pronounce it: EE-O or Eye-O). But before we look at its remarkable volcanoes, let us discuss how Io compares with the three other galilean satellites.

Interior and Energy Sources

The density of Io is 3.3 g/cm^3, slightly higher than that of Europa and indicative of a rocky object with little or no water. In fact, Io has nearly the same size and density as our Moon and therefore probably a very similar bulk composition. Like Europa, Io has a high surface reflectivity. But here the similarities end. Instead of a bland white ice surface like that of Europa, or the gray dirty-ice surfaces of Ganymede and Callisto, Nature has given Io a multihued face of materials with pastel shades of yellow, red, and brown, all related to its unique level of volcanic activity (Fig. 15.15).

Why is the face of Io constantly changing due to volcanic activity, while the Moon's volcanic fires cooled more than 3 billion years ago? What is supplying the power to maintain such a high level of activity on this one satellite? The answer is found in tidal heating, which we first discussed in Section 8.3 on Mercury.

Io owes its special character primarily to its unique location. It is about the same distance from Jupiter as is our Moon from Earth, yet Jupiter is

Figure 15.15   Io as imaged by the Galileo spacecraft in 1999 on its first close flyby. The surface looks completely different from any other members of the solar system because of Io's high level of volcanic activity. This photo reproduces many subtle shades of color, most of which are caused by sulfur from the volcanoes.

more than 300 times more massive than Earth, causing tremendous tides on Io. These tides pull the satellite into an elongated shape, with a several-kilometer-high bulge extending toward Jupiter. This tidal bulge would not contribute to heating the interior if Io always kept exactly the same face turned toward Jupiter, which is the state it would eventually occupy if there were no other nearby satellites to perturb its orbit. However, the gravitational pulls of Europa and Ganymede will not permit Io to settle into an exactly circular orbit. Instead, Io's orbit is slightly eccentric, with the result that it twists back and forth with respect to Jupiter on each orbital circuit, while at the same time moving nearer and farther from the planet. When the satellite is closest to Jupiter, it is moving more rapidly in its orbit (following Kepler's laws) and the rate of rotation falls behind the rate of revolution. The converse occurs when the satellite is farther from the planet. In each case, the tidal force from Jupiter tries to pull the bulge into exact alignment. The resulting twisting and flexing of the

tidal bulge heats Io, much as repeated flexing of a wire coat hanger heats the wire. In this way, the complex interaction of orbit and tides pumps tremendous energy into Io, melting its interior and providing power to drive its volcanic eruptions. A similar but much smaller tidal heating provides the energy to keep Europa's ocean from freezing solid.

A lower limit on the size of the tidal energy source of Io can be estimated from the heat radiated by its volcanoes, which has been measured both with telescopes on Earth and by infrared detectors on the Voyager and Galileo spacecraft. The Io power plant generates at least 100 million megawatts, more than ten times the total energy consumption of humans on Earth. Ultimately, the source of this energy is the spin of Jupiter itself, which is slowed very slightly by the interaction with Io.

After billions of years, this tidal heating has taken its toll on Io. Any water or other ices, as well as most carbon and nitrogen compounds, have long since been driven off, until now sulfur and sulfur compounds are the most volatile materials remaining. The silicate interior of the satellite must be entirely liquid, with a solid crust probably no more than 25 km thick. The crust is constantly recycled by volcanic activity, and geologic features must be short-lived.

GEOLOGY OF IO

The geology of Io is unique. There are no impact craters, indicating a geologically young surface. Even more than Europa, Io is capable of erasing its craters as fast as they are formed. Calculations of the resurfacing rate required to destroy the craters on Io indicate that hundreds of meters of new crust must be formed each million years, a value similar to that of areas with high sedimentation on Earth.

The colored materials on the surface of Io ought to be identifiable, but unfortunately the evidence from spectra is not entirely clear on this subject. The one material positively detected is sulfur dioxide (SO_2), which is an acrid volcanic gas on Earth but freezes on Io as a white frost. A thin atmosphere of SO_2 gas has also been detected, while the breakdown products of this material, sulfur and oxygen, are major contributors to the charged particle population of the jovian magnetosphere, as described in Section 13.6.

It seems probable that sulfur dioxide is the primary component of the white areas on Io, but what of the more colorful yellows, oranges, browns, and blacks (Fig. 15.16)? There is some spectral evidence for H_2S, but the most likely candidate is elemental sulfur, which has the property of taking on many colored forms depending on its temperature and the way it is cooled from the liquid state. Another possibility is provided by compounds of sulfur and sodium, which also display an appropriate diversity of hues. But even given the evidence for sulfur and sulfur compounds, we do not know whether they make up only a thin coating, perhaps disguising more ordinary rocks.

Instead of impact craters, Io displays numerous volcanic features, including layered lava plains and flow fronts, volcanic mountains, calderas, and vents. Long, twisting flows issue from some vents, while white aprons surround others, apparently deposits of sulfur dioxide frost. There are a few low shield volcanoes, and near the south pole there are also some isolated high mountains. With altitudes as high as 9 km, Io's tectonic mountains are the highest on the galilean satellites. As magma is erupted on the surface, it is being withdrawn from subsurface reservoirs. In consequence, parts of the surface collapse, creating tectonic pressures that can tilt surface blocks and occasionally even produce upthrust mountains (Fig. 15.17).

All this evidence of volcanic activity is dramatic enough to the geologist, but even more exciting was the Voyager discovery that ongoing

Figure 15.16 View of Io's colorful surface. This Galileo image shows mountains, volcanic calderas, and extensive flows of silicate lava and sulfur. The colors have been enhanced to bring out compositional differences. The mountain at the far right is 8 km high. Features as small as 1 km are visible in this image.

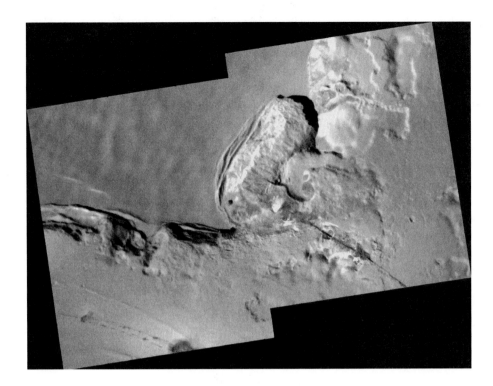

Figure 15.17 High-resolution view of an eroding plain or mesa on Io. Subsurface shifting of magma can leave isolated highlands that subsequently erode by collapse and debris flow, rather like the erosion in the walls of the Valles Marins is on Mars.

eruptions were visible as the spacecraft flew past. For the first time outside Earth, it was possible to see geologic changes as they took place, rather than being forced to infer them from evidence collected after the fact. However, the eruptions photographed by Voyager were misleading, and it was not until Galileo arrived 19 years later that the true nature of the volcanoes was revealed.

VOLCANIC ERUPTIONS

The largest volcanic events on Io are the plume eruptions, discovered by the Voyager cameras (Fig. 15.18). In March 1979, Voyager 1 photographed nine plume eruptions fountaining up to 300 km above the surface. Four months later, Voyager 2 found that one of these had ceased activity and one was unobservable, but the others were still going strong. By decision of the International Astronomical Union, the volcanoes of Io are named for volcano or fire gods from various cultures, such as Pele, the Hawaiian volcano goddess, and Loki, the Norse fire god.

Plume eruptions on Io are produced when hot liquid turns to gas as it rushes upward and then condenses again as snow when released into the cold of space. Plume eruptions resemble steam geysers on Earth, such as the famous Old Faithful in Yellowstone National Park. However, the hot fluid that drives the eruptions on Io is sulfur or sulfur dioxide, rather than water. About 100,000 tons of material is erupted each second in these plumes, enough to cover the entire surface of Io

Figure 15.18 Discovery image of Io's erupting volcanoes. In this distant and overexposed Voyager photo, we can see two huge plumes: one silhouetted against space, and one shining brightly near the edge of the lighted crescent, where the tall plume catches the light of the setting Sun.

to a depth of tens of meters in a million years, or to alter the color of an area of thousands of square kilometers in a few weeks (Fig. 15.19).

Most of the sulfur and sulfur dioxide in the eruption plumes rains back onto the surface, but a small fraction, probably amounting to 10 tons per second, escapes from Io. Quickly broken down by sunlight and the impact of energetic magnetospheric particles, this material provides most of the charged ions of sulfur and oxygen that are the primary constituent of the inner jovian magnetosphere. Thus, the volcanic plumes of Io contribute directly to maintaining the huge magnetosphere of Jupiter.

How are these spectacular plumes formed? The answer was not revealed until the Galileo camera began to photograph the surface at much higher resolution, while another on-board instrument measured the surface temperature. The new photos showed smaller but more familiar-looking eruption features, including fire fountains, lava flows, and lava lakes—all similar to the geology of volcanically active regions on Earth such as Hawaii (Fig. 15.20). Both the form of the eruptions and the temperatures of the vents and lava flows indicated that this was silicate volcanism similar to that of our planet but on a larger scale. The plumes, in contrast, did not originate in the active volcanic vents at all! Plume eruptions take place when lava flows encounter the thick deposits of frozen sulfur dioxide that cover much of the surface. They are similar to the explosive eruptions that sometimes occur when terrestrial lava flows meet bodies of ice or water. On Io, however, where the temperatures

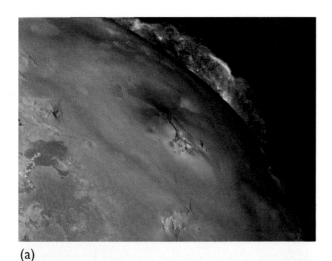

(a)

Figure 15.19   (a) Composite Voyager image of the Pele volcano. The oblique image of the surface shows the origin of the Pele plume. A longer-exposure image shows the plume itself rising 300 km above the horizon. Colors are enhanced. **(b)** Changes over 3 years produced by plume eruptions from the Pele and Pillan volcanoes. The left image was taken in April 1997, the middle 5 months later, and the right image in July 1999. The images are each about 500 km across. Colors are enhanced.

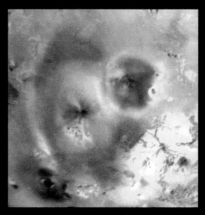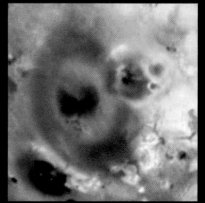

(b)

Figure 15.20   The great eruption of Tvashtar volcano on Io. On February 22, 2000, the Galileo cameras caught the first direct images of large-scale silicate volcanism on Io. At the left edge of the Tvashtar caldera (depression) is a "curtain of fire," and extending upward are two lava flows with bright fresh lava at their toes. The orange and yellow ribbon is a cooling lava flow that is more than 60 km long. Colors are enhanced, and red has been added to indicate the fresh lava that was detected by its infrared emission. The picture is 250 km across.

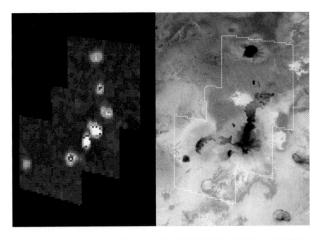

Figure 15.21   Infrared emission from Io hot spots. The right-hand image shows several active lava flows from the Amirani volcano. The left-hand image is based on thermal infrared data at a wavelength of 5 μm. The resolution of the infrared image is 6 km, and that of the visible image is about 1 km. The surface of Io is dotted with hundreds of similar volcanic hot spots, through which much of the internal heat escapes.

are lower and there is almost no atmosphere, the plumes can shoot up to tremendous heights.

Most of Io's internal heat is radiated from a few **volcanic hot spots.** Lava lakes at current eruption sites occasionally overturn, and for a few hours or days the underlying lava glows brightly. Such an "outburst" can be measured easily even with telescopes on Earth. Other hot spots are more stable. One area near the volcano Loki has dominated the heat flow from Io for 20 years. This seems to be a 200-km-diameter lake of liquid sulfur, which has a melting temperature of 385 K (Fig. 15.21).

No astronaut is likely ever to stand on the surface of Io; the intense bombardment by energetic particles from the magnetosphere would kill a human in a matter of minutes. But perhaps someday we will be able to view the scene remotely, through the cameras of a lander spacecraft. From

one hemisphere of Io, Jupiter itself would dominate the view, with 50 times the apparent diameter of the Moon in the skies of Earth. Even at night, the colored landscape of Io would be illuminated brightly by Jupiter. In the darkness of Jupiter's shadow, we might see the black sky above glowing faintly with electrical discharges in the tenuous atmosphere of Io. The most spectacular sight would surely be one of the huge plume eruptions, sending a fountain high above the surface. Anywhere within hundreds of kilometers of such a vent, we would find ourselves in a gentle snowfall of crystals of sulfur or sulfur dioxide. After a few weeks, the surrounding landscape and the spacecraft itself would acquire a fresh coating of this falling material. Meanwhile, we would have to look over our shoulders, for who knows when a tongue of molten lava from some other vent might be headed in our direction.

15.5   TITAN, THE ATMOSPHERIC MOON

With a smog-filled nitrogen atmosphere that is denser than our own, Titan is one of the special places in the planetary system. Saturn's largest satellite, Titan is essentially the same size and density (1.9 g/cm³) as Ganymede and Callisto. One might therefore expect it to look similar to these giant jovian moons, but sunlight reflected from Titan carries a message suggesting that this expectation will not be met.

As viewed with a telescope from Earth, Titan appears as a tiny, barely resolvable reddish disk, different in color from Ganymede and Callisto, which are essentially colorless. In the middle decades of the last century, Gerard Kuiper carried out a systematic spectroscopic survey of the solar system using photographic plates that were specially sensitized to record short-wavelength infrared light. When he examined Titan with a spectrograph attached to the McDonald Observatory 82-inch (2.1-m) telescope in 1944, Kuiper found that Titan's spectrum showed absorption bands of methane gas in addition to the expected solar lines. Unlike Ganymede and Callisto, Titan has an atmosphere (Fig. 15.22).

It may seem surprising that a satellite could have a substantial atmosphere; our Moon certainly doesn't. You will recall that the ability of a planet or satellite to retain an atmosphere is determined primarily by its mass and temperature. Titan is both more massive and colder than the Moon. It has a stronger gravitational field, so molecules must move faster to have sufficient energy to escape into space. Yet the lower temperature means that molecular velocities will be slower, making it more difficult for them to escape. Thus, both size and temperature favor retention of an atmosphere. But why Titan and why not Ganymede or Callisto? We shall return to this question after we learn more about Titan's atmosphere.

After Kuiper's discovery of methane, knowledge about Titan grew slowly, and it was not until nearly 25 years later that the next major discovery was made—a measurement of Titan's temperature using the Very Large Array (VLA) radio telescope

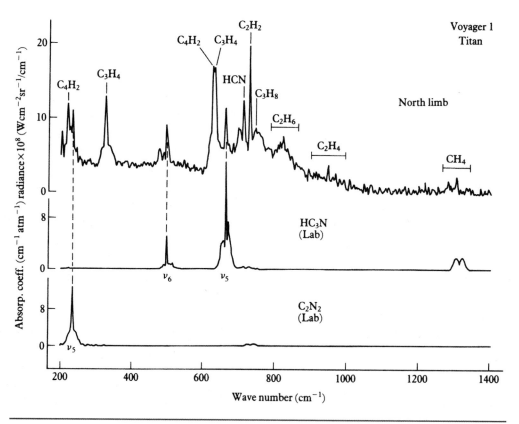

Figure 15.22 Infrared spectrum of Titan (top) showing emission bands from several gases. The two laboratory spectra of individual gases show how these substances can be identified in Titan's atmosphere.

in New Mexico. Even on the eve of the 1980 Voyager encounters with Titan, many basic questions concerning the satellite's atmosphere remained unanswered. Only methane and some other hydrocarbons had been detected spectrally, but the presence of nitrogen or argon was suspected. The composition of the clouds was unknown. An even more basic problem was the determination of the total mass of Titan's atmosphere and thus its surface pressure. Different scientists favored various possible models, in which the surface pressure ranged from a value almost as low as that on Mars up to several times the surface pressure on the Earth. Before Voyager, the information available was insufficient to distinguish among these possibilities.

VOYAGER RESULTS

The pictures of Titan sent back by the Voyager 1 spacecraft were disappointing, since the surface of the satellite was obscured by a ubiquitous smog layer (Fig. 15.23). Most of the exciting results came from the other Voyager instruments.

As the spacecraft passed behind Titan, its radio signal was attenuated by the atmosphere and ultimately blocked entirely by the satellite's solid surface. This occultation permitted Voyager scientists on Earth to determine how the density of the satellite's atmosphere varied with altitude above its surface. Assuming a nitrogen-rich composition based on results from the spectrometers on board the spacecraft, they deduced a surface pressure of 1.5 bars.

With this high surface pressure, Titan has a larger atmosphere than all of the terrestrial planets except Venus. Since its gravity is much less than that of the Earth, more gas is required on Titan to exert the same pressure as it would have here. The amount of gas above the surface of Titan is approximately ten times greater than that above an equal area on Earth, and the atmosphere of Titan also stretches nearly ten times farther into space than does our own.

Voyager made important discoveries about the composition of Titan's atmosphere as well as its structure. The methane detected by Kuiper turned out to be a minor constituent. Titan's atmosphere

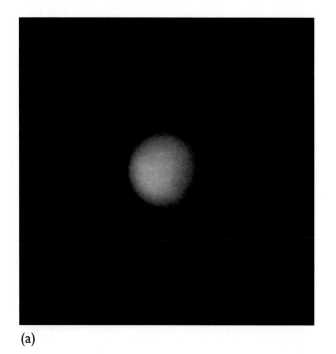

(a)

(b)

Figure 15.23 Voyager photos of Titan, showing only its ubiquitous haze-filled atmosphere: **(a)** the fully illuminated hemisphere; **(b)** the back-lit image, revealing the extended atmosphere more clearly.

is mostly nitrogen, as is ours. Impacting electrons from Saturn's magnetosphere, cosmic rays, and solar ultraviolet light have sufficient energy to break apart molecules of methane and nitrogen. As the fragments recombine, new molecules are created, and a number of these were detected in Titan's atmosphere by the Voyager spectrometers.

Table 15.2 shows that some of the chemical reactions that are taking place on Titan are interesting in the context of the **prebiotic chemistry** required for the origin of life. These reactions are producing hydrogen cyanide (HCN), a starting point for the formation of some of the components of DNA. The presence of both carbon monoxide (CO) and carbon dioxide (CO_2) in this mixture makes the formation of amino acids possible.

The presence of argon can be deduced only indirectly. There may be none at all, in which case the abundance of nitrogen would increase. The amount of methane varies with altitude and is still poorly determined.

PHOTOCHEMICAL SMOG

In Titan we have discovered a world nearly frozen in time, where we can examine a primitive environment in which chemical reactions taking place today may resemble those that preceded the evolution of life on our own planet. It would be fascinating to study the end products of this chemistry, to use Titan as a natural laboratory for testing our ideas about chemical evolution. Are certain pathways toward complexity preferred? We know that there are more complex substances on this satellite than those already found because of the existence of the global haze layer, but what are these compounds?

The haze is all that we can see in the pictures sent back by Voyager. Even close-up views that show a small area at high magnification reveal no detail at all: no gaps in the haze, no structure that could be used to map winds. The haze of Titan is even more uniform than the clouds of Venus. Only when viewed from the side does it reveal structure, in the form of a distinct layer hundreds of kilometers above the surface (Fig. 15.24). The haze is less opaque at infrared wavelengths, however, and it is possible to glimpse the surface using the infrared camera on the Hubble Space Telescope.

Observations by the Voyager ultraviolet spectrometer revealed additional absorbing layers between 300 and 500 km above the surface. Some of the organic molecules formed in Titan's atmosphere, such as HCN, can form long chains called polymers. The reddish brown haze that pervades Titan's atmosphere is probably a combination of these polymers and condensed organic compounds.

As the haze particles grow larger, they will gradually drop out of the atmosphere onto the satellite's surface, forming deposits in low-lying areas, drifts of organic material that by now may

Table 15.2 Composition of the atmosphere of Titan

Gas	Formula	Amount
Main Constituents (percent)		
Nitrogen	N_2	90–99
Argon	Ar	0–6
Methane	CH_4	2–10
Trace Constituents (parts per million)		
Hydrogen	H_2	2000
Hydrocarbons		
Ethane	C_2H_6	20
Propane	C_3H_8	20
Ethylene	C_2H_4	0.4
Diacetylene	C_4H_2	0.1–0.01
Methylacetylene	C_3H_4	0.03
Nitrogen compounds		
Hydrogen cyanide	HCN	0.2
Cyanogen	C_2N_2	0.1–0.01
Cyanoacetylene	HC_3N	0.1–0.01
Acetonitrile	CH_3CN	trace
Dycanoacetyline	C_4N_2	crystals
Oxygen compounds		
Carbon monoxide	CO	50–150
Carbon dioxide	CO_2	0.015
Water	H_2O	trace

Figure 15.24 Detached layers of haze in Titan's atmosphere. The Voyager close-up image has been enhanced and the color intensified.

be several hundred meters thick. Since the lower atmosphere and surface of Titan are colder than the upper atmosphere, most of the products of this precipitation and photochemistry cannot evaporate and mix with the atmosphere again, the way water does on Earth. The only exceptions are gases like methane and ethane, which may exhibit something similar to the terrestrial water cycle.

EVOLUTION OF THE ATMOSPHERE

Although the surface of Titan is preserved in a deep freeze, the atmosphere is evolving. As the methane is broken apart by electrons and ultraviolet photons, the hydrogen that is produced escapes because hydrogen atoms and molecules move fast enough even at Titan's low temperatures to reach escape velocity. A small amount of hydrogen is always present in Titan's atmosphere, since it is continuously being produced. Over the lifetime of the solar system, however, a large amount of hydrogen must have escaped into space. Therefore, deuterium should be enriched on Titan as it is on

Venus, although not to the same extent. There is evidence that such enrichment has occurred, in the form of a high abundance of CH_3D-methane in which one of the hydrogen atoms is replaced by an atom of deuterium.

The loss of hydrogen must be balanced by a loss of carbon; this loss is represented by the precipitation of aerosol particles and hydrocarbons onto Titan's surface. Ethane (C_2H_6) is the most abundant end product of the reactions occurring in the atmosphere, and Titan's surface is sufficiently cold that this gas would condense on it. Thus, we may imagine lakes, seas, even oceans of liquid ethane. Methane (CH_4) and propane (C_3H_8) would also form liquids, but CO_2, the other hydrocarbons, and the nitrogen compounds in Table 15.2, will solidify at these temperatures. If hydrocarbon seas are present, these solid compounds would sink rather than float, since their densities are higher than that of liquid ethane, methane, or propane.

Titan offers us the apparent paradox of an atmosphere that has undergone extensive evolution without losing its primitive characteristics. The chemistry that occurs on Titan today must be very similar (if not identical) to that of 4 billion years ago. This situation, which is totally different from the more rapidly evolving conditions found among the inner planets, is due to Titan's very low temperature.

The average surface temperature on Titan is 94 K, a frigid −179 C. Titan is so cold that liquid water—so vital to life as we know it—is an impossibility. Even water vapor will be almost totally lacking from this atmosphere. That is the reason we still find methane and other hydrogen-rich compounds. If water vapor were plentiful, Titan's atmosphere would resemble that of Mars. The water molecules would be broken apart (just as the molecules of methane and nitrogen are dissociated on the real Titan), and the liberated oxygen would convert all the hydrocarbons to carbon dioxide. However, the only sources of oxygen on Titan are the ice particles the satellite sweeps up as it orbits around Saturn and the breakup of CO molecules in the atmosphere. These sources are sufficient to form the tiny amount of CO_2 detected on Titan but are hopelessly inadequate to oxidize all of the methane.

SURFACE OF TITAN

Now we have the components of a most unusual landscape! The basic structural element should be water ice, overlain in places by wind-blown drifts of aerosol particles and crossed by gullies formed by rivers of liquid hydrocarbons. Some residual topography from the cratering events that have marked the surfaces of other solar system bodies may be present, modified by these two processes. As on Earth and Venus, the atmosphere will screen out the smaller impacting bodies. Finally, there is the distinct possibility that the crust of Titan is subject to the icy equivalents of volcanism and plate tectonics caused by mantle convection, leading to mountain chains, ridges, scarps, volcanic peaks, and blocks of ice strewn over the surface.

What would it be like to sit in a boat on an ethane ocean on Titan (Fig. 15.25)? Cold certainly, and probably gloomy. The level of illumination at Titan's surface at midday is estimated to be similar to that from a full moon at midnight on

Earth. The entire sky would appear to be glowing faintly, as sunlight filters through the orange-brown smog layer. The Sun itself would be invisible, as it is when there is heavy smog over one of our cities. Yet the horizontal visibility would be very good (unless, of course, we were caught in an ethane fog bank), since the thick part of the haze is many kilometers above the satellite's surface.

We don't expect to be rocked by big waves because global temperature differences on Titan appear to be small, and thus there should not be strong surface winds. Local weather may exist, since the ocean is not global in extent, but the surface temperature is expected to be very uniform as a result of the moderating effect of the dense atmosphere. In principle, enough ethane is produced by the atmospheric photochemistry to form a global ocean of this compound; however, observations of Titan by Earth-based radar and with infrared cameras have shown that the reflectivity of the surface at both microwave and infrared wavelengths varies with location—a result that is inconsistent with a

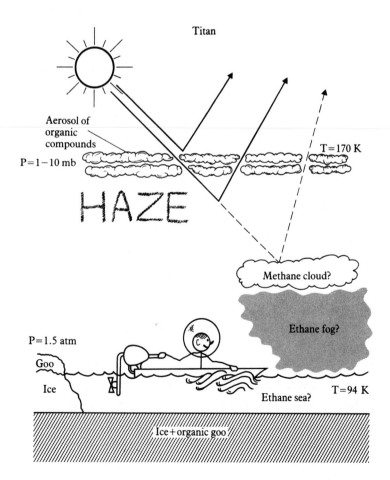

Figure 15.25 The surface of Titan may be predominantly ice, with a covering of organic matter and lakes and seas of hydrocarbons. Thus, it could be an interesting place to explore by boat.

uniform global ocean. The darkest regions are likely to be seas of ethane or other hydrocarbons, but we can't be sure; the dark areas may be caused by something else altogether (remember that early astronomers thought the dark maria of the Moon were seas!). Perhaps the brighter regions are drifts of accumulated aerosols, but it is not possible to derive surface composition simply from surface reflectivity. What we really need for exploration of Titan's surface is a spacecraft equipped with radar and an entry probe. This is exactly what the Cassini-Huygens mission is designed to provide when it arrives in 2005.

15.6 TRITON, NEPTUNE'S MAVERICK MOON

There is one more large satellite to consider: Neptune's frigid satellite Triton. Neptune was the god of the sea in the Roman pantheon, and Triton was one of his children. Triton's diameter is 2705 km, making it a little larger than half the diameter of Titan (with which it is sometimes confused because the names are so similar).

Because Triton is relatively faint, we knew very little about it until the Voyager 2 flyby in 1989. This satellite moves around Neptune in the direction opposite to the planet's rotation and opposite to the direction that all the planets follow in their orbits around the Sun. Nevertheless, its rotation is still synchronous with its period of revolution: Just as our own Moon keeps one hemisphere facing the Earth, the same side of Triton always faces Neptune. Triton must be a captured object, having formed in the solar nebula rather than in the disk of material that surrounded Neptune at the time the planet itself was forming. This suggests that Triton may be more closely related to Pluto than to Titan, since Titan appears to have formed along with Saturn.

Using the diameter of Triton and the mass determined from its gravitational effect on the trajectory of the Voyager spacecraft, we find the satellite's density to be 2.1 g/cm^3, indicating an ice-rock mixture in which rock is the dominant material. Evidently, it is deficient in ice or it has acquired some additional high-density component.

Thus the density as well as the diameter of Triton are very similar to those of Pluto.

TRITON'S SURFACE

Spectral observations from Earth in the early 1980s by astronomer Dale Cruikshank and his colleagues revealed that methane was frozen on Triton's surface, and subsequent observations revealed the presence of ices of carbon monoxide and carbon dioxide as well. In addition, the spectrum showed a feature caused by frozen molecular nitrogen (N_2). The presence of methane and nitrogen ice at Triton's temperature produces a tenuous envelope of gases. The situation is analogous to that of water on the Earth. We have liquid water or ice on the Earth's surface and water vapor in the atmosphere, vapor that occasionally makes its presence known through the formation of clouds, rain, frost, and snow.

The Voyager photos revealed Triton to be yet another unique world, fundamentally different from the other satellites surveyed by the spacecraft as it toured the outer planets. There are no large impact craters and few small ones, indicating that the surface is relatively young, perhaps as young as that of Europa. Triton also has a highly reflective surface, so that very little of the radiant energy arriving from the distant Sun is absorbed. The average surface temperature is therefore just 37° above absolute zero, or 37 K. Water ice is also present on the surface, but it is mostly covered by the more volatile ices of methane, nitrogen, CO, and CO_2 that dominate the spectrum.

Most of the surface visible to Voyager consisted of a relatively flat plain, mottled with darker material (Fig. 15.26); this appears to be an extensive cap of ice that surrounds the south pole, which was pointed toward the Sun at the time of the spacecraft flyby. Some of the dark material has been blown about by surface winds to form triangular wind streaks. These wind streaks may be caused by sublimation of surface ice deposits during the summer, leading to a flow of nitrogen gas toward the cold pole. This seasonal transport of nitrogen may be the way Triton maintains such a high albedo.

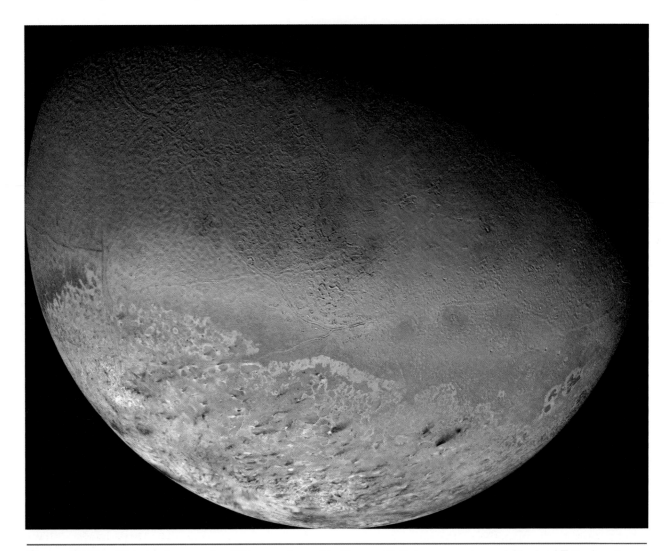

Figure 15.26 Global color mosaic of Triton as imaged by Voyager 2. Colors are enhanced. Much of Triton's unique surface appearance is linked to its extremely low temperature, since most gases will freeze onto the surface. The mottled surface in the upper part of the image is called cantaloupe terrain for obvious reasons.

TRITON'S ATMOSPHERE

In addition to the indirect evidence of an atmosphere provided by the observations of wind streaks, the Voyager cameras also photographed haze and clouds. The ultraviolet spectrometer on the spacecraft found nitrogen emission lines in the satellite's spectrum, confirming that this gas must be the major constituent of Triton's atmosphere. The total atmospheric pressure is just 16 millionths of the sea-level pressure on Earth (16 microbars). This is the maximum pressure that an atmosphere of nitrogen could have with a surface temperature of 37 K.

The composition of the *surface* of Triton is reminiscent of the *atmosphere* of Titan. The splotches of dark material on Triton's surface may therefore be the equivalent of the atmospheric smog particles found on Titan. Calculations based on the apparent escape of hydrogen suggest that a 6-m layer of hydrocarbons could have been produced on the surface of Triton over the age of the solar system. In fact, the dark splotches we see must be relatively recent because they will subside into the ice with time. Hence, a large amount of material produced by the atmospheric chemistry on Triton may be buried under the geologically recent visible surface.

Plume

Figure 15.27 Geyser on Triton. This Voyager photo shows what appears to be an eruption rising to an altitude of about 8 km, where it encounters a transverse wind in the very thin atmosphere.

Surely the most remarkable features of Triton discovered by Voyager 2 are geyserlike plumes that rise from the frigid surface into the tenuous atmosphere (Fig. 15.27). These plumes appear as narrow columns of dark material that rise about 8 km above the surface, forming small dark clouds at their summits. At this level, there are smoothly flowing horizontal winds that carry the lofted material downwind for 100–150 km, while maintaining it in streamers that are about 10 km wide.

What is causing these geyserlike plumes to occur at a local temperature of only 37 K? The structure the plumes exhibit is consistent with a warm source that produces buoyant gas at the base of the plume. We must remember that a "warm source" in this context may still be colder than −200°C! The buoyant gas then rises through the lower atmosphere, where the atmospheric measurements indicate the presence of a thermal inversion. We see a similar phenomenon on Earth, where smoke from a fire on a cold night will rise vertically through the still air, piling up at the top of the inversion layer, where it may be blown away by a horizontal wind.

What is the tritonian equivalent of that fire? We don't have a clear answer yet. But even if we don't understand them, we can see that these geysers could have parallels on cold comet nuclei and may explain some of the activity that comets exhibit at large distances from the Sun. In any case, the production of a nitrogen atmosphere on Triton also explains why many comets appear to be deficient in this element: Nitrogen will start subliming (evaporating) from comet nuclei even at distances of 40 AU, so by the time the comets reach the inner solar system where we usually first detect them, they will have lost a significant fraction of this highly volatile gas.

15.7 COMPARING THE LARGE SATELLITES

Following our practice with the planets, it is useful to compare the large satellites with one another to understand their individual peculiarities. Figure 15.28 illustrates how different their geology is, even for the three galilean satellites with icy crusts. Even more fundamental is their distinctive interior structure. In terms of bulk properties, we have three natural groupings: the outer icy "giants," Ganymede and Callisto; the inner rocky moons, Europa and Io; and Titan and Ganymede, similar in size but so different when it comes to an atmosphere.

GANYMEDE AND CALLISTO

The first problem is to determine why Ganymede differentiated and then maintained a substantial level of geologic activity for hundreds of millions of years while Callisto did not. The distinctions between these two objects are in their distances from Jupiter and in small differences in their sizes and densities. Could any of these factors be responsible for their different histories?

Because Ganymede is closer to Jupiter, it is subject to larger tidal stress, and it also experiences more impacts because of the attraction of Jupiter's gravity. Neither of these effects is large, but both could have resulted in slightly greater heating for Ganymede than Callisto early in their history.

Ganymede is larger than Callisto and also slightly denser, indicating that it contains more radioactive material and therefore had (and still has) a greater internal heat source. In addition, Ganymede's larger size means that it can hold heat slightly more easily than Callisto. These effects also point toward a higher internal temperature and slower cooling for Ganymede.

Each effect just discussed could have contributed toward a more active geology on Ganymede than on Callisto. The problem is that each individual factor is small, and even taken together they do

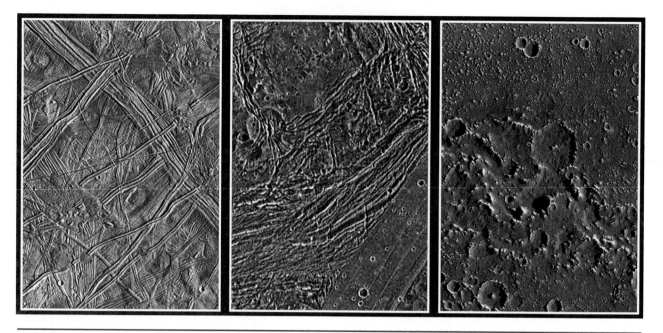

Figure 15.28 Three high-resolution images that show how different the surface geology is even for the three satellites of Jupiter with ice crusts. Here we see (left to right) Europa, Ganymede, and Callisto, scaled to the same size (about 100 km across).

not suggest that Ganymede should have differentiated and remained a great deal warmer than its twin moon. What must have happened is that these differences, even though small, managed to trigger a major change in the internal processes on Ganymede that was out of all proportion to the temperature difference. Such a major change can occur with ice because it is a substance that can take on many different forms, each with its own density, depending on the temperature and pressure.

Calculations show that just a small difference in temperature early in the history of Ganymede may have led to convection in the upper mantle, perhaps sufficient to drive some kind of plate tectonic activity for a period of the satellite's history. Subsequently, about a billion years after the formation of Ganymede, a change of phase from liquid to solid in the interior may have caused a small expansion, sufficient to crack the crust and initiate the period of mountain building. According to these calculations, Callisto, by being just a little cooler, escaped these internal events and was not able even to complete the process of differentiation before it froze up. But in all honesty, the observed differences between Callisto and Ganymede are still a mystery.

EUROPA AND IO

When we turn our attention to Europa and Io, we encounter other mysteries. Why are these two satellites so depleted in water (and other volatiles) relative to Callisto and Ganymede? Much of this difference may be a consequence of the high temperature of Jupiter early in its history, when it was contracting and forming its satellite system. Calculations of temperatures within the nebula surrounding this proto-Jupiter suggest that it may have been cool enough for ice to condense at the distances of Ganymede and Callisto but not closer to the planet, where Europa and Io formed. In that case, we should expect both Europa and Io to be dry, whereas Europa today probably contains about 10% water.

This simple interpretation fails to explain the existence of water on Europa because it neglects the role of impacting bodies from elsewhere. We know that the solar system was full of debris of various kinds early in its history and that a substantial part of that debris was cometary. The impacts of many icy comets must have supplied a great deal of water to all the galilean satellites, probably enough to account for Europa's present icy surface. In the case of Io, however, this early infusion of water was doomed by the orbital lock that developed among Io, Europa, and Ganymede. Held in the gravitational grip of the other two satellites, Io could not circularize its orbit, with the result that tidal heating drove off all of the water and other volatile substances.

DIFFERENCES BETWEEN TITAN AND GANYMEDE

Ganymede is about the same size as Titan and is also composed of a mixture that is roughly half ice and half rock. Yet Ganymede does not have an atmosphere, while Titan has an atmosphere denser than ours. What accounts for this difference? How did Titan acquire its atmosphere?

We can perform an easy test to determine whether the mixture of gases we find on Titan today is a relic of a dense, primary atmosphere captured by the satellite as it formed from the solar nebula. If that were the case, we would expect neon to be as plentiful as nitrogen in the present atmosphere, since these elements are nearly equally abundant in the Sun and hence in the solar nebula (see Table 2.2). In fact, the ultraviolet spectrometer on the Voyager 1 spacecraft detected no evidence of neon, setting an upper limit of 0.1% for the abundance of this gas in Titan's atmosphere. Thus, we conclude that the satellite did not capture its gases directly from the nebula, but instead the atmosphere was formed by the release of gases from the solid materials making up the bulk of Titan, in much the same way as the inner planets developed their atmospheres.

Like the inner planets, Titan must also have accumulated some gases as a result of cometary impacts. The big difference in the case of Titan is that approximately 50% of its initial mass was water ice, which now forms a crust and mantle around the satellite's rocky core. Thus we can say that Titan is partially made of comets, so the external cometary contribution will add to a volatile mixture contributed by the icy planetesimals that accreted to form the satellite. Like Mars, Titan must have lost most of its original atmosphere, as we find the light

isotope of nitrogen strongly depleted in Titan's HCN. Again like Mars, the carbon shows no sign of isotopic fractionation. Evidently there is a large reservoir of CH_4 that can replenish any carbon lost from the atmosphere. Such a reservoir is required in any case to maintain CH_4 in Titan's atmosphere, where the rate of conversion of CH_4 to C_2H_2 and other hydrocarbons would use up the available methane in just 20 million years. Where is this reservoir? Subsurface seas of methane have been suggested as well as the alternative of occasional contributions from impacting comets.

Water ice at low temperatures (less than 75 K) is an excellent carrier of gases. There is room in the irregular structure of this low-temperature ice for molecules of gases to be accommodated. When the ice is warmed above 135 K, the lattice structure changes, becoming regular, and most of the trapped molecules are released. The ability of water ice to capture gases in this way depends on the size and electrical properties of the gas molecules. Hydrogen, helium, and neon are not trapped except at temperatures lower than 25 K, much lower than the temperature in the vicinity of Saturn during the time the planet and its satellites were forming, but argon, methane, nitrogen, and carbon monoxide can be captured in varying amounts, depending on the temperature at which the ice formed.

These properties of ice may explain why Titan has the atmosphere it does and why Ganymede does not. The jovian satellites were simply too warm for ice to trap large amounts of gas, warmer even than their present temperatures, since they were heated during formation by Jupiter itself. At the distance from the Sun where Titan formed, however, gas-rich ice was able to condense and serve as the carrier of nitrogen and the other volatiles now found on Titan.

There is a different possibility to consider, however. Another major difference between Jupiter and Saturn is the larger mass of Jupiter. Thus, comets colliding with Callisto and Ganymede would crash with far higher energy than those hitting Titan, simply because of the greater gravitational attraction exerted by the more massive planet. The result could have been that Titan was able to retain the gases delivered by comets that originated in the

PERSPECTIVE

EXPLORATION SCIENCE

One of the most important things to learn in any science class is how scientific research is carried out. It is the *process* of science, more than specific facts, that every citizen needs to know something about.

The simplest descriptions of the scientific method portray it as a series of steps, in which a hypothesis is suggested and experiments are conducted to determine whether the hypothesis is correct. The hypothesis must be *falsifiable*; that is, it must make predictions that can actually be tested experimentally. The idea is that if the experiment yields a result that is inconsistent with the predictions, then the hypothesis is rejected. This is a fine approach in some kinds of science, especially where experiments can be performed in the laboratory. But it is not a very good description of the way most astronomers and planetary scientists work. After all, we cannot carry out experiments on the stars and planets. Along with geology and paleontology, astronomy and planetary science are often classed as *historical sciences*. We use observations made today to determine what happened in the past. The scientist functions rather like a detective, searching out clues and looking for a pattern that will explain what happened.

Science is like detective work in another way. To solve a crime, it is not enough for one detective to think he has identified the culprit. The detective must convince his police colleagues, the prosecutor, and ultimately a judge and jury of the validity of the evidence and the correctness of his conclusions. In science, also, conclusions are tested through discussion with colleagues, presentations of the results at meetings, and eventual publication in scientific journals. If the issue is controversial, there may be challenges to the reliability of the evidence or the logic of the conclusions. In the end, the new conclusions are accepted only if they can withstand all the questions and challenges that can be raised.

Let's look at some examples from Io. One of the biggest successes in predictive planetary science took place in 1979, as Voyager 1 was approaching Jupiter. Astronomers already knew that Io had a peculiar color, but most of them expected it to be mostly Moon-like, a dead, heavily cratered world. Three theorists, however, calculated the possible effect of jovian tides on Io. Just two weeks before the spacecraft arrived, they published a paper in *Science* describing the expected tidal heating and predicting that Io would be a volcanically active world. Within a few days, the first close-up photos of Io confirmed this prediction. Even though the Voyager instruments had not been designed with this in mind, they provided a good test of the volcanic hypothesis.

Finding that Io's surface showed evidence of past volcanoes did not, however, hint that we might see actual eruptions happening in front of our eyes. No one on the Voyager science team thought about looking for plume eruptions in the pictures sent back from the spacecraft. Rather, it was an engineer from the JPL navigation team who, working over a weekend, discovered the huge but faint plumes. This was a serendipitous discovery—but one that could not have been made without the Voyager spacecraft and its excellent cameras.

Following the Voyager encounters, many scientists worked out how the giant plumes might be formed, from the ejection of sulfur and sulfur dioxide. They made calculations and used the steam geysers of Yellowstone as terrestrial analogues. In previous editions of this textbook we reported that the volcanoes of Io were erupting liquid sulfur and sulfur dioxide from the interior. Then we sent the Galileo spacecraft to Jupiter with its better cameras and new instruments for imaging the surface in the thermal infrared. The plan was to take close-up photos of Io on the initial approach to Jupiter, just before firing the rocket engine to put the spacecraft into orbit. In the final week, however, the spacecraft computer developed problems, and the entire sequence of Io observations was cancelled so that full attention could be given to orbital insertion. It was not until 3 years later that Galileo again closely approached Io.

With closer flybys and better instruments, Galileo showed that most of the eruptions were silicate volcanoes like those on the Earth. The observed temperatures were too high for sulfur volcanism. The cameras photographed individual fire fountains and lava lakes similar to those in Hawaii. These had been too small to recognize at Voyager resolution. The improved photos also showed that the huge plumes did not even originate at the main volcanic vents. Thus, our picture of Io volcanism changed, and the corresponding sections of this text had to be rewritten.

The Voyager and Galileo experiences illustrate a different kind of science—what we might call *exploration science*. A successful explorer should not be burdened with too many hypotheses. Instead, we build the best instruments we can and launch them on voyages of discovery to other worlds. So far we have not been disappointed, and this approach has vastly increased our understanding of the planetary system. The interplay of data and theory is the same as in any other form of science, but the initial step is to collect as much data as possible, in an unbiased way, and see just what Nature has to tell us.

Although it can require more than a decade between the time a mission is designed and the arrival of the spacecraft at its target, once a mission like Galileo or Voyager is under way, things move quickly. The entire cycle of observation, hypothesis, and resolution with new data can be compressed into just a day or two. The mission scientists had an extraordinary opportunity—we could look at new pictures one day, discuss them with colleagues, suggest hypotheses, make predictions, maybe even hold a press conference—knowing that we might be proved right (or more likely wrong) by the next day's data. What seemed like an entire scientific career could be compressed into a few heady weeks as each day brought new and wonderful (and often baffling) revelations.

Uranus-Neptune region, while on Ganymede, the gases escaped into space after the impacts. We will not know which (if either) of these possibilities is correct until the Huygens probe reaches Titan and gives us the kind of data on isotope ratios and noble gas abundances that we use to try to understand the origins of the atmospheres of the inner planets.

Summary

Whereas the inner planets have few or no satellites, the outer planets resemble miniature solar systems. Jupiter, Saturn, Uranus, and Neptune each have many more moons than the Sun has planets. Ganymede and Callisto are larger than the planet Mercury. They are nearly the same size and density, inviting comparison between them. With densities less than 2.0 g/cm^3, they have obviously different compositions from the inner planets, which they resemble in size. They are nearly half composed of water ice, but the interior structure is quite different. Ganymede is fully differentiated, with a core of heavy material and a mantle and crust of water ice. But Callisto appears not to have completed its differentiation, and its mantle and crust are composed of a primitive mixture of ice with silicate and carbonaceous dust. Callisto's cratered icy surface shows little sign of internal activity. Ganymede, in contrast, once experienced extensive geologic activity, with resurfacing and the widespread formation of long mountain ridges and valleys.

Io and Europa differ from their larger cousins in that they are not composed of equal parts of ice and rock. Both of these satellites are about the size and density of our Moon and presumably have a similar composition. Europa has a remarkably smooth, icy surface, floating on a global ocean of water as deep as 100 km. There are few craters, indicating that the surface we see was formed recently, within the past few million years. The geology is what we might expect in a crust of ice about 10 km thick, freely floating on the ocean below. Extraordinarily long, straight ridges seem to have been formed by cracking of the icy crust, responding to global tectonic forces. The dark, mysterious ocean below is of great interest as the possible abode of alien life, but it will be difficult to access this water world.

Io is the most volcanically active body in the solar system. Its energy source is tidal heating, a result of tides raised by Jupiter acting in conjunction with a noncircular orbit. This heating is sufficient to melt the interior and power widespread surface volcanism. Over its history, Io has outgassed and lost highly volatile materials, such as water and carbon dioxide, and its surface is covered by sulfur and sulfur compounds, which are recycled through its volcanic eruptions. Today the surface is dotted with active volcanoes, which include fire fountains, lava lakes, and extensive flows of lava. Where the lava encounters thick deposits of sulfur dioxide ice, huge plumes of SO_2 gas and snow can shoot up more than 100 km. Some of the SO_2 from these eruptions also escapes, where it becomes ionized and contributes significantly to the inner magnetosphere of Jupiter.

Saturn's large satellite Titan has a nitrogen-methane atmosphere with a thick layer of photochemical smog and a surface pressure 1.5 times the sea-level pressure on Earth. The nitrogen and methane are being continuously broken apart by solar ultraviolet radiation and electron bombardment from Saturn's magnetosphere. The fragments recombine to form a rich variety of organic compounds in the atmosphere, in reactions that resemble some of the chemistry that took place on the primitive Earth, ultimately leading to the origin of life on our planet.

Neptune's satellite Triton is in a much colder environment than Titan. Yet even at 37 K, chemical reactions are taking place that slowly destroy methane and nitrogen, building up complex organic compounds that we find in the form of dark, windblown material on Triton's surface. In addition to frozen nitrogen and methane, the surface includes ices of carbon monoxide and carbon dioxide.

When we compare these satellites with one another, the differences in the interiors and surfaces of Ganymede and Callisto are especially puzzling. There are reasons to think that Ganymede has always been slightly warmer than Callisto, however, and that may have been crucial to its evolution.

Both Io and Europa may have formed without water in the warm inner region of the jovian nebula, but cometary impacts could have supplied the ocean we find on Europa today. The tidal heating of Io prevented the accumulation of a similar icy crust. Titan was able to acquire and retain an atmosphere, either because it was bombarded by cooler ices that contained abundant gas, or because its impact environment was less violent and allowed it to retain the volatiles that it collected.

REVIEW QUESTIONS

1. Describe the general characteristics of Jupiter's four largest satellites. How do they resemble the characteristics of the planets in the solar system? How would you explain this resemblance?

2. Compare the expected densities of impact craters on the satellites of outer planets with the observed values on the maria and highlands of the Moon. What accounts for the differences?

3. What is the evidence that the surfaces of Ganymede, Callisto, and Europa are covered with ice? How can impact craters be preserved in ice?

4. How do the interior and surface of Ganymede differ from those of Callisto? What accounts for these differences?

5. What distinguishes the surface of Europa from the equally icy surfaces of Callisto and Ganymede? What makes this satellite interesting to scientists (and science fiction writers—see Arthur C. Clarke's *2010*) trying to understand the origin of life?

6. What accounts for Io's remarkable geology? Why are there no impact craters? What is the source of Io's internal heat?

7. Compare the volcanoes of Io with those of the Earth.

8. How can little Titan have an atmosphere with a higher surface pressure than that on Earth?

9. Explain how Titan's atmosphere changes with time. What do you think will ultimately happen to it?

10. Perform a "thought experiment" in which you move Triton to the position of Titan. What do you think would happen? How would Triton then appear to a passing spacecraft?

QUANTITATIVE EXERCISES

1. Find a formula for the Titius-Bode rule for planetary distances from the Sun on the Internet. Compare the actual distances of the planets with the predictions of the formula. Then see if this same formula works for the galilean satellites of Jupiter or the inner satellites of Saturn. Can you develop a similar formula that works better?

2. How fast must a molecule of SO_2 travel to escape from the gravitational field of Io? Compare this with the velocity required for a molecule of H_2O. How do you then explain the absence of H_2O and the presence of SO_2 on Io?

ADDITIONAL READING

Beebe, R. 1995. *Jupiter: The Giant Planet*. Washington, DC: Smithsonian Institution Press. Excellent pre-Galileo mission overview of the planet and its satellites for the technically literate general audience, combining results from both astronomical and spacecraft investigations.

Hanlon, M. 2001. *The Worlds of Galileo: The Inside Story of NASA's Mission to Jupiter*. Beautifully illustrated popular account by a science journalist, includes interviews with mission scientists and engineers.

Lorenz, R., and J. Mitton. 2002. *Lifting Titan's Veil: Exploring the Giant Moon of Saturn*. Cambridge: Cambridge University Press. Up-to-date review of Titan on the eve of the Huygens probe.

Morrison, D., and J. Samz. 1980. *Voyage to Jupiter*. NASA Special Publication 439. Washington, DC: U.S. Government Printing Office. Popular description of the Voyager encounters with Jupiter, including an account of the discovery of volcanoes on Io.

16

SMALL SATELLITES AND PLANETARY RINGS

Saturn and its rings accompanied by two of its inner satellites.

In Chapter 15, we discussed the largest satellites of the outer solar system, completing our survey of objects with diameters larger than 2000 km. We now turn to the smaller satellites, objects of mostly icy composition that generally lack the distinctive individual personalities of their larger siblings. Recall that small objects cool quickly and usually settle into their geologic "old age" shortly after they are born. But even here we will meet with some surprises! It is also important to remember that the apparent ordinariness or normality of some of these small satellites may simply be the result of inadequate observations, as we discovered when Voyager 2 made a close flyby of Miranda, a small satellite in the Uranus system—anticipated to be bland but revealed to be one of the weirdest objects in the planetary system.

In addition to their extensive collections of satellites (Fig. 16.1), each of the four giant planets is surrounded by a system of rings made up of huge numbers of small particles, too small to be seen individually even from a close-flying spacecraft. These myriads of tiny moons follow Kepler's laws as they orbit the planet.

Why do the giant planets all have ring systems? For 300 years, this was an unasked question, as Saturn was the only planet known to have rings. With their usual thoroughness, scientists wrote learned papers explaining why Saturn had rings and the other planets did not. Then we discovered the rings of Jupiter, Uranus, and Neptune, and now the same scientists are working to explain how each of these systems formed.

We know that the history and structure of the rings are intimately connected to the satellites of these planets, as we will see later in this chapter.

16.1 SATELLITE AND RING SYSTEMS

The satellites of the outer planets are conveniently divided into two groups on the basis of their orbits. Many of them, especially the larger ones, are **regular satellites,** meaning they revolve in orbits of low eccentricity near the equatorial plane of their planet. The regular satellites are the analogues of the planets in the solar system, which revolve in nearly circular orbits close to the Sun's equatorial plane.

Figure 16.1   The larger satellites of the outer planets compared with the Moon (top) and Pluto (bottom). This composite of spacecraft images provides an overview of the satellite systems. The relative sizes of the satellites are correct, but their spacing is not to scale. Many faint outer satellites are omitted from this figure.

The second group of moons are called **irregular satellites.** Their orbits are peculiar, having either large eccentricity or high inclination, or both. Some even revolve in a retrograde direction. Neptune's satellite Triton, discussed in Section 15.6, is an irregular satellite with a retrograde orbit. The irregular satellites have orbits resembling those of short-period comets or asteroids around the Sun. Most of them are captured objects. Table 16.1 gives the numbers of each kind of satellite for each of the giant planets.

The orbits of the inner satellites of each planet overlap with the rings. The rings and moons lead synergistic lives, influencing each other in many ways. Because they are so closely bound to their planets, each regular satellite system probably formed with its planet from a **subnebula,** a disk of dust and gas around the forming giant planet, just as the planets themselves formed from the primordial solar nebula. Perhaps the rings formed in the same way, but we will later see intriguing evidence that some of the rings are much younger than the age of the solar system—and some may be the remnants of shattered satellites.

Table 16.1 gives the dimensions of the rings in terms of the radii of their respective planets. We will convert these dimensions to kilometers as we discuss each system. Despite their individual differences, all the rings are rather close to their planets, within a distance of 2.5 times the planet's radius (Fig. 16.2).

Table 16.1 Satellite and ring systems

Planet	Regular Satellites	Irregular Satellites*	Ring Radius (R_{planet})	Ring Mass (kg)
Jupiter	8	31	1.8	10^{10}
Saturn	15	15	2.3	10^{18}
Uranus	15	6	2.2	10^{14}
Neptune	6	2	2.5	10^{12}

*The number of known irregular satellites is increasing every year as new discoveries are made. The numbers given here correspond to July 2002.

We have no complete theory for understanding why these systems are so different from one another, but we do know why the inner planets have no rings. The lifetime of a small particle in orbit around one of the inner planets is just too short. Gravitational perturbations by the Sun as well as the effects of solar radiation will cause such particles to spiral into the planets on relatively short timescales. Only in the outer solar system, with massive planets at large distances from the Sun, do we find the right conditions for extensive ring and satellite systems. Yet it is still not clear just how these systems formed, or how stable they are.

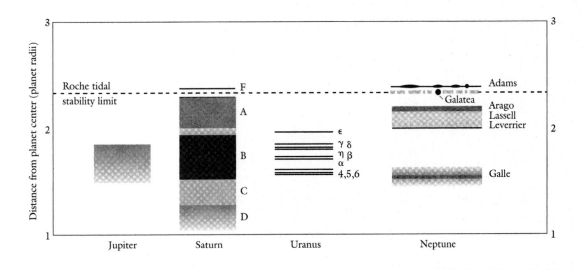

Figure 16.2 The four well-observed ring systems in the solar system are compared in this diagram that presents each one in terms of its distance from its respective planet.

SATELLITE DISCOVERIES

Galileo discovered the "galilean satellites" in 1610, on the first night he turned his newly built telescope toward Jupiter. Titan, the largest satellite of Saturn, was found in 1655 by Christian Huygens of the Netherlands. Discoveries continued over the next two centuries at a rate of nearly one satellite per decade, with Cassini, Herschel, and Nicholson each equaling Galileo's record of four.

Many additional satellites were discovered with the Voyager cameras during the first spacecraft reconnaissance of the outer solar system. From 1975 to 1990, Voyager added three new satellites of Jupiter, eleven of Saturn, ten of Uranus, and six of Neptune. Since then, the pendulum has swung back toward ground-based astronomers, especially to find very small satellites orbiting far from their planet. One single announcement in 2002 from David Jewitt at the University of Hawaii added nine additional satellites to Jupiter's family. Undoubtedly more satellites will have been discovered between the time we write these words and the time you read them.

THE JUPITER SYSTEM

The four large galilean satellites have sizes ranging from a little smaller than our Moon to larger than the planet Mercury, as we saw in Chapter 15. Four more small jovian satellites are inside Io's orbit. The others are all distant irregular satellites, with orbits of high inclination or eccentricity. Most—perhaps all—of them are captured comets or asteroids, and they are probably primitive objects rich in volatile compounds.

The very tenuous rings of Jupiter were discovered by Voyager in 1979 and confirmed within a few days by ground-based observations (it helps if an astronomer knows just what to look for!). One of us (Owen) led the Voyager effort to search for this ring, overcoming a certain amount of resistance from skeptics who thought it was impossible for Jupiter to have a ring. This negative expectation was based on the high temperatures of Jupiter early in its evolution. Ice, such as that in the rings of Saturn, could never have survived close to Jupiter.

The strategy was to take a single long-exposure picture as Voyager 1 passed through the planet's equatorial plane, centered on the distance of the Saturn rings, scaled for the larger size of Jupiter. This single frame succeeded in catching the rings, although the picture was blurred from spacecraft motion. Using the results from Voyager 1, the Voyager 2 cameras were programmed to take a variety of ring pictures, preparing the way for even better images of the ring from the Galileo mission in the late 1990s (Fig. 16.3).

The primary ring is 54,000 km from the planet and about 5000 km wide. It is much more tenuous than any of the other known rings,

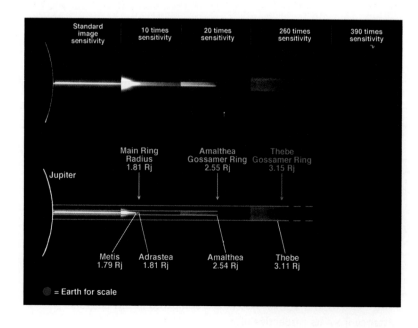

Figure 16.3 The rings of Jupiter seen edge-on by the Galileo spacecraft. Progressively longer exposures trace the rings farther and farther from the planet, showing individual components produced by dust from different inner satellites.

blocking only about 1 millionth of the light from a source seen through it. For comparison, a sheet of clear glass absorbs several percent of incident light, so one could say that the ring of Jupiter is 10,000 times more transparent than window glass! A still more insubstantial doughnut or torus of material surrounds the main ring. These are aptly called "gossamer rings."

Most of the particles in the jovian ring are very small, similar in size to the particles in cigarette smoke. They are made of dark silicate dust, eroded from the two innermost satellites by a combination of small impacts and the blasting of the satellite surfaces by energetic atomic particles of the jovian magnetosphere. Because of their small size, they easily acquire a small electric charge from interactions with the jovian magnetosphere, and the mutual repulsive force of these charged particles lifts them up out of the ring plane. The dust is continuously swept away to be replenished by new particles eroded from the small satellites. Scientists who have studied the formation of gossamer rings conclude that most of the small inner satellites may have such rings, produced by the erosion of dust from their surfaces, but we will not discuss them further in this book.

THE SATURN SYSTEM

Saturn has one of the largest regular satellite systems, just as it has the largest system of rings (and perhaps for the same reasons). As we saw in Chapter 15, Saturn's largest satellite, Titan, is about the same size as the two largest galilean moons, Ganymede and Callisto. Next in size are six satellites with diameters of 400–1500 km, corresponding to surface areas ranging from that of France up to the whole of Europe. Jupiter has no satellites this size; in fact, the only objects of this size we have encountered previously are the largest asteroids and KBOs, such as Ceres and Varuna.

The outer satellites of the Saturn system are more irregular. The most interesting is Iapetus, which has several properties that set it off from the rest of the system, as we will discuss in Section 16.2. The rest of the distant satellites, like those of Jupiter, are small, dark, and irregular.

Saturn's rings are by far the largest (most massive) and brightest in the solar system. They were first glimpsed by Galileo with his small telescopes, although he did not understand what he was seeing. Who might have expected a ring around a planet! Section 16.3 is dedicated to these marvelous rings.

THE URANUS AND NEPTUNE SYSTEMS

Uranus has 15 regular satellites, the same as Saturn. Since they orbit in the equatorial plane of the planet, they share its high inclination. The five larger uranian satellites are all in the same size range as Saturn's medium-size satellites.

Neptune is different. It has two systems of satellites, six in regular orbits and two that are highly irregular. Triton, its one large satellite (similar in many ways to the planet Pluto), is in a retrograde circular orbit. The second and much smaller irregular satellite, Nereid, has the highest eccentricity (0.75) of any satellite; it is the "comet" of the Neptune system. The regular satellites are all close to the planet, just like the inner satellites of Uranus.

Both Uranus and Neptune have substantial rings, but they are unlike those of Saturn. Rather than being broad and bright, they are narrow and dark. In part the difference is that these rings are much less massive than those of Saturn (see Table 16.1). But the distinction is deeper. The rings of Saturn are wide with a few narrow gaps. Those of Uranus and Neptune are narrow with wide gaps. We will try to explain these differences in Section 16.5.

16.2 THE SATELLITE ZOO

In this section, we will sample a few of the small satellites of Saturn and Uranus. Some will be picked because they are well behaved and typical, others because they are peculiar or pose difficult questions for us. Most of the 80 small moons are not mentioned individually at all, and many remain no more than faint points of light, barely glimpsed in the world's most powerful telescopes.

Even from a distance, we can sometimes identify fundamental characteristics that challenge our abilities to provide satisfactory explanations for what we observe. The simple distinction between high and low reflectivity is one of these characteristics, since it carries implications for the composition and evolutionary history of moons and planets. Picking one example, we note that the small satellites of Saturn are surprisingly bright. We have seen that even comet nuclei, which we know are composed largely of water ice, are covered with a layer of dark material. Exposed rocky and carbonaceous surfaces are always dark. Ice can be bright if it has a relatively recent origin, like the surface of Europa, but we will find bright, heavily cratered (and therefore old) surfaces on many of Saturn's satellites. Some kind of sorting or coating process has apparently occurred in this system that did not occur around Jupiter, Uranus, or Neptune.

SATURN'S WELL-BEHAVED MOONS: RHEA AND HER SIBLINGS

The medium-size inner satellites of Saturn, together with the rings and about a dozen small satellites, form a regular and tightly knit group (Table 16.2; Fig. 16.4). Their unusual names were taken from Greek myths that describe the establishment of the Olympian gods: Zeus, Hera, Aphrodite, and so forth. To achieve their ascendancy, the Olympians had to fight with the Titans, a family of gods that included Kronos (Saturn), Rhea, Dione, Iapetus, Tethys, and Hyperion, who were assisted by giants, such as Enceladus and Mimas.

Let us first focus on these satellites: Rhea, Dione, Tethys, and Mimas. They show less variety than the galilean satellites of Jupiter. In particular, they do not have the range of density, and hence of bulk composition, of the galilean satellites. These differences, as well as the existence of Saturn's extensive rings, clearly set this system apart from that of Jupiter. The reasons may have to do with the smaller mass of Saturn, the lower temperatures in the subnebula that surrounded it as the system formed, or perhaps other processes not yet identified.

Rhea, as the largest of these satellites, has a diameter of 1530 km, just half as big as Europa, but it is 60% larger than the largest asteroid, Ceres. Its density is only 1.3 g/cm^3, lower than that measured for any solid body discussed previously in this book. This does not mean, however, that the composition of Rhea is much different from that of Titan or the large icy satellites of Jupiter; rather, the density of Rhea is less primarily because it is small. Ice is a compressible material, resulting in a higher density for large satellites. Rhea, being smaller than Ganymede, Callisto, or Titan, has a less compressed interior. Taking compression into account, we can estimate that Rhea, like Titan, has a composition that is roughly half water ice and half silicate minerals and metal.

Rhea is highly reflective (60%), and its infrared spectrum is dominated by the absorptions of water ice. That this satellite is nearly as bright as Europa might be taken as evidence that it too has occasionally resurfaced itself with fresh ice. But the Voyager pictures tell a different story. The surface of Rhea is

---

Table 16.2 Medium-sized satellites of Saturn

Name	Period (days)	Diameter (Moon = 1)	Mass (Moon = 1)	Density (g/cm^3)
Mimas	0.94	0.11	0.0005	1.2
Enceladus	1.37	0.14	0.0011	1.1
Tethys	1.89	0.30	0.010	1.2
Dione	2.74	0.32	0.015	1.4
Rhea	4.52	0.44	0.034	1.3
Iapetus	79.3	0.41	0.026	1.2

heavily cratered, as shown in Figure 16.5. Unlike the subdued craters of Callisto and Ganymede, the craters of Rhea look remarkably lunarlike. In fact, most geologists, looking at this picture, would be hard pressed to distinguish Rhea from Mercury or the Moon, unless they were told that they were viewing brilliant white ice instead of a dark gray surface of rock. At the low temperatures (about 100 K) prevailing at the distance of Saturn, ice behaves much like rock when a crater-forming impact takes place. The colder ice is, the less plastic and the more brittle it becomes.

The crater density on Rhea is about 1000 10-km craters per 1 million km², as high as the value for the lunar highlands. Further, there is little if any indication of internal geologic activity to erase or distort craters. Lack of geologic activity

should not surprise us; after all, what is there to heat a small icy world out at the distance of Saturn?

Most of the same comments could be made about Rhea's companions: Dione, Tethys, and Mimas. They all have surfaces of relatively pure water ice, and from their densities we can infer a bulk composition that is about one-half water ice as well. It is interesting to ponder the effects of being so close to Saturn. Probably the most important influence of the giant planet is gravitational, as noted in Section 15.1. Passing comets are pulled inward, converging toward the planet and increasing both the impact rate and the impact speeds for the inner satellites. The closer a satellite is to Saturn, the larger these effects. Thus, the same flux of cometary impacts that will just build up a heavily cratered surface on Rhea

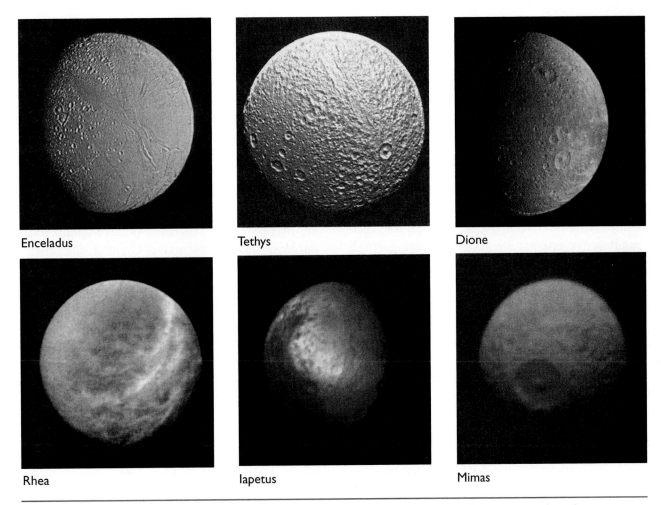

Enceladus   Tethys   Dione

Rhea   Iapetus   Mimas

Figure 16.4 The six medium-size, icy satellites of Saturn. Although shown here at the same size, their diameters range from 390 km for Mimas up to 1530 km for Rhea.

Figure 16.5   The icy surface of Rhea. Craters are shoulder to shoulder in this close-up view from Voyager. The smallest features shown are 2 km across.

will result in a much more severe pounding of Mimas, where Saturn's gravity produces a convergence of impacting objects.

ENCELADUS AND ITS RING

Two of the medium-size Saturn satellites stand out sharply from their rather bland siblings. Enceladus appears, in spite of its small size, to have remained highly active geologically. It also has a ring, the E Ring of Saturn, associated with it. Iapetus, in the outer part of the Saturn system, is a unique two-faced moon, with its leading hemisphere covered by a mysterious deposit of black material.

Even from a distance, Enceladus looks strange. Its surface is blindingly white, reflecting nearly 100% of the incident sunlight. The high albedo results in a low surface temperature, approximately 55 K. In addition, Enceladus seems to be the source of a ring around Saturn, appropriately called the E Ring (Fig. 16.6). This faint, tenuous cloud of very small, icy particles fills much of the space between the orbits of Mimas, Enceladus, and Tethys, with its maximum brightness at the orbit of Enceladus. Since the E-Ring particles are so small, they cannot

survive for long in their present orbits; instead, radiation pressure disperses them like the dust in the tail of a comet. We therefore conclude that either there is a continuing source of particles or the E Ring is young, having been formed by some recent event. Either way, Enceladus seems implicated as the most likely source of the E-Ring particles. These icy particles may provide an explanation for the surprisingly high reflectivities of Saturn's inner moons if they now coat these satellite surfaces.

Seen at closer range, Enceladus lives up to its billing as a weird place (Fig. 16.7). Over much of the surface, all impact craters have been erased, a sure sign of high levels of geologic activity. It appears that these smooth plains are no more than a few hundred million years old, about the same age as the early dinosaur fossils on Earth. Some of these smooth plains also show ridges and flow marks. Here, surely, we are seeing evidence of water volcanism or **cryovolcanism**—the term planetary geologists use to describe any process that involves the flow of partially melted ice that mimics lava flows on silicate planets. Various ices or combinations of ices might be involved, each with its own melting temperature. The flows on Enceladus seem to be water ice, but material with a lower melting temperature may have been important for cryovolcanic activity on some of the satellites of Uranus and Neptune.

It is tempting to compare Enceladus to Io, another outer planet satellite that is thought to be currently volcanically active. Granted, the activity rate on Enceladus is much lower, but then Enceladus is a much smaller body than Io. Further, the composition of Enceladus is mostly ice, while Io is a silicate object. Both objects present essentially the same problem: to find a relatively large source of internal heating that is capable of maintaining geologic activity in spite of the rapid escape of heat from the interior.

In the case of Io, that mechanism has been identified as tidal heating. Could the same thing be happening to Enceladus? The difficulty is that nearby satellites do not force Enceladus to revolve in a noncircular orbit the way Io is constrained by Europa and Ganymede. Efforts have been made to construct scenarios in which Enceladus is

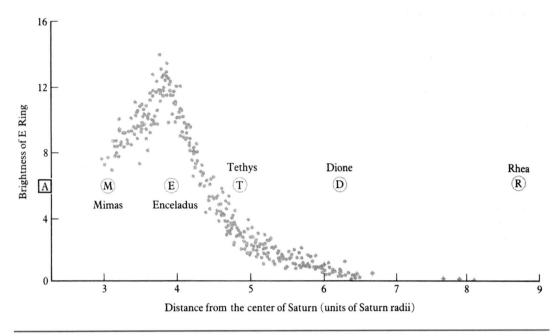

Figure 16.6 Saturn's E Ring is closely associated with the satellite Enceladus, as revealed in this plot of ring brightness versus distance from the planet. Positions of the other satellites and the A Ring are shown for scale.

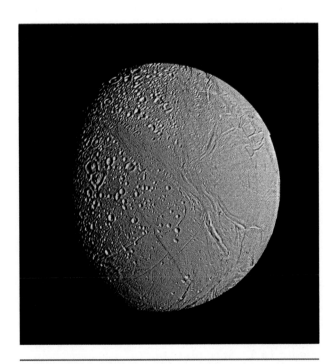

Figure 16.7 Enceladus is one of the most mysterious moons yet discovered. A large portion of the surface seems melted, with no evidence of impact craters, but the cause of this smoothing has not yet been established.

occasionally forced into an eccentric orbit, resulting in episodic heating, but to date these ideas have failed to convince skeptics. The questions of how Enceladus stayed hot and how it maintains the E Ring therefore remain open.

TWO-FACED IAPETUS

Iapetus presents us with a different set of mysteries. It is a two-faced satellite with a dark leading hemisphere and a bright trailing hemisphere (Fig 16.8). Since Iapetus, like most satellites, always keeps the same face toward its planet, we see the brightness of the satellite vary dramatically as it moves around its orbit, presenting us first with its dark side and then with its bright side. The bright side is water ice with reflectivity about 50%. The dark side, however, is covered with a reddish-black material, probably organic (carbonaceous) in composition, that reflects only 3% of the incident sunlight—among the darkest surfaces in the planetary system.

Efforts to identify this dark coating have revealed that a nitrogen-rich organic material

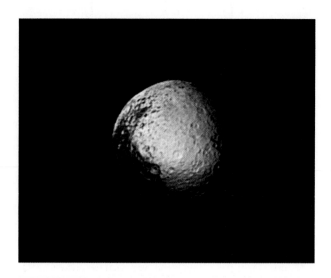

Figure 16.8 Iapetus exhibits a leading hemisphere that is roughly ten times darker than the trailing side. In this view, very little detail can be seen in the underexposed dark side.

mixed with carbon and a frosting of ice (see Section 4.4) matches the spectrum of sunlight reflected from the dark side of Iapetus. But is it really the same material? Our ability to identify solids by remote measurements is not good enough to give us a definitive answer. Similar dark, reddish material seems to be common in the outer solar system, on some asteroids, and on some comet nuclei. Recall that the object called Pholus has the reddest color in the solar system (see Section 5.4). The dark material that makes up the rings of Uranus and coats the inner satellites of both Uranus and Neptune has equally low reflectivity but lacks the reddish color of Iapetus. Perhaps all of these surfaces represent different blends of primitive, carbon- and nitrogen-rich substances from the solar nebula.

Voyager also provided the first measurement of the mass of Iapetus. In combination with the diameter of 1460 km, this mass yields a density similar to that of the other icy satellites of Saturn. Thus, Iapetus appears to be similar to the others in bulk properties, and the dark deposit seems likely to be of external origin. The zebra is revealed as a white horse with dark stripes, rather than the other way around.

The dark hemisphere is symmetric with respect to the direction of orbital motion of Iapetus, providing further support for the idea that the dark surface must be the result of some external material striking the surface. But what might this material be? The next outer satellite, Phoebe, is dark and has been suggested as a source for the dark material. But the color of Phoebe does not match that of the dark side of Iapetus, so that theory is suspect. Perhaps the Phoebe dust that is hitting Iapetus interacts with the surface material there to change its color. Alternatively, the dark material may be indigenous to Iapetus, and it is concentrated on the surface of the leading side by impacts that selectively evaporate away the water ice. Such a process would be similar to that by which a comet develops a dark surface by concentrating carbonaceous material as the ice evaporates.

These are intriguing ideas, but unfortunately they beg the question of the uniqueness of Iapetus. Why should this satellite alone have a dark leading hemisphere? A solution involving collisions—either to deposit dark material or to wear away bright material—seems to be the most reasonable possibility. But we do not know, and the Voyager data are not sufficient to provide a satisfactory answer (Fig. 16.9). We expect to get a

Figure 16.9. Scientists at work. This amusing photo was taken at JPL at the time of the Voyager 2 encounter in 1981, during a discussion of Iapetus. The most surprised looking scientist is Carl Sagan. Others (left to right) are Jim Pollack, Steven Soter, David Morrison, and Gene Shoemaker.

much closer look at the dark side of Iapetus in 2005 with the Cassini orbiter.

MIRANDA: THE "ENFANT TERRIBLE" OF THE URANUS SATELLITES

Unlike the other three giant planets, Uranus has no satellite with a diameter as large as 2000 km. Instead, it has five medium-size satellites called Miranda, Ariel, Umbriel, Titania, and Oberon—four named after Shakespearean figures and the name Umbriel taken from Alexander Pope's "Rape of the Lock" (Table 16.3).

Because Uranus is so distant, its satellites are difficult objects for astronomical investigation. Before the Voyager encounter, our meager knowledge of these frigid worlds had to be obtained from the light collected by large telescopes on the surface of Earth. It was not until 1977 that the rings were discovered, 1982 that the satellite sizes were first measured even approximately, and 1983 that the masses of two of the satellites were first determined by their mutual gravitational influence. Voyager greatly increased our knowledge, but since that brief flyby, progress in studying this system has depended on Earth-based astronomical observations, challenging the technology and ingenuity of planetary astronomers.

In size, the satellites of Uranus resemble the inner satellites of Saturn. Their densities are also fairly similar to those in the saturnian system, although at 1.3–1.6 g/cm^3 they suggest a slightly larger proportion of silicate materials and metals, with correspondingly less ice. In general terms,

however, it is fair to say that the satellites of both planets are composed of about half rock and half ice by mass, with only minor variations from one object to another. The surface compositions of the satellites of Uranus also resemble those of Saturn. Water ice is detected spectrally, but reflectivities are only 20%–30%, suggesting that the ice is relatively dirty.

Miranda provided the big surprise of the 1986 Voyager encounter. The spacecraft passed only 28,000 km from this satellite, obtaining images with resolution better than 1 km. However, the choice of Miranda to be the best-studied uranian satellite originally was not a happy one, since most geologists expected it to be a dull, cratered little world. The only reason we obtained this close view of Miranda was that Voyager had to pass through a point near Miranda's orbit to receive the gravitational boost it needed to reach Neptune in 1989.

By good fortune, however, Miranda was not dull after all (Fig 16.10). It has a surface extensively modified by internal processes. There are great valley systems up to 50 km across and 10 km deep, apparently produced by large-scale tectonic stresses. Peculiar oval or trapezoidal mountain ranges cover about half of the surface. In some areas the craters are apparently mantled and "softened" by overlying material, while elsewhere the craters are sharp and fresh looking. One huge cliff seen at the border of the illuminated area is between 10 and 15 km high.

What caused Miranda, which is smaller than 500 km in diameter, to be so much more active

Table 16.3 Large satellites of Uranus

Name	Period (days)	Diameter (Moon = 1)	Mass (Moon = 1)	Density (g/cm^3)
Miranda	1.41	0.14	0.0011	1.3
Ariel	2.52	0.33	0.018	1.6
Umbriel	4.14	0.34	0.018	1.4
Titania	8.71	0.46	0.048	1.6
Oberon	13.5	0.45	0.039	1.5

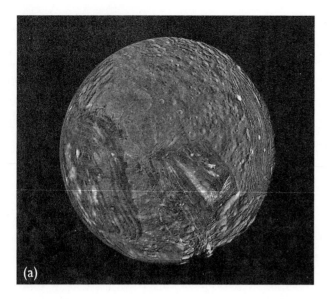

Figure 16.10 **(a)** Miranda. The surface of this satellite is like nothing seen elsewhere in the planetary system. **(b)** Close-up of Miranda.

than its larger neighbors? One suggestion is that an impact shattered Miranda after it had differentiated, and the fragments fell back together randomly like a jumbled jigsaw puzzle. Another possibility is that the satellite, too small to complete its internal mixing, froze partway through the process of differentiation. We simply don't know what happened here.

SMALL INNER SATELLITES OF SATURN

The small inner satellites of Saturn are all icy objects with bright surfaces. Figure 16.11 is a composite group photo. The most surprising aspect of these satellites is their orbits, each of which is special in some way. The three innermost satellites have orbits that are embedded within the rings, and the next three skirt the very edge of the rings and interact with them. Others share orbits with the larger satellites Dione and Tethys, providing the only known examples of Lagrangian orbits within a satellite system (see Section 5.4). But perhaps most special are the **co-orbital satellites,** Janus and Epimetheus. They were first spotted with ground-based telescopes in 1966. In 1980, Voyager 1 confirmed that these two satellites were in nearly the same orbit, about 13,000 km beyond the rings. The larger is about 200 km in its longest dimension, and the smaller is about 150 km.

If the two satellites had exactly the same orbital period, they might be able to avoid interfering by remaining on opposite sides of the planet. But the co-orbitals actually have orbits differing by about 50 km in radius. The orbital period of the inner one is 16.664 hr, and that of the outer one is 16.672 hr. This difference causes the inner one to catch up with the outer at a relative speed of 9 m/s. About once every 4 years the inner satellite laps its slower-moving sibling, but since the space between orbits is smaller than the objects, there is no room to pass. How is a collision to be avoided? What happens is that, short of a collision, the two satellites attract each other gravitationally and exchange orbits. They then slowly move apart, and the 4-year cycle starts again. This strange orbital dance is unique, as far as we know, in the planetary system.

The co-orbital satellites are elongated and irregular in appearance, looking very much like fragments from some ancient catastrophe. Many scientists think that they were once joined and that some long-ago impact fractured their parent body to leave these two pieces in nearly the same orbit. All the small inner satellites of Saturn also look as if they might be remnants of past collisions. Recall that the inner part of the Saturn system was subject to very heavy impact cratering early in its history. If we extend that argument to include these smaller bodies, we should not be

Figure 16.11   A group portrait of Saturn's tiny satellites. At left is Atlas, guarding the outer edge of the A Ring, then Prometheus (top) and Pandora (bottom), which shepherd the F Ring; Janus (top) and Epimetheus (bottom), which occupy nearly the same orbit; two Lagrangian satellites of Dione; and the single Lagrangian of Tethys. Note the crater in the latter—a slightly larger impact would have shattered it.

surprised to see a few remnants of objects that were destroyed by such impacts. Indeed, the rings themselves might have been formed in this way, as we will discuss later.

16.3   THE BROAD BRIGHT RINGS OF SATURN

Many people consider the rings of Saturn to be the most beautiful sight that can be seen through a telescope. The rings of Saturn were first seen by Galileo in 1610, but recognition of the nature of this new phenomenon did not come for another 50 years. Galileo thought he was glimpsing bumps on Saturn or perhaps a triple planet. Some other seventeenth-century observers drew the rings as handles extending on either side of the planetary disk. Not until 1659 did the Dutch astronomer Christian Huygens recognize that Saturn was surrounded by a "thin flat ring, nowhere touching" the planet and lying in its equatorial plane. In 1675, Giovanni Cassini, who also discovered the two-faced nature of Iapetus, found that there were at least two concentric rings, not a single, continuous one. The dark lane that

separates the two parts is still called the Cassini Division in his honor.

That the rings are not solid but composed of billions of tiny moons orbiting the planet was not demonstrated until the second half of the nineteenth century, although many astronomers had suspected this for some time. According to Kepler's laws of planetary motion, the particles closest to Saturn must travel faster in their orbits around the planet than the more distant ones. At the inner edge of the main rings, one circuit of Saturn requires only 5.6 hr, while the period at the outer edge is 14.2 hr.

CHARACTERIZING THE RING PARTICLES

Further understanding of the nature of the ring particles awaited the development of modern astronomical instrumentation. Infrared spectra revealed in 1970 that the particles were composed primarily of water ice. Interpretation of the first radar signals bounced from the rings in 1973 led to information on the sizes of the particles, which are typically tens of centimeters in diameter. Most of what we now know about the rings is the direct result of the Voyager encounters in 1980 and 1981. Popular interest in these encounters was so great that Saturn and its rings appeared on the covers of both *Time* and *Newsweek*, with long articles in the magazines describing the Voyager discoveries.

The rings of Saturn consist of a thin sheet of small icy particles, ranging in size from grains of sand up to house-size boulders. Just as with the asteroids and other populations of debris, there are many more small particles than large ones. Most of the individual particles are bright, reflecting 50%–60% of the incident sunlight and exhibiting the strong spectral signature of water ice. Some particles are darker, perhaps indicating the presence of organic material or silicates. An insider's view of the rings would probably resemble a bright cloud of floating hailstones with just a few snowballs and larger objects, many of them loose aggregates of smaller particles.

With the few exceptions noted later, each ring particle follows an almost perfectly circular orbit around Saturn in the equatorial plane of the planet. It is easy to see why this must be. Imagine a particle with an eccentric orbit. During each circuit of Saturn, it would move in and out, crossing the orbits of other particles. Low-speed collisions would result, and the particle would lose energy, almost as if it were rubbing against its neighbors. The same sort of "fender-benders" would be encountered by a particle in an inclined orbit, swinging back and forth across the plane of the rings. Either way, the effect of the friction is to circularize the orbit and bring the particle into the same plane with the rest of the ring particles. This is generally what we see at Saturn, and yet there remain many surprises when we look closely: narrow gaps, waves, eccentric rings, kinky rings, and even braided rings!

OVERVIEW OF THE RINGS

The rings of Saturn are very broad and very thin. Their dimensions are listed in Table 16.4. The main rings, those visible from Earth, stretch from about 7000 km above the atmosphere of the planet outward for a total span of more than 70,000 km (Fig. 16.12). The distance from one

Table 16.4 Dimensions of the main Saturn rings

Name	Outer Edge (R_{planet})	Outer Edge (km)
D	1.233	74,400
C	1.524	91,900
B	1.946	117,400
Cassini Division	2.212	133,400
A	2.265	136,600
F	2.324	140,180

edge of the rings through the planet to the opposite edge is almost as great as the distance from the Earth to the Moon. Yet the thickness of this vast expanse is only about 20 m, the width of a typical house lot. If we made a scale model of the rings out of paper the thickness of one page in this book, we would have to make the rings more than 1 km across—about eight city blocks. On this scale, the planet would loom as high as a 100-story skyscraper.

Figure 16.12 Saturn's rings as seen by Voyager. Note that the C Ring is semitransparent. A part of the Cassini Division can be seen at the extreme top and bottom. Look also at the shadow of the rings on the planet, where the Cassini Division shows as a light streak because the sunlight shines through it.

Three distinct rings, called the A, B, and C Rings, can be seen from Earth; an additional narrow F Ring was discovered in 1979 by Pioneer. Voyager, however, revealed much greater complexity. The ring material is organized into tens of thousands of ringlets visible in Voyager photographs. These ringlets are not generally separated from each other by gaps, but rather represent local enhancements or depletions of the concentration of ring particles. Only a few empty gaps exist, providing natural boundaries between parts of the rings. We will discuss the origin of these gaps in Section 16.5, but we first take a tour of the rings, noting the major features as we proceed outward from the planet.

THE INNER RINGS

Between the upper atmosphere of the planet and the inner edge of the C Ring, Voyager discovered several thin rings invisible from Earth, collectively called the D Ring. However, the substantial part of the ring—the C Ring—starts 7000 km out, where the particles are packed densely enough to reflect a fair amount of sunlight. There are two major gaps in this ring, each several hundred kilometers wide.

Inside one of the C-Ring gaps is a remarkable narrow **eccentric ring.** This entire ribbon behaves as if each of its particles orbited Saturn with the same eccentricity and with all the individual orbits exactly aligned. The color of this ring is subtly but definitely different from that of its neighbors, suggesting that the particles that make up the eccentric ring have a different composition. There are only three of these eccentric ringlets in the Saturn system, each very narrow and each lying in an empty gap. As we will see later, the rings of Uranus and Neptune are also narrow and eccentric.

THE A AND B RINGS

At a distance of 32,000 km from Saturn, the ring particles suddenly become more densely concentrated, and the structure of the concentric ringlets becomes more complex. This is the edge of the B Ring, the brightest part of the ring system and the part that contains most of the mass. Through-

out much of the B Ring, which stretches out to 57,000 km, the particles are so closely spaced that the ring is nearly opaque. Particles are typically from tens of centimeters up to meters in diameter. There are no empty gaps in the entire 25,000-km span of this ring.

At the outer edge of the B Ring lies the 3500-km–wide Cassini Division, the one break in the rings that can easily be seen from the Earth. As shown in Figure 16.13, however, the Cassini Division is by no means an empty gap. Within the division there are a number of discrete ringlets, separated by several true gaps, including one eccentric ring, and a great deal of fine structure visible in the spacecraft photos. At least one small satellite also orbits inside this division. At a time when it was still thought that the Cassini Division was empty, a proposal was considered to target the Pioneer Saturn spacecraft, which reconnoitered the planet in 1979, to pass through the rings within this division. Had this been attempted, the Pioneer Saturn mission would have come to a very sudden end!

Beyond the Cassini Division, beginning at 61,000 km above the planet, is the last major ring, the A Ring. The A Ring is intermediate in brightness and transparency between the opaque B Ring and the translucent C Ring. Its most outstanding feature is in one of its gaps, the 360-km–wide Encke Division, which contains two discontinuous, kinky ringlets and one known small satellite. These peculiar ribbons of material are only about 20 km wide and were observable from the Voyagers only when the spacecraft passed very close to the rings. The problem of the origin of kinky rings is discussed in Section 16.6, but don't expect any answers: Kinky rings remain a complex problem in celestial mechanics.

BEYOND THE MAIN RINGS

The A Ring ends abruptly 96,000 km from the planet. There is no tapering off of ring particles, no stray ringlets drifting off into space. There is, however, the fascinating F Ring 4000 km farther out. Unlike the rings discussed so far, the F Ring is an isolated bright ribbon whose width varies

Figure 16.13   The Cassini Division in Saturn's rings, viewed from its underside. In this highly magnified view, the Division, which looks dark and empty when viewed from Earth, is actually found to contain several ring structures of its own. The classical Cassini Division, just to the right of the center, contains five bands of material bounded by dark gaps.

from 30 to 500 km. It is the third eccentric ring in the Saturn system. It also has the most complex and peculiar structure of any Saturn ring, including areas where it divides into multiple strands that appear to be intertwined or braided (Fig. 16.14).

The discovery of braiding in the F Ring was one of the sensations of the Voyager encounters, leading to newspaper headlines asserting that these rings disobeyed the laws of physics. They don't, but they do make it clear that scientists are sometimes unable to interpret the laws of nature to explain very complex situations. What makes the F Ring both complicated and fascinating is the presence of two small satellites of Saturn that orbit the planet on either side of the ring. The boundaries of the ring are clearly set by these satellites, and it is their gravitational influence that generates its fine structure. We will return to this subject when we discuss satellite-ring interactions.

Continuing outward from Saturn, we encounter the tenuous E and G Rings. We already noted that the E Ring seems to originate from the

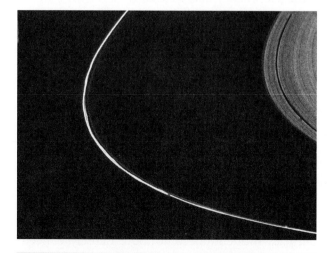

Figure 16.14   Saturn's F Ring consists of at least five separate strands, of which two are easily seen here. They exhibit a braided appearance, with knots of greater density (moonlets?) embedded within them.

satellite Enceladus (see Fig. 16.6). The G Ring, 1000 m wide and composed of very small particles, is even more mysterious. Perhaps unseen small satellites are the source of its particles, just as the ring of Jupiter is eroded from two small satellites.

16.4 THE NARROW DARK RINGS OF URANUS AND NEPTUNE

The rings of Uranus were discovered accidentally. They were "seen" not by their ability to reflect light but by their ability to block it. The technique of studying a planetary object by its blockage of the light from a distant star (or the radio transmission from a spacecraft) is called an occultation, as we have described previously. One of the early triumphs of occultation astronomy was the discovery of these rings, the first to be found since Saturn's rings were discovered three centuries earlier.

DISCOVERY OF THE RINGS OF URANUS

On March 10, 1977, an occultation of a star by Uranus occurred that was particularly favorable, except for one thing: The event could not be seen from any of the more populated parts of the Earth. Only from Antarctica and the Indian Ocean was it a "sure thing," with marginal conditions in Australia, southern India, and South Africa. Nevertheless, observers set up equipment at all the observatories that might be in the path to await this exceptional event. To supplement the ground-based observations, James Elliot and his colleagues at MIT proposed to use NASA's Kuiper Airborne Observatory to fly above the southern Indian Ocean, well within the predicted zone for the occultation. To accommodate this project, the airplane had to fly to the Southern Hemisphere and operate out of an Australian airbase.

As expected, Elliot's team succeeded in measuring the occultation of the star by Uranus, while the ground-based observers saw the planet skim by the star without blocking its light. But the most important results that night were not from the planet at all. As measured from both the airplane

and the ground, the occulted star began to wink out about 40 minutes before it should first have been affected by the upper atmosphere of the planet. Several times the star dimmed dramatically for intervals that lasted from 2 to 8 seconds, then returned just as suddenly to full brightness (Fig. 16.15).

Clearly something was briefly blocking the starlight, and at first Elliot and his colleagues on the plane thought they might be seeing a swarm of small uranian satellites. The occultation events were not randomly spaced, however. As Uranus moved beyond the star, exactly the same sequence of occultations occurred, but in reverse order. Further, the occultation pattern was the same at the different observing sites, even though some of them did not detect an occultation of the star by the planet itself. What had been discovered was a series of narrow, opaque rings, invisible in reflected sunlight but detected by their ability to block the light from a distant star. The situation resembles counting the number of cars in a train on a dark night by recording the visibility of a streetlight viewed through the passing cars.

THE ADVANTAGE OF OCCULTATIONS

The uranian rings had not been detected before because they are narrow and dark. But even if they had been seen in reflected sunlight, the occultation technique reveals much more than could be photographed with any telescope. The reason is that the occultation resolution is not limited by the size of the telescope or the shimmering of the Earth's atmosphere. The level of detail probed by an occultation depends on how rapidly the rings appear to move across the sky and on how often we can sample the changing brightness of the star. In the case of the Uranus occultation, structure as small as a few kilometers could be measured.

Since their discovery, the rings of Uranus have been repeatedly studied using occultations. This technique is very well suited to the task for three reasons: (1) Since the rings are dark, they produce almost no reflected light to interfere; (2) at near-infrared wavelengths, the planet itself

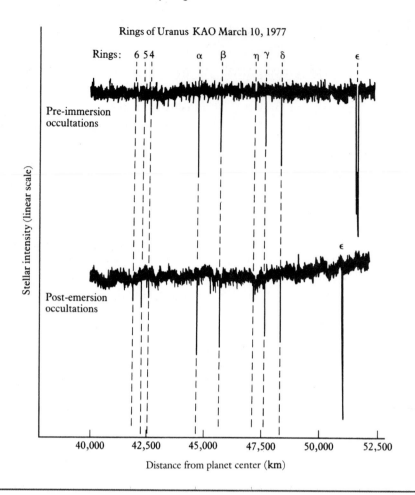

Figure 16.15 The two intensity tracings that revealed the presence of Uranus's rings. Top: As Uranus passed in front of a star, the intensity of starlight recorded by a photometer on the airborne telescope suddenly dimmed, then recovered, and repeated this behavior nine times. Bottom: The same thing happened on the other side of the planet, except that the eccentric Epsilon Ring was closer to Uranus and narrower on this side of the planet.

is also very dark, further reducing interference; and (3) because of Uranus's high tilt, the rings present a wide target. In contrast, stellar occultations seen from Earth are not very useful to study the saturnian rings, which are very bright and are viewed obliquely.

The rings of Uranus are just the opposite of the Saturn rings in several respects. At Saturn we have broad rings interrupted by a few narrow gaps; at Uranus we have very narrow rings separated by broad gaps. The Saturn ring particles are bright and composed of ice; those at Uranus are dark and apparently made of (or at least coated by) some sort of carbonaceous material. The dimensions of these rings are listed in Table 16.5.

Table 16.5 Rings of Uranus

Ring Name	Distance (km)	Width (km)	Eccentricity
6 Ring	41,850	1–3	0.0010
5 Ring	42,240	2–3	0.0019
4 Ring	42,580	2	0.0011
Alpha	44,730	8–11	0.0008
Beta	45,670	7–11	0.0004
Eta	47,180	55	0
Gamma	47,630	1–4	0
Delta	48,310	3–9	0
Lambda	50,040	1–2	?
Epsilon	51,160	22–93	0.0079

STRUCTURE OF THE RINGS OF URANUS

In 1986, the Voyager cameras photographed the rings directly (Fig. 16.16) and also observed occultations at close range, yielding ring profiles of substantially higher resolution than that obtained either by the Voyager cameras or by Earth-based occultation observations. Most of the 10 rings of Uranus are nearly circular and exceedingly narrow, no more than 10 km wide. In other words, their lengths are nearly 100,000 times their breadth, like a piece of spaghetti several city blocks long. Yet particles within each ring are very close to one another, blocking most of the starlight when the ring passes between an occulted star and us.

Two of the rings have a special structure, nearly as peculiar as that of the F Ring of Saturn. The Epsilon Ring, which probably contains as much mass as all the other rings combined, is both eccentric and variable in width (Fig. 16.17). Where it comes closest to Uranus, its width is 22 km; at the opposite side, where the ring is 700 km farther from the planet, it has a much greater width of 93 km. Like the eccentric rings of Saturn, the Epsilon Ring must be held together by some gravitational force to maintain its shape. The Eta Ring is equally misshapen. It consists of a relatively broad, low-density ring 55 km wide, with a narrow, denser component at its inner edge.

RINGS OF NEPTUNE

Once rings had been discovered around Jupiter and Uranus, it was natural to ask whether Neptune might also be a ringed planet. Prior to the Voyager 2 flyby, information about possible rings around Neptune was contradictory. Several occultations were observed for Neptune. In some cases, no evidence for rings was found. In others, the star's light disappeared briefly on one side of the planet but not on the other. These results gave rise to the idea that Neptune might be surrounded by discontinuous rings, or arcs, as they came to be called, and small satellites. An arc or a satellite would block the light of a star on only

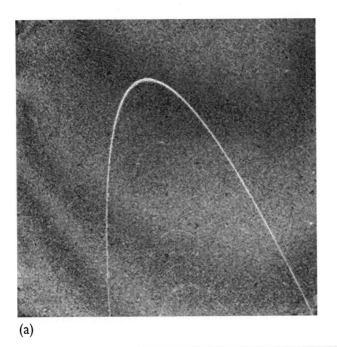

(a)

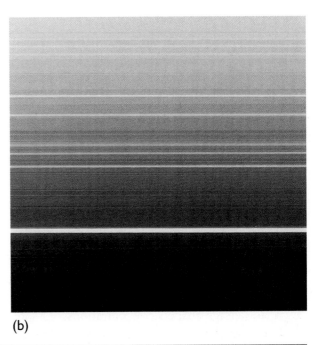

(b)

Figure 16.16 The Uranus rings as revealed by the Voyager 2 cameras corroborated and extended the occultation observations. **(a)** A distant view of the planet shows only the Epsilon Ring, revealing its variation in thickness. **(b)** Here we see nine rings in false color. Starting from the bottom, they are 6, 5, 4, α, β, η, γ, δ, ϵ.

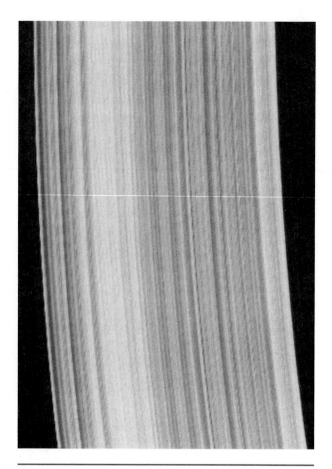

Figure 16.17 A profile of the Epsilon Ring of Uranus obtained by the Voyager 2 spacecraft as the ring occulted a star.

Figure 16.18 The ring system of Neptune recorded by Voyager 2 as the spacecraft passed behind the planet. The faintness of the rings required a long exposure, which reveals the background stars as well as intense, scattered light from the planet itself. The overexposed image of the planet is in the lower left.

one side of the planet, and only if it happened to be in the proper place in its orbit at the time.

The eagerly awaited images from Voyager 2 revealed a situation even more complex. The arcs turned out to be discrete features in a very narrow, faint, but continuous ring, which is one of three prominent, narrow rings surrounding the planet (Fig. 16.18). These rings are named Galle, Leverrier, and Adams, in order of increasing distance from the planet. (Recall that Adams and Leverrier independently calculated the position of Neptune, while Galle found it with the telescope.) The four discrete, sausage-shaped concentrations of material known as arcs are in the Adams Ring. There is still no accepted theory that explains how the material in these arcs remains concentrated in this way instead of spreading around the orbit with the rest of the ring particles.

The Leverrier and Galle rings, as well as the material in the orbit of the satellite Galatea and the wide ring stretching outward from Leverrier, consist of particles that are too widely dispersed to be detected by the occultation technique. They show up in some Voyager pictures because of an optical effect called **forward scattering.** When particles in a certain range of sizes are struck by a beam of light, most of that light is scattered forward in nearly the same direction as the incident beam. This is why the windshield of your car looks so much dirtier when you are driving toward the Sun. As Voyager 2 approached Neptune (or one of the other giant planets), the light scattered back to the cameras by the small ring particles was very faint. But after the spacecraft passed the planet, got into the planet's shadow, and was able to look back in a sunward direction, the rings just "lit up" (Fig. 16.19).

Figure 16.19 The rings of Uranus in forward-scattered light. Seen from this perspective, the fine dust in the rings "lights up" just like windshield dust when you are driving toward the Sun. The streaks are background stars that made trails during this Voyager time exposure.

16.5 STABILITY OF RING AND SATELLITE SYSTEMS

As the resolution of ring images improves, we find more and more structure. What appeared to Huygens as a single sheet of material surrounding Saturn was found by Cassini to be divided into two parts, and Johann Encke found a second gap late in the nineteenth century. The Voyager pictures demonstrated that the particles in the rings are organized into thousands of ringlets, including some within the Cassini Division, and the occultation results have shown that still more complex structures are present.

WAVES IN THE RINGS

Remarkably, the fine structure of Saturn's rings varies with time and location. Figure 16.20 illustrates the structure in the outer part of the B Ring, at the border of the Cassini Division. Four different photos, taken at different times, are shown side by side. While the major features line up well,

it is clear that the smaller ringlets do not. These ring structures are not fixed, but shift from hour to hour. We should think of them as transient waves, flowing back and forth over the rings' spinning surfaces like waves on the ocean. Many of the patterns are examples of **spiral density waves** (Fig. 16.21). As the name implies, these are waves that follow a spiral pattern, like the grooves on an old-fashioned phonograph record. Density waves had been predicted theoretically but never observed before Voyager arrived at Saturn. They are a phenomenon peculiar to a flat spinning disk in which individual particles can interact gravitationally. Even though the individual particles rarely touch, the ensemble of billions of particles behaves in many ways like a thin sheet of rubber because of mutual gravitational attractions. This is

Figure 16.20 These four radial segments of the outer edge of the B Ring of Saturn show that the fine structure in the ring is not uniform around the planet. The outer edge of the B Ring itself is eccentric, so it occurs at different distances from the planet at different places along its circumference.

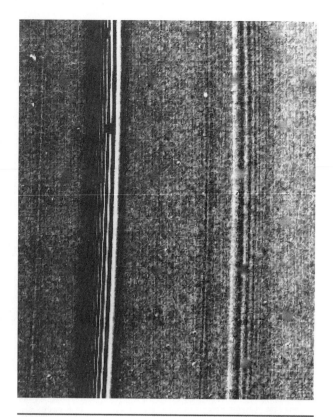

Figure 16.21 The same 5:3 gravitational resonance with the satellite Mimas generates both these systems of waves. The one on the left is a bending wave with ripples 1–2 km high, sufficient to cast shadows. On the right is a spiral density wave consisting of local concentrations of material within the ring plane.

especially true in the B Ring, where the particles are most closely packed together.

Dozens of individual spiral wave patterns have been identified, primarily in Saturn's A Ring, but a great deal of the structure, particularly that seen in the B Ring, is not yet explained. Some of this structure is probably the result of the superposition of unidentified wave patterns, but much of it may have quite different causes. We could understand the behavior of the rings better if we had moving pictures of the changing ring patterns. Unfortunately, however, there was only a single Voyager occultation, and the high-resolution images are only snapshots. A Saturn orbiter spacecraft is required to obtain the data we need, and this is one of the tasks of the Cassini-Huygens mission.

As we wait for these new data, we can try our best to understand what we have seen so far. Why are the edges of the rings so sharply defined? What causes the Cassini Division and the smaller gaps in the rings of Saturn? Why are three saturnian ringlets, as well as most of the uranian rings, eccentric? Why are the rings of Uranus so narrow? And what generates the spiral wave patterns? It turns out that the answers to all of these questions involve interactions between the rings and the satellites.

EDGES AND SHEPHERDS

Left to itself, a planetary ring will slowly spread out. Over hundreds of millions of years, the interactions between closely spaced ring particles will force them apart, with some spiraling toward the planet to disintegrate like meteors in its atmosphere and others expanding out to ever greater distances from the planet. To be stable, a ring must be bounded. The binding force that limits the outward spreading of the rings of Saturn is gravitational. The outer edge of the A Ring is in a six-sevenths resonant orbit with Janus and Epimetheus, and the gravitation of these two co-orbital satellites keeps the ring particles firmly trapped at this particular distance from Saturn.

A different kind of gravitational influence holds the thin F Ring of Saturn in place and even explains its slightly eccentric shape. In this case there are two satellites, Prometheus and Pandora, one on each side of the ring (Fig. 16.22). These are called the **shepherd satellites** for their role in keeping the particles of the F Ring narrowly confined. Probably it is the influence of the shepherd satellites that also generates the braids in the F Ring, but a complete theory for this process has not yet been worked out.

When they were discovered, the narrow rings of Uranus confounded astronomers. In response, theorists Peter Goldreich and Scott Tremaine at Caltech first developed the idea that a ring could be confined by shepherding satellites. These shepherds, first predicted for Uranus, were confirmed in concept by the discovery of Pandora and Prometheus on either side of the F Ring of Saturn.

Figure 16.22   The F Ring of Saturn appears to be held in place by the two small shepherding satellites, Prometheus and Pandora. This view shows a segment of the F Ring with Prometheus in the same field of view.

It was therefore with great interest that the Voyager cameras searched for similar shepherds at Uranus. Surprisingly, however, the search was only partially successful.

Two uranian shepherds (Cordelia and Ophelia) were discovered in association with the Epsilon Ring. Each is a small dark satellite less than 50 km in diameter. These shepherds orbit about 2000 km on either side of the ring, very much like the F Ring shepherds at Saturn. However, no other shepherd satellites were seen near the other uranian rings. Since the cameras were sensitive only to dark objects 10 km or more in diameter, smaller shepherds may be present. Thus we still have no confirmation of the postulated 22 satellites needed to explain all the other known rings of Uranus.

EMBEDDED SATELLITES

If there are satellites inside the rings, they can have gravitational effects also. Within the broad main rings of Saturn, the presence of unseen satellites is thought to produce the few genuine gaps. By its gravitational herding effect, a small satellite could sweep clear a lane much wider than its own diameter. To confirm this explanation, very careful searches were made of the Voyager images for these **embedded satellites.**

The first evidence that embedded satellites are indeed present was the discovery of scalloped edges in the Cassini Division and the Encke Gap. These scallops are the result of waves generated in the bounding ring material by the passages of the satellites, much like the wake that stretches far behind a ship moving across the oceans of Earth. From the properties of their wakes, the positions and masses of these embedded objects can be calculated, and both turn out to be smaller than 15 km in diameter. So far, only one of them, Pan, has been found in a Voyager picture.

Embedded satellites have also been suggested as an important agent for forming narrow, eccentric rings. Contradictory though it may seem, calculations have shown that under some circumstances—generally involving closely packed ring particles—an embedded satellite will produce a narrow ring rather than clear a gap. If the orbit of the embedded satellite is eccentric, so will be the path of the thin ring it controls. This is an especially attractive theory for trying to understand the kinky rings in the Encke Division of Saturn's A Ring, and it has also been suggested as an alternative way of forming some of the rings of Uranus and Neptune.

RESONANT EFFECTS OF THE LARGER SATELLITES

The larger external satellites also influence the ring structure by their gravitational effects. The Cassini Division, for example, is caused by Mimas. At the inner edge of the division, a ring particle orbits Saturn exactly twice for each orbit of Mimas, and this strong resonance is apparently responsible for the resonant gap located here, just as the resonant gaps in the asteroid belt are caused by Jupiter (see Section 5.2).

Weaker resonances with other smaller satellites do not produce gaps, but they do disturb the ring particles having orbital periods that are simple fractions of those of the satellites. The two co-orbital satellites generate a large number of these resonances. The disturbances in these cases produce the spiral density waves. Each spiral represents a wave pattern that originates at the resonant position and then winds its way outward.

By invoking the presence of satellites both seen and unseen, we can explain a great deal of the complex structure found in the rings of Saturn. We don't yet understand all of the varied phenomena that we see in the rings, but most of the prominent features seem understandable. The main problem with this hypothesis is that it depends in part on the existence of a variety of unseen satellites embedded within the rings. These satellites have been searched for but not found. If they are present, we wonder how they have eluded detection; if they are absent, then our understanding of ring structure is much poorer than we would like to think. Similarly, the presence of two shepherds for the Epsilon Ring of Uranus seems to provide an adequate explanation for the existence of this ring, but the satellites to control the other rings of Uranus remain at present only speculation.

The planetary ring systems have three features in common: They are composed of small particles of a wide variety of sizes; they are close to their planets; and much of their structure appears to be controlled by a few larger objects, including the external satellites. The interactions with satellites in particular seem to be necessary to bound a ring and keep it from gradually spreading out and fading away. But what could have produced the original swarm of billions of particles in a disk surrounding the planets?

16.6   THE ORIGIN OF RINGS AND SMALL SATELLITES

There are two basic theories of ring origin, both of them involving the small inner satellites of the giant planets. First is the breakup theory, which suggests that the rings are the remains of a shattered satellite. The second hypothesis, which takes the reverse perspective, suggests that the rings are made of particles that were never able to come together to form a satellite in the first place.

THE TIDAL STABILITY LIMIT

In both hypotheses of formation, tidal forces play an important role. When tides were discussed in Section 8.2, we noted that they are very sensitive to distance, varying as the inverse cube of the separation between two bodies. The effect of tides is to distort a satellite, raising bulges on it that face toward and away from the planet it orbits (see Section 15.4). The closer a satellite comes to its primary, the larger the tidal distortion. Ultimately, the satellite can be torn apart if its internal strength is not great enough to withstand the tidal stress.

Around each planet there exists a **tidal stability limit,** the distance within which tidal forces are stronger than self-gravitational forces, so that tides can destroy an intruding satellite. This limit was first calculated by the nineteenth-century French mathematician Edouard Roche, and it is often called the Roche limit. Its exact value depends on the density and internal strength of the satellite. If we simply consider two particles that are just touching each other, the tidal stability limit is at about 2.5 planetary radii from the center of the planet. In this case, the limit defines the distance from the planet at which the difference in the gravitational force exerted by the planet on each of the two particles is greater than their mutual gravitational attraction. Thus, they will not be able to coalesce and form a larger body. As shown in Figure 16.2, the major ring systems of each of the four giant planets lie within the respective Roche limits. The G and E Rings of Saturn are notable exceptions, their distant locations emphasizing their relatively recent origins.

We stress that the tidal stability limit is calculated for an intruding satellite with no intrinsic strength of its own. It tells us where a liquid satellite, or one made up of disconnected bits of gravel, would be pulled apart by tidal forces. The stronger a satellite, the less it is affected by tides. As an extreme example, consider the space shuttle or any other low-orbit artificial Earth satellite. These operate well inside the stability limit, yet they are not pulled apart by tides. On the other hand, a loose tool left floating in the shuttle bay will drift away under tidal forces and be lost in space. This is what it means to be inside the Roche stability limit.

Another way of looking at the stability limit is to think of it as the distance within which individual particles will not come together under their own gravity to form a larger body. If a very large disk of particles once surrounded Saturn, for instance, it is easy to imagine them coalescing to form individual satellites everywhere except inside the stability limit, where they might remain to this day as the ring particles. In such a case, the ring particles would probably all be very small, and they would retain their primitive composition.

RING FORMATION BY SATELLITE BREAKUP

In the breakup hypothesis of ring formation, we might imagine a satellite or even a passing comet coming too close and being torn apart by tidal forces. There is an even more likely scenario, however, in view of the very heavy bombardment that took place early in the history of the Saturn system. We have already suggested that a number of small inner satellites might have been broken up during this early bombardment, leaving fragments such as the co-orbital and shepherd satellites. If a satellite were disrupted inside the tidal stability limit, it would be unable to reform itself, and the fragments would spread out into a ring. In this case, we would expect a wide range of particle sizes, including a few large fragments several kilometers across. The embedded satellites found in the rings of Saturn may be examples of such fragments. If the presence of many such bodies in the size range of 1–10 km could be confirmed, this would also argue for the breakup hypothesis.

How much material is actually present in the four ring systems we are discussing? This is difficult to determine, since what we see is primarily the smaller particles, while most of the mass may be contained in just a few particles (or moonlets) near the upper end of the size distribution. This caution about unseen large particles aside, however, it has proved possible to measure the mass of the main B Ring of Saturn from Voyager data. Since the spiral density waves depend on the gravitational interactions among ring particles, their behavior is sensitive to the mass of material present. Within the B Ring, there is nearly 1 ton of material per square meter spread throughout the 20-m thickness of the ring. Summing over the entire ring system, we obtain a total mass of about 10^{18} kg. The mass of the Uranus rings is guessed to be more than a thousand times less, and that of the Jupiter rings more than a million times less, than the Saturn rings. The mass of Neptune's rings is not well determined. Estimates for the masses of the rings were given in Table 16.1.

The total mass of the saturnian rings of 10^{18} kg is equal to that of an icy satellite about 250 km in diameter—about the same size as the larger co-orbital satellite, Janus. This mass is therefore consistent with the idea that the rings might have resulted from the breakup of a typical inner satellite near the tidal stability limit. Alternatively, it is easy to imagine there being this much leftover material that never formed a satellite.

Although the presence of small moons appears to provide a gravitational explanation for much of the ring structure, it does not answer intriguing questions about the lifetime of rings. Calculations show that if the rings were made of only small particles—the kind of particles that we deduce from Voyager data—they would not survive for more than a few hundred million years. In that time the ring particles would be eroded away by meteoritic impacts and sputtering by charged particles in the magnetosphere. Thus, either the rings are relatively young or they are renewed from within by the continuing breakup of kilometer-sized particles, just as the near-Earth asteroids are renewed by collisions among larger objects in the main asteroid belt. So far, these kilometer-size ring particles remain conjectural because they are too small for the Voyager cameras to detect.

We are left with the mystery of the different chemical compositions of the four known ring systems. Saturn seems to make the most sense, since both its rings and its satellites are made of the same icy material. Jupiter can be understood also when we remember that icy bodies, either satellites or ring particles, could not have survived near the planet early in its history, when it radiated more heat than at present. The dusty rings of Jupiter represent the results of continuous erosion from rocky satellites. In fact, even the bright rings of Saturn must include some non-icy material. Voyager observations show subtle color differences between the C and B Rings as well as between discrete ringlets in the B-Ring system and their immediate neighbors. Ground-based observations that record spectra of the entire ring system demonstrate that the rings are absorbing more ultraviolet light than pure ice would. Perhaps some silicate or organic dust is mixed in with the ice, as one would indeed expect if these rings represent the disruption of an icy object like a small satellite.

Uranus and Neptune remain perhaps the most strange. Their large satellites are icy, but both the rings and the small inner satellites are made of (or at least coated by) very dark, presumably carbonaceous material. The association of the dark satellites with the dark rings suggests that the rings might have been formed by the breakup of one or more inner satellites. This possibility is supported by some models for the evolution of these systems: The inner satellites may once have been even closer to their parent planets but may have moved out to their present positions during the lifetime of the solar system in response to tidal forces.

16.7 QUANTITATIVE SUPPLEMENT:
THE TIDAL STABILITY LIMIT

The tidal stability limit, also called the Roche limit, is the distance from a planet at which an unconsolidated satellite (one made of liquid, for example) will be disrupted by tidal forces. We can calculate this distance by finding the balance between the self-gravitation of an object that holds it together and the differential gravitational forces from the planet that tend to pull it apart.

For the sake of simplicity, consider the situation for two identical spherical particles, each with mass m and radius r, that are just in contact with each other. The gravitational force between them is

$$F_g = G \frac{m^2}{4r^2}$$

These two particles will be at the tidal stability limit when the differential gravitational force on them is equal to F_g. Suppose they are in orbit about a planet of mass M at a distance d from the center of the planet. If they remain aligned as they orbit so that one is at a distance from the planet of $d + r$ and the other is at $d - r$, the differential force on them is

$$F_d = G \frac{Mm}{(d-r)^2} - G \frac{Mm}{(d+r)^2}$$

When we equate the differential force to the self-gravitational force we obtain

$$G \frac{m^2}{4r^2} = GMm \left(\frac{1}{(d-r)^2} - \frac{1}{(d+r)^2} \right)$$

Suppose that these are two particles of a planetary ring. Farther from the planet than this limit, the gravitational attraction between them can allow them to merge into a larger particle that can in turn continue to grow. But inside this tidal stability limit, the particles will not stick together because of their mutual gravitational attraction.

The precise distance at which these forces balance depends on the densities of the particles and, to some extent, on their exact shapes. The term *Roche limit* is usually applied to the case considered by Roche, in which the orbiting particles have the same density as the planet itself. Algebraically, we can derive it from the above equation by expanding the right side and simplifying, to obtain

$$G \frac{m^2}{4r^2} = GMm(4rd_R^{-3})$$

Solving for d_R, the Roche limit, for the case where the densities of planet and particle are the same, we have

$$d_R = 2.5R$$

where R is the radius of the planet.

SUMMARY

The outer planets resemble miniature solar systems, with complex systems of satellites and rings. The typical medium-size satellites of Saturn and Uranus are similar in size to the largest asteroid, Ceres, and are composed of about half rock and half ice. They show little evidence of internal geologic activity, and their surfaces have been heavily cratered by impacts. But there are interesting exceptions. Enceladus has a geologically active surface and appears to be generating the faint E Ring of Saturn. Iapetus is a two-faced satellite, with its leading hemisphere made of dark material of mysterious origin. Miranda is a mishmash of strange geologic features. Also intriguing are the co-orbital satellites Pandora and Prometheus, which perform a strange orbital dance just outside the main rings of Saturn.

Each giant planet also has a ring system made up of small particles orbiting within the planet's tidal stability limit. The particles forming the bright rings of Saturn vary in size from 10 m to dust. The rings of Neptune and Uranus are made of very dark material, and most of the particles span a more limited size range, from a few centimeters up to tens of meters, while the tenuous ring of Jupiter, the E and G Rings of Saturn, and the wide ring of Neptune are little more than bands or ribbons of smoke-size particles. In the case of Jupiter, these particles must consist of dust eroded from the inner satellites. For Saturn, water ice should be the dominant constituent of the small particles as it is for the larger ones. The dark ring particles surrounding Uranus and Neptune probably derive their low reflectivity from organic matter.

Structurally, the rings display a remarkable diversity. Within the rings of Saturn, by far the best studied, we see examples of eccentric rings, kinky rings, and braided rings. In some places the boundaries of the rings are established by shepherding or embedded satellites, and in some cases resonances with more distant satellites play this role. Other resonances stimulate spiral density waves that propagate across the rings, generating a complex structure of thousands of ringlets. All the rings of Uranus and all but one of Neptune's rings are confined to narrow ribbons. The outer Epsilon Ring of Uranus has two shepherding satellites, as does the narrow F Ring of Saturn, but no such shepherds have been found for the other narrow rings of Uranus or Neptune. The outermost, or Adams, ring of Neptune displays four concentrations of material, called arcs, that resemble sausages on a string.

The rings of Uranus were discovered by observations of a stellar occultation, and this technique has proved to be very powerful for studies of the other ring systems as well. Occultation observations from and with spacecraft have permitted resolution of detail within the rings to scales of a few tens of meters.

We are still not sure of the ways in which the various rings originated, but tidal forces and impacts must have played important roles. At least two possibilities exist. Perhaps the rings within the tidal stability limit represent material that was prevented from forming a large satellite by the presence of strong tidal forces. The alternative option, which is currently more favored, suggests that a satellite near or within the tidal stability limit was broken apart by one or more impacts and the fragments were unable to re-accrete owing to tidal forces. In the case of distant rings such as Saturn's E and G Rings, one must invoke a local source that produces very fine particles because these rings must be relatively young.

KEY TERMS

co-orbital satellite	forward scattering
cryovolcanism	shepherd satellites
eccentric ring	spiral density wave
embedded satellite	tidal stability (Roche) limit

REVIEW QUESTIONS

1. Make four separate charts showing the satellite systems of the giant planets. Use the radius of each planet as the unit of distance and plot the positions of the satellites relative to their planets. What similarities and differences do you see in the different systems?

2. What is so strange about Enceladus? How would you try to explain this strangeness? Consider both Io and Europa as you develop your answer.

3. Describe the appearance of Iapetus. Our own Moon also exhibits a hemispherical asymmetry. How does the variation in the appearance of the surface of our own Moon differ from the asymmetry on Iapetus?

4. Compare the four outer planet ring systems. What do they have in common? What are the most striking differences?

5. Describe the individual particles in the rings of Saturn. Do they all have the same composition? How do these particles compare with those found in the other three systems? Why are they different?

6. Compare the discoveries of the rings of Uranus and Neptune. Why was the interpretation of the Neptune results so difficult? What are "arcs" and where are they found?

7. Explain how the occultation technique is used to probe ring structure. What exactly is measured? What determines the resolution? How do ground-based and spacecraft data compare?

8. Ring structure is determined in part by satellite interactions. Explain the roles of shepherd satellites, embedded satellites, and satellite resonances. Which of these produces what structures in the known ring systems?

9. What is the tidal stability limit? How is it possible for artificial satellites to circle the Earth inside this limit?

10. Describe alternative ways in which rings can form. Would you expect frequent formation of short-lived ring systems early in solar system history? Explain.

Quantitative Exercises

1. Find the distance in kilometers above the surfaces (or cloud tops) corresponding to the tidal stability limits for Mars, Earth, Saturn, and Neptune.

2. Make a chart showing the positions of the A, F, G, and E Rings of Saturn and the orbits of Atlas, Pan, Pandora, and Prometheus with respect to the Saturn tidal stability limit. What happens if you use a density of 0.7 g/cm³ for Saturn and 1.4 g/cm³ for the satellites?

3. Repeat Exercise 2 for the Neptune system, plotting rings and inner satellites. Do you think there may be some significance to the fact that arcs are found only in the Adams Ring and not in Leverrier or Galle? Explain.

4. The NASA Space Shuttle typically orbits the Earth at an altitude of about 200 km. Is this inside or outside the Earth's Roche limit? How does the orbit of Phobos relate to the Roche limit of Mars?

5. Compare the angular speed of Jupiter's small satellite Leda in its orbit around Jupiter with the angular speed of the entire Jupiter system as it moves in its orbit about the Sun. Will these two motions ever cancel each other so that Leda appears stationary? Explain.

Additional Reading

Elliot, J., and R. Kerr. 1984. *Rings: Discoveries from Galileo to Voyager.* Cambridge, MA: MIT Press. A well-written survey of planetary rings prior to the Voyager 2 encounters with Uranus and Neptune; the product of a collaboration between a scientist and a science journalist.

Miner, E. 1990. *Uranus: The Planet, Rings and Satellites.* New York: Wiley. Comprehensive review of the Voyager results, at a moderately technical level.

Miner, E., and R. R. Wessen. 2002. *Neptune: The Planet, Rings, and Satellites.* New York: Springer Verlag. Companion volume to *Uranus* (above).

Rothery, D. A. 1999. *Satellites of the Outer Planets: Worlds in Their Own Right,* 2nd ed. Oxford: Clarendon Press. A well-illustrated compendium of information about the larger satellites of the outer planets.

17 ORIGIN OF THE PLANETARY SYSTEM

Shiva, the Hindu god of destruction and rebirth, is shown here in his form as the cosmic dancer, Nataraja. The Hindu concept of the recurring cycle of life, death, and rebirth is a good analogy for the formation of a planetary system from the gas and dust produced in the interiors of multiple generations of stars.

In this book we have discussed several dozen worlds, some in considerable detail. These planets, satellites, asteroids, KBOs, and comets display an incredible diversity of composition and history. Yet they were all apparently formed at about the same time, from the same primordial solar nebula that gave birth to the Sun. Exactly how this happened is not yet clear, but there is nothing in our current models that would make the origin of planets a rare event. It appears that the formation of planetary systems is a natural part of star formation, opening the possibility that there may be many such systems around other stars and that some of these systems may be similar to our own. These ideas are supported by recent observations of disks around many young stars, disks that are probably even today forming their own planetary systems.

If we could understand the fundamental processes that formed the planets we know, we could move beyond these speculations to a rigorous prediction of how probable it is that these same processes produced other planetary systems like ours. Conversely, if we could find Earth-like planets orbiting other stars, their properties would help us to understand the origin and evolution of our own system.

Having stated what we would like to do, we must admit right away that it is not yet possible to do it. We are unable to work backward from the wealth of data on the present state of the solar system to derive a unique, detailed description of how the system began. Neither can we work forward from a theory of star formation to the production of a solar system with all the properties we find in ours today. Even the most basic part of the process, the formation of the Sun itself, is only partly understood. Instead of a unique and all-encompassing theory, we must work with a collection of reasonable explanations for those properties of our solar system that seem especially basic. As we will discuss in Chapter 18, we can finally begin to test these ideas against the proliferating discoveries of other planetary systems, most of which are startlingly different from our solar system.

17.1 BASIC PROPERTIES OF THE PLANETARY SYSTEM

No theory of planetary formation can deal with the wealth of detail about individual objects presented in this book. To begin, therefore, we should try to identify the really basic properties of the system that must be explained by an acceptable theory of its origin. We have made a list of 13 of these properties in Table 17.1. In the subsequent discussion, we refer to them as Fact 1, Fact 2, and so forth.

The "facts" from Table 17.1 are used in the following sections to constrain theories of the origin and early evolution of our planetary system. Of course, the concepts that emerge must be consistent with other data as well, particularly the detailed chemical and isotopic composition of the planets and the primitive objects in the system, such as the comets and meteorites. In addition, any theory of origins must make use of the increasingly detailed models for star formation that astronomers are developing from improved observations and applications of theoretical astrophysics.

There are other items that we could add to the list. We may wonder about the large mass of Jupiter, for example, or the fact that the distances between pairs of outer planets are so large. You should think about whether the model we develop is capable of addressing other phenomena that may have seemed especially puzzling when they were presented in earlier chapters.

17.2 THE LIFE OF A STAR

At the beginning of the twentieth century, many astronomers thought that the planets had been formed through a remarkable accident: the near-collision of the Sun with another star. Since then, we have come to realize that stars form in a spinning cloud of dust and gas that we have called the *solar nebula* in the case of our own system. Furthermore, we have located many companions, both dwarf stars and giant planets, orbiting nearby stars. The abundance of such companions strengthens the idea that single stars do not form alone. Finally, we now know that the ages of the Sun and of the planetary system are approximately the same. As we have seen, the Moon, the meteorites, and Earth all formed 4.5 billion years ago. Astrophysicists who study stellar evolution give this same value for the age of the Sun. From all of these arguments, we conclude that the Sun and the planets probably formed together from a common source of material (Fact 1).

STELLAR NURSERIES

On a galactic timescale, the Sun is a relative newcomer. We are not among the latest arrivals, however. Stars that are more massive than the Sun have much shorter lifetimes. Their internal temperatures are hotter, and their nuclear fires burn with greater intensity. The bright blue-white stars that dazzle us at night are all younger than the Sun. Some of them are only a few million years old. For example, the stars in the Pleiades, a familiar cluster of stars in winter skies, came into existence after the dinosaurs had disappeared from Earth (Fig. 17.1).

The most luminous and massive of these stars are destined to explode as **supernovas,** generating a very special group of elements that can be created only under the unique conditions that briefly occur during these cosmic cataclysms. Atoms of

Table 17.1   Facts that any theory of origins should explain

1. The most ancient age recorded in the solar system is less than 4.6 billion years, even though the Galaxy that contains the solar system is much older than this. Furthermore, many meteorites share this common age (what we have called "the age of the solar system").

2. The planets all move around the Sun in the same direction that the Sun rotates and nearly in the plane that passes through the Sun's equator.

3. Although the Sun has 99.9% of the mass in the solar system, the planets have 98% of the system's angular momentum.

4. The inner planets, which are composed primarily of the cosmically rare silicates and metals, are smaller and denser than the outer planets. The giant planets, in contrast, have a more nearly cosmic (or solar) composition, and their satellites, with the exception of Io and Europa, have lower densities, indicative of water ice and other volatiles.

5. The asteroids, which represent a composition intermediate between the metal-rich inner planets and the volatile-rich outer solar system, are located primarily between the orbits of Mars and Jupiter. Within the main asteroid belt, there is a distinct gradient in composition, with the outermost asteroids being richest in volatile elements and compounds.

6. The primitive meteorites are composed of compounds representative of solid grains that probably formed in a cooling gas cloud of cosmic (solar) abundance at temperatures of a few hundred Kelvins. Some of these grains even predate the formation of the solar system.

7. Comets, like some outer-planet satellites, appear to be composed largely of water ice, with significant quantities of trapped or frozen gases like carbon dioxide and methane, plus silicate dust and dark carbonaceous material.

8. Specific isotopic ratios established prior to solar system formation are preserved in objects formed at different distances from the Sun.

9. Volatile compounds (such as water) have reached the inner planets even though the bulk composition of these objects suggests formation at temperatures too high for these volatiles to form solid grains.

10. Despite the general regularity of planetary revolution and rotation, Venus, Uranus, and Pluto all rotate in a retrograde direction (although their revolution is normal).

11. All the giant planets have systems of regular satellites orbiting in their equatorial planes. They resemble miniature versions of the solar system itself, with the addition of rings, which the Sun does not possess, unless you consider the asteroid belt and the Kuiper belt as ring systems circling the Sun.

12. All four of the giant planets, Jupiter, Saturn, Uranus and Neptune, have one or more highly irregular satellites—either in retrograde orbits or with high eccentricities.

13. All of the giant planets are enriched in heavy elements equivalent to 10–15 Earth masses, all have atmospheres rich in hydrogen and helium, and all except Uranus are radiating substantial quantities of heat from their interiors.

these elements are shot into interstellar space and become incorporated in the next generation of stars and planets. All the elements we have found elsewhere in the cosmos are present on Earth, but the proportion in the interstellar gas and dust of these manufactured heavy elements increases over time.

Even younger stars exist, and some are being formed today. Star formation takes place in clouds of interstellar gas and dust, the most famous of which is in the direction of the constellation Orion (Fig. 17.2). A typical interstellar cloud in which star formation is occurring has a mass hundreds of times greater than that of the Sun. It is composed primarily of hydrogen and helium, the predominant elements in the stars that it will spawn. The other elements that can be studied appear to be present in roughly the same proportions as they are found in the Sun and other young stars, exactly as one would expect (see Table 2.2).

What may seem surprising, however, is the richness of the molecular chemistry that takes place in these clouds. In addition to simple compounds like methane and ammonia, a large array

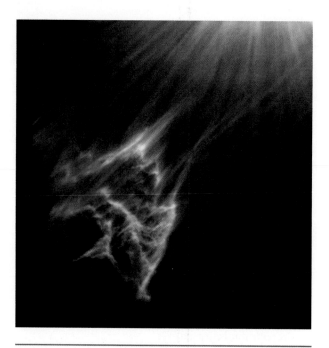

Figure 17.1 Ghostly nebula in the Pleiades star cluster. In this HST image, bright young stars are illuminating clouds of interstellar dust and gas as they move from the place where they formed a few million years ago.

A STAR IS BORN

One of the first things to notice about stars is that many of them are members of multiple systems. Doubles are almost as common as singles. In principle, a double star could have planets in stable orbits if those orbits either are close to one of the components or are at a large distance from both of them. In practice, we don't know whether Earth-like planets will form under such conditions, so we will concentrate our attention on the much simpler case of single stars, like the Sun.

We start with a slowly rotating cloud that may itself be part of a much larger complex such as one of the giant clouds in Orion. At some point the cloud begins to collapse (Fig. 17.2). Perhaps some gravitational instability has been created in its interior by a random coming together of some of the material, or possibly a nearby star has exploded as a supernova, seeding the cloud with short-lived radioactive elements and sending out shock waves that begin to compress the cloud.

of molecular species is continually being formed. A list of those known at the time of this writing would include more than 120 entries. Among the more interesting interstellar compounds are ethyl alcohol (CH_3CH_2OH), the essential ingredient of liquor, beer, and wine; formaldehyde (CH_2O), good for preserving corpses; and hydrogen cyanide (HCN), a deadly poison that is also a vital compound in experiments designed to simulate chemical evolution on the primitive Earth. New molecules are constantly being discovered as astronomers use more sensitive radio and submillimeter telescopes and study new segments of the radio spectrum.

It is interesting to compare a list of interstellar molecules with the molecules found in comets (see Tables 6.2 and 6.3). They have many species in common: H_2O, CO_2, HCN, CH_3OH, OH, and others. As we noted in Chapter 6, some scientists think that the icy nuclei of comets contain unaltered interstellar material, trapped from the interstellar cloud during the earliest stages of the formation of the solar system.

Figure 17.2 Star-formation region. This HST photo of the Triffid Nebula shows dense areas of dust and gas that are bathed in the ultraviolet light of nearby stars. The thin protuberances are collapsing clouds that may be forming both a star and a circumstellar disk.

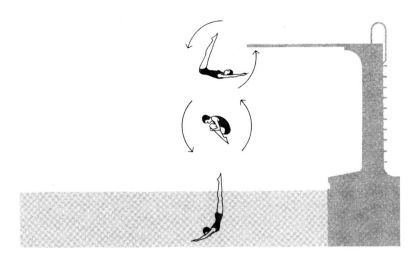

Figure 17.3 Conservation of angular momentum. A high-diver who contracts her body in the direction perpendicular to her spin axis will increase her rate of spin. Her angular momentum remains constant as she falls, and the angular momentum depends on the product of her rate of spin and the square of her size in the direction perpendicular to her axis of spin.

The collapse is possible as long as the energy of motion (internal pressure) of the gas in the cloud is less than the gravitational energy represented by the mass of the cloud and the distance through which it collapses.

As the collapsing cloud becomes smaller and denser, three things happen: (1) its rate of rotation increases; (2) it flattens into a disk; and (3) it heats up, especially near the center. The heating is the conversion of gravitational energy to thermal energy (like the main source of internal energy for Jupiter). The increase in rotation rate results from conservation of angular momentum (see Section 4.1): As the mass of the cloud comes closer to the center, the angular velocity must increase to keep the momentum constant (Fig. 17.3). The more rapid spin in turn causes the material to flatten into a disk. This disk is the critical feature of the classical model for planet formation.

The time for the collapse to form a disk is remarkably short, less than a million years. At first, only the gas and dust that start with relatively little angular momentum fall directly to the center. Material that is on orbits that do not intersect the central portion of the collapsing cloud falls toward the equatorial plane, where it meets material coming in from the other direction. The kinetic energy of this colliding material is dissipated as heat. The disk then heats up to an extent that is dependent on the local opacity. In the thin outer parts of the disk, the gas and dust can radiate energy to space much more easily than in the center, where a spherical condensation forms. Thus, a temperature gradient develops in the disk, with the center growing hotter than the outer regions.

At the center of the disk, the temperature continues to rise until it finally reaches approximately one million degrees, where nuclear fusion of deuterium into helium can occur (see Section 2.4). Now compression stops until the deuterium is exhausted. For stars with masses above about 70 times the mass of Jupiter, the collapse then continues until the internal temperature rises to about 10 million degrees and the complete proton–proton chain fusing hydrogen into helium is initiated. At this stage, this spherical assemblage of matter "turns on" and begins its life as a star. Close to the forming star, the opacity is high because of the high density, and the dissipated energy is also high because of the gravitational attraction of the central condensation, which is on its way to becoming a star. Thus, disk temperatures near 1 AU may reach 1500 K. Stars with masses less than 70 Jupiters simply begin a slow process of cooling down. We shall discuss these objects (called brown dwarfs) in the next chapter.

MASS AND DIMENSIONS OF THE DISK

While we are rapidly gaining a good idea about the dimensions of the original solar nebula through observations of disks around forming stars, there is some dispute about the mass of the original solar nebula and of the disk itself. The central condensation must have had a mass approximately equal to that of the present Sun, but what about the disk?

We can gain an idea of the minimum amount of mass that must have been present if we ask how much material of cosmic composition would be required to make all of the present planets. The idea behind this calculation is that the solar nebula started with cosmic composition and then the individual planets formed from it, with compositions that depended on the local temperature and the availability of solar nebula gas. For example, if it had been cooler close to the Sun, and if enough hydrogen and helium had been present, perhaps massive planets like Jupiter and Saturn might have formed there.

To discover what these hypothetical planets would have been like, we can perform the thought experiment of adding hydrogen and helium to the existing planets until the ratio of these light elements to a key heavy element like silicon or iron is the same as it is in the Sun. The masses of the resulting planets are indeed similar to those of Jupiter and Saturn. Thus, we conclude that the initial disk must have had a mass roughly equal to at least 3% of the solar mass. In this case, the minimum mass for the entire nebula is about 1.03 solar masses.

More likely, planet formation is not so efficient, and there was probably much more material available that has since been lost by the blowing away of the nebula or by gravitational ejection of larger objects after the planets formed. Scientists generally adopt a value for the entire nebula of about 1.1 times the mass of the Sun (corresponding to a disk mass of 0.1 solar mass). The maximum mass of a disk before it becomes unstable is about 0.3 solar masses.

DIRECT OBSERVATIONS OF CIRCUMSTELLAR DISKS

Throughout the twentieth century, while astronomers tried to understand the processes that took place when our solar system formed, they were frustrated by their inability to discover and study the circumstellar disks that they assumed must exist around stars that are just being born today. The closest active star-formation regions are associated with the nearby Orion arm of our Galaxy (see Fig. 4.1). The distance of the Orion arm is about 1300 LY (light years) away. At that distance, the angular diameter of our solar system (including the Kuiper belt) would be less than 1 arcsecond, which has been until recently the best angular resolution obtainable with ground-based telescopes. Thus circumstellar disks, just like craters on Mars, were beyond the reach of conventional astronomy.

The advent of the Hubble Space Telescope (HST) in the 1990s brought the Orion cloud and other star-forming regions within reach, and today ground-based telescopes can achieve a resolution as good as or better than HST. Astronomers can also observe at infrared and millimeter wavelengths, which are more effective at penetrating the thick dust that shrouds newborn stars.

Figure 17.4 illustrates several of the newly discovered circumstellar disks in the Orion region. The diameters of these disks range from 50 to 1000 AU, embracing the dimensions of the Kuiper belt in our system. Presumably they include material near the outer edges that will eventually be blown away from the mature system. Some also show intriguing hints of dust depletion within the inner 50 AU or so of their star. It is tempting to hypothesize that planets (or at least planetesimals) are already forming in the inner regions and thereby sweeping up most of the dust, to leave a donut-shaped disk. Motions in the disks have also been inferred from infrared and submillimeter observations, and the majority of the disks have masses estimated as 0.01–0.1 solar mass—consistent with the estimate given above for the mass of the disk that gave birth to our own solar system.

17.3 THE PROBLEM OF ANGULAR MOMENTUM

Stars are born at the centers of spinning disks of gas and dust. The disk is the endpoint of what we call the primordial solar nebula, while deep in the interior of the central condensation, nuclear reactions

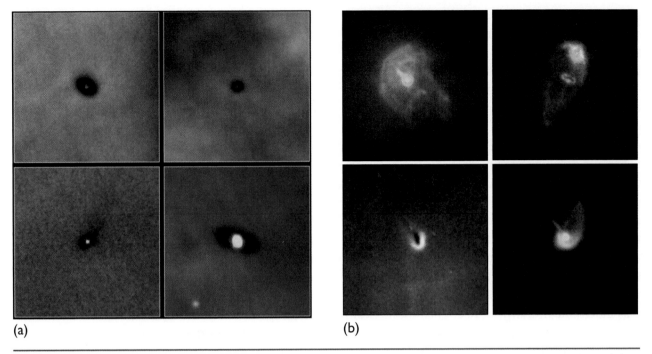

(a)　　　　　　　　　　　　　　　　　　　　(b)

Figure 17.4 **(a)** Circumstellar disks photographed by HST. **(b)** Disks in destruction. These HST photos show disks that are being blasted by ultraviolet light from nearby hot stars; probably the disks will be destroyed before a planetary system can form.

are beginning to convert hydrogen to helium. This is probably a good place to stress again that the exact sequence in which events occurred in the early nebula is not known. Thus, some of the processes we will describe in the next section must have started before the angular momentum transfer that we discuss here was completed.

ANGULAR MOTION OF THE SOLAR NEBULA

The disk of material revolves around the forming star in the same sense that the star itself spins on its axis—both reflect the original slow spin of the interstellar cloud as it began to contract. The disk must also be symmetric about a plane passing through the star's equator. This configuration provides a ready explanation for Fact 2: All the planets revolve around the Sun in the same sense in which the Sun rotates, and they move approximately in the Sun's equatorial plane. Planets forming out of this disk will have exactly that configuration and motion. It was considerations such

as these that led Immanuel Kant and Pierre Simon Laplace to propose (independently) nebular theories for the origin of the solar system as early as the eighteenth century.

The main reason these early nebular theories were challenged was the attention given to Fact 3: Why do the planets have more angular momentum than the Sun? Or, to pose this question in the framework of the nebula theory: How could the Sun slow down by transferring angular momentum to the disk and hence the planets?

It is enlightening to put this problem in context. It turns out that all stars with masses at least 15% greater than the Sun's rotate much more rapidly than our star, whereas all stars with comparable or smaller masses exhibits slow rotation rates like our Sun. Furthermore, if one calculates the rotation rate the Sun should have, given conservation of angular momentum as the original cloud collapsed, this rate turns out to be similar to that of the more massive and rapidly spinning stars. It is also the rate that would result if the present angular momentum of the planets were put back into the Sun. We conclude that the planets have the correct angular momentum for the system as it formed, and that some process must have slowed down the rotation of the Sun.

There is a second aspect of this angular momentum transfer. The protostar in the center quickly attracts all the gas and dust with low angular momentum, but there are parts of the collapsing cloud following orbits that do not intersect the center as we described already. This gas and dust fall onto the disk instead, providing both mass and angular momentum to the disk. Yet we know that in the end, most of the mass will be in the star. The problem is to understand how some of the disk material loses angular momentum and "falls" into the central condensation. The process is as much one of transport of angular momentum within the disk as it is of transfer of momentum from the star to the disk.

MAGNETIC BRAKING

One solution to this angular momentum problem invokes magnetic braking. Here one imagines the magnetic field of the Sun (or proto-Sun) moving through the disk of material around it as the Sun rotates. Since the material in the disk is following orbits defined by Kepler's laws, it is orbiting the Sun more slowly than the Sun is rotating. The situation is analogous to the interaction of the plasma in the Io torus with the rapidly spinning magnetic field of Jupiter (see Sections 13.6 and 15.4). The material in the inner part of the disk is ionized. The Sun's magnetic field encounters stiff resistance from this plasma as the Sun rotates, slowing down the spin. When the material in the disk is later dissipated, it takes the excess angular momentum with it.

This hypothesis has several problems. It does not explain why only stars with low masses exhibit slow rotation, since some stars with large masses have strong magnetic fields, far stronger than the Sun's. A way out of this dilemma would be to postulate that only low-mass stars form with disks. But we now know that disks are not uncommon around young stars with masses greater than the Sun's. For example, the star Beta Pictoris, which we will discuss in Section 17.6, has twice the mass of the Sun and yet is surrounded by a clearly visible disk of material. Furthermore, this process does not explain how material in the inner disk loses angular momentum, as it must if it is to make its way into the center of the nebula.

THE SOLAR WIND: BLOWING THE PROBLEM AWAY

A second hypothesis for slowing down the solar rotation involves the solar wind. Recall that the gases in the outer fringes of the solar atmosphere have enough energy to escape into space, flowing steadily outward through the solar system at speeds of about 400 km/s. The amount of matter lost in this way is a tiny fraction of the Sun's total mass, yet it is carrying angular momentum with it, and this momentum is lost by the Sun. Furthermore, observations of very young stars with masses like the Sun's indicate that they generate particularly intense stellar winds shortly after they form. While it is difficult to make an accurate estimate of this effect, it may be sufficient to account for the present slow rotation of the Sun.

An especially appealing aspect of this hypothesis of solar wind braking is its natural explanation of the mass dependence of stellar rotation. Only stars with masses comparable to or less than that of the Sun have the proper atmospheric structures to produce steady stellar winds. Hence, the same strong wind that ultimately clears residual gas and dust from the disk can slow down the rapidly rotating star that generates it. More massive stars will not produce such winds, so they will continue to exhibit rapid rotation.

Attractive as this idea is, it is still just a hypothesis. We need more observations of real disks to see what actually takes place in them. For example, in a real disk of gas and dust, there will be complex gravitational interactions among different components as soon as any clumping of material destroys the disk's homogeneity. Large protoplanets forming in the disk can clear annular gaps in the regions where they form. Remembering the gravitational interactions of satellites with the material in Saturn's rings, we can expect similar resonance clearing and generation of spiral density waves in the disk of material surrounding the star. These interactions themselves can transfer angular momentum, as indeed they must, if the disk itself is to evolve.

TURBULENCE IN THE DISK

Scientists today who are trying to understand the formation of the Sun and planets have concluded that much of the required angular momentum transfer must take place within the disk, not just between the disk and the protostar. The orbiting molecules of gas can transfer mass and angular momentum within the disk, although it was not possible to calculate detailed models until the advent of modern computers.

Convection in the nebular gas, the same process that causes winds on Earth, can produce turbulent eddies that can be particularly important for momentum transfer in the inner part of the disk where densities are relatively high. Just how important this process is will depend again on the local opacity to thermal radiation. A low opacity in directions perpendicular to the disk will allow rapid cooling and hence damping of convection.

Other processes are primarily gravitational. Clumps of material within the disk interact with each other and with their surroundings in complicated ways that are able to transfer angular momentum outward, thus allowing the gas and dust to move inward. As (or if) planets or protoplanets start to form in the disk, the situation becomes even more complex. Figure 17.5 illustrates a computational model for a disk with one large protoplanet. The newly forming planet sweeps out a clear lane and induces waves in the surrounding disk. These waves contribute to the transfer of angular momentum within the disk.

A complete model for the disk/nebula must take all these effects into account. It will trace the flow of momentum as well as mass and energy

Figure 17.5 Model for a circumstellar disk. This computer simulation shows the empty lane and spiral density waves that result as one giant planet is forming. The false colors represent regions of differing density in the disk.

within the disk and determine the degree of mixing from the hot inner nebula to the cold outer regions. The effects should be revealed in the composition of comets, about which we still know far too little to draw any definite conclusions.

17.4 EVOLUTION OF THE DISK: CONDENSATION, ACCRETION, AND DISSIPATION

We return now to the evolution of the material in the disk. How is it possible to get from this flattened mixture of dust and gas to a system of planets, satellites, and comets moving through nearly empty space? Of course, the space is not completely empty, since we have the solar wind and the myriad dust grains produced by collisions in the asteroid belt and the dissipation of comets. Compared with the solar nebula, however, the present interplanetary space is a pretty good vacuum, better in fact than the best vacuum we can produce in our laboratories. We must try to reconstruct the processes of agglomeration and dispersal that led to this condition.

EARLY HISTORY

The first thing to remember is that temperatures in the disk were not uniform. The initial condensation would generate heat as the gravitational energy of the extended cloud was converted to thermal energy. Most of the mass in the collapsing cloud became concentrated toward the center, converting enough potential energy to raise the internal temperature of this central condensation above the ignition point for thermonuclear reactions. The material in the outer part of the disk was collapsing only in the vertical direction, as particles in the cloud sank to the midplane (the plane that ran through the middle of the disk), forming a dense sheet of material. The disk was heated by the gravitational energy of the infalling material. Additional local sources of heat from short-lived radioactive elements and electromagnetic discharges were also available in the disk.

As the central condensation increased and the dust in the nebula gravitated to the midplane, the entire nebula warmed up. Temperatures above 2000 K were reached in the disk near the Sun, according to present estimates. This means that the dust in this region was vaporized and the inner nebula was totally gaseous. Subsequent cooling of this hot gas allowed the condensation of molecules and the formation of grains again, but with some sorting of condensates according to the decrease of temperature with distance from the Sun (Fig. 17.6). This temperature-dependent condensation was one of the primary causes of the chemical fractionation of material in the planetary system. Calculations indicate that the collapse of the cloud to a disk with an infant Sun took about 10 million years. At this point, the young Sun provided a source of energy to keep the temperatures high in the inner parts of the disk, inhibiting the condensation of most volatiles in the region now occupied by the inner planets.

This scenario provides a partial explanation for Facts 4 through 7. The inner planets are deficient in light elements compared with the cosmic distribution. Our model suggests that the inner planets formed at relatively high temperatures, at which volatile compounds of light elements

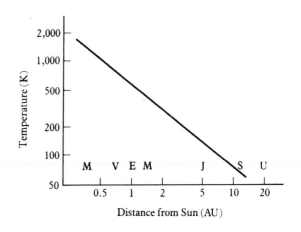

Figure 17.6 Condensation temperatures in the solar nebula disk. This graph illustrates one model for the decrease in temperature with increasing distance from the Sun. The positions at which the various planets formed are marked.

(H_2O, CO_2, etc.) could not condense. Acquisition of these volatiles by adsorption is much less efficient, although still adequate to produce oceans of water, for example. The increasing fraction of volatiles in the solid objects we find today as we move through the asteroids to the outer solar system and the comets roughly follows the temperature gradient in the original solar nebula.

When we reach Uranus, a further change in the composition of solid objects occurs. The larger satellites of Uranus, Neptune's Triton, and the Pluto-Charon system all exhibit densities significantly higher than those of the icy satellites of Saturn. It is not yet clear what change in composition is responsible for this. Perhaps the more distant objects contain a higher proportion of oxidized carbon, especially CO_2, than the Saturn satellites, which formed at relatively higher temperatures. Solid CO_2 has a higher density than H_2O, and it would also use up some of the available oxygen that might otherwise form water. The net result would be a higher ratio of rock to ice than we find in the Saturn system and hence satellites with higher densities.

There is thus a reasonably good agreement between the decrease in temperature with increasing distance from the Sun in the original nebula and the chemical fractionation we observe today, as we move from the rocky inner planets to the ice-rich and gas-rich outer planets. This agreement is attractive, but it implies a segregation of material in the early nebula that was not rigorously maintained. There must have been some mixing, as solid, condensed material from the outer part of the disk (large orbital radius) reached the inner part (small orbital radius) because it was moving in eccentric orbits. This allowed material formed under one set of pressure and temperature conditions to be mixed with material from another environment.

This radial mixing explains Fact 9: Despite the temperature gradient in the solar nebula, some volatile constituents found their way to the surfaces of the inner planets, yet this is consistent with Fact 8. Although solids were mixed to some extent, the high-temperature chemistry that took place in the inner nebula did not wipe out isotope anomalies throughout the disk. Condensed matter from the outer parts of the nebula has preserved its original isotopic signatures.

The early vaporization of the interstellar grains in the inner part of the nebula helps us to understand the age cutoff in Fact 1: Despite the comparatively young age of the Sun, compared to about 10 billion years for the Galaxy, no objects large enough to have their ages determined are older than 4.6 billion years. In other words, we have not yet detected solid interstellar material in sufficient quantity to determine its age, although there are clear indications of the presence of interstellar carbon and silicon compounds as microscopic grains in some meteorites and interplanetary particles. As mentioned earlier, interstellar grains may also be present in KBOs and comets, which formed (and remained) at low temperatures in the outer nebula. If we could obtain large enough samples from comets for study in the laboratory, we should find ages older than the 4.6-billion-year limit.

CONDENSATION IN THE SOLAR NEBULA

It is possible to calculate what compounds and minerals would form from this cooling cloud by investigating various possible chemical reactions among the elements that were present. Such calculations indicate that there will be a temperature sequence in which various compounds form (Fig. 17.7).

As the hot gas cools, one of the first things to condense is spinel, a magnesium-aluminum compound with the formula $MgOAl_2O_3$. Further cooling leads to the condensation of a wide variety of silicates, which must have appeared as dust

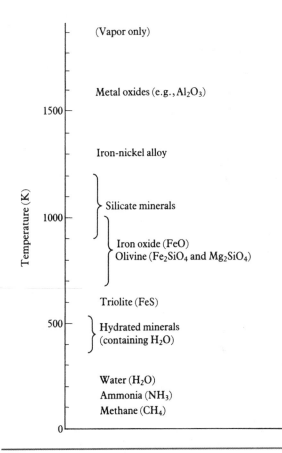

Figure 17.7 The chemical condensation sequence in the solar nebula, showing the primary chemical species that would be expected to form in a cooling gas cloud of cosmic (solar) composition. As the temperature drops, each material forms and condenses into liquid droplets or solid grains. The calculations upon which this diagram is based, originally carried out by John Lewis of the University of Arizona, presume that the grains remain in chemical equilibrium with the residual gas. A slightly different condensation sequence results if the grains are immediately swept up into planetesimals and taken out of contact with the gas, but the general results are the same. These solid grains then become the building blocks of the planetesimals and ultimately of the planets themselves.

grains in the inner disk. Since silicon and oxygen are both abundant elements, silicates should have become the dominant minerals in the inner disk, where temperatures were too high for the condensation of volatiles.

The inner boundary for the condensation of H_2O ice in the solar nebula moved steadily inward as the nebula cooled, ending at a distance of about 4 to 5 AU from the Sun called the **snowline.** The snowline is a major threshold, since in a cosmic mixture, H_2O ice is potentially far more abundant than rock, even though the reverse is true on Earth and the other inner planets. Recall that in the Sun and stars oxygen is approximately 18 times as abundant as silicon. This means that in the solar nebula, H_2O should be approximately nine times as abundant as SiO_2. After all possible oxides and silicates are formed, there is still plenty of oxygen left to make ice. The mass of this ice is about 1.5 times the mass of rock and heavy elements in a solar mixture. As we saw in Section 13.7, crossing the threshold for the formation of ice enables the growth of giant planets. But how are planets formed from the tiny grains of ice and dust that condense in the cooling nebula?

AGGREGATION OF GRAINS

The next step toward making planets after condensing grains from the gases in the solar nebula is to get these grains to accumulate into larger aggregates. This process is aided by the fact that at any given point in the nebula, most of the material should be in nearly identical orbits, so the relative velocities between nearby grains will be small. This enhances the chances that two colliding grains will be able to stick together, although it tends to reduce the total number of impacts. Sticking is especially likely if the grains are fluffy.

The evidence we have for the appearance of grains in the solar nebula suggests that they indeed had this desirable fluffy characteristic. Figure 4.11 shows one of the grains, probably of cometary origin, collected in high-altitude flights by a project designed to capture interplanetary particles. These should be the most unmodified

grains remaining in the solar system, since they have never been heated. These fluffy particles form clumps that will gradually grow through additional collisions. While this growth is occurring, the solid material continues to settle toward the midplane of the nebula.

Within the disk, the fluffy clumps of grains stick together to form loosely bound objects as large as 10 km across; we call these objects *planetesimals*, the building blocks of the planets (see Section 4.1). The planetesimals are large enough to exert a significant gravitational attraction on each other, leading to further rapid growth.

TIMESCALES FOR THE FORMATION OF PLANETESIMALS

The hypothesis that the planets formed from intermediate-size objects called planetesimals was originally the work of Russian theorist V. A. Safronov, who investigated this problem in the 1950s and 1960s. Previously, many scientists had thought that planets probably had formed directly from the solar nebula and its tiny dust grains, but we now recognize in the asteroids and comets surviving examples of planetesimals, and increasingly sophisticated calculations support Safronov's hypothesis.

It is possible to estimate the length of time required for the various stages in the formation of planetesimals. The theoretical models suggest that it may have taken as long as 100,000 years for 1-mm particles to reach the midplane of the disk, whereas 1-cm particles would settle out in only 10 years. The initial loose condensations of dust should evolve into solid objects with diameters of about 10 km in roughly 1000 years.

Confirmation of these surprisingly short timescales is provided by investigations of the abundances of isotopes formed from radioactive parents with very short half-lives. Such studies indicate that the solid material that produced the original asteroid parent bodies formed from the solar nebula in less than a million years. Most of the meteorites that scientists have examined were formed from these grains within the first several million years after the nebula collapsed to a disk.

The Earth and the other solid planets were essentially complete in less than 50 million years from the time the solar nebula formed.

FROM PLANETESIMALS TO PLANETS

The inner planets and the cores of the giant outer planets formed from the continued accretion of planetesimals. Rapid gravitational growth quickly used up much of the solid material in the disk to form thousands of objects with diameters of a few hundred to a few thousand kilometers. These accumulations, spanning the size range from the largest asteroids up to objects larger than the Moon, are called **planetary embryos.**

Once most of the mass was in the form of planetary embryos, collisions were much rarer. But they were also more violent, with impact speeds of several kilometers per second. As the gravitational reach of the embryos extended farther and farther, the violence of the impacts increased. When a small embryo struck a large one, the material was accreted and the large embryo grew, but when two embryos of comparable size collided, they were likely to be fragmented into thousands of smaller pieces. A competition developed between the processes of accretion and the processes that break growing planets apart (Fig. 17.8).

Several scenarios have been calculated for the evolution of the embryos in the inner solar system. The models of George Wetherill, in particular, suggest that a great many objects of lunar size or larger were formed in addition to the growing protoplanets themselves. According to this picture, the late stages of the accretionary process were characterized by impacts of incredible violence as these lunar-size objects bombarded the planets (or, in some cases, were gravitationally ejected from the solar system). The possibility of a few discrete, random impacts of this magnitude can explain some of the peculiarities of the planets, as we shall see in the next section.

Gradually, over tens of millions of years, the protoplanets emerged from this chaotic situation. It seems inevitable that they were heated by the energy of these collisions once again above the melting point of silicates. Thus, the planets probably differentiated as they formed, and no primitive material survived the violence of the accretionary process.

Figure 17.8   Mimas, a satellite that was almost destroyed by an impact. This inner satellite of Saturn bears a crater so large that, if the impact had been only a little more energetic, the satellite would likely have been broken up and dispersed.

In the outer part of the solar system, more mass was present in the form of condensed water ice, and the velocities of the planetary embryos were slower. As a result, the level of violence was less and the protoplanets grew larger. Calculations indicate that within less than a million years, cores as large as 10 Earth masses could form and begin to attract the gas as well as the solid matter in the disk. At first these protoplanets were too hot to hold much gas, but after several million years they cooled to the point where large quantities of hydrogen and helium were acquired, at least by Jupiter and Saturn. In the case of Uranus and Neptune, it appears that the nebular gas dissipated before much of it could be captured by these planetary cores. This process of planetary formation, in which a solid core forms first and then captures an envelope of gas from the surrounding disk, is often referred to as a "bottom up" process, to distinguish it from the "top down" process by which stars are formed. Detailed calculations show that the accretion of planetary embryos must have taken place while the disk was still present, and when perhaps most of the mass was still in the form of dust and gas.

DISSIPATION OF THE NEBULA

At some point after the planets formed, a short-lived but highly intense blast from the solar wind cleaned out the remaining gas and small dust. Observations of young stars have revealed that such intense winds are a natural part of the star-formation process. In some cases, they take the form of bipolar outflows, in which two streams of gas oriented in opposite directions are seen to emanate from young stars. Most of the observations revealing these winds are indirect, involving the interpretation of Doppler-shifted lines in stellar spectra that indicate the presence of moving clouds of gas. Occasionally, however, it is possible to record the effects of jets from a young star as they clear away material from the larger cloud of dust and gas from which the star (and associated planets, if they are present) has formed (Fig. 17.9). The total lifetime of the gaseous phase of the solar nebula is estimated to be about 10 million years.

17.5 TERRESTRIAL PLANETS AND THE IMPORTANCE OF IMPACTS

We know from our detailed study of the impact history of the lunar surface that all of the inner planets must have continued to accumulate mass for the next 600 million years. Although this late bombardment effectively created the surfaces we see today on all small objects, it contributed only a small percentage of the total mass of an individual object. It is just the icing on the cake. In contrast, massive impacts that took place in the first 100 million years had a profound effect on the ultimate structure of the solar system.

A RETURN TO CATASTROPHISM

The science of terrestrial geology took a great leap forward in the nineteenth century by adopting the principle of **uniformitarianism.** As stated by one of

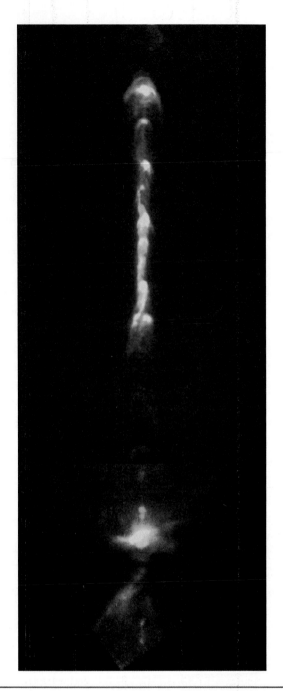

Figure 17.9 Jets of material from a young star. This HST photo shows a pair of 12-LY-long jets of gas blasted into space from a system of three young stars. At late stages in the star-formation process, most stars expel material, often in such narrow jets.

its foremost promoters, English geologist Charles Lyell, this principle holds that the landscapes we find around us were formed by processes we see operating today. This idea, which was so important for understanding the great age of the Earth and the slowness with which it changes, failed to account for the landscape we find on the Moon, as discussed in Chapter 7. The processes acting today could not supply the observed number of craters in the total history of the solar system. It is necessary to invoke bombardment by the fragments left over from the formation of the planets to account for the craters we see on every ancient solid surface in the solar system, from Mercury out to Triton.

A careful reader of this book will have realized that impacts have had even greater effects than the formation of craters and basins on planetary and satellite surfaces. A giant impact between Earth and a Mars-size planetary embryo was probably the reason we have the Moon. A massive impact with Mercury apparently stripped away a large fraction of that planet's mantle, leaving Mercury with a disproportionately massive metallic core. Giant impacts on Venus, Uranus, and Pluto are the most likely causes of their strange rotations (Fact 10). At least some of the planetary rings must have resulted from impacts with objects already in orbit about the planets.

Contrary to normal human intuition, which likes to think of nature as smoothly following fundamental laws, it is becoming clear that the early solar system was a chaotic place, with objects randomly smashing into each other—sometimes destructively, sometimes creatively, and often with long-lasting and fundamental effects. We regard the recognition of the importance of impacts for the origin and evolution of planets and of terrestrial life as a major contribution of solar system exploration to the history of ideas.

This bombardment is not yet over, as the extinction of the dinosaurs, the lunar crater Tycho, and Saturn's G Ring make abundantly clear. To this extent, impacts do not flagrantly violate Lyell's uniformitarian hypothesis; they too are processes that are ongoing today, but fortunately they don't happen very frequently. They also no longer occur on the large scales that formed the Moon or tipped Uranus on its side. Asteroids and comets can still alarm us, however, which is one reason we should try to understand how they formed.

ORIGIN OF ASTEROIDS

The small objects now in independent orbits about the Sun represent material that was left over from the planet-formation process. The asteroids appear to have several sources. Some may simply represent planetesimal growth that was halted at an early stage, when planetesimals had reached a maximum size of 100–1000 km. The gravitational effect of nearby Jupiter removed most of the planet-forming material from the region of the asteroid belt before it could accumulate into a full-size planet (Fact 5). Perhaps the gravitational influence of Jupiter was also responsible for Mars being so much smaller than Earth and Venus.

Some of the dark C-type asteroids might actually be the carbon-rich remains of huge comet nuclei, whose volatiles were gradually exhausted once they were captured into their present orbits. There are indications that some of these may still have water ice in them. Other objects in the asteroid belt may be the remains of some of the giant collisions that seem to have been an important part of the history of the inner planets. In other words, the main asteroid belt may be a kind of quasi-stable dumping ground, rather like the little pile of debris that sometimes accumulates at the exact center of the intersection of two busy highways.

WAS THERE A LATE HEAVY BOMBARDMENT?

The lunar cratering record suggests to many geologists that there was a discrete burst of impacts—the late heavy bombardment—about 600 million years after the formation of the Moon. Where did these projectiles come from, at a time when the inner solar system had been largely cleared of debris left over from the accretionary process? Many scientists think they may have originated near the orbits of Uranus and Neptune.

Because the distances between the outer planets are so great, it required a much longer time to clear the debris from this region. Quite possibly, the late heavy bombardment consisted largely of icy planetesimals, gravitationally scattered inward by Uranus and Neptune. Perhaps this was the time when most of the volatiles that make up the terrestrial planet atmospheres arrived (Fact 9 again). At the same time, an even larger number of such objects should have been scattered outward, where they became the comets of the Oort comet cloud.

In addition to contributing volatiles to the atmospheres of the terrestrial planets, the cometary ices would also have been added to the "primordial soup" of organic chemicals accumulating in the oceans of Earth, ultimately to become part of organisms as life began. You may be pleased to realize that some of the carbon, nitrogen, and oxygen atoms in your cells were formed in stellar interiors, shot into interstellar space by supernova explosions, trapped in the ices of comets and delivered to the surface of the primitive Earth. Some of the water you drank today may also have come from comets, but the alcohol in your beer was produced by local yeast.

This apparently benign contribution of comets to the course of life on our planet must be contrasted with a darker side of the bombardment of the early Earth. During the first 600 million years of our planet's history, impacts by objects of 10-km diameter or larger were far more frequent than they are today. Recall that one 10-km object striking our planet 65 million years ago caused widespread extinctions of life on Earth. Now imagine the planet being bashed by still larger projectiles. Some scientists think that the results of such impacts would have been sufficiently severe that they speak of the "impact frustration" of the origin of life during this period. We will return to this hypothesis in Chapter 18.

17.6   GIANT PLANETS AND THEIR RINGS AND MOONS

From the model we are presenting for the formation of the solar system, it seems natural to expect that the giant planets with their systems of satellites and rings must have formed in a manner similar to that of the Sun and its retinue of planets. In the outer reaches of the solar nebula, enough matter was present to form giant protoplanets surrounded by disks that we call *subnebulae*. The satellites evidently developed from these disks, followed or accompanied by the formation of rings. Unlike stars, however, the giant planets formed "from the bottom up" as we have seen. This is an important difference because it leads to an enrichment in heavy elements in the planets compared to the stars.

GROWTH OF THE GIANT PLANETS

The giant planets consist of cores of rocky and icy material surrounded by huge envelopes of gas. All four of these planets have cores amounting to 10–15 Earth masses. Calculations show that in the case of Jupiter, this core could have formed in less than 1 million years. These massive cores are apparently the key to the further attraction of gas from the solar nebula to make up the hydrogen and helium atmospheres observed today.

As the rock and ice cores of the outer planets approached their maximum masses (Fact 13), they caused the surrounding gas in the nebula to collapse, simply from the strength of their gravitational attractions. Thus, the cores are now surrounded by atmospheres that are a mixture from two sources: the gases released by the solid material that formed the core and a primordial mix of gases collected from the solar nebula.

It is this mixture that explains the observation that all four giant planets have atmospheres that exhibit a solar abundance of hydrogen and helium, while methane is enriched to various degrees. The hydrogen and helium were contributed by the solar nebula, while the extra methane was produced by the growing core as one of the components of its outgassed atmosphere. On Saturn, we find a slightly greater enrichment of methane than on Jupiter, where it is about three times the value expected from a solar mixture of the elements. Jupiter is the only planet where we know the global, well-mixed abundances of heavy elements other than carbon, and they, like the carbon in methane, are also enriched by the same factor of three. On Uranus and Neptune, methane is 25 to 35 times more abundant than a solar mixture would predict, in keeping with the much higher ratio of core mass to total mass exhibited by those two planets. For reasons not yet understood, Uranus and Neptune were not able to capture as much hydrogen and helium from the solar nebula as their more massive cousins closer to the Sun.

FORMATION OF SATELLITE SYSTEMS

Just as a disk of gas was left around the forming Sun, the matter that formed the giant planets produced both central condensations and disks. The disks shared the sense of rotation of the forming planets, so satellites forming in these disks orbited their planets in the same direction that the planets rotated. These became the regular satellites that we find today (Fact 11).

Within one of these disks (subnebulae), one could expect a temperature and pressure gradient similar to the one we just discussed for the solar disk (nebula). Near the planet, the density of gas would be greater and the temperature higher than at large distances. Obviously these protoplanetary disks never got as hot as the solar nebula near the Sun. Nevertheless, this change in physical properties with distance from the planet should have left some signature on the forming satellites.

In the case of Jupiter, the hottest of these planets because it was the largest, we can see the effects of this heat on the forming satellites when we examine them today. Recall that at Jupiter's distance from the Sun, water ice is stable, even over the lifetime of the solar system. Thus, we expect to find satellites that reflect the cosmic proportions of the elements, roughly half ice and half rock by mass, leading to densities of 1.9 g/cm^3. This is indeed the case with the two outer satellites, Callisto and Ganymede (Fig. 17.10).

Io and Europa, closer to Jupiter, have higher densities (3.0–3.6 g/cm^3), exactly as this scenario would predict. The corresponding proportion of water is no more than 10% on Europa, while Io has been baked out completely by tidal heating and the resulting volcanic activity. This low proportion of water on Europa can be understood in terms of higher subnebular temperatures than those in the regions where Callisto and Ganymede formed.

Figure 17.10 Collage of Jupiter and its four galilean satellites as photographed by the Galileo spacecraft. From top to bottom, the satellites are Io, Europa, Ganymede, and Callisto.

The satellite systems of Saturn and Uranus do not show a similar variation of satellite density with distance from the planet. Several of the inner satellites of Saturn actually appear even less dense than the half-ice, half-rock composition of Callisto and Ganymede. These satellites may be composed of fragments of the icy mantles of disrupted larger objects and therefore may be nearly pure ice. Another contributor to their low densities may be increased pore space resulting from reassembly of fragments. There may also be variations in other volatiles trapped in the ice. For example, such variations provide the most probable explanations for the absence of methane-nitrogen atmospheres on Ganymede and Callisto and the presence of such an atmosphere on Titan (see Section 15.7).

While Ganymede and Callisto formed at temperatures low enough for water ice to condense, they may have been too warm to trap methane and ammonia.

The irregular satellites must have formed differently. Early in the history of the solar system, some planetesimals could have been captured into stable orbits when they passed close to a giant planet. These captures could occur as a result of frictional drag when these objects penetrated the extended atmospheres (or subnebular disks) of the forming planets. The friction between the moving planetesimal and the gas that still existed in the planetary subnebula would cause free-roaming planetesimals to lose energy and enter orbits around the planets, providing a natural explanation for Fact 12: All the outer planets have one or more satellites in retrograde orbits. Satellites forming with the planets would not assume such orbits, but captured satellites could. In the inner solar system, the two satellites of Mars may also have been captured by the same process.

FORMATION OF RINGS

Why do some planets have rings while others don't? What determines the kinds of rings a given planet may have? These are questions that are now under active investigation. In Section 16.6, we discussed several scenarios for the origin and evolution of ring systems.

The general picture of rings as debris that did not accrete into satellites provides a possible explanation for the existence of the rings of Saturn. However, the close association of this ring system and all of the others with a number of inner satellites has led scientists to argue in favor of the impact-induced breakup of one or more satellites as a more likely cause. Since it is clear that the bombardment of inner satellites by infalling debris must have been very heavy in the first few hundred million years of planetary history, it seems inevitable that satellites close to the planet, if they existed, were at least heavily eroded if not completely disrupted by large impacts. The debris of such events, if it were inside the tidal stability limit, could not have pulled itself together gravitationally to reform the satellite.

If these ideas are correct, the rings did not form directly from a subnebula surrounding the giant planets. Instead, the ring systems are an indirect result of the formation of satellites, combined with the high intensity of impacts produced close to the planet by its massive gravitational field. In other words, they are collisional debris. This process may still be taking place. For example, if a passing comet broke up as it made a close approach to a giant planet, it might lose enough energy to be captured into an orbit about the planet (as happened in 1992 for Comet Shoemaker-Levy 9). Under some circumstances, this cloud of orbiting debris might evolve into a short-lived ring. The nongravitational forces produced as gas was suddenly liberated from freshly exposed surfaces could perhaps slow the comet sufficiently to allow the capture.

FORMATION OF THE COMETS AND KBOS

A large percentage of the icy planetesimals that were not incorporated into the giant planets were forced into orbits with very high eccentricities that either expelled them from the solar system or took them out to a radial distance of 50,000 AU from the Sun to form the Oort cloud of comets, as discussed in Section 6.5. The long-period comets we discover from time to time are coming from this distant region, where gravitational perturbations from the galactic tide, giant molecular clouds, and passing stars occasionally cause these icy nuclei to change their orbits, bringing them close enough to the Sun for us to discover them.

A second group of icy planetesimals exists just beyond the orbit of Pluto in the disk-shaped Kuiper belt, which is the main source of the short-period comets (see Section 6.5). The largest of the KBOs we have discovered are bigger than the largest asteroid, Ceres—and we have only begun to explore the Kuiper belt, so we really don't know what still awaits us at these great distances from the Sun. In October 2002, a new KBO named Quaoar with a diameter of 1200 km was announced. It is the largest object discovered in the solar system since Pluto. The KBOs apparently formed in the same region of space in which we find them today. Like the asteroid belt, these objects also have their own history of mutual collisions, so that we see a population in which most of the members are the survivors of such impact catastrophes. They represent the outer fringe of planetesimal formation in the solar nebula.

The KBOs are important not only as a source of the short-period comets but also as a proving ground for ideas about the evolution of the outer edge of the solar nebula. Why is there such an abrupt decline in mass going from Neptune to Pluto and the inner edge of the Kuiper belt? Was there much more material in the Kuiper belt originally than we find there today? These and other basic questions are under investigation as data come in slowly from observations of these faint, distant members of our system.

EVIDENCE FROM OTHER STARS

In addition to the disks that occur around young stars as part of the star formation process (see Fig. 17.4), there is another class known as **debris disks.** These are found around older stars that have already become stable members of the stellar community. The most visible example surrounds the nearby star Beta Pictoris, which has an age of about 20 million years (Fig. 17.11). Other less prominent disks have been detected around more familiar stars such as Vega, which is part of the famous triangle of bright stars (the other two are Deneb and Altair) that decorate the summer sky in the northern hemisphere. These stars are substantially closer and hence easier to study than the newly forming systems in the Orion arm.

These disks are composed of material produced by the mutual collisions of planetesimals. They are thus analogous to the collisional fragments we expect to exist in the Kuiper belt. The masses of such disks are typically quite small—on the order of a few lunar masses—although their dimensions may be a few hundred AU to 1500 AU in the case of Beta Pic. There are indications of density flutuations in the Beta Pic disk that may indicate the presence of planets. The complete ring, with a diameter of 80 AU, that surrounds the

Figure 17.11 The Beta Pictoris disk. The bright image of the star itself has been blocked leaving only a residual crescent of starlight.

star HR 4796A provides more convincing evidence for planets, as they are needed to shepherd the ring material, thereby preventing it from spreading (Fig. 17.12).

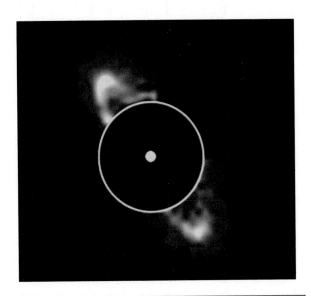

Figure 17.12 The star HR4795A is surrounded by a ring of debris that is about 10 AU wide, at a distance of approximately 40 AU from the star. This image was recorded with a special infrared camera mounted on the Hubble Space Telescope with a device to block most of the light of the central star.

17.7 QUANTITATIVE SUPPLEMENT:
ANGULAR MOMENTUM IN THE SOLAR NEBULA

The distribution of angular momentum in the solar system has always challenged astronomers investigating the origin of the planets. To see why, it is instructive to calculate a table of the angular momenta associated with the Sun, Jupiter, Saturn, and Earth, relative to the center of the solar system. For the Sun, this momentum is rotational and is given approximately by $0.1mR^2/p$, where m is the mass of the Sun, R is its radius, and p is its rotation period. For the individual planets, almost all of the angular momentum is contained in their orbital motion, given approximately by mr^2/p, where r is the radius of the orbit and p is the period required for one orbital revolution. With these formulas and the data given in the endpaper of this book, we calculate the numbers shown in Table 17.2. Examining these values, we see that there is much more angular momentum in the planets, especially Jupiter, than in the Sun. The disparity is even more dramatic if we consider the angular momentum per kilogram, obtained by dividing the numbers in the last column by the mass: For the Sun this is 2×10^{10}, while for Jupiter it is 2×10^{15} m^{2}s, or 100,000 times greater.

For comparison, we can calculate the angular momentum for a condensing solar nebula of about 1 solar mass (2×10^{30} kg). Let us assume that the original nebula has dimensions typical of interstellar distances—say 1 light year or about 10^{16} m. Its minimum rotation period corresponds to one orbit around the center of the Galaxy in 200 million years, or 6×10^{15} s. Then its angular momentum is that given by the formula above for the Sun, or about 3×10^{45} kg m^{2}s^{-1}, which is 100,000 times greater than the angular momentum of the Sun today. The angular momentum of the solar nebula per kilogram is, however, about 10^{15} m^{2}s, similar to that of Jupiter. Therefore, we can conclude that the planets have about the correct quantity of angular momentum, while the Sun is highly deficient. Hence, some process in the past must have robbed the Sun of its angular momentum, but most of this momentum was not transferred to the planets; it was simply lost from the system. As we have discussed, the solar wind seems likely to be the responsible agent.

SUMMARY

There is a consensus among astronomers that the solar system was born about 4.5 billion years ago from a collapsed fragment of an interstellar cloud. The collapse caused a disk to form, which is called the solar nebula. Dense interstellar clouds of gas and dust are observed widely in the Galaxy where they are recognized to be the birthplaces of stars. One approach to understanding the origin of the solar system is to study the young stars surrounded by disks that we can see around us today.

The solar nebula was the nursery of both the Sun and the planetary system. The Sun, forming in the center, attracted most of the mass and evolved into a self-sustaining star. The rest of the gas and dust collapsed into a turbulent disk with a strong

Table 17.2   Angular momenta of the Sun, Jupiter, Saturn, and Earth

	m (kg)	r (m)	p (s)	Momentum (kg m^{2}s^{-1})
Sun	2×10^{30}	7.0×10^{8}	2×10^{6}	5×10^{40}
Jupiter	2×10^{27}	7.8×10^{11}	4×10^{8}	3×10^{42}
Saturn	6×10^{26}	1.4×10^{12}	9×10^{8}	1×10^{42}
Earth	6×10^{24}	1.5×10^{11}	3×10^{7}	5×10^{39}

temperature gradient, hot at the center and cooling toward its outer rim. As the proto-Sun and its disk developed, angular momentum was transferred outward while mass flowed inward to the growing core. Modern telescopes can observe analogous circumstellar disks in regions where the formation of stars (and probably planets) is occurring today.

As temperatures dropped in the disk, solid grains condensed, sorted compositionally to some extent by the temperature structure in the disk. These grains rather quickly aggregated to build up solid objects called planetesimals. These planetesimals, tens to hundreds of kilometers in diameter, became the building blocks of the planets. The planets, growing by accretion and heated by impacts, differentiated as they formed.

For the first few hundred million years of solar system history, a great deal of solid debris remained between the planets, and impacts were much more frequent than they are today. The heavily cratered surfaces of the Moon and other solid objects throughout the system date from this time. A small fraction of the leftover icy planetesimals from the outer solar system contributed to the atmospheres of the terrestrial planets, while others were gravitationally ejected from the solar system or into the Oort comet cloud. The Kuiper belt, just beyond the orbit of Pluto, is a second reservoir of icy planetesimals. Both reservoirs still contribute comets that occasionally strike the inner planets today.

In the outer solar system, solid planetary cores formed that were large enough (10–15 Earth masses) to attract and hold part of the remaining gases of the solar nebula, thus creating the giant planets. These planets formed within their own subnebulae, which gave rise to their regular satellite systems. The rings may have originated in the same way, but it seems more likely that they are the product of later impact fragmentation of inner satellites. Ultimately the remnants of the original nebula were blown away by strong solar winds from the young Sun.

This picture of the origin of our system is consistent with the basic properties listed in Table 17.1. It also suggests that the formation of other planetary systems should be a common aspect of the origin of single stars.

KEY TERMS

debris disks	supernova
planetary embryos	uniformitarianism
snowline	

REVIEW QUESTIONS

1. Review Table 17.1. Do you agree that all of the "facts" listed there are fundamental in the sense that a theory of solar system origin must explain them before we can believe it? What facts would you add or delete?

2. Describe the current model for star formation. Do you expect stars to form soon near the Sun—just outside the Oort cloud, for example? Why or why not?

3. Explain how angular momentum is conserved as an interstellar cloud collapses to form the solar nebula. Why is there a problem in accounting for the difference in the angular momentum carried by the Sun and by the planets? What solutions could you suggest for this problem?

4. Make a chart depicting the various steps in the evolution of matter in the solar nebula disk, from its initial dispersed state to the formation of planets. Describe what happens in each step and give the length of time required.

5. What role (if any) did icy planetesimals play in the formation of the inner planets?

6. How did comets form? What are the two major reservoirs of comets in the solar system and how do we distinguish between them? Or, to put it differently, when a comet is discovered, how do we know from which reservoir it originated?

7. Explain the steps in the formation of satellites and rings. Why doesn't Venus or Earth have rings or extensive satellite systems?

8. What kind of observational evidence exists to support the "nebula hypothesis" for star and planet formation? What kind of evidence is likely to accrue in the next decade?

QUANTITATIVE EXERCISES

1. The outer planets with their regular satellite systems offer us many opportunities for comparison with the solar system as a whole. Calculate the rotational angular momentum of Jupiter and the orbital angular momentum of each of the four galilean satellites. How do these values compare with the situation for the Sun and planets?

2. Jupiter and its galilean satellites are thought to have formed from the collapse of a protojovian nebula. Suppose this nebula had an initial radius of 2 AU and shared the orbital motion of Jupiter around the Sun. Use the same methods described for the solar nebula to calculate the angular momentum of this protojovian nebula, and compare your results with the angular momenta found in Exercise 1. What can you conclude about the loss of angular momentum in the jovian system? Why might the situation here differ from that of the Sun and planets?

ADDITIONAL READING

Hutchison, R. 1983. *The Search for Our Beginnings.* New York: Oxford University Press. Highly readable discussion of solar system origin with emphasis on cosmochemical issues.

Jakosky, B. 1998. *The Search for Life on Other Planets.* Cambridge: Cambridge University Press. An excellent introduction to both planetary science and astrobiology, this book includes several chapters on the origin of the solar system.

McSween, H. Y., Jr. 1997. *Fanfare for Earth.* New York: St. Martin's Press. A nontechnical summary of our understanding of the formation of the solar system and the origin and evolution of the Earth.

Ward, P., and D. Browlee. 2000. *Rare Earth: Why Complex Life Is Uncommon in the Universe.* New York: Copernicus. Provocative discussion by a geologist and an astronomer of what makes the Earth special, from its origin through the development of advanced life.

18 DISTANT WORLDS

This statue of Giordano Bruno is in the Campo de Fiori in Rome, the place where he was burned alive by the Catholic Church in 1600. Among the cited "heresies" for which he lost his life, Bruno believed in the plurality of worlds. He was right.

Comparison is one of the most powerful methods available to scientists in their quest to understand nature. By examining different manifestations of the same phenomenon, we can derive the general laws and principles that are at work. For example, finding that craters on the Moon had the same depth/diameter ratios as bomb craters on Earth helped scientists to deduce that explosive impacts rather than volcanism were responsible for these features of the lunar landscape. Newton discovered the law of gravity by noticing the effects of this force on the motions of the legendary apple and the Moon. The comparative approach to planetary science has been a central theme of this book.

For the first time in history, we are finally in a position to apply this same approach in our efforts to understand the origin and nature of the solar system. For decades scientists felt frustrated by the fact that ours was the only planetary system they knew. The danger in this situation is the seductive assumption that our family of planets is typical, that all other planetary systems must be like ours. It was obvious that we would never be able to gain a full understanding until we could compare different planetary systems to see what common properties they exhibited. These properties would help us determine which aspects of our own system are truly fundamental, requiring a common theory for their explanation, and which are random, the result of chance events. Do all systems consist of small planets close to their stars and massive planets farther out? Do the planets always revolve around the star in the same plane? Are comets and asteroids always "left behind" when solar systems form? There are certainly many questions we can ask. As we begin the twenty-first century, we can at last begin to find some answers to these questions, and the answers are very surprising!

18.1 THE SEARCH FOR PLANETARY SYSTEMS

The discovery of other planetary systems was one of the most important astronomical developments of the last decade of the twentieth century. Until then, we had to face the possibility that our solar system is a fluke of nature. It seems that there have always been two competing schools of thought about our existence. On one hand, people we shall call pessimists believed that our system of planets was rare or even unique in the universe, with Earth the only inhabited world. On the other hand, the optimistic scientists were convinced that planetary systems must be common, and many went on to suggest that some of these planets must host other forms of intelligent life. Both sides supported their positions with observational evidence, which became increasingly sophisticated with time.

A decade ago, the pessimists could point out that despite many clever and careful attempts to find them, no planets had been discovered around any other star besides the Sun. However, this apparently definitive negative result was not as conclusive as it sounded.

CLASSICAL METHODS AND ARGUMENTS

The simplest method to search for other planets would be to look at stars through a powerful telescope and see whether they had any planets. This

was how Galileo found the moons of Jupiter, but this direct approach is unfortunately infeasible when it comes to other planetary systems. Even the nearest star is so far away that its planets—if it had any—would appear so close to the star from our distant vantage point that they would be lost in the glare from the star itself. A star is typically more than a million times brighter than any of its planets, and hence it is impossible to detect these planets with conventional telescopes. The pessimists readily conceded this fact but called attention to other telescopic searches that could detect planets indirectly without actually seeing them. These techniques had also failed to yield results.

The most widely used of these techniques is called the **astrometric method,** based on very careful measurements of a star's apparent position in the sky as the star moves through interstellar space in its orbit about the center of the Milky Way Galaxy. If it is accompanied by a giant planet like Jupiter, a star will follow a wavy trajectory because of the changing direction of the gravitational attraction of the planet as the star and planet move around their common center of mass. The **center of mass** of a system of two bodies is a familiar concept that governs the action of a child's teeter-totter. It is the point on a line connecting two masses from which the product of mass and distance is the same for both objects (Fig. 18.1). Decades of searches for tiny wobbles in the paths traced by stars had failed to find a single convincing example of a star whose motion was influenced by an unseen planet (Fig. 18.2). The pessimists concluded that the absence of wobbles indicated an absence of planets.

The optimists conceded the absence of an observable effect but suggested that the tiny displacement of the star caused by an orbiting planet might be too small to measure with the available technology. Instead, they used the theory for the formation of stars to suggest that stars normally form with a disk of material around them, and planets would then be expected to form within these disks just as our planets formed from the solar nebula. Sure enough, observations of young stars in dense interstellar clouds showed that

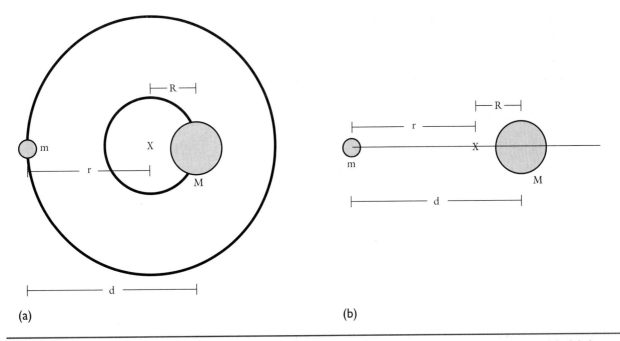

(a)

(b)

Figure 18.1 A planet and its star are both moving around the center of mass of the system (marked X): **(a)** face-on view, corresponding to 0° inclination of the orbits; **(b)** edge-on view, inclination is 90°. Note that the scale is grossly exaggerated!

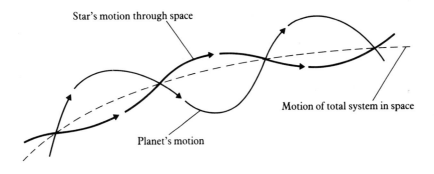

Star's motion through space

Motion of total system in space

Planet's motion

Figure 18.2 If a star has one or more sufficiently massive planets, it will move in a small orbit about the center of mass of this system. The center of mass will move along a smooth, elliptical orbit about the center of the Galaxy. However, the star, which is what we observe, will follow a wavy path back and forth across the orbit of the center of mass in response to the gravitational attraction of its planet.

almost all of these stars are surrounded by disks (Fig. 18.3). The optimists concluded that the formation of planets should indeed be a natural part of star formation, just as the theory predicted, and therefore planets should be rather common.

DISCOVERING PLANETS WITH THE DOPPLER EFFECT

This is how matters stood in the early 1990s, when a new technique was brought to bear on the problem. This technique uses the *Doppler effect* to search for unseen companions to stars (see Section

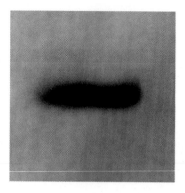

Figure 18.3 Here we see two views of the same protoplanetary disk (Orion 114-426) viewed edge-on against the bright background of the Orion nebula. The left hand image gives a view in visible light, emphasizing the dust in the disk. The right hand image in infrared light penetrates the dust somewhat, revealing the star forming in the center of the disk. The disk is nearly 1000 AU across

8.2). The spectrum of the star is recorded at high spectral resolution, so the absorption lines in the stellar spectrum are spread far apart. Then the wavelengths of these lines are measured with high precision relative to the wavelengths of absorption lines from a gas contained in a tube through which the star's light must pass (Fig. 18.4). If the star is moving freely through space by itself, the wavelengths of the absorption lines in its spectrum will not change with time. But if an unseen companion is orbiting the star, then the slight periodic motion of the star toward and away from the observer as the star and the planet pursue their orbits around their common center of mass will cause the star's absorption lines to shift periodically to shorter and longer wavelengths. These tiny changes in wavelength can be measured with respect to the stationary absorption lines of the reference gas.

The first attempt to use this method led to negative results on some 30 stars, further reinforcing the position of the pessimists. But subsequent efforts with higher precision have been spectacularly successful. The first extrasolar planet, orbiting star number 51 in the constellation Pegasus (abbreviated 51 Peg), was discovered in this way by Michel Mayor and Didier Queloz in 1995. Subsequent discoveries followed rapidly, with most coming from the team of Geoff Marcy and Paul Butler (Fig. 18.5). As of October 2002, over 100 new planets have been detected.

The Doppler technique can supply an impressive amount of information about any planetary system it discovers. First, by observing the repetitive cycles in which the absorption lines in the star's spectrum shift to longer and shorter wavelengths, astronomers can determine the planet's orbital period directly by measuring how long it takes for the changes in wavelengths to pass through a single cycle (Fig. 18.6). Second, as Isaac Newton showed long ago, the orbital period depends only on the planet's average distance from the star and the star's mass. Since the mass of a star can be determined from its spectrum, astronomers can then find the

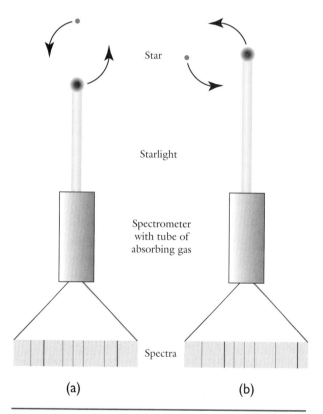

Figure 18.4 The combined spectrum of the reference gas (blue lines) and the star (black lines) is shown below the diagrams of the planetary system. When the star is not moving relative to the observer **(a)**, no shift in its absorption lines occurs. But when it moves away from the observer **(b)**, its absorption lines are shifted to longer wavelengths (highly exaggerated!).

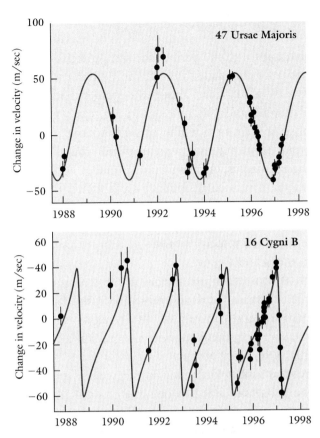

Figure 18.5 Michel Mayor (right) who discovered the first extrasolar planet around the star 51 Peg, together with Geoff Marcy (left) who immediately confirmed it. Both men have gone on to discover many more wonderful worlds.

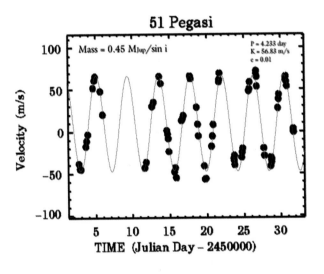

Figure 18.6 The radial velocity curve of the star 51 Peg. The sinusoidal shape provides clear evidence that a giant planet was circling it at a very close distance.

Figure 18.7 Velocity curves for two stars with planets in relatively long-period orbits. The vertical lines through the observed points show the measurement errors. The curve that best fits the data for 47 U Ma is nearly sinusoidal, revealing that the planet is in a nearly circular orbit. The curve for 16 Cygni B shows the distinctive signature of a much more elliptical orbit.

average planet-star distance using Kepler's third law. Third, by examining the rhythms with which the star's velocity changes, astronomers can determine whether the planet moves in a circular orbit; if it does not, they can measure the *eccentricity* of the orbit (how much it deviates from circularity).

The shape of the planet's orbit reveals itself in the rhythms by which the star's velocity varies during each orbital cycle. A planet in a circular orbit produces velocity changes at a steady rate, so that a graph of these changes versus time shows a sinusoidal pattern. In contrast, a planet moving in an elongated orbit makes the star change its velocity at different rates, first more rapidly (when the planet orbits closer to the star and perturbs the star with greater-than-average amounts of force, which produce larger velocity changes) and then more slowly (when the planet orbits at greater distances from the star and therefore perturbs the star with less-than-average amounts of force) (Fig. 18.7).

FINDING THE MINIMUM MASS

In addition to the period, the size, and the eccentricity of the planet's orbit, the Doppler method reveals a key fact about the planet itself: the planet's mass. The changes in the star's velocity observed along our line of sight to the star depend upon the mass of the planet that induces these changes. A more massive planet orbiting a star at a particular distance (a distance that we know from determining the planet's orbital period) will produce proportionately larger changes in the star's velocity, producing a larger amplitude in the velocity curve.

A complicating factor in determining the planet's mass, however, arises from the fact that astronomers who rely on the Doppler shift can measure only the component of the star's velocity that is directed toward us or away from us, and not the total velocity, which may well be greater if the orbit is inclined to our line of sight. As a result, the Doppler method measures only the *minimum mass* of the perturbing planet. The method determines the planet's mass multiplied by a factor that depends on the inclination of the planet's orbit, the angle by which the plane containing the orbit is inclined to the plane of the sky. This is the same as the angle between our line of sight and a line drawn perpendicular to the plane of the orbit (Fig. 18.8).

This factor (which is equal to the trigonometric sine of the angle of tilt) varies from 1, which occurs when the plane containing the orbit coincides with our line of sight (edge-on), all the way down to 0, when the orbital plane is perpendicular to the line of sight (face-on). The amplitude of the velocity curve is thus not directly equal to the mass of the planet, m. Instead, it is proportional to $m \sin i$, where sin is the sine function, which varies from 0 to 1, and i is the inclination angle of the orbit. It is $m \sin i$ that is derived from the Doppler shift observations. When $i = 90°$, $\sin i = 1$ and we can determine the mass exactly. When $i = 0°$, $\sin i = 0$, so we can't even detect the planet by this method because there is no component of its orbital velocity in our line of sight and hence no Doppler shift.

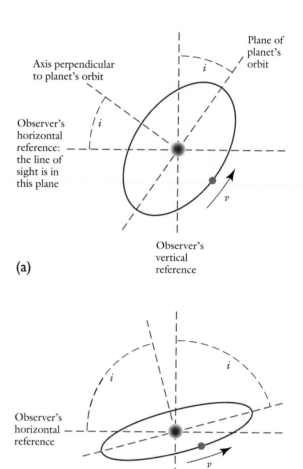

(a)

(b)

Figure 18.8 The orbits of the star and its planet are inclined with respect to the plane of the sky by an angle i, which we do not know. When i is small, as in **(a)**, the projected velocity that we measure, $v \sin i$, is a small fraction of the true velocity, v. When i approaches 90° as in **(b)**, $v \sin i$ approaches the true velocity, v.

If we assume that the inclinations of these systems are distributed randomly over the sky, we can show that statistically we expect that the vast majority of planets detected by the Doppler shift technique will have masses within a factor of two of the minimum mass, when $\sin i = 1$. Expressed in terms of the inclination angle, this is equivalent to concluding that the vast majority of these systems will have inclinations of 30° or greater. For example, an inclination of 45° would give a true mass equal to 1.4 times the derived minimum mass.

18.2 MEETING BRAVE NEW WORLDS

Table 18.1 lists the first 101 extrasolar planets that have been discovered. They are illustrated graphically in Figure 18.9. The table shows their minimum masses, their orbital periods, their average distances from their stars (orbital semimajor axes), and their orbital eccentricities. The full list, which contained no entries at the beginning of 1995, will be well over 100 planets by the time you read this. Almost all of these extrasolar planets orbit Sun-like stars that lie within 100 light years of the solar system. They all have minimum masses comparable to Jupiter's (about 300 Earth masses) or, for the least massive of them, to Saturn's mass (95 Earth masses).

Obviously we cannot know as much about these unseen worlds as we know about the giant planets in our own solar system. Yet even a cursory glance at Table 18.1 and Figure 18.9 reveals that these new planetary systems are radically different from the one we inhabit. Why do so many enormous planets occupy orbits that are closer to their stars than the distance of Mercury to the Sun? Indeed, the space within 4 AU around these stars that should be populated by small rocky inner planets like ours is instead the realm of giants!

Experience with our own system suggests that these planets, massive or not, should move in orbits that are nearly circular. Instead, we find many of them in elongated elliptical orbits, some with eccentricities approaching those of short-period comets!

THE RIDDLE OF ECCENTRICITY

The fact that so many of the new planets follow elongated paths has surprised astronomers. If objects form from condensations within a rotating disk of gas and dust (as described in Chapter 17), we would expect them to move in almost circular orbits. If, on the other hand, two or more objects form almost independently, still acquiring orbits around their common center of mass, then the orbital eccentricities could easily range from zero up to quite large values. This second scenario, which describes how astronomers think double-star and multiple-star systems formed, would lead to more elongated orbits than the rotating-disk model.

Some scientists argue that the large eccentricities for many of the extrasolar planets support the idea that the icy planetesimals in their cores did not form out of a circumstellar disk but rather had their beginnings in the interstellar cloud that preceded the formation of this disk. Most scientists prefer the dominance of the disk, however, and look for dynamical explanations for these highly elongated orbits. One possibility is that these eccentricities result from perturbations caused by the gravitational forces exerted in close encounters between the forming planets, one of which is then ejected from the system or forced into the star, while the other assumes a highly eccentric orbit. Recall that on a much smaller scale, such a two-body encounter has been invoked to explain the retrograde orbit of Triton (see Section 15.6).

ORBITAL INCLINATIONS

In discussing the determination of planetary masses with the Doppler method, we pointed out that there is no information about the inclination of the planetary orbits to our line of sight, which means that we can determine only a minimum mass for the planet. In that case, we are speaking of the inclination of the entire system. The orbital inclinations within each system, which we also don't know, are interesting in their own right.

In our solar system, all the planets except Mercury and Pluto have orbits lying in the same plane, and these two outliers have inclinations that are still rather small. As we saw in Chapter 17, the fact that all our planets have orbits in nearly the same plane has been one of the strongest arguments for the formation of planets from a circumstellar disk. When we consider other systems, we assume that here too the planets orbit their stars in the plane of the star's equator. When there are multiple planets in one of these systems, we again assume they all move in the same plane. Neither of these assumptions may be correct.

Table 18.1 Masses and orbital characteristics of extrasolar planets

Star Name	$M \sin i$ (M_J)	Period (days)	Semimajor Axis (AU)	Eccentricity
HD83443	0.35	3.0	0.038	0.00
HD46375	0.25	3.0	0.041	0.02
HD187123	0.54	3.1	0.042	0.01
HD 179949	0.93	3.1	0.045	0.00
BD-103166	0.48	3.5	0.046	0.05
HD209458	0.63	3.5	0.046	0.02
Tau Boo	4.14	3.3	0.047	0.04
HD75289	0.46	3.5	0.047	0.01
HD76700	0.19	4.0	0.049	0.00
51 Peg	0.46	4.2	0.052	0.01
HD49674	0.12	4.9	0.057	0.00
Ups And b	0.68	4.6	0.059	0.01
HD168746	0.24	6.4	0.066	0.00
HD68988	1.90	6.3	0.071	0.14
HD217107	1.29	7.1	0.072	0.14
HD162020	13.73	8.4	0.072	0.28
HD108147	0.41	10.9	0.079	0.20
HD130322	1.15	10.7	0.092	0.05
55 Cnc b	0.84	14.7	0.12	0.02
GJ86	4.23	15.8	0.12	0.04
HD38529b	0.78	14.3	0.13	0.28
GJ876 c	0.56	30.1	0.13	0.27
HD195019	3.55	18.2	0.14	0.02
HD6434	0.48	22.1	0.15	0.30
GJ876b	1.89	61.0	0.21	0.10
Rho Crb	0.99	39.8	0.22	0.07
55 Cnc c	0.21	44.3	0.24	0.34
HD74156b	1.55	51.6	0.28	0.65
HD168443b	7.64	58.1	0.30	0.53
HD121504	0.89	64.6	0.32	0.13
HD178911	6.46	71.5	0.33	0.14
HD16141	0.22	75.8	0.35	0.00
HD114762	10.96	84.0	0.35	0.33
Mercury	0.0002	88.0	0.39	0.21
HD80606	3.43	112	0.44	0.93
70 Vir	7.41	117	0.48	0.40
HD52265	1.14	119	0.49	0.29
HD1237	3.45	134	0.50	0.51
HD37124b	0.86	153	0.54	0.2
HD73526	3.63	188	0.65	0.52
Venus	0.0025	225	0.72	0.01
HD82943c	0.88	222	0.72	0.54
HD202206	14.68	259	0.77	0.42
HD8574	2.08	229	0.77	0.30
HD40979	3.16	260	0.82	0.26
HD150706	1.0	265	0.82	0.38
HD134987	1.63	265	0.82	0.37
HD169830	2.95	230	0.82	0.34
HD12661b	2.30	263	0.82	0.35
Ups Andc	1.90	241	0.83	0.28
HD89744	7.17	256	0.88	0.70
HD17051	2.12	312	0.91	0.15
HD92788	3.88	337	0.97	0.28
HD142	1.36	338	0.98	0.37

Star Name	$M \sin i$ (M_J)	Period (days)	Semimajor Axis (AU)	Eccentricity
Earth	0.0030	365	1.00	0.02
HD128311	2.63	414	1.01	0.21
HD28185	5.70	383	1.03	0.07
HD108874	1.65	401	1.07	0.20
HD4203	1.64	406	1.09	0.53
HD177830	1.24	391	1.10	0.40
HD210277	1.29	437	1.12	0.45
HD27442	1.32	415	1.16	0.06
HD82943b	1.63	445	1.16	0.41
HD19994	1.66	454	1.19	0.20
HD114783	0.99	501	1.20	0.10
HD147513	1.00	540	1.26	0.52
HD20367	1.12	500	1.28	0.23
Hip75458	8.68	550	1.34	0.71
HD222582	5.20	577	1.36	0.76
HD23079	2.76	628	1.48	0.14
HD160691	1.74	637	1.48	0.31
HD141937	9.67	659	1.48	0.40
Mars	0.0003	687	1.52	0.09
HD114386	0.99	872	1.62	0.28
16 Cyg b	1.68	798	1.69	0.68
HD4208	0.81	829	1.69	0.04
HD190228	3.44	1112	1.98	0.52
HD213240	4.49	951	2.02	0.45
HD114729	0.88	1136	2.08	0.33
47 Uma b	2.56	1090	2.09	0.06
HD10697	6.08	1074	2.12	0.11
HD2039	5.1	1190	2.2	0.69
HD13507	3.19	1318	2.30	0.13
HD50554	3.72	1254	2.32	0.51
HD216437	2.09	1293	2.38	0.34
HD136118	11.91	1209	2.39	0.37
HD196050	2.81	1300	2.41	0.20
HD37124c	1.00	1550	2.5	0.40
Ups And d	3.75	1284	2.52	0.27
HD106252	6.79	1503	2.53	0.57
HD12661c	1.56	1444	2.56	0.20
HD216435	1.23	1326	2.6	0.14
HD33636	7.71	1553	2.62	0.39
HD30177	7.64	1620	2.65	0.21
HD23596	8.00	1558	2.87	0.31
HD168443c	16.96	17770	2.87	0.20
HD145675=14Her	3.90	1775	2.87	0.37
HD72659	2.54	2185	3.24	0.18
HD74156c	7.46	2300	3.47	0.40
HD38529c	12.78	2207	3.71	0.33
Eps Eri	0.92	2550	3.39	0.43
HD39091	10.39	2280	3.50	0.63
GJ777A	1.15	2613	3.65	0.00
47 Uma c	0.76	2640	3.78	0.00
Jupiter	1.00	4345	5.2	0.05
55 Cnc d	4.05	5360	5.9	0.16

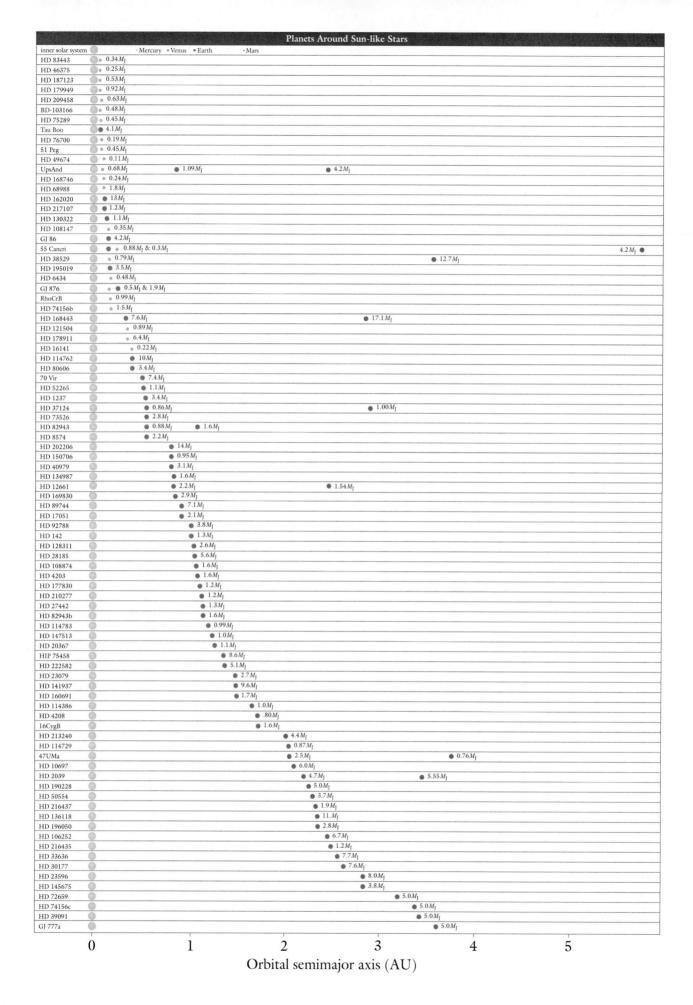

Planets Around Sun-like Stars

The system of HD 168443 provides an illuminating example (Fig. 18.10). At least two objects are orbiting this star, one of them at an average distance of 0.3 AU and the other at an average distance of 2.9 AU (see Table 18.1). The inner object has at least 7.6 times Jupiter's mass, making it an impressively large giant planet. The outer object, however, has at least 17 times Jupiter's mass and is probably a **brown dwarf,** an object too big to be a planet but with too little mass to ignite the nuclear reaction that would allow it to become a full-fledged star. In this case, one would expect that the brown dwarf was formed from the top down, like other stars, and therefore its presence in the system may be unrelated to the existence of a disk around the star. On the other hand, the giant planet should have formed from the bottom up, in just such a disk. So do these two objects have orbits that are in the same plane? It is not obvious that they should.

Looking for guidance in our own system (always somewhat dangerous!), we find that when objects have highly eccentric orbits (some asteroids, the long-period comets, some irregular satellites), these orbits are also often highly inclined to the equatorial plane of the Sun or the planet around which the small object revolves. So the highly eccentric orbits in Table 18.1 may also be highly inclined, which would again challenge classical models. On the other hand, they may not be so inclined! At this point, we simply don't know.

STARS AND PLANETS: A MATTER OF MASS

Now that we have introduced the observation of a brown dwarf, we can ask whether there is a continuous distribution of masses that reaches from the most massive stars down through stars like the Sun to dwarf stars and brown dwarfs and finally down to true planets. Brown dwarfs had been predicted long before they were found, as they are

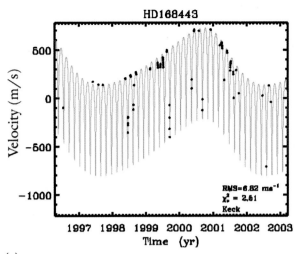

(a)

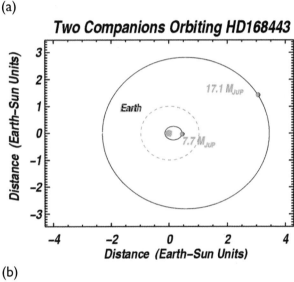

(b)

Figure 18.10 The star HD 168443 has two massive companions. The inner giant planet's $m \sin i = 7.5$ Jupiters, while the outer object's $m \sin i = 17$ Jupiters. The outer object is thus a brown dwarf, not a planet. The presence of these two fundamenatally different objects in the same system poses a serious puzzle (see text). **(a)** The velocity curve of the star, showing two superimposed sinusoidal patterns: The large amplitude, tightly spaced one is from the short period planet (P = 58.1 days, Semimajor axis = 0.3 AU) while the gently undulating pattern with a smaller amplitude and larger period is produced by the brown dwarf (P = 4.85 years, Semimajor axis = 2.87 AU). **(b)** An artist's impression of the system, with the Earth's orbit sketched in for reference.

Figure 18.9 (facing page) Graphical display of the extrasolar planetary systems known as of the middle of 2002. This diagram gives the orbital distances and approximate masses of the new planets ordered by the distance of the planet. (or of the nearest planet if there are several) from the star.

very dim objects, not easy to discover. But now many are known, and they have been observed in star-forming regions where the statistics for their occurrence can be compared with the statistics for ordinary stars. Similarly, we now know enough planetary systems to be able to assess the numbers of planets of different masses.

Remember that the basic difference between a star and a planet is their mass. Stars are sufficiently massive ($M > 70 \ M_J$) to produce central temperatures that are so high that nuclear reactions begin to convert hydrogen to helium, as described for the Sun in Section 2.4. A brown dwarf is not that hot but becomes warm enough to convert the deuterium it acquired from the interstellar cloud that spawned it into helium. This process occurs at lower temperatures than the first step in the proton-proton chain; it is in fact the second step (see Fig. 2.7). Every star will pass through this stage as it heats up, proceeding to the proton-proton chain if it has sufficient mass. That is why the Sun, for example, no longer contains any deuterium in its outer layers; the deuterium was consumed long ago. A planet is still less massive ($M < 13 \ M_J$) and is unable to extract energy from any nuclear reactions. That is why we find the original abundance ratio of deuterium to hydrogen in Jupiter's atmosphere (see Section 13.3).

If stars and planets formed in exactly the same way, we would expect the distribution of masses to be continuous as we move from larger to smaller objects. However, there is a distinct gap between the smallest brown dwarfs and the largest giant planets (Fig. 18.11). The classical concept that stars form directly from collapsing fragments of interstellar clouds while planets form from circumstellar disks is consistent with this distribution if the mass of material in the disk is not sufficient to produce a second star. One of the most important results from the discovery of new planetary systems is the observational confirmation of this fundamental difference between stars and planets. It is now clear that there truly is an upper limit to the size of planets at a few times the mass of Jupiter, and that the numbers of planets rise rapidly with decreasing mass.

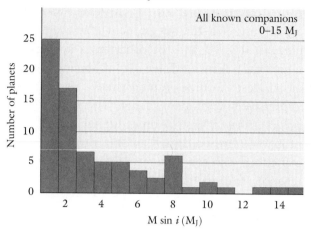

Figure 18.11   The number of planets of a given mass is plotted against the mass. The mass boundary between planets and brown dwarfs occurs at 13 M_J, a region of the diagram that is thinly populated. It appears that there is a pronounced increase in the number of planets with smaller masses, suggesting different formation processes for planets and brown dwarfs. The large number of small planets is especially striking when we consider that the Doppler technique is biased against discovery of low-mass planets.

THE HOST STARS

With a hundred planetary systems known, we can look at the nature of the stars that have planets and ask whether they are unusual in any way. First note, however, that not all stars have been searched for possible planetary systems. Stars with rapid rotation or turbulent atmospheres have spectral absorption lines that are too blurred for the Doppler technique to work. This restriction effectively excludes stars of high mass and luminosity. In practice, solar-type stars seem to be the best candidates for finding planets.

There is one distinguishing factor, however, that does seem to separate the solar-type stars with planets from others of the same mass and luminosity. The majority of stars with planets have unusually high abundances of the heavier elements, as revealed by their spectra. This difference has led to two interesting hypotheses. One is that planets are more likely to be born in disks that contain more of the planet-forming elements: silicon, iron, oxygen, and others. The other suggestion is that in many of these systems, a large number of the planets that originally formed have since fallen into the star. If the central star is "eating planets," then the planetary debris might remain in the star's atmosphere as an enhancement in heavy elements.

18.3 THE CHALLENGE TO STANDARD MODELS

The discovery of other planetary systems has challenged our deep-seated Ptolemaic tendency to see ourselves as setting the standard for the universe. Ours may not be a "typical" solar system after all. Furthermore, the theories we have developed to explain the architecture of our system, with giant planets beyond 5 AU and rocky planets closer to the Sun, must be at least incomplete and at worst fundamentally flawed.

When new discoveries challenge existing theories, we encounter some of the most exciting moments in science. The new planets described in Table 18.1 have given us a refreshing opportunity to revise our ideas about a basic subject: the origin and evolution of the planetary system we inhabit.

As we attempt to understand these new discoveries, the first thing we need to realize is that the Doppler technique works better and yields results more swiftly when a massive planet causes its star to move rapidly, which in turn occurs when the planet orbits the star at close range. Hence, the technique we are using to find extrasolar planets carries a strong bias toward detecting large, close-in planets.

The Doppler method does not have the sensitivity to find low mass planets—that is, Earth-like planets—in orbits of any size. We should therefore not worry that there are no Earth-like planets in Table 18.1. We will return to the challenge of detecting other Earths later in this chapter. The real issue is why so many of the giant planets in Table 18.1 have close-in orbits: Is this a result of the biases toward short orbital periods, or it is a real effect? The Doppler technique can also find, and has

found, giant planets that orbit their parent stars at larger distances, provided that we observe over time periods sufficiently long to reveal these changes. Indeed, the most recent entries in Table 18.1 (September 2002) include the star with the most distant known planet (at 5.9 AU), star number 55 in the constellation Cancer. It has taken more than a decade of observations to establish the 14.7-year period of this planet's orbit. We expect additional distant giant planets to be added to the list as time goes on. In this sense, we are just beginning to open the window on systems like our own—systems with giant planets at the same distances from their stars as Jupiter and Saturn in our own solar system.

This observational bias explains why the giant planets closest to their stars were discovered first. It is also important to note that fewer than 10% of solar-type stars have such hot giants orbiting close to the star. These are relatively unusual, and the majority of planetary systems might turn out to be like ours. But why are there *any* giants in close orbits? These had not been predicted by the models for solar system formation discussed in Chapter 17. Let's look at some intriguing explanations that may help us to understand this new phenomenon but still require additional testing before a definitive conclusion is reached.

HOW TO MAKE HOT GIANTS

Starting from ideas about the formation of our solar system, most scientists assume that these close-in giant planets must have formed farther from their stars than the positions in which we find them today. Conventional theories suggest that the heat radiated by the parent star, plus the heat generated by the formation of the circumstellar disk, will prevent any giant planet from forming closer to the star than about 5 AU. The reason for this is that stellar heating at distances less than 5 AU will prevent ice from condensing and will make gases like hydrogen and helium evaporate into interstellar space. The importance of ice stems from its cosmic abundance. Hydrogen and oxygen are the two most abundant chemically active elements in the universe. (Helium, second in abundance to hydrogen, is an inert noble gas.) Thus, ice is potentially the most abundant solid material. If it is possible for ice to condense with rock in an environment that also contains cosmic proportions of hydrogen and helium, there is the potential to build the massive giant planets we find beyond 5 AU in our solar system.

This simple physical argument provides the standard explanation for why the four inner planets in our solar system are rocky, while the four giant planets have orbits beyond five times the Earth-Sun distance. The same gradient in composition with distance from the center of the system is found in the galilean satellites of Jupiter.

This property of planetary and satellite systems led some scientists to suggest that the newly discovered planets might actually be giant rocky objects rather than fluid planets made of hydrogen and helium like Jupiter. To test this radical interpretation, we need to know the density of one of these planets. To derive the density, we need an exact value of the planet's mass and volume, so we have to determine the planet's diameter as well as its mass. If one of these giant planets were to pass in front of its star as seen from Earth, an event astronomers call a **transit,** we could determine both the mass and the diameter. For the transit to take place, the planet's orbit must have an apparent inclination close to 90°, so the sine of the angle to the plane of its orbit is 1 and we get the exact mass. As the transit proceeds, the light from the star will dim slightly, by an amount depending only on the diameter of the planet (Fig. 18.12).

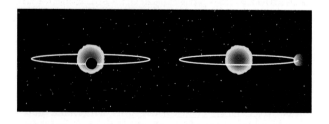

Figure 18.12 We observe the transit of a planet if our line of sight is in the plane of the planet's orbit about the star (*i* = 90°). In that case, the transit causes a modest diminution in the star's brightness that lasts for a few hours.

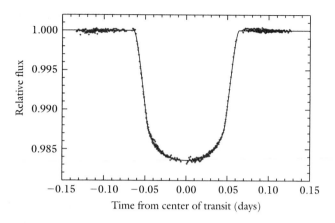

Figure 18.13 High-precision measurement of a planetary transit. These observations of a transit by the giant planet orbiting star HD 209548 were made by the Hubble Space Telescope. The diameter of the planet is 1.35 Jupiter diameters, yielding a density slightly less than 1 g/cm^3.

In fact, one such transit has been observed at high precision using the Hubble Space Telescope, and the results confirmed conventional expectations (Fig. 18.13). The planet's diameter was found to be 1.35 times that of Jupiter, while its mass is only 0.6 that of Jupiter. The implied density is roughly equal to that of Saturn, a little less than 1 g/cm^3. The low density means that this planet cannot be a huge, rocky Earth-like planet; it must be a hydrogen-helium-dominated planet like Jupiter. The reason it has a density lower than that of Jupiter is probably the extra heating it receives so close to its star, since the orbital period is less than 4 days.

If this one example is typical, these are really giant planets of jovian composition. We are thus led back to the problem of producing giant Jupiters closer to their stars than 5 AU, the limit set by the distance at which ice begins to condense in the standard model of solar system formation, a boundary often called the snowline. But perhaps this snowline is not so important as the standard model suggests.

Recall that the Galileo probe discovered that the chemically inert element argon has an abundance in Jupiter's atmosphere that is about three times greater than its abundance in the Sun and other stars (see Section 13.3). It is just as enriched as the highly reactive element carbon. This factor of three enrichment, which was also found for all the other elements that could be measured except helium and neon, must have been embodied in the icy planetesimals that produced Jupiter's core and original atmosphere as the planet formed. Yet argon, an inert gas, could not be incorporated in such a high abundance into the icy planetesimals unless these cometlike objects formed at temperatures below 30–40 K. Such temperatures are far colder than the approximately 150 K corresponding to the snowline that would allow ice to form at Jupiter's distance from the Sun. Thus, this result implies that the planetesimals formed somewhere else, perhaps even in the interstellar cloud that produced the solar nebula. In that case, it may have been possible to assemble giant planets faster and earlier than conventional models suggest, allowing them to form closer to their stars than 5 AU. If these giants reached their present masses before the Sun got hot enough to blow off the surrounding gas, their gravitational fields would have been sufficiently strong to enable them to retain hydrogen and helium despite the high temperatures caused by their proximity to their stars.

Migrating Planets

There is another way to produce the new extrasolar planets that does not require such a radical departure from conventional wisdom. Assuming that these planets did indeed form at much larger distances, they must have migrated inward. What could have caused the changes in their orbits that brought them to their present positions?

The answer may lie in the disk of material from which the planets originated. Even after the planets had formed, some of the material in the protoplanetary disk remained in orbit around the star. This material would have exerted a small amount of gravitational force on any planet, with subtle but important effects. The attraction of the material farther from the star than the planet, moving more slowly than the newborn planet in concentric orbits around the star, would drag energy from the planet, while the matter closer to the star, moving more rapidly than the planet,

would transfer energy to the planet. So long as roughly equal amounts of matter move around the star on either side of the planet's orbit, these effects would cancel each other.

Consider, however, what would happen if the material in the disk began to spiral into the star and disappear. As this occurred, the planet would experience a weaker gravitational tug from the depleted material closer to the star compared with the attraction from the amount farther out. As a result, the planet would lose orbital energy and move into a smaller orbit (Fig. 18.14). This new, smaller orbit would allow the planet to "catch up" to the material spiraling into the star, but as the material continued to move inward, so too would the planet.

If this were so (and many scientists think it explains in outline how planets can migrate into

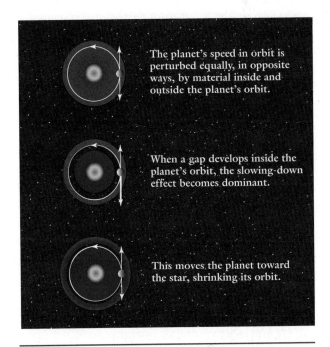

Figure 18.14 Gravitational forces from matter in a circumstellar disk can change the orbits of newly formed planets nearby. Matter farther from the star than the planet tends to slow the planet and make it move inward, while matter closer to the star tends to accelerate the planet's motion and make it move outward. If the amounts of matter on either side of the planet are roughly equal, the planet will not move, but if significantly more matter exists just outside the planet's orbit than just inside it, the planet will "migrate" inward.

smaller orbits), why would the process ever stop? In other words, why do the planets now orbit their stars at different close-in distances, instead of having followed the material in the disk all the way into the stars? One possibility is this: As the planet came close to its star, it experienced a significant tidal effect, similar to the ones that have caused the Moon's rotation rate to become equal to its orbital period, or have likewise locked the rotation rates of Jupiter's large moons, so that they too equal the times that the moons take to orbit the planet. Once the planet's rotation period became equal to its orbital period, the tidal interaction would prevent any further inward migration of the planet. Moving into a still smaller orbit would require the planet to take less time to complete an orbit than to rotate once, and this is prevented by the tides.

According to this interpretation, the inward migration of the gas giants depends on, and follows, the similar migration of the material in the circumstellar disk, but the migration ceases because of the tidal interaction between the star and its planet. Although we cannot measure them now, astronomers expect the rotation periods of all extrasolar planets that orbit their stars at distances of 0.1 AU or less to equal their orbital periods.

This seemed like a fine idea when these very close companions were the only ones we knew. However, it does not provide a satisfactory model for the numerous giant planets with orbits larger than 0.1 AU yet still inside the 5 AU snowline (Fig. 18.10). The two objects orbiting the star HD 168443 provide a major challenge. If the giant planet in this system was formed beyond 5 AU, how did it migrate to its present position at 0.3 AU with the brown dwarf at 2.9 AU apparently blocking its way? Astronomers currently have no good answer to this question.

There is another possibility: Perhaps many giant planets actually did fall all the way into their stars, and only a comparative few remained behind when the disk cleared. If this were true, it might help explain the enhanced heavy elements in many of the stars with inner giant planets. But this hypothesis requires that a great many more giant planets formed than is predicted by normal star-formation models, so this too seems like an unsatisfactory solution to the mystery. As more planetary systems are found, our current ideas about star formation are likely to continue to be challenged by unexpected discoveries.

One consequence of any model in which giant planets migrate inward from a starting point at several AU is that planets such as the terrestrial planets in our solar system will be destroyed—swept away as the larger mass migrates inward.

18.4 SEEKING OTHER EARTHS

From our parochial point of view, we want to know the probability of finding other planets like our own—rocky objects with approximately the same mass as Earth, moving in orbits near 1 AU from their stars. These are beyond reach of the Doppler method, so we must develop some alternative search technique.

GIANT NEIGHBORS MAY BE HAZARDOUS TO YOUR HEALTH

Recall that the only star in Table 18.1 with a planet in a Jupiter-like orbit is 55 Cancri, which actually has three companions. The two closest to the star have orbits within Mercury's distance from the Sun and minimum masses approximating those of Jupiter and Saturn. The third has a mass four times that of Jupiter and caught our attention because its orbit is at 5.9 AU. The geometry of this system allows an (undetectable) Earth-like planet at 1 AU, but whether or not such a planet could exist there depends on where the giant planets in the system formed and whether and how they moved to occupy their present positions.

Accepting the inability of the Doppler method to detect Earth-like planets, we are still missing a system with only one giant planet in a nearly circular orbit some 5 AU from its star. This is how our solar system would appear to an observer in the 55 Cancri system (for example), using the Doppler technique to examine our Sun. Given the long times required to establish such orbits, we have only just entered the epoch when we can expect these systems to be discovered, so we should not be put off by their absence from Table 18.1.

As we anticipate such discoveries, it is useful to consider another consequence of Jupiter's distance of 5 AU from the Sun. If a giant planet orbits its star at a distance significantly less than this, its gravitational effects will prevent rocky planets from forming near 1 AU, just as Jupiter prevented the material in the asteroid belt, orbiting at distances between about 2 and 3 AU from the Sun, from coalescing into a sizable planet. It is possible that Jupiter's nearby presence also kept Mars from growing as massive as Earth or Venus. This inhibiting influence will be especially effective if the giant planet moves in a highly elliptical orbit, as many of the newly discovered giant planets actually do.

Aside from the 55 Cancri system, the extrasolar giant planets with the greatest orbital distances discovered so far move in orbits at about 4 AU from the stars. The destructive effects of the Jupiter-size planets around 14 Herculis and Epsilon Eridani are maximized by the large eccentricities of these planets' orbits, which bring the planets significantly closer to their stars than their average distances suggest. We may conclude that rocky, Earth-like planets are highly unlikely to exist in orbits at distances near 1 AU from any of the stars listed in Table 18.1. Only when we discover Jupiter-size extrasolar planets that orbit at Jupiter-like distances from their stars in nearly circular orbits (and with no giants inside) can we feel confident that we have found planetary systems that might contain one or more Earth-like planets.

GUESSING THE NUMBER OF EARTHS IN THE MILKY WAY

Given that we do not yet understand how life developed on Earth, we can't expect to deduce a meaningful value for the number of inhabited planets in the Milky Way Galaxy. However, we can at least try to estimate the number of planets that could be inhabited if life began and survived on them.

In the (to us) exotic planetary systems that have been discovered so far, we can always imagine the existence of some equally exotic worlds, such as Earth-size satellites in orbit around giant planets that are themselves in circular orbits approximately 1 AU from their stars. Earth-like

planets trapped in the stable Lagrangian points in the orbits of such planets (like Jupiter's Trojan asteroids; see Fig. 5.9) might offer another ecological niche. But let us concentrate on planets more like ours. How many does our Galaxy contain?

The first step in this calculation is relatively straightforward: How many stars are like the Sun? They must be single, sufficiently rich in heavy elements to have rocky planets, and located in a safe region of the Galaxy where supernovas are rare, with masses and thus luminosities similar to the Sun's. This last point ensures that they spend at least 5 billion years steadily burning their nuclear fuel at nearly constant luminosity. Adopting what astronomers Joseph Shklovskii and Carl Sagan have called the *principle of mediocrity*, we assume (always dangerous!) that life on Earth is an average example of life in the universe. In that case, the planets we seek need 4 billion years to produce moderately intelligent life. Two out of the 34 stars closest to the Earth satisfy all of these criteria. If this ratio holds for all safely located stars in the Galaxy, we find that approximately 15 billion stars would qualify.

Table 18.1 tells us that planetary systems are relatively common. What it doesn't tell us is whether planetary systems like ours are common. Given the bias of the Doppler method toward discovering close-in, massive planets, we can estimate from the percentage already discovered that approximately 10% of all Sun-like stars have giant planets. We could then guess (since the data are not yet in) that perhaps 10% to 50% of these giants are in orbits like Jupiter's. What would this imply for the chances of finding Earth-like planets around these stars?

Again we are forced (with the usual reservations) to rely on our own solar system for insight. All of our giant planets have families of one or more relatively large satellites with smaller ones closer to the planet. This hierarchical distribution imitates the distribution of our planets with respect to the Sun. We thus tentatively conclude that a planetary system that has at least one giant planet beyond 5 AU and none inside that distance is likely to have rocky planets distributed between 5 AU and the star. We conclude that the number of habitable planets in the Galaxy is somewhere between one (ours) and several hundred million. But this is not much of a conclusion. To do better, we need to find some of these other Earths and establish empirically how often planetary systems occur that include terrestrial planets like ours.

THE PHOTOMETRIC METHOD

Fortunately, another technique is being developed to detect planets like the Earth in orbit about stars like the Sun: highly accurate **photometry,** the measurement of the intensity of light from the star. Imagine how our solar system would appear to an observer on a planet somewhere else in the Galaxy. If this vantage point were located in the plane of our planetary orbits, then once every year, Earth would pass between the Sun and the observer, blocking a small fraction of the sunlight for a period of about 10 hours. This same technique was used to determine the size of a giant planet as shown in Figure 18.13. For our distant observer of the Sun, however, the effect would be extremely small because of the much smaller size of Earth. The decrease in the light from the Sun would be only 1/10,000 of the Sun's brightness, but if the solar brightness were monitored with sufficient accuracy, this photometric signature of the presence of Earth could be detected. From the time between transits, the orbital period of Earth could be derived, and from the amount of light lost during the transit, the size of our planet could be estimated.

Bill Borucki of NASA's Ames Research Center has shown that current technology is capable of measuring stellar brightness with the requisite precision, and he has successfully proposed building a small space telescope called Kepler that will monitor the light of 100,000 stars simultaneously for 4 years. The telescope must be in space, since only above the atmosphere can the requisite precision be achieved. We need to look at so many stars because we can expect only about 1% of them to be oriented so that transits can be seen from our direction in space. When astronomers first detect the diminution of a star's light that appears to indicate a planetary transit, they cannot be sure that this is really a planet and not some

problem with the data. A second transit, however, will increase their confidence and yield the period of revolution of the planet and hence its distance from the star. The third time is the charm: If a third transit occurs after the same period of time that was deduced from the second transit, it will confirm the existence of the planet.

The Kepler mission is presently scheduled for launch in 2007. During its nominal 4-year lifetime, it is expected to discover one hundred or more Earth-like planets, if our present ideas about planetary formation are valid. In the process, it will of course also identify more giant planets in close-in orbits, adding to the list of those discovered by the Doppler method. If the system works well but finds many fewer terrestrial planets, or even none at all, then we must consider a major revision in our ideas about planet formation. Either way, the results will be exciting.

GOING FOR BROKE: INTERFEROMETRY FROM SPACE

The next step in extrasolar planet detection will be a major advance. It will not only enable us to detect Earth-like planets regardless of the orientation of their orbits; it will also allow us to determine the composition of their atmospheres. We will therefore be able to find out exactly what kind of planets we are detecting. We will even be able to search for the signature of gaseous oxygen in the atmosphere of an inner planet that would reveal the presence of life (more on this topic later).

This breakthrough requires the use of **interferometry,** the simultaneous observation of an object with two or more telescopes (Fig. 18.15). Radio astronomers have been using interferometry for decades, but it is just becoming feasible at optical wavelengths. The great power of this technique is that it allows an investigation of the target object with the same angular resolution that would be provided by a single mirror as large as the separation of the individual telescopes. It is therefore possible to detect the planets directly, the method described as impossible with conventional telescopes at the beginning of this chapter.

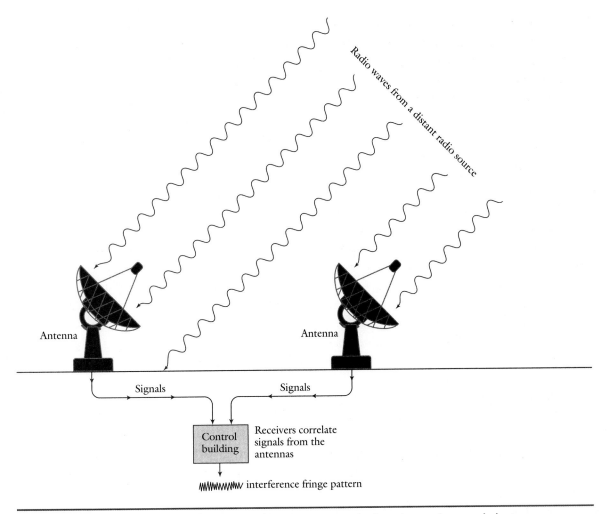

Figure 18.15   The basic principle of an interferometer system is that observations made by two or more individual instruments can be linked, so that light or radio waves reaching each instrument can be compared with those reaching the others at the same time. This comparison allows the system to achieve an angular resolution equal to that of a single large telescope whose diameter equals the maximum separation of the individual telescopes in the interferometer.

The first of these next-generation instruments, called SIM (for Space Interferometry Mission), would orbit the Sun at roughly Earth's distance, always remaining well away from our planet. It is necessary to get into space to avoid the blurring effects of our planet's atmosphere. SIM would carry an array of small telescopes that would use interferometry to combine the images that each produces, thus achieving the resolving power of a single, much larger telescope.

Although SIM's superior resolving power would allow it to detect giant planets orbiting nearby stars, this instrument could not observe Earth-size planets at distances from their stars roughly equal to the Earth-Sun distance. It will constitute what engineers like to call a "proof-of-concept" mission, demonstrating that optical interferometry from space is feasible. If we hope to observe Earth-like planets directly, we must send a much larger interferometric array of telescopes to greater distances from Earth.

Astronomers hope to do just that, perhaps in two decades, with a U.S. mission called the Terrestrial Planet Finder (TPF) and a similar mission called Darwin under development by the European Space Agency. As shown in Figure 18.16, these interferometry missions will consist of an array of four or five telescopes orbiting the Sun

Figure 18.16 By the start of the second decade of the twenty-first century, we may be ready to begin construction of TPF, the Terrestrial Planet Finder. TPF will be an interferometry system millions of kilometers from Earth that is able to secure direct images of Earth-like planets orbiting other stars. By studying the composition of these planets' atmospheres, TPF should be able to determine which planets are most likely to have already produced life.

millions of kilometers from Earth and maintaining their relative positions with an accuracy precise to distances less than one wavelength of visible light. Achieving this feat is now technologically impossible. But if we hope to make direct observations of Earth-like planets, including spectral observations that can determine the major constituents of their atmospheres, we shall require a mission something like the one envisioned for Darwin or TPF. There are obvious advantages to combining the two concepts into a single mission, one of which would be the sharing of costs and thus the acceleration of the launch schedule. We may thus hope one day to see in operation a single, international spacecraft facility carefully designed to achieve our ultimate goal in planet detection: the identification of Earth-like planets that can be examined to see whether they show evidence of life.

18.5 LIFE IN OTHER PLANETARY SYSTEMS

The discovery of other planetary systems raises the important question whether any of these planets might support life. We can divide the question into two parts. The first part is how to define what environmental conditions we should look for as being capable of supporting life. The second part addresses the more difficult challenge of planning how to actually detect evidence of life on such distant worlds.

CONDITIONS FOR LIFE

What kind of life are we looking for? Since we know only one form of life, scientists are unable to answer some basic questions about the general nature of life and the range of conditions that might support it. We are limited to the kind of life we have on Earth, which is based on complex carbon chemistry and requires liquid water. Carbon is unique among the elements in its ability to form varieties of complex compounds that can store information and participate freely in chemical reactions, while water is an excellent solvent with many biologically helpful properties, such as the low density of its solid form. We conclude that life on other planets will probably be based on carbon and water, although its basic chemical components and structures could be very different from the amino acids, proteins, and DNA we know. In any case, life on Earth is the only life that we can deal with in a scientific way. Of necessity, when we speak of planetary habitability, we are focusing our attention on environments that might support life as we know it.

The basic requirements are the presence of liquid water, access to the biogenic elements that are the building blocks of life, and a source of energy to drive the chemical reactions that life requires. Only if there is a continuing stable source of energy will the complex molecules of life be able to grow and reproduce, a process that creates ordered systems within a universe that otherwise tends toward disorder.

We find the same chemical elements throughout the universe. Organic (carbon-based) chemistry is common in space, as astronomers have found from their studies of the clouds of gas and dust between the stars. The biogenic elements (primarily carbon, hydrogen, oxygen, and nitrogen) are available almost everywhere, and as long as temperatures are not too high (as they are on stars themselves), we can anticipate an abundance of the chemical building blocks of life.

Accessible energy is a bit harder to find. There is plenty of energy throughout the universe: Every object with a temperature above absolute zero is radiating some energy into its surroundings. The challenge is to find energy in a useable form, capable of driving chemical reactions that are essential to life. On Earth, we have two primary sources of useable energy. First, and still important for some microbial communities, is chemical energy generated by reactions that take place most commonly when minerals dissolved in water are heated—for example, the hot plumes of mineral-laden water that are emitted at submarine volcanic hotspots (Fig. 18.17). The second and much more powerful source of energy is sunlight. In order to utilize this source, however, life had to develop the means to extract chemical energy from sunlight through photosynthesis. Photosynthesis involves sophisticated reactions that probably required hundreds of millions of years of evolution.

The condition of liquid water is the most demanding. Liquid water requires an environment with a restricted range of temperatures, roughly between −10°C (for a brine) and +100°C. Although water can remain liquid at higher temperatures when the pressure is also high, the complicated chemistry of life starts to break down for temperatures much above 100°C. The highest temperature at which life has been found on Earth is 113°C. Liquid water also requires pressures that are greater than the triple point for water—that is, pressures above about 0.06 bar. As we saw when studying conditions on Mars, most of the martian surface is

Figure 18.17 Hydrothermal vent on the ocean floor. Hot water laden with minerals provides an environment that can sustain many microorganisms.

at an atmospheric pressure too low for liquid water to be stable. On the other hand, our studies of Mars (and of life in the cold deserts of Earth) has also taught us that liquid water can exist in favorable microenvironments, such as the interiors of rocks, or tens to hundreds of meters below a planet's surface, even when the outside conditions are unsuitable. In any case, the requirement of large amounts of liquid water for life to get started suggests temperate environments on or near the surfaces of planets.

What kinds of planets meet these requirements? Certainly a terrestrial-type planet with oceans of liquid water would be a suitable place for the formation and persistence of life. Presumably we also need a degree of environmental stability—as would be found on a planet with a nearly circular orbit about a nonvariable star.

What about giant planets? There are regions within their atmospheres that have both liquid water clouds and the chemical building blocks for life, but there may be problems resulting from the absence of a stable surface and from the fact that little sunlight penetrates to the depths where liquid water exists. On Earth, one of the few environments that life has not colonized is the clouds. In spite of the presence of both liquid water and ample sunlight, there are no microbes that carry out their life cycle floating in the air.

The satellites of giant planets offer other possibilities. We have seen that Europa, within the jovian system, is considered a prime candidate for future astrobiology investigation. It is not too difficult to imagine habitable environments on giant planetary satellites at distances from the central star where liquid water is stable on the surface.

THE HABITABLE ZONE

These environmental requirements, especially the need for liquid water, focus our attention on Earth-like planets at distances from their parent star where the surface temperatures are between the freezing and boiling points of water. A planet of roughly terrestrial size is the most likely habitat to have both a solid crust and a substantial atmosphere, with the possibility of liquid water on or near the surface. The surface temperatures will fall within the correct range when there are appropriate combinations of stellar luminosity, distance of the planet from the star, and nature of its atmosphere (which determines the magnitude of the greenhouse effect).

These considerations led scientists to define a **habitable zone** as a range of distances from the central star for any planetary system where liquid water could exist on the surfaces of terrestrial-type planets. The actual dimensions of the habitable zone depend on the luminosity of the star. We have referred to distances of approximately 1 AU in the above discussion about searching for Earth-like planets, but let us examine this optimum distance more closely.

We can start by looking at the habitable zone within our own solar system. As we discussed in Section 12.6, the Earth is today the only planet with the proper surface temperature for liquid water. Remember, however, that without its atmospheric greenhouse effect, Earth temperatures would be lower and most of the water on Earth would be frozen solid. Venus is far too hot, and we think that the runaway greenhouse effect that led to its current inhospitable conditions was inevitable for a planet this near the Sun, at 0.7 AU. As we look outward, however, Mars is a marginal case. Today liquid water cannot exist at the surface except in unusual and short-lived circumstances. But if Mars had a larger mass (similar to Earth and Venus), it would have retained a larger atmosphere with a greater greenhouse effect. Both the higher temperatures from the greenhouse effect and the greater surface pressures would have favored liquid water on the surface. In addition, even for Mars with its small mass, there is ample evidence of past surface water, even perhaps large lakes or seas in which water might have persisted under ice caps. We must therefore allow the habitable zone for our solar system to extend at least as far as Mars.

Detailed calculations that consider the likely evolution of climate and atmospheres yield an estimate for the inner edge of the solar system habitable zone at about 0.8 AU, between Venus and Earth. The outer limit is between 1.4 AU (just beyond the orbit of Mars) and 1.8 AU, depending on which atmospheric model is adopted.

Since temperature varies as the luminosity of the star to the 1/4 power, it is a simple matter to scale the habitable zone boundaries for stars of different types. Stars of lower mass and lower luminosity have smaller habitable zones, while massive, highly luminous stars have larger habitable zones. Planets circling massive stars probably don't qualify for life, however, because such very bright stars are more variable than our Sun and have much shorter lifetimes. Long-term stability is one of the requirements for the development of advanced life.

For stars like the Sun, the dimensions of the habitable zones are similar to those in our solar system. The planets most likely to support life are roughly Earth-mass planets (say from 0.3 to 3 times the mass of the Earth) at distances near 1 AU. We hope the Kepler mission will discover many such Earth-like planets within the habitable zones of their parent stars.

Habitability does not necessarily imply the presence of life, however. We know too little about the origin and history of life to estimate how many habitable planets will actually harbor life. To find the answer, we must have a way of detecting the presence of life on a planet.

BIOMARKERS

We are unlikely ever to send spacecraft to provide a direct reconnaissance of the planets around other stars. The detection of life will depend on remote sensing. Our main hope of success rests on the ability of life to alter the environment on a planetary scale. Such a global signature of life is often called a **biomarker.**

Our own planet Earth provides examples of such biomarkers. The photosynthetic life that dominates our biosphere utilizes solar energy to produce complex organic structures, some of which in turn emit unexpected gases into the atmosphere. These gases represent a state of chemical disequilibrium. They would not be present were it not for their continuing production in the biosphere.

The most obvious biomarker today is the molecular oxygen that constitutes 21% of Earth's atmosphere. Because oxygen is highly reactive chemically, it quickly recombines with other atoms in the environment to produce oxidized surface minerals or atmospheric water and carbon dioxide. Without a continuing source, most of the free oxygen in our atmosphere would disappear in a few thousand years. As we discussed in Chapter 12, such a large abundance of atmospheric oxygen would be a clear indication of the presence of life on any planet.

There are other even more dramatic examples of chemical disequilibrium caused by life on Earth. In the presence of atmospheric oxygen, methane is quickly oxidized and converted to carbon dioxide. Yet we have a small quantity of methane in our atmosphere thanks to its continuous production by microbes—primarily in swampy areas (methane used to be called "marsh gas") and in the digestive systems of some animals (including humans!). The co-existence of oxygen and methane is a powerful biomarker, since we cannot imagine any nonbiological process that could produce such a striking chemical disequilibrium.

Let us turn the clock back, however, and ask whether the atmosphere of the ancient Earth displayed such obvious biomarkers. As we saw in Section 9.5, our planet acquired its atmospheric oxygen only during the past 2 billion years, and it is probably only within the past 1 billion years that oxygen exceeded a concentration of 10%. Before the evolution of oxygen photosynthesis, terrestrial life may have extracted energy from sulfur photosynthesis or from chemicals dissolved in water as a by-product of volcanic activity (as certain microbes on Earth still do). In these latter circumstances, life does not generate chemical disequilibrium; quite the opposite. In extracting energy from its surroundings (rather than from sunlight), life drives the environment toward chemical equilibrium. The unfortunate result is that no strong chemical biomarkers are generated.

Research on biomarkers has only begun, but already we realize that there are many circumstances where it will be difficult to detect any global consequences of life. The presence of abundant oxygen and the simultaneous existence of atmospheric methane would be an excellent indicator of life. But for much of its history, the terrestrial biosphere would have been undetectable using such criteria. The age of the star in the system will be an important piece of information to use to discern whether or not a given gas should be considered a sign of life, as opposed to an early episode of disequilibrium caused by impacts, out-gassing, or an ongoing runaway greenhouse effect. Scientists will need to consider carefully a number of possible paths for the co-evolution of a planet and its biosphere in order to evaluate the usefulness of particular biomarkers.

SEARCHING FOR BIOMARKERS

We have already described the Darwin and Terrestrial Planet Finder (TPF) space telescopes, which will be the first instruments that have much hope of detecting the biospheres of distant planets. They will be designed to isolate individual planets, including terrestrial-size planets, circling nearby stars. Note that we will not resolve these planets—no detail can be seen on their disks. The best we can do is separate some of their light from that from the central star and from other planets in the system.

Infrared spectra can then be obtained to search for the disequilibrium gases. The strategy is to look for atmospheric oxygen (O_2) not by detecting its spectrum directly but by looking for the spectrum of ozone (O_3), which is much easier. The concept is that ozone will indicate plentiful oxygen, which is in turn a reliable biomarker. Other gases including methane are also detectable by their infrared spectra. The challenge is to derive the maximum useful information from these extremely faint astronomical sources.

While we wait for these future space missions, here is one way that we could detect life unambiguously using currently available telescopes. If the life on a distant planet has followed the same path as ours, evolving intelligence and ultimately sending out radio broadcasts, then we could detect it today.

18.6 THE SEARCH FOR EXTRATERRESTRIAL INTELLIGENCE

An obvious way to accelerate the process of finding other planets like Earth is to try to make contact with intelligent, technologically advanced civilizations that could be living on them. If such places exist, and if the inhabitants are transmitting powerful radio signals in our general direction, then it is within our present capabilities to detect such signals. For example, the giant radio telescope in Arecibo, Puerto Rico, could easily exchange messages with another telescope of the same size on the other side of the Galaxy (Fig. 18.18). No new developments of technology are needed.

The discovery of such signals would not only demonstrate the existence of other planets like ours. Once two-way communication had been established, there would be an opportunity to learn in detail about this other system, plus vast amounts of additional knowledge that we can only speculate about at this point.

The Search for Extraterrestrial Intelligence, usually abbreviated as **SETI,** has been described by astronomer Frank Drake, one of its leading proponents, as "the last great adventure left to humanity." It is indeed, and well worth spending a few minutes to consider.

FINDING ANOTHER CIVILIZATION

How likely is it we will find another advanced civilization in the Galaxy? This question can be answered only by doing the experiment—trying to find them. Our knowledge about such civilizations is limited to our own, and we must be among the most primitive. We have discussed above the probability of finding other planets like ours. We can conclude from this discussion that nothing we know now would exclude the existence of such worlds. Instead of waiting for other investigations to find these planets and then listening for possible signals coming from civilizations that may inhabit them, SETI strategists have opted to go straight after the signals. They have adopted a dual approach that consists of an *all-sky survey* and a *targeted search.* In both cases, the radio telescope is equipped with a sensitive detector and a spectrum analyzer backed up by a powerful computer that allows the astronomers to analyze over 100 million frequencies in the radio spectrum simultaneously.

This brute force approach is required because we have no idea which frequency the hypothetical civilization is using to transmit its signal. We also don't know where they are. In the all-sky survey, the telescope slowly and systematically surveys the

Figure 18.18   Arecibo telescope. This instrument could communicate with a telescope of the same size across most of the Milky Way Galaxy if we only knew where to point it and what radio frequency to use.

entire sky, looking for evidence of an artificial signal. In this case, there is no judgment about civilizations living on Earth-like planets; it is a basic search for signs of intelligence in the Galaxy. These signs could be signals coming from the civilizations themselves, from their spacecraft, or from beacons they may have established throughout the Galaxy. A minimum of four telescopes are required for this huge project, two in the northern hemisphere and two below the equator.

The targeted search focuses on some 1000 relatively nearby stars that have been chosen for their similarity to the Sun. Once Earth-like planets have been found by Kepler or other dedicated observatories, they will receive special attention.

IMPROVING THE ODDS

The efficiency of these searches would be greatly improved if the searchers knew which frequency their quarry was using to communicate. Bernard Oliver, one of the pioneers in the field, made an

intriguing suggestion about this. Suppose water is indeed the universal solvent for life, used again and again by life forms throughout the universe. Water is broken apart by ultraviolet radiation into H and OH. Each of these molecular fragments radiates at a specific frequency, H at 1428 megahertz (the famous 21-cm line of atomic hydrogen), while OH produces a cluster of lines near 1665 megahertz. Oliver called the frequency range between these two signposts the "water hole" and suggested that this is where galactic civilizations would meet.

This range of frequencies has another advantage. An enormous amount of noise is generated by the Galaxy at radio frequencies. Most of the low-frequency noise is coming from electrons spiraling around magnetic field lines in various sources such as supernova remnants, producing the synchrotron radiation we met in studying the radiation belts of Jupiter. At the high-frequency end of the radio spectrum, gases in any planetary atmosphere absorb incoming radiation. The "water hole" occurs near

---

PERSPECTIVE

UFOS AND ANCIENT ASTRONAUTS: TWO MODERN MYTHS

Why should we spend so much effort to find Earth-like planets and try to make contact with their inhabitants when these intriguing aliens may already be visiting the Earth? This question arises from the intense popular interest in UFOs (unidentified flying objects), believed by some to be extraterrestrial spaceships on mysterious missions to Earth.

Complementing these contemporary visits, many people believe that thoughtful astronauts from advanced civilizations provided guidance to our early, primitive ancestors. An international audience has bought more than 7 million copies of a book by Erich von Däniken that purports to prove that ancient monuments such as the Great Pyramids of Egypt were either built directly by extraterrestrials or at least constructed with the assistance of their superior wisdom and technology.

It's a pity that this instructive activity no longer takes place. There is so much these aliens could tell us! After all, by their standards we are still extremely primitive. We cannot reach even the nearest star with our technology, and on our own planet, people are still starving and engaging in homicidal combat to resolve political and religious differences. Instead of advice from advanced civilizations, today we have only reports of lights moving strangely in our night skies, with perhaps an occasional landing and abduction.

The idea that we humans periodically receive visitors from the sky is not new. We seem to have a deeply rooted yearning for some personal connection with the cosmos, an indication that somebody out there cares enough about us to guide us along our way or at least to stop in to see how we're doing from time to time. Today we talk of aliens from interstellar spaceships, but in the past our visitors were various kinds of spirits, ghosts, avatars, devils, and angels.

The authors of this book and their scientist colleagues would be thrilled and delighted if any of these stories were true. Unfortunately, there is not a shred of evidence to support them. Instead, serious investigators have found only examples of simple mistakes and outright fraud, compounded by wishful thinking.

In the case of UFOs, ordinarily reliable people like airplane pilots and police who are simply not accustomed to observing the variety of natural phenomena that take place in the sky have reported sightings of spacecraft that are easily shown to be something else. Venus heads the list. When it is near the Earth and thus unusually bright, this planet can startle people who are unaware of its existence. Seen among moving clouds, Venus is frequently mistaken for an alien spaceship maneuvering through our skies. Former President Jimmy Carter is one of many who have contributed such reports. Venus has been chased by police squad cars and fired at by naval cannons. Military planes have been scrambled to shoot it down. Fortunately, it is sufficiently distant to survive such indignities.

Weather balloons, lenticular clouds, airplanes, even flocks of migrating birds have all been reported as UFOs. And in fact they *are* UFOs to the people who make these reports because to them, these are *unidentified* flying objects. But they are not (alas!) extraterrestrial spaceships. Amateur astronomers who study the sky more intensively than most of their professional colleagues have discovered many comets (e.g., Hyakutake and Hale Bopp), but they have never recorded a sighting of a spaceship. We have marvelous videos of rare wild animals in unusual activities and more mundane scenes like car chases and police brutality, even videos of a meteorite entering the Earth's atmosphere that are so good that astronomers were able to trace the orbit back to the asteroid belt. Yet we have no such sharp pictures of spaceships, only grainy black and white stills and shaky out-of-focus videos, often of dubious authenticity.

What about a conspiracy? The government knows that spaceships have landed, but it refuses to tell us. Think about it! We can't imagine that these aliens would come all the way from some distant star and visit only the United States. Surely they would whiz around the world, enjoying every continent and culture. Thus, we would have to believe that there is a giant conspiracy of all the various governments; they have jointly agreed to keep this mighty secret from their constituents, sharing it only among themselves or among their militaries. Consider just the I's: Iceland, India, Indonesia, Iran, Iraq, Ireland, and Israel all working together in this plot. It is simply not credible!

What about extraterrestrial encounters in earlier times? Ancient visits by aliens are supported by the idea that our primitive ancestors were incapable of building the monuments that appeared during their lifetimes. What an insult! The problem here is the reluctance or inability of the average modern mind to appreciate what can be accomplished without the benefit of modern technology by the combined efforts of many skilled people working together for long periods of time. The Egyptians certainly built the Great Pyramids themselves. We can see the quarries from which they took the stones, inspect the tools they used, even admire the drawings they left to show how teams of workers could pull heavy stones and statues into position.

We have found marvelous things as we look back in time, well beyond von Däniken's modest horizon of human activity. We find the bones and shells of vanished species, rocks laid down before the atmosphere had so much oxygen, and still older pebbles rounded by the erosive action of ancient seas. Yet nowhere in this 4-billion-year record contained in the rocks do we find evidence of a visit by alien astronauts.

Could it be that we are actually alone, the only intelligent life in a vast, uncaring universe? Some scientists seriously think so, and there is no hard evidence to refute them. Others, including the authors of this book, prefer to believe that the same natural processes that led to the origin and evolution of life on Earth are operating throughout the universe and have surely produced intelligent life on other worlds.

If so, we are very recent arrivals on the scene. Given that the age of our Galaxy is more than twice the age of our solar system, there could be many civilizations that are millions or even billions of years older than ours. So, as Nobelist Enrico Fermi once famously asked, "Where is everybody?"

The great distances between neighboring stars make interstellar travel extremely difficult, just in terms of fundamental physics, never mind the engineering problems or the costs. Even in the most optimistic case, matter-antimatter annihilation is required to reach the relativistic velocities needed for trip times of less than millennia. In contrast, communication by electromagnetic radiation is easy and cheap and proceeds at the velocity of light, the universal speed limit. SETI, using radio and optical wavelengths, remains our best hope of making contact with other forms of intelligence, thereby finally relieving that great sense of cosmic loneliness that seems to be an intrinsic component of the human condition.

the minimum in this noise spectrum, where interstellar transmission could proceed most clearly.

Scientists are becoming increasingly worried about interference arising from human broadcasts of radio and television signals on Earth. If this interference becomes sufficiently bad, it may become necessary eventually to resort to construction of a huge antenna on the farside of the Moon, where radiation from Earth would be completely blocked.

Given all this noise, how can SETI researchers be sure they have found a true signal from outer space and not some secret military broadcast or some stray signal from an amateur radio buff? This is a real problem and one that has been met head-on. To qualify as a detection worth verifying, an incoming signal has to trigger a second detector simultaneously on another telescope pointing in the same direction but located far from the main telescope.

Should the signal survive that test, requests are made for other telescopes to search for it and a concerted effort is made to track the signal as Earth turns and different observatories come into its path. No signal has yet survived this close scrutiny.

INTELLIGENT LIFE ON EARTH

The largest (most expensive) SETI program, involving both a targeted search and an all-sky survey, was originally funded by NASA and began on October 12, 1992, commemorating the 500th anniversary of the voyage of Columbus. The program was terminated one year later by the U.S. Congress in a Senate vote of 77 to 23. The leading congressional antagonist, Senator Richard Bryan of Nevada, referred to this project as "The Great Martian Chase," derisively pointing out that "not a single Martian has said 'take me to your leader,' and not a single flying saucer has applied for FAA approval."

Fortunately, the search for intelligent life on Earth is not compelled to stop with a study of the U.S. Congress. In the aftermath of this cancellation, the SETI program has risen like a phoenix from the ashes of senatorial ridicule. A nonprofit corporation known as the SETI Institute has raised over $20 million in private and corporate dona-

Figure 18.19 Astronomer Jill Tarter, Director of Project Phoenix and one of the leading figures in SETI research.

tions. Renamed Project Phoenix, the search is under way as you read this (Fig. 18.19).

Project Phoenix is not the only effort being made to detect signals from other civilizations. A project called Serendip, run from the University of California at Berkeley, piggybacks on ordinary astrophysical observing programs at radio telescopes, searching for background signals while the astronomers record the data they want on their chosen objects. The Planetary Society is sponsoring a program at the Harvard College Observatory under the direction of Paul Horowitz. Funding for the radio telescope used in this survey was provided by filmmaker Stephen Spielberg, who directed such popular movies as *ET, the Extraterrestrial*. Horowitz is also exploring optical SETI, in which the extraterrestrials are assumed to attempt communication using optical rather than radio frequencies. A second optical SETI project at the University of California has been developed by Shelley Wright, an undergraduate student, working with Frank Drake of the SETI Institute. They have been observing nearby stars regularly since 2000 in a search for very rapid light pulses (Fig. 18.20).

Serendip offers a novel solution to one of the greatest handicaps this work has always faced: the need to get observing time on a sufficiently large

Figure 18.20 Optical SETI. Shelley Wright, a senior physics major at the University of California, Santa Cruz, with the optical SETI detector she built as part of a prize-winning senior thesis project. The search uses a 1.1-m telescope at Lick Observatory to search for very rapid pulses of laser light.

radio telescope. Such telescopes are in great demand to carry out a variety of astrophysical investigations, so the amount of time available for SETI is very small. The SETI Institute was therefore delighted to receive another $20 million from Paul Allen of the Microsoft Company to build a dedicated telescope for this project, with a collecting area of 0.1 hectare. The amazingly weak radio signals from some distant civilization that may be passing through this page right now could be detected by this telescope within the next 10 years.

18.7 QUANTITATIVE SUPPLEMENT:
MOTIONS IN PLANETARY SYSTEMS

We refer back to Chapter 1 and consider the simplest case of a small planet of mass m moving in a circular orbit with radius D around a star of mass M. Then the gravitational force between the star and the planet is

$$F_g = \frac{GMm}{D^2} \qquad (1)$$

where G is the gravitational constant:

$$G = 6.672 \times 10^{-8} \text{ N m}^2/\text{kg}^2$$

As Newton recognized, the gravitational force on the planet must be balanced by a centrifugal force:

$$F_c = \frac{mv^2}{D} \qquad (2)$$

where v is the velocity of the planet in its orbit. If $F_c = 0$, the planet would be drawn into the star; if $F_g = 0$, the planet would fly off into space in a straight line. To remain in orbit, we must satisfy the condition

$$F_g = F_c \qquad (3)$$

Substituting from equations (1) and (2), we find

$$\frac{GMm}{D^2} = \frac{mv^2}{D} \qquad (4)$$

The velocity of the planet, v, is simply the circumference of its orbit, $2\pi D$, divided by its period of revolution, P. Substituting into equation (4) and collecting terms, we have

$$\frac{GM}{D} = \frac{(2\pi D)^2}{P^2} \qquad (5)$$

Rearranging the terms in this equation, we find

$$\frac{P^2}{D^3} = \frac{4\pi^2}{GM} \qquad (6)$$

which is Kepler's third law.

This derivation of Kepler's third law actually works quite well in our solar system, where the largest planet, Jupiter, is a thousand times less massive than the Sun and in an orbit with $D = 5.2$ AU.

To gain an appreciation of the conditions in the new planetary systems with close-in giant planets, we consider both the planet and the star to move in circular orbits (still an approximation) about their common center of mass, X (see Fig. 18.1). Now the planet's distance to X is r, while the star's distance to X is R. Then, by definition,

$$MR = mr \quad \text{and} \quad r + R = D \quad (7)$$

Using these two equations, we can solve for r and R:

$$m(D - R) = MR \quad \text{so} \quad mD = R(M + m)$$

and

$$R = \frac{m}{(M + m)} D \quad (8)$$

Similarly, we can establish that

$$r = \frac{M}{(M + m)} D \quad (9)$$

Therefore,

$$mr = MR = \frac{Mm}{(M + m)} D \quad (10)$$

Now for each of the two masses, the gravitational force must be balanced by the centrifugal force, as in equations (3) and (4). Both the planet and the star move around the center of mass with the same period, P, but their velocities, v and V, will be different because the sizes of their orbits are different:

$$P = \frac{2\pi r}{v} = \frac{2\pi R}{V} \quad (11)$$

It is then easy to show that the relationship between period P and the separation of the planet and star D is

$$\frac{P^2}{D^2} = \frac{4\pi^2}{G} \left(\frac{1}{(m + M)} \right) \quad (12)$$

This is the more accurate form of Kepler's law when the mass of the planet is a more significant fraction of the mass of the star.

SUMMARY

The last years of the twentieth century saw the first discovery of planets orbiting stars beyond the Sun. To make these discoveries, astronomers study the positions of the absorption lines in a star's spectrum with amazing precision, seeking to find small, repetitive changes in the wavelengths of these features. These changes arise from the Doppler effect—that is, from changes in the star's velocity as it moves in response to the gravitational pull from an unseen planet. This Doppler technique works best for more massive planets and for planets orbiting close to their stars, because such planets produce the greatest changes in a star's velocity. Not surprisingly, the approximately 100 planets already found with this technique all have masses larger than or comparable to those of Jupiter and Saturn, the Sun's most massive planets. Like the Sun's giant planets, all the newfound extrasolar planets should be fluid and gaseous, composed primarily of hydrogen and helium.

What surprised astronomers most about these planets was that 25% of them move in orbits smaller than those of any planet in the solar system. Most scientists conclude that these giants could not have formed at these relatively tiny distances from their stars because the heat would have prevented ice from condensing and would have evaporated hydrogen and helium before it could be attracted gravitationally to form a planet. Instead, these scientists think that the giant planets most probably formed at great distances from their stars and then migrated inward to separations, in some cases, of only 1/20 of Earth's distance from the Sun. Yet some of the newly discovered systems with multiple planets or a planet and a brown dwarf pose severe challenges to this model for the formation of these new systems.

The Doppler discovery technique is limited to giant planets. We know nothing yet about the presence or absence of Earth-like planets orbiting other stars. Fortunately, high-precision photometric observations offer the promise of detecting other Earth-size planets in the habitable zones of their stars, and an instrument called Kepler is being built to carry out this search. Farther in the future, orbiting optical interferometers will allow a direct search for evidence of oxygen or other biomarkers in the atmospheres of these planets.

Meanwhile, a continuing search for signals from advanced civilizations is going on at optical and radio frequencies. Known as SETI, this search may provide a dramatic shortcut to the knowledge we seek about the origin and evolution of planetary systems, if contact is ever established. Communication with an advanced civilization on another world would have profound effects on our own culture. Humans are thus engaged in an extraordinary adventure whose ultimate outcome is impossible to predict.

KEY TERMS

astrometric method	habitable zone
biomarker	interferometry
brown dwarf	SETI
center of mass	transit

REVIEW QUESTIONS

1. In applying the Doppler-shift method to the discovery of extrasolar planets, what do astronomers actually observe? Why does this allow them to deduce the existence of a planet or planets?

2. What property of the Doppler-shift observations allows astronomers to deduce the orbital period of an extrasolar planet? What property allows them to estimate the planet's mass?

3. Why do Doppler-shift observations provide only a lower bound on the mass of an extrasolar planet?

4. What characteristics of the planets discovered around other stars during the late 1990s surprised astronomers? What made them so surprising?

5. What makes the detection of Earth-like planets much more difficult than finding planets like those discovered so far?

6. What instrument is being developed to search for Earth-like planets? How does it work? What characteristics can be deduced about a planet by this technique?

7. Describe the plans for searching for life on Earth-like planets. Can you suggest other approaches?

QUANTITATIVE EXERCISES

1. Derive equation (12) in Section 18.7.

2. Find the location of the center of mass in the Jupiter-Sun system, assuming these are the only two bodies involved. Compare the distance of the center of mass from the center of the Sun with the radius of the Sun.

3. What would be the apparent separation of Jupiter and the Sun in seconds of arc as seen from Alpha Centauri, the closest star? How does this compare with the maximum separation of Ganymede and Jupiter seen from Earth?

4. What is the orbital velocity of the planetary companion to the star HD 83443? How does that compare with the rotational velocity of the Sun at its surface?

ADDITIONAL READING

Achenbach, J. 1999. *Captured by Aliens: The Search for Life and Truth in a Very Large Universe.* New York: Simon & Schuster. Delightful romp through both science and pseudoscience, contrasting the different ways people approach the search for life beyond the Earth.

Boss, A. 1998. *Looking for Earths: The Race to Find New Solar Systems.* New York: Wiley. Historical discussion of half a century of efforts to find extrasolar planets, written by one of the astronomers who participated in this often frustrating search.

Dick, S. J. 1996. *The Biological Universe: The Twentieth-Century Extraterrestrial Life Debate and the Limits of Science.* New York: Cambridge University Press. A distinguished historian of science looks at these issues from both the scientific and the popular perspectives.

Goldsmith, D. 1997. *Worlds Unnumbered: The Search for Extrasolar Planets.* Sausalito, CA: University Science Books. A popular account of the discovery of extrasolar planets and their significance.

Goldsmith, D., and T. Owen. 2002. *The Search for Life in the Universe,* 3d ed. Sausalito, CA: University Science Books. An entry-level textbook for nonscience majors discussing aspects of astronomy, biology, and planetary science relevant to the search for life and strategies for carrying out that search.

Koerner, D. W., and S. LeVay. 2000. *Here Be Dragons: The Scientific Quest for Extraterrestrial Life*. New York: Oxford University Press. A popular discussion of astrobiology, from life's origin on Earth to extrasolar planets, SETI, and UFOs.

Shapiro, R. 1999. *Planetary Dreams: The Quest to Discover Life Beyond the Earth*. New York: Wiley. Entertaining and informative popular account of our progress in finding habitats for life beyond the Earth.

APPENDIX: UNITS AND DIMENSIONS

Throughout this book, we have tried to emphasize the metric system, since it is the system of measurements commonly used in science at the present time. It is also the most widely used system in the world, with only the stubborn Americans adhering to the foot-pound system. That adherence makes metric values unfamiliar to many of us, however, so we present tables below with some equivalences.

Note that in the metric system, units of measurement increase by factors of 10, 100, 1000, and so on:

1 centimeter = 10 millimeters

1 meter = 100 centimeters
       = 1000 millimeters

1 kilometer = 1000 meters
          = 100,000 centimeters
          = 1,000,000 millimeters

American Unit	Metric Equivalent
1 inch	25.4 millimeters 2.54 centimeters
1 foot	30.48 centimeters
1 yard	91.44 centimeters 0.9144 meter
1 mile	1609.3 meters 1.6093 kilometers
1 ounce	28.4 grams
1 pound	453.6 grams 0.454 kilogram
1 quart	1136.5 cubic centimeters 1.137 liters
1 mile/hour	1.609 kilometers/hour 1609 meters/hour 0.45 meter/second

Metric Unit	American Equivalent
1 millimeter	0.04 inch
1 centimeter	0.39 inch
1 meter	39.37 inches
1 kilometer	0.62 mile
1 gram	0.04 ounce
1 kilogram	35.27 ounces 2.2 pounds
1 liter	0.88 quart
1 kilometer/hour	0.62 mile/hour
1 meter/second	2.2 miles/hour

In discussing the dimensions and masses of atoms, planets, and satellites, as well as distances between objects in the solar system and between waves of electromagnetic radiation, we encounter some very large and very small numbers. Instead of coping with a lot of zeros, it is easier to resort to what is called exponential notation, or powers of 10. This way we can write one million as 10^6 instead of 1,000,000. If this is unfamiliar, you can think of the exponent as giving you the number of zeros following the 1. Thus $10^0 = 1$.

Similarly for small numbers: $\frac{1}{10}$ is simply 10^{-1}; $\frac{1}{10,000}$ is 10^{-4}. Now the negative exponent tells you the number of places to the right of the decimal point. The diameter of a hydrogen atom is $0.00000001 = 10^{-8}$ cm.

Still further simplifications are used. The distance from the Earth to the Sun is 93 million miles or 93×10^6 miles or 149.6 million kilometers or 149.6×10^6 kilometers or 1.5×10^{13} centimeters or 1 astronomical unit. This last convenient unit is often used to give other distances in the solar system. For example, Jupiter is 5.2 astronomical units (5.2 AU) from the Sun.

At the short end of the scale, we need units for describing wavelengths of electromagnetic radiation. The common unit for visible and ultraviolet wavelengths is the nanometer (nm), where 1 nm = 10^{-9} meters. The longer wavelengths of infrared light are often given in terms of micrometers, where 1 micrometer (abbreviated μm) equals 10^3 nm or 10^{-6} meters.

GLOSSARY

absorption line Removal of energy from a narrow range of wavelengths or frequencies in the electromagnetic spectrum due to atomic or molecular absorption. (Ch. 2)

accretion Gravitational accumulation of mass in a planet or protoplanet. (Ch. 4)

albedo The reflectance or reflectivity of an object; specifically the ratio of reflected energy to incident energy (also called the Bond albedo). (Ch. 3)

angular momentum A measure of the momentum associated with rotational motion about an axis. (Ch. 4)

anorthosite Primary igneous rocks of the lunar highlands, a type of rock that is very rare on Earth. (Ch 7)

anticyclonic Rotation induced in a high-pressure mass of atmosphere by Coriolis forces; clockwise in the northern hemisphere of a directly rotating planet like Earth. (Ch. 9)

aperture The diameter of the primary lens or mirror of a telescope and hence the best single measure of the light-gathering power of a telescope. (Ch. 3)

aphelion For an object orbiting the Sun, the point in the orbit that is farthest from the Sun. (Ch. 1)

asteroid Any small body (less than 1000 km in diameter) orbiting the Sun that does not display the atmosphere or tail associated with a comet; formerly called minor planets. (Ch. 4)

astrometry Precise measurement of positions in astronomy; hence an approach to searching for external planets by measuring the tiny periodic shifts in a star's position caused by the gravitational pull of invisible objects in orbit about it. (Ch. 19)

astronomical unit (AU) The semimajor axis of the Earth's orbit, or equivalently the average distance of the Earth from the Sun; approximately 150 million km. (Ch. 1)

atmospheric window The part of the electromagnetic spectrum within which a planetary atmosphere is more or less transparent; for Earth, the wavelength regions where astronomical observations can be carried out from the ground. (Ch. 3)

aurora Light emitted by atoms and ions in a planetary ionosphere, mostly in the magnetic polar regions; also called polar lights. (Ch. 9)

bar Unit of pressure equal to 10^5 pascals, or approximately the atmospheric pressure at the surface of the Earth. (Ch. 9)

basalt Common igneous rock, composed primarily of silicon, oxygen, iron, aluminum, and magnesium, produced by the rapid cooling of lava. Basalts make up most of Earth's oceanic crust and are also found on other planets that have experienced volcanic activity. (Ch. 3)

biomarker An indicator of life; specifically, any constituent of a planetary atmosphere or of sedimentary rocks that would be unlikely to form from non-biological processes, and that therefore strongly suggests past or present biological activity. (Ch. 18)

Bode's law See Titius-Bode rule.

breccia Any rock made up of recemented fragments of material, usually the result of extensive impact cratering. (Ch. 4)

brown dwarf An object intermediate between a planet and a star; roughly, with mass less than 0.07 solar masses and greater than 13 times the mass of Jupiter. A brown dwarf does not have enough mass to generate self-sustaining thermonuclear fusion reactions in its core. (Ch. 18)

C-type asteroid One of the most populous group of main belt asteroids, characterized by dark, spectrally neutral or slightly red surfaces; thought to be primitive, similar in composition to the carbonaceous meteorites; also sometimes applied to any dark primitive asteroid. (Ch. 5)

caldera Volcanic crater resulting from collapse following withdrawal of subsurface magma, often found at the summit of a shield volcano. (Ch. 9)

carbonaceous meteorite A primitive meteorite made primarily of silicates but including chemically bound water, free carbon, and complex organic compounds; also called carbonaceous chondrites. (Ch. 4)

carbonate A chemical compound that contains CO_2, such as calcium carbonate ($CaCO_2$), the primary constituent of the shells of marine organisms. When decomposed by heating, carbonates release carbon dioxide. (Ch. 9)

catalyst An atom or molecule that enables or accelerates a chemical reaction without itself being altered or consumed. (Ch. 14)

catastrophism The concept that the geology of the Earth and planets has been greatly influenced by rare events of large magnitude, such as impacts by asteroids. In the eighteenth and nineteenth centuries, catastrophism was associated specifically with attempts to explain geologic features as results of the biblical flood, but today the term is used in the much broader sense given above. (Ch. 7)

center of mass The average position of the various mass elements of a body or system, weighted according to their distances from that center of mass; also called center of gravity. (Ch. 18)

CFC Chlorofluorocarbons, nontoxic industrial chemicals used in refrigerators and air conditioners, as propellants in spray cans, and for cleaning electronics—but with the unintended consequence of destroying stratospheric ozone. Manufacture of most CFCs has now been halted. (Ch. 9)

chaotic terrain Irregular, jumbled terrain; for example, regions of the martian uplands consisting of jumbled depressions and isolated hills, thought to have been produced by collapse induced by the withdrawal of subsurface water or ice; also areas of Europa where the icy crust has been partially melted and disrupted. (Ch. 11)

chemical equilibrium Composition that reflects a chemical balance of different atoms and molecules, without any active sources or sinks of new material. (Ch. 14)

chondrite The most common type of stony meteorite, with primitive composition and usually containing chondrules (see below). (Ch. 4)

chondrule A small silicate spherule (typically a few millimeters in diameter) commonly found in primitive meteorites. (Ch. 4)

coma Atmosphere or head of a comet, forming a visible halo around the nucleus. (Ch. 6)

comet The most primitive Sun-orbiting members of the solar system, consisting of a small nucleus composed of ices, silicates, and carbonaceous material, which when heated generates a tenuous temporary atmosphere as its volatiles evaporate. Generally these icy objects are called comets only when they are close enough to the Sun to outgas a visible coma or atmosphere. (Ch. 4)

comet nucleus The solid part of a comet, typically a few kilometers (but perhaps up to hundreds of kilometers) in diameter. The nucleus consists primarily of a mixture of ices and solid grains of silicate and carbonaceous composition. (Ch. 6)

composite volcano Common type of terrestrial volcano, with a cone built up by repeated (and sometimes explosive) fountains of lava mixed with hot gas. (Ch. 9)

compound A substance composed of two or more chemical elements, such as H_2O (formed from hydrogen and oxygen). (Ch. 2)

conduction One of the basic ways of transferring energy, caused by the motion of atoms and electrons in a solid. Although it is not as intrinsically efficient as convection or radiation, conduction is usually the dominant means to transfer energy in a solid. (Ch. 2)

constellation Originally a configuration of stars; now one of 88 specific areas of the sky with internationally agreed-upon boundaries. (Ch. 1)

continental drift Name originally proposed for the observed gradual motion of the continents over the surface of the Earth due to the motion of lithospheric plates, as later described by the theory of plate tectonics. (Ch. 9)

convection One of the basic ways of transferring energy, caused by the large-scale (macroscopic) motion of material, such as the rising of pockets of hot gas and the sinking of cooler gas. When convection occurs, it is usually more effective at transferring energy than either radiation or conduction. (Ch. 2)

co-orbital satellite Informal term for the two Saturn satellites Janus and Epimetheus, which share almost the same orbit, or for any other satellites found in similar dynamical situations. (Ch. 16)

Coriolis effect The deflection of material moving across the surface of a rotating planet, producing in the case of the Earth's atmosphere the familiar cyclonic and anticyclonic patterns that characterize our midlatitude weather. (Ch. 9)

corona (Sun) The tenuous outer atmosphere of the Sun, consisting of gas at temperatures higher than a million degrees. (Ch. 2)

corona (Venus) Large, circular tectonic features unique to Venus, apparently caused by a rising plume of mantle material, often with associated volcanic activity. (Ch. 10)

cosmic rays Atomic nuclei (mostly protons) that strike the Earth's atmosphere with exceedingly high energies. Some originate in the Sun, but most cosmic rays have a galactic origin. (Ch. 9)

crater A circular depression (from the Greek word for "bowl" or "cup"), generally of impact origin. (Ch. 7)

crater density The degree to which impact craters are packed together on a planetary surface, measured in units of the number of craters of a given size per unit area (for example, the number of craters larger than 10 km in diameter per million square kilometers of surface). (Ch. 7)

crater-retention age The time over which a planetary surface has accumulated impact craters; hence, the time since the surface was mobile or extensive erosional or tectonic forces were acting to destroy craters. (Ch. 7)

crust The outer solid layer of a planet; on Earth, roughly the upper 30 kilometers. (Ch. 9)

cryovolcanism Volcano-like eruptions of material other than silicate magma, generally applied to the features seen on outer planet satellites that appear to represent flows of liquid water or other volatiles. (Ch. 16)

cyclonic The rotation induced in a low-pressure mass of atmosphere by Coriolis forces; counterclockwise in the northern hemisphere of a directly rotating planet. (Ch. 9)

debris disk Circumstellar disk that survives beyond the star-formation period, perhaps consisting of debris formed by collisions among "asteroids" in the system. (Ch. 17)

differentiated meteorite A meteorite from a differentiated parent body, as contrasted with a primitive meteorite. (Ch. 4)

differentiation The gravitational separation or segregation of different densities of material into different layers in the interior of a planet, as a result of heating. (Ch. 3)

Doppler effect Apparent change in wavelength (and frequency) of the radiation from a source due to its relative motion toward or away from the observer. (Ch. 8)

dust tail A cometary tail, usually broad, somewhat curved, and yellow-white, made of dust grains released from the nucleus of the comet. (Ch. 6)

Earth-approaching asteroid See near-Earth asteroid.

eccentric ring A planetary ring that has the form of an ellipse, as a result of the alignment of the individual elliptical orbits of the particles that comprise it. (Ch. 16)

eccentricity (of ellipse) The degree to which an orbit is noncircular; specifically, the ratio of the distance between the foci to the major axis of the ellipse. (Ch. 1)

eclipse of the Moon The phenomenon visible when the Moon passes wholly or in part through the shadow of the Earth. (Ch. 1)

eclipse of the Sun The blocking of all or part of the light of the Sun by the Moon. (Ch. 1)

ecliptic The apparent annual path of the Sun on the celestial sphere. (Ch. 1)

ejecta blanket Rough, hilly region surrounding an impact crater made up of ejecta that has fallen back to the surface; usually extending one to three crater radii from the rim. (Ch. 7)

electromagnetic radiation Radiation consisting of electric and magnetic waves; they include radio, infrared, visible light, ultraviolet, x rays, and gamma rays. (Ch. 2)

electromagnetic spectrum The whole array or family of electromagnetic waves, usually ordered by wavelength (or equivalently by frequency or energy); often called simply the "spectrum". (Ch. 2)

electron Basic subatomic particle with a negative electric charge. The number and configuration of electrons in an atom or molecule are critical to both its chemical properties and its spectrum. (Ch. 2)

element Basic form of matter; the smallest unit (atom) that retains the chemical properties of an elemental substance. There are 92 naturally occurring elements, with numbers of protons (and electrons) from 1 (hydrogen) to 92 (uranium). (Ch. 2)

ellipse A closed curve (one of the conic sections) that describes the orbit of one object about another subject only to their mutual gravitational attraction. (Ch. 1)

embedded satellite A small (perhaps invisible) satellite orbiting within a ring system and gravitationally influencing the structure of the ring. (Ch. 16)

equinox Either of the two intersections of the ecliptic and the celestial equator; occupied by the Sun on about March 21 and September 21, when the day and night are of equal length all over the planet. (Ch. 1)

escape velocity The minimum upward speed required to escape entirely from the gravitational attraction of a body. (Ch. 1)

eucrite One of a class of basaltic meteorites believed to have originated on the asteroid Vesta. (Ch. 4)

exosphere The part of the upper atmosphere of a planet from which atoms or molecules of gas can escape into space. (Ch. 3)

falls Meteorites that are seen to fall, as opposed to "finds". (Ch. 4)

family of asteroids A group of asteroids with similar orbital elements, indicating a probable common origin in a past collision. (Ch. 5)

fault In geology, a tectonic crack or break in the crust of a planet along which slippage or movement can take place, accompanied by seismic activity. (Ch. 9)

finds Meteorites that are found long after they fell, identified from their distinctive appearance or composition. (Ch. 4)

fluidized ejecta Ejecta from an impact crater that flows along the surface like a liquid rather than arching freely through space, apparently a common phenomenon in the past on Mars. (Ch. 11)

fluorescence Stimulated emission of light that was absorbed at wavelengths different from those at which it is emitted (Ch. 6)

forward scattering The tendency of small particles to reflect or scatter light primarily in a forward direction—that is, in nearly the same direction from which light is incident on the particle. (Ch. 16)

fractionation A process that changes the relative abundances of elements or isotopes in a planetary object (or atmosphere), such as by the selective loss of one component. (Ch. 4)

fragmentation A process that breaks up objects, usually as the result of high-speed collisions. (Ch. 4)

gas-retention age A measure of the age of a rock, defined in terms of its ability to retain radioactive argon. (Ch. 4)

geocentric Earth-centered, as in the pre-Copernican idea that the Sun, Moon, and planets all circled the Earth, which was thought to be located at the center of the universe. (Ch. 1)

geologic timescale The history of the Earth over the past 4.5 billion years, as determined from the rocks deposited in its crust. (Ch. 9)

granite Igneous rock associated primarily with the Earth's continental crust, composed chiefly of the minerals quartz and alkali feldspar. (Ch. 9)

greenhouse effect The blanketing of infrared radiation near the surface of a planet by infrared-opaque gases in the atmosphere (for example, carbon dioxide), producing an elevated surface temperature. (Ch. 9)

habitable zone The range of distances from a star within which liquid water can exist on the surface of an earthlike planet. In our system, the habitable zone stretches from approximately 0.8 to 1.8 AU from the Sun. (Ch. 18)

Hadley cell A theoretical model for the circulation of a planetary atmosphere, in which air heated near the equator rises and moves toward the poles, where it descends and flows back

toward the equator. The circulation of the atmosphere of Venus approximates this situation. (Ch. 10)

half-life The time required for half of the radioactive atoms in a sample to disintegrate. (Ch. 4)

heliocentric Centered on the Sun; specifically, the Copernican theory that the planets are in orbit around the Sun rather than around the Earth. (Ch. 1)

highlands (lunar) The older, heavily cratered crust of the Moon, covering 83% of its surface and composed in large part of anorthositic breccias. (Ch. 7)

ice age One of the periods in the Earth's climatic history when global cooling led to the formation of extensive ice sheets over polar and even temperate landmasses. (Ch. 9)

igneous rock Any rock produced by cooling from a molten magma. (Ch. 3)

imaging radar Radar carried on a moving platform (airplane or spacecraft) that can, with suitable data processing, produce images of the ground beneath; also called synthetic aperture radar (SAR). (Ch. 10)

impact basin A very large impact feature, usually 300 km or larger in diameter. (Ch. 7)

impact erosion Loss of atmosphere produced by repeated, large impacts. (Ch. 12)

impact frustration (of the origin of life) Environmental conditions unfavorable to the survival of life on a planet caused by large impacts in its early history. (Ch. 12)

inclination (of an orbit) The angle between the orbital plane of a revolving body and some fundamental plane—usually the plane of the celestial equator or of the ecliptic. (Ch. 1)

interferometry The technique of combining electromagnetic radiation (light or radio waves) collected by multiple telescopes in order to achieve much higher resolution than would be possible with single telescopes. (Ch. 18)

Io plasma torus Prominent feature of the jovian magnetosphere located near the orbit of Io, consisting of relatively dense concentrations of sulfur and oxygen ions trapped in the planet's rapidly rotating magnetic field. (Ch. 13)

ion An atom that has gained or (more usually) lost one or more electrons and thus has a net electric charge. (Ch. 2)

ionize To add or subtract one or more electrons from an atom, making it an ion. (Ch. 2)

ion tail See plasma tail.

ionosphere The upper region of the Earth's atmosphere in which many of the atoms are ionized, or any similar feature of the atmosphere of another planet. (Ch. 9)

iron meteorite A meteorite composed primarily of metallic iron and nickel and thought to represent material from the core of a differentiated parent body. (Ch. 4)

irregular satellite A planetary satellite with an orbit that either is retrograde or has high inclination or eccentricity. (Ch. 16)

isotope Any of two or more forms of the same element, whose atoms all have the same number of protons but different numbers of neutrons. (Ch. 2)

K/T event A major break in the history of life on Earth (a mass extinction) that occurred 65 million years ago, between the Cretaceous and Tertiary periods, due to the impact of a comet or asteroid in the Yucatan region of Mexico. (Ch. 12)

Kelvin temperature scale The absolute temperature, which is measured from absolute zero ($-273°$C). (Ch. 2)

Kuiper belt Disk-shaped region beyond the orbit of Neptune that contains many icy objects and is the main source of the short-period comets. (Ch. 6)

Kuiper belt object (KBO) An object that is resident in the Kuiper belt. (Ch 6)

Lagrangian asteroid An asteroid that is caught in the stable Langrangian orbits that lead and trail a planet by 60°. The Lagrangian asteroids associated with Jupiter are the "Trojan asteroids". (Ch. 5)

leading side The hemisphere of a synchronously rotating satellite that always faces forward in the direction of its orbital motion. (Ch. 16)

lithosphere The upper layer of the Earth, to a depth of 50 to 100 km, which is involved in plate tectonics. (Ch. 9)

long-period comet A comet with a period of revolution of approximately 200 years or more. (Ch. 6)

luminosity Intrinsic brightness; specifically, the total power output of the Sun or a star. (Ch. 2)

M-type asteroid Asteroid composed primarily of metal, presumably related to the iron meteorites. (Ch. 5)

magma Melted rock in the interior of a planet; called lava when it is erupted on the surface. (Ch. 7)

magnetic braking Process proposed to account for the slow rotation of some stars, in which angular momentum is transferred from the star to the surrounding plasma through its magnetic field. (Ch. 17)

magnetic flux tube Feature of the jovian magnetosphere consisting of a loop of electric current connecting Io and the planet. (Ch. 13)

magnetosphere The region around the Earth (or other planet) occupied by its magnetic field, and within which the planetary field dominates over the interplanetary field associated with the solar wind. (Ch. 9)

magnetotail The region of a planetary or cometary magnetosphere in which the magnetic field lines stream away from the object as they are carried "downstream" by the solar wind. (Ch. 13)

main belt asteroids Asteroids that occupy the main asteroid belt between Mars and Jupiter, sometimes limited specifically to the most populous parts of the belt, from 2.2 to 3.3 AU from the Sun. (Ch. 5)

mantle The part of a planet or satellite between its crust and core; on Earth, the mantle is the largest part of the planet, with about 65% of the mass. (Ch. 9)

mare (plural: maria) Latin for "sea"; name applied to the dark, relatively smooth features consisting of basaltic lava flows that cover 17% of the Moon. (Ch. 7)

mass extinction The sudden disappearance in the fossil record of a large number of species of life, to be replaced by new species in subsequent layers. Mass extinctions are indications of catastrophic changes in the environment, such as might be produced by a large impact on Earth. (Ch. 12)

metamorphic rock Any rock produced by the physical and chemical alteration (without melting) of another rock that has been subjected to high temperature and pressure. (Ch. 3)

meteor The luminous phenomenon observed when a bit of material (cosmic dust) enters the Earth's atmosphere and burns up; popularly called a falling star or shooting star. (Ch. 4)

meteor shower Many meteors appearing to radiate from a common point in the sky, caused by the intersection of Earth with a swarm of meteoric particles. (Ch. 6)

meteorite A rock that survives its fiery passage through the atmosphere (as a meteor) and strikes the ground. (Ch. 4)

midplane The central plane of rotation of a disk, such as the early solar nebula. (Ch. 17)

mineral A solid compound (often primarily silicon and oxygen) that forms rocks; the term could also be applied to condensed volatiles such as ice. (Ch. 3)

model See scientific model.

molecule A combination of two or more atoms bound together; the smallest particle of a chemical compound or substance that exhibits the chemical properties of that substance. (Ch. 2)

near-Earth asteroid (NEA) An asteroid with an orbit that crosses the Earth's orbit or that will at some time cross the Earth's orbit as it evolves. (Ch. 5)

neutrino A subatomic elementary particle with zero electric charge and extremely small mass, which interacts only weakly with matter. (Ch. 3)

neutron Basic subatomic particle with zero electric charge. The neutrons in an atom are located in the nucleus, and their number determines the isotope of the element. (Ch. 2)

noble gases The chemically inactive or inert elements: helium, neon, argon, krypton, xenon, and radon. (Ch. 12)

nonthermal radiation Any electromagnetic radiation from an astronomical source that is not thermal in origin, such as radio emission (synchrotron radiation) from electrons spiraling in the magnetosphere of Jupiter. (Ch. 13)

occultation The passage of an object of large angular size in front of a smaller object, such as the Moon in front of a distant star, or the rings of Saturn in front of the Voyager spacecraft. (Ch. 11)

Oort comet cloud The spherical region around the Sun from which most long-period comets come, representing objects with aphelia at about 50,000 AU, or extending about a third of the way to the nearest other stars. (Ch. 6)

opposition The position of a planet when it is opposite the Sun in the sky, rising at sunset and setting at sunrise. (Ch. 1)

orbital velocity The minimum speed necessary for a satellite to remain in a low orbit around a planet. (Ch. 1)

organic compound A compound containing carbon, especially a complex carbon compound. Organic materials are essential to life, but they can be produced non-biologically as well as biologically. (Ch. 4)

outflow channel Martian channel, typically several kilometers wide and hundreds of kilometers long, that once drained large floods of water from the southern uplands to the northern lowlands. (Ch. 11)

oxidizing Chemically dominated by oxygen; tending to form compounds of oxygen rather than hydrogen; opposite of reducing. (Ch. 3)

ozone Molecule of oxygen with three atoms rather than the more common two (symbol O_3). (Ch. 3)

parent body In planetary science, any larger original object that is the source of other objects, usually through breakup or ejection by impact cratering—for example, asteroid Vesta is thought to be the parent body of the eucrite meteorites. (Ch. 4)

perihelion For an object orbiting the Sun, the point in the orbit that is closest to the Sun. (Ch. 1)

perturbation The small gravitational effect of one object on the orbit of another. (Ch. 14)

Phanerozoic period The most recent eon in the Earth's history, covering the past 590 million years. It is divided into the Paleozoic, Mesozoic, and Cenozoic eras. (Ch. 9)

photochemical smog Aerosols of complex organic materials produced in a reducing atmosphere by various photochemical reactions. (Ch. 13)

photochemistry Chemical reactions that are caused or promoted by the action of light—usually ultraviolet light that excites or dissociates some compounds and leads to the formation of new compounds. (Ch. 10)

photodissociation The breakup of molecules by ultraviolet light. (Ch. 12)

photometry Precise astronomical measurement of brightness and fluctuations in brightness. Careful photometry of a star can reveal the presence of planets when the planets pass in front of the star as seen from the Earth. (Ch. 18)

photon A discrete unit of electromagnetic energy. (Ch. 2)

photosphere The part of the Sun from which the visible light originates; hence, the apparent surface of the Sun. (Ch. 2)

planetary embryos Hypothetical objects of roughly lunar mass proposed as an intermediate step between planetesimals and the final formation of the terrestrial planets. (Ch. 17)

planetesimals Hypothetical objects, from tens to hundreds of kilometers in diameter, formed in the solar nebula as an intermediate step between tiny grains and the larger planetary objects (or planetary embryos). The comets and primitive asteroids may be leftover planetesimals. (Ch. 4)

plasma A hot gas consisting in whole or in part of ions (charged atoms and electrons). (Ch. 2)

plasma tail A cometary tail, usually narrow and bluish in color, extending straight away from the Sun and consisting of plasma streaming away from the comet's head under the influence of the solar wind; also called an ion tail. (Ch. 6)

plate tectonics The motion of segments or plates of the outer layer of the Earth (the lithosphere), driven by slow convection in the underlying mantle. (Ch. 9)

polar cap A permanent or periodic deposit of ice or other volatiles near the polar region of a planet. (Ch. 8)

prebiotic chemistry Organic chemical reactions that take place in the absence of life and are thought to have been essential to produce the chemical building blocks for life itself. (Ch. 15)

Precambrian period The period of Earth history from the formation of the planet until the beginning of the Phanerozoic, 590 million years ago. The Precambrian is divided into the Priscoan, Archean, and Proterozoic eons. (Ch. 9)

primitive In planetary science, an object or rock that is little changed, chemically, since its formation and hence is representative of the conditions in the solar nebula at the time of formation of the solar system. Also used to refer to the chemical composition of an atmosphere that has not undergone extensive chemical evolution. (Ch. 3)

primitive meteorite A meteorite that has not been greatly altered chemically since its condensation from the solar nebula; also called a chondrite (either ordinary chondrite or carbonaceous chondrite). (Ch. 4)

primitive rock Any rock that has not experienced great heat or pressure and therefore remains representative of the original condensates from the solar nebula. (Ch. 3)

proton Basic subatomic particle with positive electric charge. The protons in an atom are located in the nucleus, and their number specifies the atomic number of the element. (Ch. 2)

proton-proton chain reaction The most important process for generating the energy (luminosity) of the Sun through the fusion of hydrogen nuclei to form helium. (Ch. 2)

protoplanet The hypothetical precursors of the planets, or young planets in the process of formation. (Ch. 4)

radar Measurement of reflected electromagnetic energy (usually microwaves) from an object (or planet) under investigation. (Ch. 5)

radial mixing Condition in the early solar nebula that allowed materials (e.g., planetesimals) formed at different distances from the Sun to intermix. (Ch. 17)

radiation Electromagnetic radiation; also one of the three basic ways of transferring energy from one location to another. Since radiation travels through a vacuum, it is the primary means of transferring energy from one object in space to another, such as from the Sun to Earth. (Ch. 2)

radioactive half-life See half-life.

radioactivity The process by which certain kinds of atomic nuclei naturally decompose with the spontaneous emission of subatomic particles and gamma rays. (Ch. 4)

reducing Chemically dominated by hydrogen; tending to form compounds of hydrogen rather than oxygen; opposite of oxidizing. (Ch. 3)

regolith The broken or pulverized upper layers of a planetary surface, fragmented by impacts; specifically, the lunar soil. (Ch. 7)

regular satellite A planetary satellite that has an orbit of low or moderate eccentricity lying approximately in the plane of the planet's equator. (Ch. 16)

remote sensing Any technique for measuring properties of an object from a distance; used particularly to refer to measurements (imaging, spectrometry, radar, etc.) of a planet carried out from terrestrial observatories or an orbiting spacecraft. (Ch. 3)

resolution The degree to which fine details in an image are separated or visible. Resolution can be specified in either angular or linear units, but is used here usually in the sense of the linear dimensions (in km) of the smallest features that can be studied on a planet or satellite. (Ch. 7)

resonance An orbital condition in which one object is subject to periodic gravitational perturbations by another; most commonly arising when two objects orbiting a third have periods of revolution that are simple multiples or fractions of each other. (Ch. 5)

resonance gap A location in an ensemble of orbiting particles that is empty because it corresponds to periods that are simple fractions of those of a perturbing external body, such as the Kirkwood gaps in the main asteroid belt. (Ch. 5)

retrograde motion An apparent westward motion of a planet with respect to the stars, caused by the motion of the Earth. (Ch. 1)

retrograde rotation (or revolution) Movement that is backward with respect to the common direction of motion in the solar system; counterclockwise as viewed from the north. (Ch. 1)

rift In geology, a place where the crust is being torn apart by tectonic forces; generally associated with the injection of magma from the mantle and with the slow separation of lithospheric plates. (Ch. 9)

runaway greenhouse effect A process whereby the heating of a planet leads to an increase in its atmospheric greenhouse effect and thus to further heating, quickly and irreversibly altering the composition of its atmosphere and the temperature of its surface. (Ch. 10)

runoff channel A branching river channel with many tributaries, found in the old martian uplands and presumably formed at a time when higher temperatures and a more massive atmosphere permitted surface water from rain or near-surface springs. (Ch. 11)

S-type asteroid The second most common class of asteroids, located primarily in the inner part of the main belt and characterized by moderate reflectivities (20%) and spectra that indicate the presence of silicate minerals similar to the ordinary chondrite meteorites. (Ch. 5)

scientific model A description (usually mathematical) of processes in nature. It should be self-consistent, provide a framework for understanding observations and experiments already carried out, and predict additional observations or experiments with which the model can be tested against alternatives. (Ch. 8)

secondary crater An impact crater produced by ejecta from a primary impact. (Ch. 7)

sedimentary rock Any rock formed by the deposition and cementing of fine grains of material. On Earth, sedimentary rocks are usually the result of erosion and weathering, followed by deposition in lakes or oceans, but breccias formed on the Moon by impact processes should also be considered as sedimentary rocks. (Ch. 3)

sedimentation The process on Earth in which eroded material is transported (by water or wind) to lower areas, often the sea floors, where it slowly accumulates to form new sedimentary rock. (Ch. 9)

seismic waves Waves in the solid Earth (or other planet) caused by earthquakes or impacts, which can be used to probe the structure of the interior. (Ch. 9)

seismometer A sensitive device for measuring seismic waves. (Ch. 9)

SETI The Search for Extraterrestrial Intelligence, carried out by looking for radio or optical signals from other civilizations in the Galaxy. (Ch. 18)

shepherd satellite Informal term for a satellite that maintains the structure of a planetary ring through its close gravitational influence—specifically, the two Saturn satellites, Prometheus and Pandora, that orbit just inside and outside of the F Ring, and the two small satellites of Uranus that orbit on either side of its Epsilon Ring. (Ch. 16)

shield volcano A broad volcano built up through the repeated nonexplosive eruption of fluid basalts to form a low dome of shield shape, typically with slopes of only 4–6 degrees, often with a large caldera at the summit. Examples include the Hawaiian volcanoes on Earth and the Tharsis volcanoes on Mars. (Ch. 9)

short-period comet A comet with a period of revolution of less than 200 years. Most known short-period comets have periods less than 15 years and are associated with Jupiter (Jupiter-family comets). (Ch. 6)

silicate Minerals containing silicon, oxygen, and one or more metals; the most common constituents of ordinary terrestrial rock. (Ch. 3)

snowline Informal term for the distance from the Sun (near 4 AU) within which water ice did not condense in the solar nebula. (Ch. 17)

solar activity cycle The 22-year cycle in which the solar magnetic field reverses direction, consisting of two 11-year sunspot cycles. (Ch. 2)

solar day The "day" as we usually think of it, the average interval (24 hours on the Earth) between the times when the Sun rises or crosses the meridian. (Ch. 8)

solar nebula The disk-shaped cloud of gas and dust from which the solar system formed. (Ch. 3)

solar wind A radial flow of plasma (mostly protons and electrons) leaving the Sun at an average speed of about 450 km/s and spreading through the solar system. (Ch. 2)

solidification age The most common age determined by radioactive dating techniques; the time since the rock or mineral grain being tested solidified from the molten state, thus isolating itself from further chemical changes. (Ch. 4)

solstice The position in the Sun's apparent path when it is farthest south (around December 21) and farthest north (around June 21). (Ch. 1)

spectrometer An instrument for forming and recording a part of the electromagnetic spectrum. In astronomy, a spectrometer is usually attached to a telescope to record the spectrum of an individual star or planet. (Ch. 2)

spiral density waves Waves produced in a self-gravitating rotating sheet of material. In the rings of Saturn, such waves (generated by resonant gravitational effects of the inner satellites) produce spiral patterns that wind around the rings like grooves in a phonograph record. (Ch. 16)

sputtering The process by which energetic atomic particles striking a solid alter its chemistry and eject additional atoms or molecular fragments from the surface. (Ch. 13)

stony meteorite A meteorite composed mostly of stony (silicate) minerals. The term can be applied to either primitive or differentiated meteorites if they are made of silicates. (Ch. 4)

stony-iron meteorite A fairly rare kind of differentiated meteorite composed of a mixture of silicates with metallic iron-nickel, thought to have originated near the core-mantle boundary of a differentiated parent body. (Ch. 4)

stratigraphy The study of rock strata, particularly the sequence of layers and the information this provides on the geologic history of a region. (Ch. 7)

stratosphere The cold, stable (nonconvective) layer of the Earth's atmosphere above the troposphere and below the ionosphere; also, the similar layer above the troposphere on any planet. (Ch. 9)

subduction In terrestrial geology, the tectonic process whereby one lithospheric plate is forced under another; generally associated with earthquakes, volcanic activity, and formation of deep ocean trenches. (Ch. 9)

subnebula A miniature version of the disk-shaped solar nebula that is hypothesized to have formed around each of the giant planets at the time of their formation, giving rise to their regular satellite systems. (Ch. 16)

sunspot A magnetically active region of the solar photosphere that is cooler than its surroundings and therefore appears dark. (Ch. 2)

supernova The final stage in the life of a massive star, when the star explodes, generating heavy elements and dispersing these elements into the interstellar gas clouds. (Ch. 17)

synchronous rotation Rotation of a body so that it always keeps the same face toward another object; the situation where the periods of rotation and revolution of an orbiting body are equal. (Ch. 8)

tectonic In geology, associated with stresses in the crust of a planet, often leading to the formation of faults (cracks) and folded ridges; in the case of Earth, associated with even the large-scale motion of lithospheric plates (called plate tectonics). (Ch. 9)

thermal inversion Situation in a planetary atmosphere where local heating causes the temperature to increase with altitude, rather than decreasing as it normally does. (Ch. 13)

thermonuclear fusion The joining or fusion of atomic nuclei at high temperatures to create a new, more massive atom with the simultaneous release of energy; the main source of energy in the Sun and stars. (Ch. 2)

tidal force A differential gravitational force that tends to deform an object or to pull it apart due to the tidal effect of its neighbor. (Ch. 8)

tidal heating Heating of one object by tidal friction or repeated stressing resulting from its motion within the strong tidal field of its neighbor, as in the tidal heating of Io. (Ch. 8)

tidal stability limit The distance—approximately 2.5 planetary radii from the center—within which differential gravitational forces (or tides) are stronger than the mutual gravitational attraction between two adjacent orbiting objects. Within this limit, fragments are not likely to accrete or assemble themselves into a larger object. Also called the Roche limit. (Ch. 16)

Titius-Bode rule A numerical scheme by which a sequence of numbers can be obtained that give the approximate distances of the planets from the Sun in astronomical units. Also called Bode's law. (Ch. 14)

trailing side The hemisphere of a synchronously rotating satellite that always faces backward, away from the direction of its orbital motion. (Ch. 16)

transit The passage of a small object in front of a larger one; compare with "occultation." (Ch. 18)

Trojan asteroid An asteroid that is caught in the stable Langrangian points that lead and trail Jupiter in its orbit by 60°. (Ch. 5)

troposphere Lowest level of the Earth's atmosphere, where most weather takes place; any region in a planetary atmosphere where convection normally takes place. (Ch. 9)

uncompressed density The density that a planetary object would have if it were not subject to self-compression from its own gravity—hence, the density that is characteristic of its bulk material independent of the size of the object. (Ch. 8)

uniformitarianism The principle in geology that the landforms we see were formed through the action of the same processes that are acting today, continuing over very long spans of time. In the nineteenth century this approach, as opposed to the biblical catastrophism of previous generations, laid the foundations for modern geology. (Ch. 7)

volatile A substance that has a relatively low boiling temperature. Although usage of this term depends on the context, volatile in this text usually refers to substances that are gaseous at temperatures above about 400 K to 500 K, such as water or carbon dioxide. (Ch. 3)

volcanic hot spot Location of a particularly large flow of heat from the interior of a planet. On Io, the name given to the areas covering about 1% of the surface from which most of the internal heat escapes to space. (Ch. 15)

zodiac A belt around the sky 18° wide and centered on the ecliptic, within which the Moon and planets move through the sky. (Ch. 1)

CREDITS

NASA/JPL/Caltech **11.6:** Courtesy of NASA/JPL/Caltech **11.8:** NASA **11.9:** NASA **11.10:** NASA **11.11:** NASA **11.12:** NASA **11.13:** NASA **11.14:** Adapted from data given by Michael H. Carr in *The Surface of Mars* (Yale University Press, 1981) **11.15:** NASA, Courtesy Michael Carr **11.16:** Courtesy of NASA/JPL/Caltech **11.17:** Courtesy of NASA/JPL/Caltech **11.18:** (a & b) NASA **11.19:** Courtesy of NASA/JPL/Caltech **11.20:** NASA **11.21:** NASA **11.22:** NASA **11.23:** Courtesy of NASA/JPL/Caltech **11.24:** Courtesy of NASA/JPL/Caltech **11.25:** Courtesy of NASA/JPL/Caltech **11.26:** NASA **11.27:** NASA **11.28:** NASA **11.29:** Courtesy of NASA/JPL/Caltech **11.30:** NASA **11.31:** Courtesy of NASA/JPL/Caltech **Perspective Figure 1:** NASA **Perspective Figure 2:** NASA **11.32:** NASA **11.33:** Courtesy of NASA/JPL/Caltech **11.34:** NASA **11.35:** G. Shirah, Goddard Space Flight Center; W. Boynton and W. Feldman **11.36:** Courtesy of NASA/JPL/Caltech **11.37:** Based on *Moons & Planets, Second Edition*, by William K. Hartmann, © 1983 by Wadsworth, Inc. Adapted by permission. **11.38:** NASA **11.39:** Courtesy of NASA/JPL/Caltech **11.40:** Courtesy of NASA/JPL/Caltech

Chapter 12

Opener: © Scala/Art Resource, NY **12.1:** Goldsmith, D. and Tobias Owen, *The Search for Life in the Universe*, © 1992 Addison-Wesley Publishing Company. Reprinted with permission. **12.2:** J. William Schopf, UCLA **12.3:** Goldsmith, D. and Tobias Owen, *The Search for Life in the Universe*, © 1992 Addison-Wesley Publishing Company. Reprinted with permission. **12.4:** J. William Schopf, UCLA **12.6:** Don Davis painting for NASA **12.7:** Kirk Johnson/Denver Museum of Natural History **12.9:** Courtesy of NASA/JPL/Caltech **12.10:** NASA **12.11:** Goldsmith, D. and Tobias Owen, *The Search for Life in the Universe*, © 1992 Addison-Wesley Publishing Company. Reprinted with permission. **12.12:** NASA **12.14:** Courtesy of NASA/JPL/Caltech **12.15:** (a & b) NASA **12.16:** Courtesy of NASA/JPL/Caltech **12.17:** Courtesy of NASA/JPL/Caltech **12.18:** NASA

Chapter 13

Opener: Courtesy of NASA/JPL/Caltech **13.1:** NASA **13.2:** NASA **13.3:** NASA **13.4:** NASA **13.5:** NASA **13.6:** NASA **13.7:** Tobias Owen **13.8:** Paul Mahaffy **13.9:** (a) The Galileo Project, NASA **13.10:** Adapted from Andrew Ingersoll **13.11:** NASA/JPL **13.12:** Tobias Owen **13.13:** NASA **13.14:** NASA **13.15:** Adapted from Andrew Ingersoll **13.16:** NASA **13.17:** NASA **13.18:** NASA **13.19:** (a, b, & c) Courtesy of NASA/JPL/Caltech **13.20:** (b) STScI and R. Beebe and A. Simon (NMSU); (c) NASA Infrared Telescope Facility and Glenn Orton **13.21:** Courtesy of NASA/JPL/Caltech **13.22:** Courtesy of NASA/JPL/Caltech **13.23:** Courtesy of NASA/JPL/Caltech **13.24:** Courtesy of NASA/JPL/Caltech **13.25:** Courtesy of NASA/JPL/Caltech_13.26: NASA **13.27:** NASA

Chapter 14

Opener: NASA **14.1:** Courtesy of NASA/JPL/Caltech **14.2:** Charles T. Kowal **14.3:** (a, b, & c) Lowell Observatory photographs **14.6:** NASA **14.7:** NASA **14.8:** Robert Danehy

and Tobias Owen **14.9:** (a, b, c, & d) Adapted from Andrew Ingersoll **14.10:** NASA/JPL **14.11:** Adapted from Andrew Ingersoll **14.12:** NASA **14.13:** NASA **14.16:** Tobias Owen

Chapter 15

Opener: Courtesy of NASA/JPL/Caltech **15.1:** Roger Clark **15.2:** NASA **15.3:** Courtesy of NASA/JPL/Caltech **15.4:** Courtesy of NASA/JPL/Caltech **15.5:** Courtesy of NASA/JPL/Caltech **15.6:** NASA/JPL and Arizona State University **15.7:** NASA/JPL **15.8:** Courtesy of NASA/JPL/Caltech **15.9:** Courtesy of NASA/JPL/Caltech **15.10:** Courtesy of NASA/JPL/Caltech **15.11:** Courtesy of NASA/JPL/Caltech **15.12:** Courtesy of NASA/JPL/Caltech **15.13:** Courtesy of NASA/JPL/Caltech **15.14:** Jeff Moore/NASA **15.15:** Courtesy of NASA/JPL/Caltech **15.16:** Courtesy of NASA/JPL/Caltech **15.17:** Courtesy of NASA/JPL/Caltech **15.18:** Courtesy of NASA/JPL/Caltech **15.19:** (a & b) Courtesy of NASA/JPL/Caltech **15.20:** Courtesy of NASA/JPL/Caltech **15.21:** Courtesy of NASA/JPL/Caltech **15.22:** Virgil Kunde, from Kunde *et al.*, *Nature 292*, 686 (1981) **15.23:** Courtesy of NASA/JPL/Caltech **15.24:** Courtesy of NASA/JPL/Caltech **15.26:** Courtesy of NASA/JPL/Caltech **15.27:** NASA **15.28:** Courtesy of NASA/JPL/Caltech

Chapter 16

Opener: NASA **16.1:** NASA **16.3:** Copyright © Cornell University, Ithaca, New York. All Rights Reserved. **16.4:** NASA/JPL **16.5:** Courtesy of NASA/JPL/Caltech **16.6:** From W.A. Baum *et al.*, *Icarus 47*, 84–96 (1981) **16.7:** Courtesy of NASA/JPL/Caltech **16.8:** Courtesy of NASA/JPL/Caltech **16.9:** NASA/JPL **16.10:** (a & b) Courtesy of NASA/JPL/Caltech **16.11:** NASA **16.12:** Courtesy of NASA/JPL/Caltech **16.13:** Courtesy of NASA/JPL/Caltech **16.14:** Courtesy of NASA/JPL/Caltech **16.16:** (a & b) NASA **16.17:** NASA **16.18:** Courtesy of NASA/JPL/Caltech **16.19:** Courtesy of NASA/JPL/Caltech **16.20:** NASA **16.21:** NASA **16.22:** Courtesy of NASA/JPL/Caltech

Chapter 17

Opener: © Giraudon/Art Resource, NY **17.1:** NASA and The Hubble Heritage Team (STScI/AURA) **17.2:** NASA and Jeff Hester (Arizona State University) **17.3:** Donald Goldsmith **17.4:** STScI/NASA **17.05:** Geoffrey Bryden/JPL/NASA **17.08:** NASA **17.09:** STScI/NASA **17.10:** Courtesy of NASA/JPL/Caltech **17.11:** B. A. Smith, Santa Fe and R. J. Terrile, JPL, Las Campanas Observatory **17.12:** Schneider, G., Smith, B.A., Becklin, E. E. *et al.*, 1999, "NICMOS imaging of the HR 4796A Circumstellar Disk," ApJ Letters, pp. L127–L130. Reprinted by permission.

Chapter 18

Opener: Corbis **18.3:** NASA/Mark McCaughrean and Robert O'Dell **18.5:** Steve Maran **18.6:** Geoff Marcy **18.11:** NASA **18.16:** NASA **18.17:** DAMTP, University of Cambridge **18.18:** National Ionosphere and Astronomy Observatory **18.19:** Seth Shostak, SETI Institute **18.20:** Seth Shostak, SETI Institute

INDEX

Planetary Exploration Spacecraft: Most Important Historical Missions

Spacecraft	Country of Origin	Launch Date
Sputnik 1	USSR	4 October 1957
Sputnik 2	USSR	3 November 1957
Explorer 1	U.S.	31 January 1958
Luna 2	USSR	12 September 1959
Luna 3	USSR	4 October 1959
Mariner 2	U.S.	27 August 1962
Mars 1	USSR	1 November 1962
Ranger 7	U.S.	18 July 1964
Mariner 4	U.S.	28 November 1964
Mariner 5	U.S.	14 June 1965
Venera 3	USSR	16 November 1965
Luna 9	USSR	31 January 1966
Luna 10	USSR	31 March 1966
Surveyor 1	U.S.	30 May 1966
Lunar Orbiter 1	U.S.	10 August 1966
Venera 4	USSR	12 June 1967
Mariner 6	U.S.	24 February 1969
Venera 7	USSR	17 August 1970
Luna 16	USSR	12 September 1970
Mars 2	USSR	19 May 1971
Mariner 9	U.S.	30 May 1971
Pioneer 10	U.S.	2 March 1972
Venera 8	USSR	26 March 1972
Pioneer 11	U.S.	5 April 1973
Mariner 10	U.S.	3 November 1973
Venera 9	USSR	8 June 1975
Venera 10	USSR	14 June 1975
Viking 1	U.S.	20 August 1975
Viking 2	U.S.	9 September 1975
Voyager 1	U.S.	5 September 1977
Voyager 2	U.S.	20 August 1977
Pioneer Venus Orbiter	U.S.	20 May 1978
Pioneer Venus Probe Carrier	U.S.	8 August 1978
Venera 11	USSR	9 September 1978
Venera 12	USSR	12 September 1978
Venera 13	USSR	30 October 1981
Venera 14	USSR	4 November 1981
Venera 15	USSR	2 June 1983
Venera 16	USSR	7 June 1983
VEGA 1	USSR	15 December 1984
VEGA 2	USSR	21 December 1984
Giotto	ESA*	2 July 1985
Magellan	U.S.	4 May 1989
Galileo	U.S.	18 October 1989
NEAR-Shoemaker	U.S	17 February 1996
Mars Global Surveyor	U.S.	7 November 1996
Mars Pathfinder	U.S.	4 December 1996
Cassini-Huygens	U.S.-ESA*	15 October 1997
Stardust	U.S.	7 February 1999
Mars Odyssey	U.S.	7 April 2001

* ESA = European Space Agency, a consortium of European countries.